REAL-TIME OBJECT UNIFORM DESIGN METHODOLOGY WITH UML

Real-Time Object Uniform Design Methodology with UML

BUI MINH DUC
Laval University, Sainte-Foy, QC, Canada

A C.I.P. Catalogue record for this book is available from the Library of Congress.

ISBN-13 978-90-481-7494-2
ISBN-13 978-1-4020-5977-3 (e-book)

Published by Springer,
P.O. Box 17, 3300 AA Dordrecht, The Netherlands.

www.springer.com

Printed on acid-free paper

Contents

List of Figures xi

List of Tables xxv

Preface xxix

1
Introduction to the world of systems, problems, and solutions 1
- 1.1 The world of systems, problems, and solutions 1
  - 1.1.1 System: problem 1
  - 1.1.2 Tools for handling complex systems 4
  - 1.1.3 Intelligent systems: hybrid systems, multidisciplinary systems 5
  - 1.1.4 Component-based systems 7
  - 1.1.5 Solutions 8
- 1.2 Real-time and embedded systems: disaster control and quality of service 10
  - 1.2.1 Real-time and embedded systems 10
  - 1.2.2 Disaster control: quality of service 12
- 1.3 Human organizations and structures: how software can simulate and study them 13
  - 1.3.1 Human organizations and structures 13
  - 1.3.2 Simulation of human organizations 14
- 1.4 This book is also a system 15

2
Coping with complexity 19
- 2.1 Visual formalisms for handling complexity: a picture is worth a thousand words 19
  - 2.1.1 Free diagrams 19
  - 2.1.2 Technical diagrams 20
- 2.2 Object/function decomposition: layering, hierarchy, collaboration 22
- 2.3 Functional decomposition to handle complexity 23
  - 2.3.1 Functional modeling is based on processes 23
  - 2.3.2 How process is converted to UML? 24
  - 2.3.3 The limit of functional decomposition: half a century of modular programming 25
- 2.4 How object technology handles complexity 28

2.4.1 Object paradigm 28
2.4.2 Information hiding, encapsulation 30
2.4.3 Composite objects are built from aggregates 30
2.4.4 Class factorization as complexity reduction mechanism 31
2.4.5 Unique messaging mechanism for object interaction 32
2.4.6 Inheritance as an economy of new class description 33
2.5 Object paradigm applied to control applications: reactive systems, control arrangement 33
2.6 Object paradigm applied to database applications: data modeling, handling database complexity 36
2.7 Objects and databases: bridging or mingling? 39
2.8 Object paradigm and component reuse 40
2.9 Mastering development complexity: a roadmap towards Model Driven Architecture 42

3
UML Infrastructure: Core concepts 47
3.1 Core concepts compliant to MOF: Modeling levels 48
3.2 Infrastructure: Core package 50
3.2.1 Core::PrimitiveTypes Package 52
3.2.2 Core::Abstractions Package 53
3.2.3 Core::Basic Package 65
3.2.4 Core::Constructs Package 67
3.2.5 Core::Profiles Package 72

4
UML Superstructure: language definition and diagrams 77
4.1 Making our own development methodology with UML 77
4.2 Structure and behavior: 13 UML diagrams classified along Structure and Behavior axes 79
4.3 Hierarchy of UML metaclasses 83
4.3.1 UML and "L" as language 83
4.3.2 Package import and merge in metamodeling 85
4.3.3 Comment on inheritance relationship at the metalevel of UML 86
4.4 Superstructure and compliance levels 86
4.5 Reuse of the Infrastructure: generalization hierarchy of metaclasses in the Structure part 88
4.6 Structure: class diagram 93
4.6.1 Example of user class diagram 95
4.6.2 Attribute 96
4.6.3 Association end, navigability, role, and qualifier 98
4.6.4 Operation, parameter list, return type, signature 100
4.6.5 Interface 103
4.6.6 Association class 104
4.6.7 Graphical notations 105
4.6.8 Dependencies 108
4.7 Structure: object diagram 109

4.7.1 Object, instance of a class 110
4.7.2 Link, instance of an association 111
4.7.3 Graphical notations 113
4.8 Structure: package diagram 114
4.9 Structure: composite structure diagram 118
4.9.1 Part 120
4.9.2 Port 122
4.9.3 Connector 123
4.9.4 Collaboration and collaboration use 123
4.9.5 Graphical notations 126
4.10 Structure: component diagram 128
4.11 Structure: deployment diagram 132
4.12 Hierarchy of metaclasses defined in the Behavior part: abstract metaclass behavior 135
4.13 Behavior: state machine diagram 139
4.13.1 State 142
4.13.2 State transition 143
4.13.3 Trigger and event 145
4.13.4 Simple, composite, and submachine states 145
4.13.5 Graphical notations 147
4.14 Behavior: activity diagram 150
4.14.1 Activity and Action metaclasses 150
4.14.2 Activity diagram 155
4.14.3 Data and control flow: execution of an activity diagram 158
4.14.4 Object nodes and storage of tokens 160
4.14.5 Control nodes 161
4.14.6 Exception handling: partition 163
4.14.7 Graphical notations 164
4.15 Behavior: interaction suite 169
4.15.1 Example of sequence diagram 169
4.15.2 Example of communication diagram 171
4.15.3 Example of Interaction Overview Diagram 173
4.15.4 Example of timing diagram 174
4.15.5 Metaclasses of the Interaction Suite 175
4.15.6 Graphical notations 176
4.16 Behavior: Use case diagram 181
4.16.1 Example 183
4.16.2 Metaclasses 183
4.16.3 Graphical notations 184
4.17 Auxiliary constructs: profiles 186
4.17.1 Example of profile 187
4.17.2 Graphical notations 188

5
Fundamental concepts of the real world and their mapping in UML 191
5.1 Abstraction, concept, domain, ontology, model 191
5.1.1 Abstraction 191

5.1.2 Concept 193
5.1.3 Domain 194
5.1.4 Ontology 196
5.1.5 Model 198
5.2 Structural, functional, and dynamic views of systems 200
5.3 Concepts of the functional view: Process and Business Process Modeling 201
5.3.1 Concepts used in Process Domain 202
5.3.2 Examples of Process Modeling and their conversions into UML diagrams 207
5.3.3 Building a proprietary methodology 211
5.3.4 Business process domain 211
5.3.5 Business Process Management and Workflow 216
5.3.6 Business Motivation Model 217
5.3.7 Mapping a process 220
5.3.8 Mapping a datastore 225
5.3.9 Mapping control flows and data flows 227
5.4 Fundamental concepts of the dynamic view 230
5.4.1 States and pseudostates 231
5.4.2 From elementary state towards global system state 234
5.4.3 Actions and activities 238
5.4.4 Events 240
5.4.5 Condition 241
5.4.6 Messages 245
5.4.7 Relations between action, activity, event, state, condition and message 248
5.4.8 Transitions: relationship with control flow 253
5.5 Fundamental concepts of the structural view 255
5.5.1 Class, type, object, and set of objects 258
5.5.2 Database objects, real-time objects, and their relationships 268
5.5.3 Dependencies, relationships, associations, links 270
5.5.4 Association class: n-ary association 272
5.5.5 Aggregation/composition in the object/component hierarchy 276
5.5.6 Inheritance 280
5.5.7 Multiple inheritances 281
5.5.8 Objects and roles 285

6
The Uniform concept 289
6.1 Elements of the uniform concept 290
6.1.1 Elements of the real world: objects and their interactions 290
6.1.2 Extension of the message interaction model 293
6.1.3 Induced energy 296
6.1.4 Use of a Monitor object 297
6.1.5 All objects are made independent even with tightly coupled interaction scheme 301
6.1.6 Cause–effect chain 303

6.2 Requirement analysis model 305
6.2.1 "What to do?" phase 305
6.2.2 Parameters of the requirements engineering step 306
6.2.3 Requirements engineering 308
6.3 Requirements analysis with technical diagrams 311
6.3.1 Requirements analysis with use case diagrams 313
6.3.2 Conducting the requirements analysis phase 318
6.3.3 Use case modeling advices 324
6.3.4 Using business process diagrams 326
6.4 System Requirements specifications 327
6.4.1 Requirements Engineering Model in the Model Driven Architecture 329
6.4.2 Dealing with platform dependent components in PIM 331
6.4.3 Various steps in the design process 331
6.4.4 Hierarchical decomposition or bottom-up method? 334
6.5 Designing real-time applications with MDA 335
6.5.1 Project breakdown structure into packages 337
6.5.2 Project scheduling 340
6.5.3 Starting a real-time system with the UML sequence diagram 340
6.5.4 What does a logical model mean? 344
6.5.5 Starting real-time systems with the UML interaction overview, activity, or state machine diagrams 347
6.5.6 Building real-time classes and object diagrams 350
6.5.7 Completing dynamic studies with UML state machine and activity diagrams 357
6.5.8 Assigning task responsibility 362
6.6 Designing reusable database models 364
6.6.1 Categorization and classification: reuse of static structures 365
6.6.2 Poor reusability of some conceptual database models 365
6.6.3 Conceptualizing domain and modeling reality: database models 367
6.6.4 Generic Classes Model 369
6.6.5 Intermediate and Application Classes Models 374
6.7 Unifying real-time and database applications 374

7
Real-time behavioral study beyond UML 381
7.1 SEN or State-Event Net diagram 383
7.1.1 Mathematical foundations of Petri network 385
7.1.2 Using Petri nets in modeling 386
7.1.3 State-Event Net abstractions 387
7.1.4 SEN subnets 389
7.1.5 SEN control structures 389
7.2 UML diagrams mapped into SEN 394
7.2.1 State machine diagram mapped into SEN 394
7.2.2 Activity diagram mapped into SEN 396
7.2.3 Sequence diagram mapped into SEN 397

7.2.4 Exception handling 397
7.2.5 Competition for common resource 399
7.3 Timing constraints with SEN 399
7.4 Case study with SEN 402
7.5 Safety-critical systems 405
7.5.1 State space search with combinatorial method: image attribute methodology 408
7.5.2 Developing safety-critical systems 411
7.5.3 Method of image attributes applied to human and social organizations 414

8
Real time case study 417
8.1 Design of an inclined elevator or track-lift tram 417
8.1.1 Early requirements 417
8.1.2 Requirements engineering: text analysis phase 420
8.1.3 Building classes for instantiating actors in use case diagrams 423
8.1.4 PIM of the inclined elevator system: creating the Generic Classes package 430
8.1.5 PIM of the Inclined Elevator System: creating the Generic Architectures package 434
8.1.6 PIM of the Inclined Elevator System: instantiating a working system 439
8.1.7 PIM of the Inclined Elevator System: behavioral study of the control unit 441
8.2 Emergency service in a hospital and design of a database coupled with a real-time system 448
8.2.1 Early requirements 450
8.2.2 Use case diagrams for System Requirements Specifications Document 452
8.2.3 Designing the patient database coupled with a real-time emergency system 452

References 467
Index 475

List of Figures

2.1.1.1	Odd shapes and free forms	20
2.1.1.2	Objects with/without volume, decorations and adornments	21
2.3.1.1	Gane-Sarson or Yourdon-DeMarco bubble representing a process in functional modeling	24
2.3.2.1	Same Process as Figure 2.3.1.1 with an UML activity diagram	25
2.5.1.1	Typical arrangement of a reactive system	34
2.5.1.2	Thermostatic control built on the principle of a reactive system	35
2.9.0.1	Classic development roadmap	43
2.9.0.2	Instance of development scenario compliant to MDA (to be compared with the classic development roadmap) that enacts high reuse	44
2.9.0.3	Transformations and bridges between models	45
3.2.0.1	Reuse of Common Core Package by UML and MOF	51
3.2.0.2	Four Core packages defined in the Infrastructure	51
3.2.1.1	Core::PrimitiveTypes Package contains four classes	52
3.2.2.1	Contents of Core::Abstractions. This package is composed of 20 elementary packages connected together via a network structure (Fig. 12, Infrastructure Book)	53
3.2.2.2	The Core::Abstractions::Elements Package contains the metaclass Element (Fig. 27 & 28, Infrastructure Book)	53
3.2.2.3	The Core::Abstractions::Ownerships Package expresses the fact that an element can contain other elements (Fig. 50, Infrastructure Book)	54
3.2.2.4	Core::Abstractions::Comments Package (Fig. 20, Infrastructure Book)	55

3.2.2.5	Core::Abstractions::Namespaces Package. A namespace provides a container for named elements (Fig. 48, Infrastructure Book)	56
3.2.2.6	Core::Abstractions::Visibilities Package. This package defines how an element can be visible to other modeling elements (Fig. 62 & 63, Infrastructure Book)	56
3.2.2.7	Core::Abstractions::Expressions Package. Expressions are used to support structured expression trees (Fig. 30, Infrastructure Book)	57
3.2.2.8	The Core::Abstractions::Literals Package specifies metaclasses for specifying literal values. A literal Boolean is a specification of Boolean value. A literal integer contains an integer-valued attribute, etc. (Fig. 40, Infrastructure Book)	58
3.2.2.9	Core::Abstractions::Constraints Package. Constraints use Expressions to define textually conditions that must be satisfied (expressions evaluated to true) in a model (Fig. 23, Infrastructure Book)	58
3.2.2.10	Core::Abstractions::Classifiers Package (Fig. 18, Infrastructure Book)	59
3.2.2.11	Core::Abstractions::Super Package (Fig. 58, Infrastructure Book)	59
3.2.2.12	Core::Abstractions::TypeElements Package (Fig. 61, Infrastructure Book)	59
3.2.2.13	Core::Abstractions::Relationships Package (Fig. 54, Infrastructure Book)	60
3.2.2.14	Core::Abstractions::Generalizations Package (Fig. 32, Infrastructure Book)	61
3.2.2.15	Core::Abstractions::Redeflnitions Package (Fig. 52, Infrastructure Book)	61
3.2.2.16	Core::Abstractions::StructuralFeatures Package (Fig. 56, Infrastructure Book)	62
3.2.2.17	Core::Abstractions::Changeabilities Package (Fig. 16, Infrastructure Book)	62
3.2.2.18	Core::Abstractions::Instances Package (Fig. 35, Infrastructure Book)	62
3.2.2.19	Core::Abstractions::BehavioralFeatures Package (Fig. 14, Infrastructure Book)	63

3.2.2.20 Core::Abstractions::Multiplicities Package (Fig. 42, Infrastructure Book) 63

3.2.2.21 Core::Abstractions::MultiplicityExpressions Package (Fig. 46, Infrastructure Book) 63

3.2.3.1 Core::Basic::Types Package (Fig. 65, Infrastructure Book) 64

3.2.3.2 Core::Basic::Classes Package (reduced version of Fig. 66, Infrastructure Book) 64

3.2.3.3 Data types in the Core::Basic::DataTypes Package (Fig. 67, Infrastructure Book) 65

3.2.3.4 Core::Basic::Packages Package (Fig. 68, Infrastructure Book) 65

3.2.4.1 Core::Constructs::Root Package (Fig. 71, Infrastructure Book) 66

3.2.4.2 Core::Constructs::Classifiers Package (Fig. 84, Infrastructure Book) 67

3.2.4.3 Core::Constructs::Classes Package (Fig. 73, Infrastructure Book) 68

3.2.4.4 Core::Constructs::Operations Package (Fig. 93, Infrastructure Book) 69

3.2.4.5 Core::Constructs::Datatypes Package (Fig. 86, Infrastructure Book) 70

3.2.4.6 Core::Constructs::Packages Package (Fig. 94, Infrastructure Book) 71

3.2.4.7 Core::Constructs::Namespaces Package (Fig. 89, Infrastructure Book) 72

3.2.4.8 Core::Constructs::Expressions Package (Fig. 72, Infrastructure Book) 73

3.2.4.9 Core::Constructs::Constraints Package (Fig. 85, Infrastructure Book) 73

3.2.5.1 Core::Profiles Package (Fig. 109, Infrastructure Book) 74

4.2.0.1 13 diagrams of UML shown as containers 79

4.4.0.1 Top level package dependencies. Relationships are "import" and/or "merge" 88

4.5.0.1 Infrastructure packages merged into the Kernel of UML 88

4.5.0.2 Flat 16 packages of the Kernel merged from the Infrastructure 89

4.5.1.1 Generalization hierarchy of the UML Structure Part showing principal metaclasses and their structure in the metalanguage (Part1) 90

4.5.1.2 Generalization hierarchy of the UML Structure Part (Part2). Operation is a BehavioralFeature and Property is a StructuralFeature 91

4.5.1.3 Generalization hierarchy of the UML Structure Part (Part3) 91

4.6.0.1 Classifier concept of the Kernel package. (Figure 7.9, Superstructure Book) 94

4.6.0 2 Class and Property concepts of the Kernel package. (simplified version of Figure 7.12, Superstructure Book) 95

4.6.1.1 Example of Class Diagram showing a temperature regulation 96

4.6.3.1 Role, multiplicity, navigability, qualifier attached to an association 98

4.6.3.2 E-R diagram replacement 100

4.6.5.1 Using interfaces to describe systems (example 7.57 of Superstructure Book) 104

4.7.0.1 InstanceSpecification concept of the Kernel package (figure 7.14, Superstructure Book) used to instantiate both objects ands links 110

4.7.2.1 Class diagram and Object diagram 112

4.8.0.1 Example of package diagram and relationships defined with packages 115

4.8.0.2 Package concept of the Kernel package (figure 7.14, Superstructure Book) 116

4.9.0.1 StructuredClassifier of the InternalStructures package (figure 9.2, Superstructure Book) 119

4.9.1.1 Differences between a class diagram, a composite structure diagram and an object diagram 121

4.9.2.1 Port metaclass defined in CompositeStructures::Ports package (figure 9.4, Superstructure Book) 122

4.9.3.1 Connector and ConnectorEnd metaclasses as defined in CompositeStructures::InternalStructures package (figure 9.3, Superstructure Book) 123

4.9.4.1 Collaboration and CollaborationUse metaclasses in CompositeStructures::Collaborations package (figure 9.6 & 9.7, Superstructure Book) 124

4.9.4.2 Example in the mechanical domain that illustrates the use of "Collaboration" and "CollaborationUse" modeling concepts and three associated kinds of dependency 125

4.9.4.3 Example elaborated in the house construction domain 126

4.10.0.1 Component concept as defined in BasicComponents package (Fig. 8.2, Superstructure Book) 129

4.10.0.2 Extension of the Component in PackagingComponents package 129

4.11.0.2 Deployment diagram showing components and nodes of a web database 135

4.12.0.1 Generalization hierarchy of the UML Behavior (part1) 136

4.12.0.2 Generalization hierarchy of the UML Behavior (part2) 137

4.12.0.3 Behavior metaclass as defined in BasicBehaviors package 138

4.13.0.1 Hair Dryer State Machine 140

4.13.0.2 Metaclasses StateMachine, State, Region, Transition, Pseudostate, FinalState defined in BehaviorStateMachines package 141

4.13.2.1 Abstractions defined in a state and on a transition 144

4.13.3.1 Trigger and Event metaclasses defines in Communications package 145

4.13.3.2 Subclasses of Event metaclass defined in Communications package 146

4.14.1.1 Activity package dependencies 152

4.14.1.2 Activity metaclass defined in FundamentalActivities (Fig. 12.3 Superstructure, Book) 152

4.14.1.3 Activity metaclass refined in BasicActivities (Fig. 12.5 Superstructure, Book) 153

4.14.1.4 Action metaclass as defined in BasicActions 153

4.14.1.5 Pin metaclass defined in BasicActions package 154

4.14.2.1 Example of activity diagram with activities, decision nodes with conditions, fork and join nodes, initial activity and final activity nodes 156

4.14.2.2 Activity diagrams with partitions and identification of participating objects 157

4.14.3.1 Data flows, action pins and activity parameter nodes 158

4.14.4.1 ObjectNode and ControlNode metaclasses defined in BasicActivities package (Fig. 12.4 Superstructure Book) 160

4.14.4.2 CentralBufferNode defined in JntermediateActivities package and DataStoreNode defined in CompleteActivities package 161
4.14.5.1 Control nodes and their corresponding packages. ActivityFinalNode defined in BasicActivities is enriched in IntermediateActivities 162
4.14.6.1 InterruptibleActivityRegion, ExceptionHandler and ActivityPartition metaclasses 164
4.15.1.1 Main constituents of a sequence diagram: lifelines and messages 170
4.15.2.1 Communication diagram, a replica of the previous sequence diagram 171
4.15.2.2 Partial class diagram of the Car 172
4.15.3.1 Example of Interaction Overview Diagram 173
4.15.4.1 Timing diagram for time specification 175
4.15.5.1 Metaclasses of the Interaction Suite 176
4.16.1.1 Concepts used in a use case diagram 182
4.16.2.1 Concepts used for modeling uses cases (Fig. 16.2, Superstructure Book) 183
4.17.1.1 Example of UML profile 188
5.3.0.1 Mapping of the Business Process Model into UML compatible BP model 203
5.3.1.1 Mapping of the Process Model directly into UML compatible Process model 203
5.3.2.1 Bubble process model of an Order Processing 208
5.3.2.2 Conversion of the DFD "Order Processing" into an activity diagram 210
5.3.3.1 Compatible strategy for adapting an existing methodology to UML 212
5.3.4.1 Overview of Business Process Modeling Notation (BPMN) of an Order Processing in the site of www.bpmi.org 214
5.3.4.2 Order Processing diagram converted from Business Process Modeling Notation (BPMN) to activity diagram notation 215
5.3.6.1 Over simplified version of the Business Organization as inspired from drafts on the OMG site to explain the mapping of business processes 218

5.3.6.2 Reduced version of the Influencer as taken from the OMG site 219

5.3.7.1 Mapping Business Processes in an Enterprise. (See Text) 223

5.3.8.1 Datastores used in the past with DFD. There is no datastore defined in the BPMN and BMM 225

5.3.9.1 Control flows and data flows in a DFD diagram 227

5.3.9.2 Conversion of the DFD of figure 5.3.8.1 into activity diagram 230

5.4.2.1 States of the Hair Dryer not decomposed. 64 states are possible at the beginning, based on 6 binary operations (activated or not) 235

5.4.2.2 States of the Hair Dryer decomposed into two separate objects. The number of states is 16 (8+8) 235

5.4.2.3 Conventional state used with UML is not elementary state (figure 15.32 Superstructure book) 237

5.4.8.1 Transition characteristics in state machine diagram. Transitions are made of known concepts exposed in previous chapters 254

5.5.1.1 Car class declared in three different contexts 265

5.5.3.1 Relationship between Relationship, Association, Link, Connector and Dependency 271

5.5.4.1 Attribute of relation in ER diagram 272

5.5.4.2 Association class. When a person takes a job in a company, the two sides are tied by a contract that is represented as a class association 272

5.5.4.3 Association Class. Association class represents a contract in business domain 273

5.5.4.4 The first example shows a misuse of 4-ary relationship, corrected in the first step to a 3-ary relationship, then corrected in a second step to transform the ternary association into two binary associations (see text for discussion) 276

5.5.5.1 Same objects but two different aggregation criteria that are "Administrative Structure" and "Geographical Localization" 278

5.5.5.2 Decomposition that respects the level semantic coherence 279

5.5.6.1 Example of inheritance hierarchy 280

5.5.7.1 In object technology, multiple inheritance use union of attributes and operations (not intersection), so the child class is essentially the union of the behavior of the two parent classes 282
5.5.7.2 Classical problem known as the "diamond" problem [Snyder, 1986] in multiple inheritance 283
5.5.7.3 Inheritance hierarchy containing simple and multiple inheritance 284
5.5.7.4 Example of questionable multiple inheritance hierarchy 285
5.5.7.5 Composition or multiple inheritance? The first solution is a multiple inheritance with some minor conflict resolution and the second is a team of three engineers 286
5.5.8.1 Representation of role in UML as name of an association end 287
5.5.8.2 Roles are modeled as separate classes 287
6.1.1.1 Object diagram of a Home Alarm System showing the modeling of visible and invisible, human and non human objects 293
6.1.2.1 Sequence and class diagrams showing the arrangement of a parking ticket delivery system from the user viewpoint 295
6.1.3.1 At the starting point, we have 4 independent pieces Mechanical_Piece_1, Mechanical_Piece_2, Bolt and Nut At the end, we have a monolithic piece shown in the figure 297
6.1.3.2 Translation of mechanical assembly into objects and messages 298
6.1.4.1 Variable speed electric fan with its power supply 299
6.1.4.2 Object diagram representing a control model 299
6.1.4.3 Use of a Monitor object 302
6.2.3.1 UML activities diagram showing activities of Requirements Engineering phase 309
6.3.2.1 Class diagram of the RMSHC system 318
6.3.2.2 Use case diagram depicting goals enumerated for the RMSHC 321
6.3.2.3 Activities of the Blood Pressure Monitor 323
6.3 2.4 Use case diagram showing activities related to the management of the Electronic Pifi Box 323
6.3.3.1 Package diagram of the Requirements Analysis of the Remote Monitoring System for Health Conditions 324

6.3.3.2 Actors are connected directly through inheritance only 324

6.3.3.3 When actors are connected together, we have an object style connection. In a use case diagram, a use case is inserted between two actors meaning that actors are involved in the activity 325

6.3.3.4 Inheritance between roles. The left diagram is replaced by the right diagram 325

6.3.4.1 Business Process Diagram showing the functional aspects of the Pill Management Software Component of the RSMHC project 327

6.4.1.1 A link Model traces all mappings between the SRS elements and components of PIM set and PSM set 330

6.4.3.1 Design process roadmap represented by a business process diagram combined with an object diagram 332

6.5.1.1 Decomposition of a Security System in an Airport 338

6.5.1.2 Re-specs with use cases and packages illustrated via the RMSHC example 339

6.5.3.1 Example of call graph showing all execution paths in a program 341

6.5.3.2 User's Instruction "How to arm a home alarm system" expressed by a sequence diagram 342

6.5.3.3 Corresponding classes identified while drawing the sequence diagram 343

6.5.3.4 Sequence diagram for developing the Home Alarm System 344

6.5.3.5 The status of a passive component in a real time system can be detected by reading the status of the component with a busy wait loop or mounting appropriate hardnre to send an interrupting signal to the Controller 346

6.5.5.1 Interaction Overview Diagram packing many references to sequence diagrams inside an activity diagram 348

6.5.5.2 Activity diagram as a starting method for the design 349

6.5.5.3 State machine diagram as a starting method for designing 350

6.5.6.1 Domain engineering with two activity flows (Domain Engineering & Client System Development) that may be conducted in parallel or not 352

6.5.6.2 Classes of Alarm System Domain (Generic Classes Package) 353

6.5.6.3 Contents of Generic Architectures Package (Alarm System) 355

6.5.6.4 Contents of Reusable Components Package (Alarm System) 356

6.5.6.5 Instance of Customer application (Alarm System). A Customer application contains objects instead of classes 358

6.5.7.1 State machine diagram of the Panel Controller showing activities accompanying the arming process 361

6.6.0.1 Physical, logical, conceptual and semantic models on a user perception axis of low or high level of data representation 364

6.6.2.1 Classic conceptual Person class often seen in database model 366

6.6.2.2 Slight correction of the model of figure 6.6.2.1 366

6.6.3.1 Organization of data models, From Generic Classes Model to the Physical Models 368

6.6.4.1 Elementary GCM elements of Humans, Identifications, Locations, Immobilizations packages illustrating the contents of Generic Classes Model 370

6.7.0.1 Deployment diagram explaining that many bagRT are created temporarily (they can be made persistent locally) 378

7.0.0.1 The Dynamic Model developed with SEN (State Event Net) acts as intermediate model between PSM and Code 383

7.1.1.1 Petri net with 4 places P1-P4 and 2 transitions T1-T2 386

7.1.2.1 Modeling task execution with a Petri net 387

7.1.3.1 Various graphic representations of SEN places 389

7.1.4.1 SEN subnets 390

7.1.5.1 Various End of Activity representation and Interruptible Activity 391

7.1.5.2 Decision nodes (a) simple IF, (b) CASE with one outgoing edge, (c) handling complex case with overlapping conditions and possibility of parallel split 392

7.1.5.3 Example of DO loop represented by a SEN at two levels, first at programming language level and at design level (as a sub SEN) 393

7.1.5.4 Fork, Join and Merge nodes 394

7.2.1.1 Mapping of a state diagram into SEN 395

7.2.2.1 Conversion of activity diagram with object flow into SEN 396

7.3.2.1 Intra-object interactions and inter-object interactions 398

7.2.4.1 Exception handling in a SEN 399

7.2.5.1 Competition for a common resource 400

7.3.0.1 Representation on the End of Activity (EOA) and Before End of Activity (BEOA) conditions 401
7.3.0.2 Timing specification with variable "end" 401
7.3.0.3 Timing dependence between two objects A and B with EOA and BEOA 402
7.4.0.1 Use case diagram representing activities of the Tamagotchi, Person and Timer 404
7.4.0.2 SEN representing the state evolution of the Tamagotchi 404
7.4.0.3 Class and object diagrams of the Tamagotchi 405
7.4.0.4 SEN versions of the Tamagotchi application with separate SEN. Cross-interactions between objects are not shown but they are parts of class and object diagrams 406
7.5.1.1 Attributes and image attributes of an object 409
7.5.1.2 SEN of the Tamagotchi established after an extensive study with the method of image attributes 412
8.1.1.1 Free form diagram showing main components of the inclined elevator 419
8.1.1.2 Control Panel of the inclined elevator 419
8.1.3.1 Classes of the Inclined Elevator identified by Text Analysis of Early Requirements Specification 424
8.1.3.1 Thirteen use case diagrams showing activities of the Inclined Elevator System from which CRT parameters must be evaluated 430
8.1.4.1 Generic Classes Package in use in LifiTram Company 431
8.1.4.2 Details of the Generic Mechanical Classes Package. These classes are specific to the Elevator Domain 431
8.1.4.3 Details of the Generic Electromechanical Classes Package 432
8.1.4.4 Details of the Generic Sensor Classes Package 433
8.1.4.5 Details of the General Inteffigent Devices Classes Package 434
8.1.5.1 For one shot deal project with small chance of recurrence (small scale reuse), components can be built directly from objects instantiated from the Generic Classes package. This diagram is incomplete and is targeted to show small-scale reuse process 435
8.1.5.2 Four components are defined in Generic Architectures Package 436

8.1.5.3 Generic Architectures Package contains the definition of four composite components 436

8.1.5.4 Composition of the Cable-Transmission-Tension component 437

8.1.5.5 Composition of the Track-Limit–Sensors component 437

8.1.5.6 Composition of the Cabin-Door-Detector-Castor-Overspeed-Obstruction-TrackBrake component 438

8.1.5.7 Composition of the Control Unit component 439

8.1.6.1 The Package Application Objects/Components contains the PIM built from merging classes and components from other packages. Only Application Classes Package is not reusable 440

8.1.6.2 The final system is composed of objects/components instantiated from reusable Generic Classes Package, reusable Generic Architecture Package 440

8.1.7.1 Object diagram of the Inclined Elevator Controller with all inputs/outputs of its first neighbors 444

8.1.7.2 When the power is on, the Controller undergoes a system check routine then powers the VFD. If faults are declared, the Controller requests that all faults must be cleared before enabling any movement 445

8.1.7.3 This routine details "System Check & Power on VFD" subroutine. It sets the main EM brake on, verifies track jam, slack cable and overspeed sensors and set lights accordingly 446

8.1.7.4 The subroutine "Update Fault Indicators" reads all sensors and buttons then updates all signal lights (Lights are exclusive, Red or Green only) 447

8.1.7.5 The main subroutine supports the normal operation mode of the cabin 449

8.2.2.1 Classes of the Hospital with the Emergency Classification Problem. They are used to instantiate objects for specification purpose only, not for design 453

8.2.2.2 UC0 to UC8 specifying the hospital emergency problem 457

8.2.3.1 Contents of the Generic Classes Model (GCM). Figures of this GCM suite must be completed with Figure 6.6.4.1 of chapter 6 459

8.2.3.2 Contents of the ICM package that contains classes belonging to the Health Care System 462

8.2.3.3 Contents of the ACM package (Application Classes Model). This package relates to all relationships specific to the Hospital X 464

List of Tables

1.1.1.1 The table shows the similarity of approaches and steps used to design a system or solve a problem 4
1.2.1.1 Characteristics of real time applications 11
1.3.1.1 Depletion of vocabulary from high to lower models 14
2.8.0.1 Survey questions to test the commitment to reuse concept in an organization 42
3.1.0.1 4-layer architecture of UML 48
4.2.0.1 Six structural diagrams of UML standard 81
4.2.0.2 Seven behavioral diagrams of UML 2 standard. Behavioral includes both functional and dynamic aspects 82
4.3.1.1 Parallel between the natural language and UML 84
4.4.0.1 Four compliance levels of UML L0 to L3 defined with regard to the abstract syntax dimension 87
4.6.7.1 Modeling concepts mostly used in a class diagram 105
4.6.8.1 Standard dependencies defined in UML. Users can freely define their proper set of dependency stereotypes 108
4.7.3.1 Modeling concepts found in an object diagram 113
4.8.1.1 Modeling concepts found in a package diagram. — 117
4.9.5.1 Modeling concepts in a composite structure diagram 126
4.10.1.1 Modeling concepts of a component diagram 130
4.11.0.1 Modeling concepts used in a deployment diagram 132
4.13.5.1 Modeling concepts used in a State Machine diagram 147
4.14.1.1 Seven levels of Activities definitions 155
4.14.7.1 Modeling concepts used in an Activity diagram 164
4.15.6.1 Modeling concepts of the Interaction Suite 176
4.16.3.1 Concepts used in a use case diagram 184

4.17.2.1 Modeling concepts used in profiles 188
5.1.1.1 Mapping of Abstraction concept and different meanings in the domain of modeling 192
5.1.2.1 Mapping of a simple concept that necessitates just a categorization process with a Class 194
5.1.2.2 Mapping of a "Complex Concept" concept and example 195
5.1.3.1 Mapping of a Domain concept in UML 19
5.1.4.1 Mapping of an Ontology concept in UML 198
5.1.5.1 Mapping of a "model" concept into UML 199
5.2.0.1 Vocabulary depletion when evolving from application domains towards programming domain 202
5.3.2.1 Mapping rules of a simple Process Model into UML 209
5.3.7.1 Mapping of process into UML metaclasses 221
5.3.8.1 Mapping of a datastore into UML concepts 226
5.3.9.1 Meaning of UML control flow, object flow, interrupt edge 229
5.4.1.1 Description of UML concepts related to the real world concept of "state" 232
5.4.4.1 Support for concept of event in UML 242
5.4.5.1 This table explains how UML deals with conditions at various levels 244
5.4.6.1 Support of Message concepts in UML 249
5.4.8.1 Mapping of system change, system evolution, system transition or system movement into UML transition and UML control flow 254
5.5.4.1 Table containing all possible entries of a ternary association between Participant, TradeShow and Location 275
5.5.5.1 Aggregation is considered as the loosely coupled form and composition the highly coupled form between parts and whole 277
6.2.3.1 Sub phases of the Requirements Engineering process 312
6.3.1.1 Convention used in this book: basic concepts in use case diagram and extensions of concept interpretation 314
6.3.2.1 Example of Text Content Analysis in order to establish use case diagrams structuring requirements 319
7.4.0.1 Text Content Analysis of the Tamagotchi specification 403
7.5.1.1 Table of attributes and image attributes of the Tamagotchi 411

7.5.1.2 Table of Sequences built from variations of state values of Tamagotchi regular attributes and image attributes 413

8.1.2.1 Text analysis of Early Requirements (Inclined Elevator). First step of Requirements Engineering phase 420

8.1.7.1 This table lists all objects/components constituting the first neighbors of the Controller and identifies the nature of communication channels exchanged with the Controller 442

Preface

Book Description

Real-Time Object Uniform Design Methodology with UML is a theoretical and practical book written for busy people who want to untangle the complex world of system development, find essential materials without digging in UML standard documentation, grasp subtle concepts of object orientation, practice the new Model Driven Architecture (MDA), experience the reuse mechanism, and transform the bare metal programming of real-time and embedded products into more handsome platform-independent and platform-specific components. With this rapid methodology of development, practitioners can spare time, avoid tons of written documentation by relieving this tedious task to smart CASE (computer-aided software engineering) tools, and have a quick and synthetic view of any system through a well-built set of pictures and blueprints.

The methodology presented in this book is a *neutral methodology* based on a thorough study of fundamental modeling concepts and then a temporary mapping of these concepts on current available standards and tools. We say "temporary" because research is in fact a never-ending activity. Good standards are evolving standards and the truth is always questionable. We are not pretending to add a new methodology to the numerous existent or in-house methodologies. We hope that the reader is able to catch the thoughts presented in this book to have a more critical view on any future methodology (a kind of meta "methodology"). So, feel free to prune off parts that you do not feel comfortable with. On the other hand, give us also the opportunity to correct any inaccuracy or vagueness in future versions of this book to always keep it at the leading edge of knowledge.

Why do we need object technology and an innovative uniform view of the message concept? Neil Armstrong had already walked on the Moon in 1969 before the maturity of object technology. Most real-time methodologies in the past were good guidelines at that time. Today, we know that every system has exactly three views: structural, functional, and dynamic ("behavioral" combines "functional" and "dynamic"). We must know all about systems before building them. At the implementation phase, algorithms are inferred from dynamic analyses. Before serious dynamic investigations, we must first structure the systems. In the past, real-time practitioners tackled systems through functions, along the functional axis. Database developers unveiled the first structural aspects with entities and relationships, and so approached problems from the structural axis. Mixed and multidisciplinary software, hardware, physical, biological, and social systems suffered from the use of different tools to explore and design multidomain systems. Object methodologists, who made the proposal of class abstraction, which contains functions, structures, and dynamic contents, claimed that every software system can be modeled with object paradigm and blueprinted with a tool like the UML. Effectively, we observe that object technology can be used to study any complex multidisciplinary system if we supply a more uniform view of the message concept to represent interaction between objects of very different nature. We therefore propose some "patches" in object paradigm, based on "induced energy," "cause–effect chain," and the generalization of the "message concept." They open unsuspected doors for the modeling of a richer set of multidomain applications.

Is object technology compatible with business processes? If everyone is convinced that object technology is the right way to go, in university school benches, why do students still make classes from each function they discover even in the presence of an object environment like Java, Ada, C++, or C# and UML modeling tools? Why do people get stuck with process? In fact, even with object technology, a process that is a real working entity in an object system still plays exactly the same role as it played in the past; object technology only dictates how to build and organize them smartly to satisfy other software engineering goals and deal with complexity. The functional view of systems remains at the foreground contrary to some prognostics of object theorists of the first wave. Processes become more structured and find their strength mainly in business models. After an exposé on business process model, we connect business models to the UML and give an answer to the question as to whether a process could be mapped into a full class or not.

Why MDA? Large-scale and multidisciplinary systems need complementary techniques. A biometric measurement system is, for instance, a multidisciplinary embedded system. An airport management is a large-scale mixing of databases, humans, and real-time components. It is well known that a large project life cycle that needs an analysis, design, implementation, and validating phases must be subdivided into sub- and sub-subprojects. This hierarchical view of the development model is not sufficient and should be revised and adapted to the convergence point of the black art of design and the recent MDA concept. The MDA deals with "models" and begins with a *Platform Independent Model* (PIM) that represents the system in terms understood by model practitioners who are often not computer specialists or implementers. An object-oriented and logical architecture is then built to represent the planned solution. Communication between domain experts and system architects is conducted to validate the PIM that can be simulated and tested. *Platform Specific Model* or PSM signs up as the intermediate phase before coding or implementing. Models facilitate communication, reuse, and structure the development process instead of the product itself. This book has clearly a convergence point with the MDA proposal.

What is the double problem of a system designer? When we develop a system, we are facing a double problem: building the right system, deploying appropriate human and technological resources to realize this system in a reasonable time frame and at an acceptable cost. In an organization business model, the managerial structure should be geared to the development process that involves a plethora of recurrent problems including management of the development process, looking for the right individuals, finding available resources, etc. Development of a product or a design of a solution for a complex system is not simple at all. The batch of stress, bad luck, or mitigated success, at personal or organizational levels, are clues showing the complexity of our daily business. Things may appear "simple" because we have experience, we react by reflex, and we apply consciously or unconsciously known solutions or resolution patterns that transform all complex processes into automated daily acts or reactions. As problems evolve slightly with time and context, systematical patterns, unified way of doing things replicated in "similar" situations, based on the erroneous thinking that they do save cost, may propose approximate solution or inadequate systems, blur nuances, and finally may create more problems in the longer term than they can solve. Moreover, patterns or ready-to-use solutions targeted to have rapid results, if applied without any methodology, shortcuts inevitably the analysis and design processes, and may impair the activity of finding the optimized solution or system. Another problem is unexpected constraints that may be added to the requirement analysis and alter the original problem. For instance, in political systems, under the pressure of the media, some hasty decisions based on commonly admitted patterns, taken under the pressure of the population in catastrophic circumstances, can calm down the media pressure but may represent atrocious examples of resource wasting. From a perspective of system and solution design, we are first solving the media problem by applying a pattern to the original problem quickly without any detailed analysis of the main problem.

Facing the MDA and current research trend, we can say that a methodology is not a mechanical act of applying patterns shortcutting analysis and design phase, but it proposes a way of quickly creating appropriate patterns for each new problem or system, reusing models. A methodology does not give up the notion of patterns or existing generic models. By organizing the development process into a multitude of models, we can easily distill universal patterns applicable to several domains, those that are specific to a particular domain and finally those that must be generated on the fly for a given situation. Experience does not mean ready-to-use solutions and patterns shortcutting analysis and design phases (they simplify them considerably instead) but suggests a process of reusing both universal and specific domain knowledge to design an optimized solution taking into account requirements of the current context. Parts of the solution patterns can be returned to the universal or domain knowledge to enrich the overall experience that constitutes the business assets of an organization or individual. This process is the essence of the reuse concept exposed in this work, and constitutes the immediate benefit of object technology and the MDA.

Audience

Designed for analysts, object programmers, system architects, application developers, project managers, and researchers, this book contributes an effort to implement multidomain and complex systems mixing databases, real-time controls, humans, etc. with object technologies. Large-scale applications are handled with complexity reduction techniques so, at any stage of system development, individual projects are self-contained and easily manageable. It is very difficult to find a complete and well-managed object project in the literature. Undoubtedly, they do exist and belong to well-kept industrial secrets. Well-built object software is still the black art of few specialists. Could this bunch become more crowded in the future?

This book can also be used as a graduate-level text for students at any domains involved in modern software development to inoculate intelligence to their biological, mechanical, electrical, physical, and social systems. They can discover that the UML and the MDA can be used in their fields unexpectedly. Specifically for computer science and software engineering domains, this book can be taught starting at the 4th trimester of undergraduate level when students have enough background on programming (an object programming course and a relational database course are necessary requisites). The reader should be exposed to difficult problem-solving situations and confronted with a large-scale project to really appreciate this neutral and uniform methodology.

Book Organization

From the scientific literature production domain, *this book is designed with our best knowledge about learning object research domain*, so individual knowledge chunks are often self-contained and self-documenting. If their knowledge dependency hierarchy and matrix cannot be avoided, superfluous cross-references/deferments are pruned to minimize learning effort.

Chapter 1 introduces modern software/hardware, database/real-time control, computer/intelligent, hybrid systems, and multidisciplinary systems. Chapter 2 explains techniques to cope with the complexity of large-scale systems. Chapter 3 introduces the metalanguage of the UML as a lesson of rigorous and metasystem development. This helps developers to acquire a deep understanding of metaelements of the UML. This chapter can be skipped at first reading. Chapter 4 is the main chapter for learning how to use the 13 diagrams of the UML; it examines the nature of each UML concept. One or many short examples accompany each theoretical exposé. Chapter 5 exposes fundamental concepts of the real world and their mappings into UML modeling elements. Their semantics are taken from problem solving and system development viewpoint, independently from the way the UML considers them. Business process diagrams in business domains are interesting examples for mapping UML diagrams. Chapter 6 presents the uniform methodology targeted to study complex multidisciplinary systems. A uniform view of the message concept is needed to represent interactions between objects of different nature. Some "patches" in object paradigm, based on "induced energy," "cause–effect chain," and the generalization of the "message concept" are introduced in order to model a richer set of applications. In Chapter 7, two advanced research topics beyond the UML are exposed to complement the arsenal of development tools. First, a state-event network (SEN), a new diagram based on Petri net, supports dynamical studies and refines the UML behavioral diagrams before implementation. Second, the "image attribute method" is exposed to systematically build sequences of a real-time system. All sequences melted together constitute the algorithm of the system or the solution. This method is a combinatorial technique deployed to study dynamic behavior of safety-critical systems. Chapter 8 contains case studies, both in technical and social domains.

Informative, actual, useful, returns from intellectual investment are targeted along this book-designing process.

Acknowledgments

All diagrams in this book are drawn with the version 6.5 of Enterprise Architect implementing UML 2.1, graciously donated by Sparx Systems Inc. SEN diagrams are drawn with Visio 2003. We would like to thank Rational, IBM, and Sparx Systems for supporting computer science and software engineering courses at Laval University, Quebec, Canada, with their software products year after year. We would like to thank OMG (Object Management Group) organization for providing the scientific community with invaluable object standards, for being the opposite of a conservative organization, and for graciously making numerous documentations available on their site. This work could not have been completed without the constant support of my wife Moc Mien, friends and relatives who were always there when positive morale, food, and scientific supports must be combined to produce words. Finally, a special thanks to the Chief Editor Mark De Jongh, and Cindy Zitter for accompanying us at every step along this arduous project.

Chapter 1

Introduction to the World of Systems, Problems, and Solutions

1.1 The World of Systems, Problems, and Solutions

1.1.1 System: Problem

Complex systems are the central theme that emerges from many disciplines and domains. Addressing the new challenge of complex system solving requires new research and educational initiatives based on the synergy of many disciplines as complex systems are often multidisciplinary and require a total integration of concepts, methodologies, and tools.

In everyday life, human activities turn around *system design* and *problem solving*. We are biological systems, we live in political and social systems and everyday, we build pieces of projects, which are parts of industrial, private, or governmental systems. According to our individual skill and expertise, we are civil, construction, mechanical, electrical, software engineers/workers, medical professionals, or businessmen involved in production or distribution systems. In order to coordinate individual or group activities, to design communications and coordinate workflows, we need management systems. Humans need social, artistic, sportive, and ludic systems to liberate their work pressure. In order to regulate relationships between humans, business corporations, workers, and administrators, we need legal, political systems. When problems cannot be solved inside the frontier of a country, we need international systems to rule exchanges.

At every stage of the human activity sphere, the work that steers the course of the society and its economical, governmental organizations is *building systems*, making *decisions*, and *solving problems*. Solving a problem impacts the system where this problem belongs to as the solution will modify its functionality. Serious analysis of the current situation and design of the correct solution need time and effort. If those phases are not executed correctly, hasty or arbitrary modification does not always mean improvement. Shortcuts, tyrannical and random decisions must be avoided as humans can build completely useless

D. M. Bui, Real Time Object Uniform Design Methodology with UML, 1–17.

systems hosting thousands of souls, gobbling huge resources in political, military, and management domains. So, the problem-solving domain is of highest importance in human organizations.

Solving problems and making decisions are not only privileges of administrators but they are also current burdens of people at any level. Budget, trade deficit, military actions, national security, and mitigation of tsunami and natural disasters are problems of all governments. Production efficiency, profit earning, product improvement, and investment choice are business organization problems. Marriage, having children, choosing a career, buying a house are individual problems for everybody.

The well-being of a society dictates that every system must be performed as effectively as it is designed. But, in practice, systems entailing problems after a certain time and new problems call for designing and implementing new systems. Even if those two notions are tied together, a system is not a problem. If a system is perceived clearly as an object or a collection of objects linked together through a bunch of relationships and involved in collaborative tasks, the problem is perceived as an undesired state of this object or group of objects. Common sense gives to the term "problem" a challenging situation or a difficulty to overcome, but a problem announces generally a dysfunction of a system or of a group of systems. However, a state may be occasionally considered as an object. For instance, in political or social systems, people often assimilate a problem (state) to a person or a group (object) to find out scapegoats.

Could a same problem be modeled inside a same system as a state and later as an object? The answer is "yes."

> Consider a car and for an unknown reason, its motor cannot start up. When the car is taken as a whole object, the state of the car switches to "malfunctioning." The problem is viewed at this high level as a state. Now, after some diagnoses, the mechanic detects a broken metal piece in the carburetor. The piece is therefore a problem but this knowledge needs a detailed investigation to find out the defective piece and reach this broken piece.

According to the audience, the same reality may be seen sometimes as a state, sometimes as an object, and sometimes even as a function. A problem could be assimilated to a function when people say "that is your problem" meaning a burdensome regular task we must carry out.

Poorly designed systems are subject naturally to recurrent problem occurrence but well-designed systems are not completely safe as all systems are made from engineering trade-offs. Moreover, evolution is a natural fact and systems are validated only inside a time frame. Today systems can be subjects to problems in the future as its environment evolves. Hence, system maintenance concept guarantees that systems will continuously keep their quality over time.

All systems and solutions are often scrutinized through a quality scale. As perfection is not a very common qualifier, every system exposes, as a normal fact, small problems or failures with time or that facing hostile environment. When the number of failures arrives at an intolerable threshold, it would be questionable to demolish an entire system to rebuild a completely new one. But, such a radical move must be dictated by a thorough analysis, a clear demonstration that old system is not upgradeable within minor modifications. There must be evidence that building a completely new system justifies the investment and finally the move.

The cornerstones for such developments and answers to all those inquiries are *modeling*, approach *methodology*, optimization, or *trade-offs* that are time-consuming activities but there is no other option. Only, methodology and *experience* can shorten the delay and therefore cost. Methodology means no trial-and-error cost, experience means existence of solution in a similar situation in the past. If we have both, methodology and experience, we can save investment, build the right system, or sometimes get a mere Boolean answer, positive or negative.

Theoretically, if a template or a schema for a given type of problem exists, then solving a new problem is simply a matter of adapting this schema to a new context, adjusting some data. Schemas are results of previous experience, so we can jump directly to the implementation stage, short-circuiting or reducing the time devoted to the analysis and design phases. Experts are good problem solvers because they are either experienced and/or they hold a good methodology. This book deals with "methodology" because novices, who do not possess problem schemas, are able to, with a good methodology, recognize problem types, model a new solution, and finally approach the expert schemas. Experts may be subject to biased solutions as they are inclined to bring everything back to older schemas. Good experts know how to adapt existing schemas to new situations.

The every day vocabulary used to describe systems and problems is very imprecise. To design well-made systems, to reach the silver bullet, we must avoid hazy meanings and imprecise descriptions. How to avoid misunderstanding on communication vehicle? Standards, metamodels developed with the UML (Unified Modeling Language) by the OMG (Object Management Group) address this concern.

When solving problems, we build a new system or modify existing system functionality. The design strategy is the same when handling the whole system or one of its functionalities. *This uniform view of system design and problem solving is very useful as we do not have to distinguish the problem from its hosting system.* We can now say that solving a problem means that we engineer the way a system works or the way many systems interact to get a common work done. Decision making is only a subphase of problem solving that implies the choice of one solution among many available solutions thoroughly studied.

Table 1.1.1.1 The table shows the similarity of approaches and steps used to design a system or solve a problem

Problem solving	*System development in computer sciences*
State the problem, clarify it, discuss it with someone, obtain needed information, look for different viewpoints, examine its history and its environment, take into account constraints, establish goals, etc.	Analysis phase
Generate idea, build many ways of solving problem, evaluate all the possibilities, choose one solution, etc.	Design phase
Try the solution, make adjustments, determine whether the solution works, etc.	Implementation/test phase

1.1.2 Tools for Handling Complex Systems

The last century was undoubtedly the starting point of modern science and technology. If electromechanical devices such as robots extend mechanical function of human arms and mobile robots extend locomotion function of human legs, computers extend abilities of the human brain. The quality of solutions and decisions nowadays depend upon electronic machines called computers. These machines have established the superiority of the US economy in the last century and are vital tools for our modern economy. Computers already help us in manipulating large volumes of data, controlling devices, and automating most processes. Promising scientific research goals target how human minds solve and make decisions. With the collaboration of psychology, statistics, operational research, political and social science, research on artificial and cognitive science must help us in handling more and more complex, hybrid, and multidisciplinary systems.

Before giving the machine the opportunity to assist us, problems must be formulated appropriately and transposed into a machine understandable state so that the machine can deploy its extraordinary computing resources. In fact, we must build a "model of computing" that is a replica of the problem or system to be solved. The process of designing and building automated information systems is greatly influenced by the tools used to assist in the process. Systems analysts, designers, and developers turn to Computer-Aided Software Engineering (CASE) software to capture information about business requirements, create design to fulfill these requirements, and generate a framework inside which programmers feed the code.

The most elementary CASE tool associates drawing software to a DBMS (database management system). It generates automatic documentation and aims to generate code in the near future. Automatic code generation is still an immature technology as it lacks a good formalism to express system dynamics.

Nowadays, code generators can easily create classes, attributes, and method headers in a given language (for instance Java, C++, C#, etc.) from UML compatible CASE tools. Programmers still have to "fill in the gaps" and manually put code inside bodies of class methods.

Roughly speaking, CASE tools assist us in handling complex systems, drawing graphics, putting an end to laborious manual writing of tons of documentation, and greatly increasing overall productivity. It is now nearly impossible to separate the modeling process from its corresponding tools. Tools available for object modeling and compliant to UML standards can be found on the OMG web site at http://www.omg.org.

1.1.3 Intelligent Systems: Hybrid Systems, Multidisciplinary Systems

There is no clear definition of complexity and degree of complexity in sciences, except in very well-targeted domains. Complexity intuitively refers to systems that have a large number of components, need important mass of specifications, exhibit thousands of model elements, require concurrency and interprocess communication, must react to external environments, etc. Complexity in the information and computation sciences is roughly based on the fact that the problem is either tractable or not inside a time frame. This complexity may be measured in terms of algorithms or steps required to compute a solution. A problem is tractable if it needs polynomial-time algorithms, whose computation time equals the problem size raised to some power, for instance n^2 for a size n problem. It is intractable if it has only exponential-time algorithms, whose computation time becomes exponential, for instance 2^n. The intractability comes from the fact that the amount of computation required exceeds any practical time limit. People talk of combinatorial explosion. Without browsing through all categories of complex systems in general sciences, not limited only to algorithmic aspects of computer sciences, we observe that they are often intelligent systems, typically nonlinear and associate many domains or disciplines.

Without drowning into a war on the definition of the term "intelligence," the first expression of intelligence for a physical system is its ability to reproduce a human behavior for problem solving, system control, and/or decision making. In that sense, an electronic (or even mechanical) thermostat is already an intelligent system/device since it can monitor the temperature of a room without requiring a person to do it manually from time to time. Some scientists insist on a learning capacity to issue an intelligence label. If a system can learn from experiences, it would be a "very" intelligent system. So there would be a lengthy scale in intelligence measurement. On the high end, research on neural networks, fuzzy logic, knowledge-based systems, learning process, genetic algorithms, evolutionary computation, and mathematical and chaos models is one of the key issues of developing intelligent systems that can assist us in making decisions and solving problems.

Hybrid systems (Alur et al., 1995) are formally defined in computer sciences. Essentially, they are heterogeneous systems that include continuous-time subsystems interacting with discrete events.

> The dynamic state of a car changes when a gear shift occurs, either because a driver moves the gear lever (input discrete event) in a manual car model or because the "speed" (continuous time) state variable reaches a specified threshold (state event) in a car with an automatic transmission.

A hybrid system can have both continuous and discrete states. If a digital computer (discrete states only) is used to control a physical system (typically with continuous states), we have a hybrid system. Controlling a hybrid system is a challenging task.

> At any moment, continuous variables as throttle angle of the car and logic decisions (gear selection, brake action) have to be adjusted in order to achieve a desired output (in-demand car speed and driver comfort).

Hybrid systems are found frequently in *multidisciplinary systems* that are found in large projects mixing many technological domains.

> A robot project currently involves mechanical, electrical, and computer engineers. A heart monitoring medical project associates biological patients to electronic equipments. An airport security system with various sensors is a multidisciplinary project.

Multidisciplinary Design Optimization (MDO) is a research branch that decomposes a large multidisciplinary system into smaller, more tractable, coupled subsystems. Decomposition results are often nonhierarchic and are a mixture of both collaborative and hierarchic components. They require iterative techniques to attain a converging system and its associated optimal design point. In automotive and aircraft industries, optimal designs may require hundreds or even thousands of iterative cycles to attain convergence. A survey article on MDO can be found in Sobieszczanski-Sobieski and Haftka (1997).

With object technology, developers address the formidable organizational challenges of multidisciplinary systems and the design concepts applied to minimize the coupling of objects at all levels. Multidisciplinary domains are difficult issues to address as current generation of scientists/technologists tends towards increasing depth of knowledge and narrowness of interest. Moreover, it is hard to find superhumans with multidisciplinary competence so when designing multidisciplinary systems, a project manager must conduct the project in such a way that the analysis phase must take into account this reality as an initial constraint. Therefore, multidisciplinary projects necessitate team work in most cases.

1.1.4 Component-based Systems

As system complexity continually tends to increase with computer and high-tech advances, compensatory mechanisms such as component development (Sametinger, 1997; Heineman and Councill, 2001) must be introduced to counterbalance the negative effect of the nonstop trend towards complexity. This kind of compensatory mechanism is essential because humans have a limited ability to deal with complexity. These components are nowadays called multidisciplinary components.

> We can buy a robot arm, choose its brain, its vision system, the tool needed to perform a specific task (assembly, welding, etc.). Many software companies sell web components (sending mail, uploading files, etc.) to be integrated in a web site as web components.

This component movement is perceived in languages, in software development platforms such as .NET (Microsoft Basic Development Platform) and J2EE (SUN Java 2 Enterprise Edition). A software component is a unit of well-specified interface and functionality, intended to be part of a third party system. A component hides implementation details and exposes only its simple interface to OEM (Original Equipment Manufacturers).

Although the notion of component/black box is a very old programming paradigm, its real implementation has changed considerably through time. While in the early days of programming, software components were libraries and source code modules, nowadays, plug-and-play drivers, web services, and text editors to be included in our final applications are real components.

One of the directions of component-based software research was to standardize the interface offered by independent developers. The first goal of such an approach is application development using commercial COTS (Component Off The Shelf), analogous to hardware electronic assembly using integrated circuits as building blocks.

To use component software, a solid foundation in computer science concepts is required and significant effort must be deployed to understand the specifics of whatever component framework we want to use. A second goal is thus the simplification and the reduction of of learning curve required for using components.

> Nearly everyone who uses a Web browser has installed plug-ins, add-ons, and extensions. The situation is similar for the Office Suite. They are examples of component software. Practically, unless an installation error is returned by the browser, most of us just have to know vaguely what a component is supposed to do to deploy it.

Very often, we simply unplug the component in case of installation errors and few of us are really interested to undertake a debugging phase. The challenge for system integrators is how to assemble components and deploy this new technology. Getting experience with components can help us in finding a good job than developing the component itself.

As we consider multidisciplinary systems, components are not only limited to software. Millions of people rely on medical devices for their survival.

> Those who suffer from certain cardiac ailments depend on pacemakers (ICD: Implantable Cardioverter Defibrillator) to regulate the beating of their hearts and to prevent sudden cardiac arrest. If ICD emits electromagnetic waves to feed in centralized monitoring system, we witness a small multidisciplinary system that is composed of components originated from various scientific disciplines for the benefit of medical therapy domain. So, modern components are often embedded hardware/software intelligent components.

1.1.5 Solutions

The theory of problem solving (Reitman, 1965; Simon, 1973; Jonassen, 1997), regarded as the most important cognitive human activity, is a research branch devoted to the exploration of mechanisms used by humans to find out solutions to various types of problems. From the simplest logical problem to assess mental acuity and logical reasoning, through algorithmic, rule-based, decision making, strategic troubleshooting, to dilemmas towards a design problem (the most complex and ill-structured kind), researchers try to discover mental mechanisms in order to automate and assist humans in their daily activities.

Problem typology has been found in the literature as:

1. *Algorithmic* – Find a set of procedural steps to get a specified goal or result.
2. *Rule-based* – In a context of ill-structured problem and in the presence of a set of heuristics, rules, and constraints, how to apply rules to get the optimal solution, for instance, moves in playing chess game?
3. *Decision making* – Select a single option from a set of alternatives based on a set of criteria.
4. *Troubleshooting* – In the presence of a faulty system or dysfunction, analyze symptoms, emit hypotheses, and perform limited tests to find out faults and their sources, suggest repair solutions.
5. *Diagnosis* – Idem to troubleshooting without any solution proposal.
6. *Strategic performance* – Apply in real time a set of known tactics to meet strategic objectives, e.g. flying an F2 jet fighter in a combat mission or playing hockey.

7. *Dilemmas* – No solution is acceptable by everybody. This kind of vexing problems is mostly found in politics when we must choose a candidate who does not represent a solution but offers the least dangerous perspectives.
8. *Case analysis* – For training, elaborate an imaginary situation inspired from a real situation, study and structure it.
9. *Design* – Our problem.

The first snag of design problem is ambiguous specifications of goals, lack of information about environment of a project followed by tremendous effort expressing and communicating conceptual intentions into blueprints, elaborating working strategy, choosing teams, coordinating groups, execution of goals and sub goals, testing, etc. Whether designs are embedded electronic devices, houses, books, or political campaigns, the process requires deployment of domain-specific knowledge, artifacts, clear standards/laws, and human skill that is as yet irreplaceable. Moreover, the design problem seems to be an ill-structured problem when starting a project.

Problems solved in schools and universities are generally well-structured. They are mostly mathematical, logical, case study types. They contain clearly defined and finite number of elements and constraints, goals are unequivocal, the convergent solution exists and can be found when applying concepts and rules that are themselves well-structured in a domain of well-structured knowledge.

Ill-structured problems contain opposite statements: unclear goals, multiple or no solution at all, imprecise criteria to evaluate the degree of confidence of the solution or of the constituent elements. Dilemmas are used as decision-making mechanism (election). Judgment is influenced by emotional parameters. The person who makes a decision on an ill-structured problem must defend it and generally, only one portion of the audience agrees with his choice.

The challenge of this book would be *bringing all design problems from an ill-structured state towards a well-structured state*. Most people seem very excited about new projects but once the subject takes shape and becomes clearer, they often abdicate it, hence, problems cannot be solved; but when passing from the ill-structured state to a well-structured state, systems can give us a lot of useful information. A negative decision does not mean a useless endeavor.

The roadmap towards structuring any system, the research of the optimal compromise/solution/conclusion passes through the deployment of development schemas/patterns, methodological framework, well-structured tools (in our case, UML, MDA), well-trained and domain-competent teams and managers as behind every project, there must be a team of right persons.

1.2 Real Time and Embedded Systems: Disaster Control and Quality of Service

1.2.1 Real-Time and Embedded Systems

Real-time and embedded systems constitute a class of systems that need very precise models by their application fields. Roughly speaking, in a real-time system, things must be done in time. Late answer, though correct, is wrong and could be catastrophic.

> Consider a cap putting machine on a bottling plant. Each bottle must be capped as it passes continuously on a conveyor belt. The speed of the conveyor can be slow, but stopping the conveyor to put a cap is a very costly operation as it can slow down considerably the production line. Therefore, the motion of the cap is coupled with the speed of the conveyor belt and the cap has only a very tiny window of opportunity to cap the bottle. Timing constraints are characteristics of real-time systems and they must be met most of the time or every time if human life is concerned. Other examples of real-time systems include: laboratory experiment control, automobile engine control, nuclear power plant control, flight control system, components in spatial domain, avionics and robotics.

Hard real-time system does not accept any delay since catastrophic failure would result. This kind of system is omnipresent in industrial control and automation, aeronautics, military, and spatial applications. *Soft real-time systems* tolerate late answers, deadlines may be missed, but the commercial value of the system diminishes with the frequency of missed answers. There is no specification of the absolute scale for time despite the fact that hard real-time systems works mostly from fractions of nanosecond to fractions of second and soft real time covers longer delay scale compared to human inertial reflexes that spread from 10th of second for human body parts and 100th of second for eye retina perception. But, those values are only indicative as the limit cannot be ascertained.

Embedded systems are systems constructed with minimum resources needed to lower the manufacturing cost of the equipment. Embedded systems associate hardware and software in a unique design so they are often hardware/software codesign problems (Wolf, 1994). From microwave oven to aircraft control system through mobile phone, MP3 player, and car antilock brake controller, embedded systems are often microprocessor based and their software is becoming more and more sophisticated (Lee, 2000).

In the past, embedded systems were programmed with assembly language. They needed a small quantity of memory, pertained to industrial domain so computer scientists largely ignored embedded software. Embedded systems are nowadays real technological pieces of jewelry. Small digital cameras nesting inside the palm of our hand are able to process the densely populated (about 10 mega photosensitive cells or more) cells, control the movement of tiny zoom actuator in real time to shoot still images, or record scenes like a classic camcorder. Small programs in small space may contain highly sophisticated signal processing algorithms.

Embedded software components are emerging technology. If, in the past, developers of this technology were bare metal assembly programmers, if they can neglect object technology, AI (artificial intelligence) foundations, multiprocessing operating systems, networking, and web programming, things are mutating speedily and in the near future, embedded developers must be aware of object/component development as embedded software are becoming more and more complex. *Reusable software components* for embedded systems are already under way, specifically in signal processing and in multimedia applications.

Embedded systems are often *embarked systems* as the clear tendency towards miniaturization is of topical interest. Moreover, they are mostly soft real-time systems. A mobile phone must sustain any conversation without delay. From the modeling and development viewpoint, the "embarkable" property or the real-time attributes are converted into requirements or constraints. There is a subtle difference between requirement, a "must have" feature, and constraint, a "not to be violated" clause. For instance, a house must be built with a patio (requirement) and it must be localized at least 20 ft from the road (constraint).

The following table summarizes fundamental concepts in real time.***

Table 1.2.1.1 Characteristics of real-time applications

Characteristics of real-time applications	
Application areas	Control and automation, avionics, military and spatial applications, scientific measurements, telecommunications, domestic appliances, automotive applications, medical instrumentation, process control, robotics, automotive control, computer domain, real time graphic simulation, etc.
Characteristics of software	Timing constraints (deadlines for tasks) Applications are said to be responsive Deterministic and predictable responses and behavior Interaction with the environment (reactive system) High reliability
Options	Embedded components (Efficiency of code, energy saving) Interaction with the environment High safety critical software
General techniques deployed	Hierarchical hardware interrupts and exception handling Task queue reschedulability (preemption, priority) First quality components to insure high reliability Extensive testing in hardest conditions Redundancy to enhance safety Program correctness checking Document and software traceability High concurrency (to satisfy timing constraints) Reentrancy (to save memory in embedded applications) Deadlock removal techniques

From the modeling viewpoint, the study of real-time and embedded systems are interesting for many reasons. First, most modern systems are soft real-time systems so time dimension is ubiquitous. For instance, if the climate change is really due to certain gases responsible for global warming and does not receive a correct reaction on time, we can assist to economic, human and ecological catastrophes at short term. Second, once logical model is established, time constraints may dictate the confection of intermediate model between logical model and implementation model. Real-time scheduling really enriches our experience when designing systems and offers an opportunity to explore modeling ideas. Third, embedded systems may add physical, electrical, power consumption, and environmental constraints and may greatly influence the choice of the initial approach at the early stage of logical model building in a MDA (model driven architecture).

Real-time computing is fast and many interesting theoretical problems (scheduling, competing for resource, etc.) are brought to light while studying real-time systems. Given a set of demanding real-time constraints and an implementation using a fastest processor and hardware, how can we be certain that a specified timing behavior is indeed achieved since even testing is not a correct answer? Speed does not always mean predictability. Real-time engineering is labor-intensive performance and optimization exercise so the interaction of many research domains can be applied in real-time context. Real-time systems offer a great challenge for system modeling, specification, verification techniques, scheduling theory, and AI.

1.2.2 Disaster Control: Quality of Service

Besides the relevant question of how to build systems to meet requirements at acceptable costs, hard real-time systems, by their implication in high security transportation, medical domain, avionics, and spatial applications must deal with possible disaster situation in case of component dysfunction. Soft real-time systems, by their usage in economical environment (banking, ticket reservation, credit card approval at boxing day, etc.), social systems (tsunami, storm alert, etc.), and political organizations (Kyoto Protocol, military response, etc.), aroused in the past lot of economical and social concerns such as how to manage hardware/software failure and how to insure quality of service (QoS) even in high-load situations.

Disaster control of real-time systems passed through redundancy and replication of spare components. (Pierce, 1965; Randell, 1975) investigate fault-tolerant systems, which can tolerate faults and continue to provide functionalities despite occasional failures, transient or permanent, of internal components. If we can replicate merely hardware units to realize redundancy, software redundancy cannot be merely a program replication but multiplication of independent designs. In case of failure, control is switched to spare components, whether hard and/or soft. Reconfiguration may be needed. Disaster control in database systems practice data redundancy and checkpoint layouts.

If failure to meet deadlines in soft real-time systems does not create disaster right away, it may affect the QoS concept developed originally in the context of network bandwidth utilization and packet loss management (Aurrecoechea et al., 1998). As examples, waiting queues in health system, traffic control, reservation and banking systems, multimedia distribution, internet server access, etc. are domains sensitive to QoS, which is directly a real-time constraint as service time is perceived as the principal factor of quality. QoS is not a functionality, but characterize how fast this functionality is performed. If QoS is taking into account early in the design process, it may transform the overall architecture of the design. The MDA paradigm fits perfectly with QoS as it adds a supplementary QoS-aware model transformation and does not modify the original model addressing the functionality itself.

1.3 Human Organizations and Structures: How Software can Simulate and Study Them

1.3.1 Human Organizations and Structures

Human organizations have been considered as distributed intelligence (Lewis, 1988). Another direction of research considers agents and multiagent systems (Flores et al., 1988; Jennings, 2000; Odell et al., 2000; Luck et al., 2003; Epstein et al., 2001). A social system or a human organization is considered as a multiagent system, and computing results are extracted mainly by simulation (MABS: Multi Agent Based Simulation). Traditionally, agent and object technologies are similar (Davidsson, 2001) and MABS should overlap with object-oriented simulation (OOS). What is referred as agent in the context of MABS covers a spectrum ranging from ordinary objects to full agents with autonomy, proactiveness, mobility, adaptability, ability to learn, and faculty of coordination, briefly speaking, all skills and aptitudes that normal human has. An agent is therefore a human model in an organization.

What is really important to notice is that agent technology naturally extends an object component-based approach. The agent technology already supports already applications like automation of information-gathering activities, purchase transactions over the Internet, etc. In fact, agent technology relies on object/component technology as its implementation model. The difference is only at *development model* metaphor level and lies in various terminologies used in different technological domains or, to use a newer language, an ontological issue in the recent MDA proposal.

Agent-based technology is thus derived from the practice of computer, AI, object/component technology, and modeling of human organizations. Roughly

Table 1.3.1.1 Depletion of vocabulary from high to low models

Concept	*Application level*	*Design/modeling*	*Implementation level*
Entity	System, architecture, prototype, product, organization, structure, actor, agent, object, component/tool, simulator/emulator web service, database, data schema, code, etc.	Component/aggregate, package, module, model, class, object, association, relationship, node	Class, object, component, interface
Data	Property, data, variable, parameter, state, knowledge, etc.	Attribute, variable, state	Attribute, constant
Function	Plan, function, task, operation, activity, action, process, procedure, script, scenario, formula, equation, algorithm, inference rule, service, etc.	Operation, task, action, activity service, method, function, process	Method, thread, task

speaking, agents are social objects or components. According to the level of discussion and the audience, we can use terms fitting inside a cultural, social context to help people understand the model; but in the implementation model, everything are objects for programmers. Assimilating agents to objects means simply adopting the implementation ontology to describe a richer social ontology (that is fundamentally a communication error). Developers must switch easily between domain terminologies as modern development like MDA evolves through many domains when navigating among models. Table 1.3.1.1 shows that the general vocabulary used to identify things, data, and functionalities, depletes and differs when evolving from higher models towards lower models.

1.3.2 Simulation of Human Organizations

According to Bradley et al. (1988), "Simulation means driving a model of a system with suitable inputs and observing the corresponding outputs." To complete this definition, first, we operate on a model, so the validity of the results depends upon the exactness of the model and how close the model matches the reality. If we pass this basic premise, according to the purpose of the simulation, we can predict (weather forecast, prediction of interest rate for the next month), train (flight simulator), entertain (electronic games), educate (create imaginary worlds to pass concepts), study the performance (computer simulation), prove, discover hidden behavior, and correct the original model. In the last case, simulation can be assimilated to testing.

Results obtained with simulation (Axelrod, 1997) are based mainly on the two concepts, *induction* (discovery of laws or patterns from empirical data or reasoning from detailed facts to general principles) and *deduction* (proof from a set of axioms or reasoning from the general to the particular). Simulation is a third research methodology used mainly in computer sciences. We cannot prove theorems with simulation. Simulation generates data that can be used inductively with the difference that data come from the model but not the real world as measurement does with scientific experiments. Simulation stimulates our intuition to find natural laws in the simulated discipline.

Gilbert and Troitzsch (1999) recognized that simulation is a particular case of modeling. Scientists build simplified, less complex models to understand the world. They hope that conclusions drawn about the model will also apply to the target because the two are sufficiently similar. A model can be a formula to predict the interest rate of the next month, a logical statement saying for instance that "if some meteorological conditions are met in winter, there will be risk of black ice formation on the highways." Models can be complex and costly objects with simulated functionalities. A proportional and reduced robot model, representing a real and more heavy robot, or a complex computer program of many thousand lines of code that predicts, are considered as models of the real world. Statistical models and simulation models both predict or explain outputs. Statistical models are mathematics and they typically extract correlations between variables measured at one single point in time. Simulation is rather process-based and a simulation program has to run many times in order to observe or get the correlations depicted by statistical laws or formulas.

1.4 This Book is also a System

To establish the unified view of system building and problem solving, this book project can be stated as "how to produce a book at the frontier of graduate/undergraduate level, teaching modeling of real time systems, reaching a very large audience and having a relative long life despite the fact that it adheres to an object and rapidly evolving standard like UML." The system is a real human social interaction system with three main actors/agents: the *author, publisher*, and the *public*.

If the *author* is alone, the *publisher* is a huge organization of more than 5,000 employees working in 19 countries around the world. So this organization delegates a person (*publisher contact* agent) at the interface to manage the book project. A *public* agent does not play a visible role at the elaboration phase but, he/she influences greatly the decision of the *publisher* to go ahead with the project and invest. The *publisher* has great experience about the *public* behavior and the *public* expectancy in hi-tech learning. Once the book is published, the *public* agent will decide on the actual success of the project but this matter should be another concern. This kind of inquiry, out of scope, must be pruned off at the starting point to avoid blurring the current analysis.

Is there any relationship between the *public* agent and the *publisher* agent at this conception stage? One possible answer could be the *publisher* belief parameters based on market trends, research interest, and data available for similar books, inside or outside the frontier of the *publisher* organization. Moreover, a *public* agent may influence the *author* agent during his writing since meticulous *Author* could ask, during his writing, some agents belonging to the *public* group to read the proof and give him useful feedback.

Other important objects in this system are documents exchanged among agents: *proposal, contract, schedule, book template, guidelines, author proof* with all their versions, etc.

In the second stage, *author proof* is composed of chapters and each chapter is an aggregate of learning objects. States of *author proof* is a multivalued attribute qualified as *first draft, draft reviewed, typeset proof, final proof*, etc.

When starting this analysis, secondary agents like the *technical proof reader* and people behind the scenes who belong to the *publisher* organization are not identified. But, if we establish a UML sequence diagram to follow the evolution of the writing and publishing process, all the secondary agents will appear successively as we examine our system in more details.

Is this system real time?

An *author* agent must end his schedule at the date that the contract stated for completion of the first draft. The system is in fact a soft real-time system. The *publisher* agent has a stake in making sure that the *author* schedule is reasonable because they have dependent services and resources that they need to schedule such as layout people, publicity material for conferences, and mass mailing advertising. If the schedule slips, the schedule of the *publisher* may suffer and deadlines may cascade through all related areas. Moreover, racing against a technology may outdate the book before the draft is finished. So time is an important constraint for such a system.

To put the investigation a notch further, the very first question does not specify the time frame and voluntarily, we terminated the system analysis at the end of the first production process. If we consider a longer time frame including several years with possible reprints or new editions, if we want to monitor *public* reactions, the problem would be more complicated. In this case, to manage complexity, the question should be reformulated into several simpler problems, considering the already studied production phase as system S1 and others new systems to come as S2, S3, and so on. All these small systems, hooked together, will provide insights about this publishing system dynamics over a longer time frame.

A *publisher* agent can monitor the book distribution, feed data inside a simulation program, and get advice from this simulator on how to proceed in the future with the evolution of the book revenue. In the absence of the simulator, the editor staff plays this role. In the latter case, results would be variable according to who makes the decision at time T in the *publisher* organization (biological intelligence at the rescue of AI).

This problem can be studied in more detail with all the formalisms, tools, concepts, and models exposed in this book and the design/implementation models will produce a set of blue prints with suggestive diagrams replacing this boring text. But, at this stage, we are faced with the question of how to interpret UML diagrams correctly to understand the communication material transmitted through blueprints. So, the analysis process brings us back to a very elementary issue "we still need text to start somewhere the learning process." This is why you should read this book completely.

Chapter 2

Coping with Complexity

2.1 Visual Formalisms for Handling Complexity: A Picture is Worth a Thousand Words

2.1.1 Free Diagrams

"Seeing is believing" or "A picture is worth a thousand words" stress the importance of visual formalisms in learning and communication. Every time, we want to draw a system, unconsciously, we draw a circle or a rectangle. A closed contour delimits a small surface that materializes an abstraction, in our case, a system, an object or a component. Children draw forms before language. Primates such as chimpanzees produce paintings with a striking similarity to drawings made by children at the early stages. The world seems to have an iconic representation before we can abstract it with language. Six recurring "diagrams" were identified by psychologists with children: circles/ovals, squares, rectangles, triangles, crosses, and irregular odd shapes.

Irregular odd shapes are difficult to draw (Fig. 2.1.1.1). In the early stage of object technology, Booch (1993) used a cloud or blob notation to emphasize abstractions. Venn diagrams use a very simple form of blob. Odd shape is dangerous because people draw it differently. Since it conveys artistic flavor, it may be appealing to someone and may create an opposite reaction to another and this fact can impair the message we want to pass. Moreover, the natural tendency when drawing odd shape is to include technological details in shapes and then suggest precociously an implementation detail in a logical model. For instance, when drawing a security lamp, if we represent it as a rectangle, we let to the implementer enough margin to decide and operate a technically optimized choice, but if we represent it with a pear form, it suggests an early incandescent lamp, so we have unconsciously passed a message to the implementer. Unless, this choice is an initial requirement of the contract, this design "bug" should be avoided while modeling systems. So, to avoid cloudy and hazy messages, irregular odd shapes must be banned from technical diagrams.

D. M. Bui, *Real Time Object Uniform Design Methodology with UML*, 19–45.

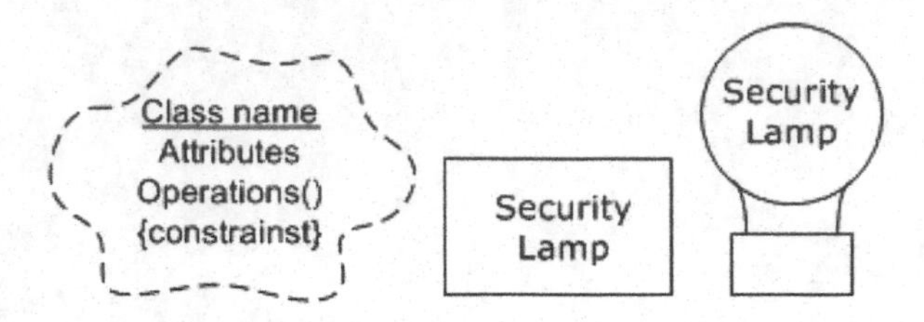

Fig. 2.1.1.1 Odd shapes and free forms. Despite its artistic flavor, the cloud representation of object is difficult to draw. The pear form of the security lamp passes an unclear and unattended message to the implementer. A simpler and neutral shape such as a rectangle is more appropriate to represent object at the logical phase

However, free diagrams or nontechnical diagrams that contain odd shapes are currently used by many people, incidentally those who give contracts. In fact, we cannot command everybody to know how to interpret technical diagrams. These persons come from disciplines distinct from Sciences or Computer Sciences (that is why they need us), they try to communicate with us and express their vision of things differently with nontechnical diagrams, sometimes with an artistic flavor. We must clarify things and identify true requirements with them in the analysis phase. Sometimes, it would be advisable to insert nontechnical and artistic diagrams accidentally in a conference or a meeting to capture the attention of the audience. So, do not underestimate the accidental power of unconsciousness.

2.1.2 Technical Diagrams

Technical diagrams are iconic by nature. Icons accelerate the recognition of the abstract concept. They must be self-explaining, suggestive, and should not, as said before, contain any detail that could be interpreted as implementation constraint. Each profession has its own set of icons. Electrical engineering for instance has a large set of icons to represent real objects at various scales (op-amp, resistor, capacitor, transistor, etc.). Computer analysis and logical models make use of standard and neutral forms like circles/ovals, squares, rectangles, and triangles which are connected by lines and arrows (single or double oriented lines). Except for circle/oval, these forms are made from a succession of straight lines (edges) and vertices forming a close contour, colored or not, delimiting an abstraction. They are *objects with volume*, meaning that they fill up a given space on a drawing area. Lines and arrows that are lightweight *objects without volume* may then connect objects with volume to establish relationships or communication channels between them.

The number of objects with volume laid on an $8.5'' \times 11''$ document must be limited to avoid *visual surcharge*. Psychologists think that visual

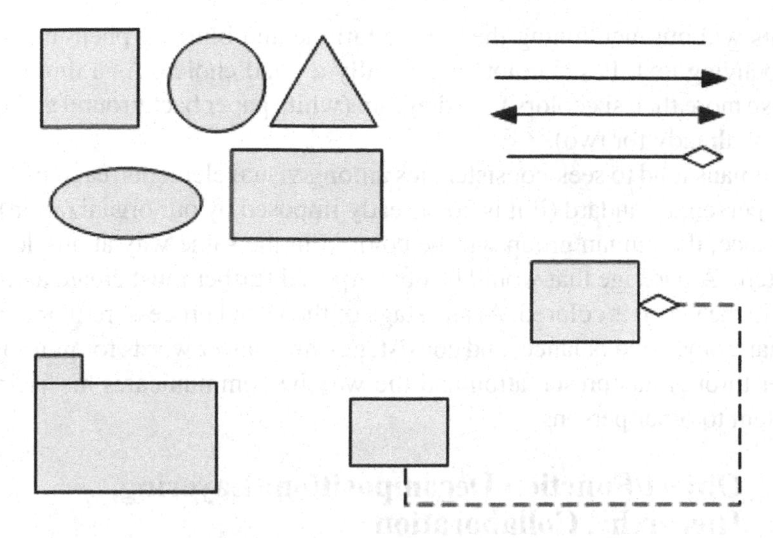

Fig. 2.1.2.1 Objects with/without volume, decorations, and adornments. Primary shapes with volume occupy space in an area. Arrows, connectors are object without volume. If connector makes a roundabout way to reach two objects, it may be equivalent to an object with volume and may create an unattended visual surcharge. A rectangle with a smaller rectangle as adornment is considered as a unique object as long as the adornment volume stays reasonable

short-term memory is able to store 7 ± 2 objects only (Miller, 1956) as a population average. Even if the theory of short-term memory and their empiri cal result are controversial, it would be common sense that there must be some limit for the number of objects with volume we can put on a diagram, by the simple fact that identification lettering of objects must remain visible. Objects without volume, such as lines and arrows, do not contribute to the visual surcharge at first approximation but, if a line follows a complex way to join two graphical objects with volume, it mimics a visual object with volume and by the way, may be assimilated to a pseudoobject with volume and could increase the visual overload as an object with volume does. Figure 2.1.2.1 summarizes these notions.

Must we fill objects with volume with color? The theory and treatment of color is a huge topic that has spawned massive discussions with conflicting opinions. Color sends a very strong signal and bear cultural values. If color can be used effectively to differentiate efficiently model elements, there is no single accepted theory about what color combinations are most effective. Color can make a dramatic difference in the aesthetic qualities of a diagram and as consequence can influence our desire to go further with its content. Ill-colored diagram can disturb, create confusion, and discomfort the viewer. So, be careful. A diagram without color (or white colored) can be more interesting than a diagram colored with a doubtful combination. Heavily saturated or primary colors should be used sparingly because they draw attention away from other elements without mentioning the viewer fatigue and blurring phenomenon of accompanying text. Pastel colors are usually a good choice. As a thumb rule, never use more than six colors for a diagram (white paper background and black ink count already for two).

As humans tend to seek consistencies among visual elements, each of us can adopt a personal standard (if it is not already imposed by our organization). So, for instance, the human group will be colored in the same way at any level of the system. A package that would be decomposed further must create a kind of anchor in the way it is colored. At any stage of the visual process, remember that unity, harmony, good balance, and consistency are master words for guiding the modeler through his presentation and the way he communicates his vision of the system to other persons.

2.2 Object/Function Decomposition: Layering, Hierarchy, Collaboration

In the past, when modeling was in process, designs are conducted in a *top-down* manner, so the methodology of complexity reduction was called functional *decomposition*. Combining elementary objects to form a larger object is the *bottom-up* method and the process is called *composition*. Although composition/decomposition has been used for a long time in conventional modularization techniques, they are still valid mechanism of complexity reduction in the context of object-orientation.

When decomposing, objects identified at each level of decomposition compose a layer at that level. This process thus creates a hierarchy of objects with the most complex object lying on the top of the hierarchy. Elementary objects are found at the bottom layer.

> The car system can be broken down into major interacting subsystems, such as the engine, the transmission, the cockpit . . . that are made up of more primitive components. For instance, the engine consists of carburetion and cooling subsystems. At the most detailed level (bottom layer), subsystems are primitive physical component like screws, metal sheets, molded parts . . . The car's cooling subsystem can be described in terms of radiator, water reservoir, rubber tubing, and channels through the engine block. The car stays alone at the topmost layer. At intermediate layers, we found engine, transmission, brake, direction, passenger cockpit. At lowest level, a car repair shop can list a computerized inventory of thousands of parts.

Hierarchy and layering technique correspond to aggregation/composition counterpart in object technology. But this is not the only structuring mechanism governing the world of complexity. As applications become more and more sophisticated, they associate many objects coming from various disciplines. Tasks in everyday life are executed with the contribution of many objects which cannot be put inside a hierarchy. In this case, the term *collaboration* is used to describe such association, and objects are collaborative objects/agents assuming some roles while achieving a common goal. The same object may have multiple roles if it can execute more than one task. If aggregation/composition hierarchy is seen as a *vertical* mechanism structuring the world, collaboration is the *horizontal* mechanism. In fact, modern systems call for these two mechanisms simultaneously at various levels.

If we consider now the previous car with a conductor inside interacting with the car and looking at the highway, we have both a hierarchical system (the car) and collaborative objects (conductor, car, local geometry of the highway, other vehicles, local meteorological objects like fog or snow, mobile phone, etc.). All these objects cannot be aggregated inside the car hierarchy, so, only collaboration could be the most appropriate description at the driving level. In a normal situation, all the car objects are hidden from the car hierarchy to simplify the discussion on driver reactions, but in case of an accident caused by the driver grasping his mobile phone, the quality of a hidden component, its airbag, will be of highest importance.

Collaboration is also the main working mechanism inside human organizations. Roughly speaking, the hierarchical vision is more appropriate for dumb objects. If we take individually molded parts of a car engine, we can do nothing with them, they must be mounted together to accomplish a given task. But, when objects become more and more intelligent, they can execute several functions besides those specific to collaborative tasks. They are autonomous *agents*; they can share their knowledge and capacity in diversified collaborative tasks involving many agents.

When designing the structure of a company, people mostly see it as a hierarchical organization with a president, departments, managers, engineers, workers. A containment relationship is often proposed to put workers under identified teams and humans inside departments. But, the way we model the system is not static and depends upon the current context, time, space and culture. There is no universal valid model. Must we model the interaction model of this enterprise with a horizontal or a vertical scheme? With a tribal organization in undeveloped countries, the company is a real hierarchy as the head acts as a tribe chief or a king, but in democratic countries, a president of a company can be replaced by its administration board without affecting the functionalities of the company. So the model depends on space, time and culture. In North America, we can still have a tribal organization but, generally, workers/engineers/managers can be replaced so a collaborative model is more appropriate to account for its operation. Sometimes, we must choose a mixed model to account for the reality.

2.3 Functional Decomposition to Handle Complexity

2.3.1 Functional Modeling is Based on Processes

A *process* can be seen as a set of ordered and procedural steps (atomic actions or activities) intended to reach a goal (Fig. 2.3.1.1). A process is present in the UML standard which defines it, within the current grounds, as a heavyweight

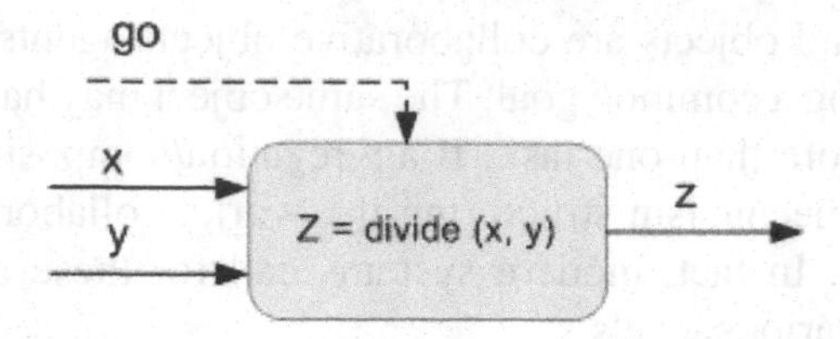

Fig. 2.3.1.1 Gane-Sarson or Yourdon-DeMarco bubble representing a process in functional modeling. Control is based on process/functional modeling. Arrows arriving on the process materialize input data and arrows leaving the process are output data. Dotted arrows are interpreted in DeMarco methodology as control signals. In the figure, the "go" signal fires the division and yields z data output when x and y data are made available at the inputs

unit of concurrency in an operating system and reserves the term *thread* to characterize a lightweight process or simply a single path of execution through a program. Function, process, procedure, thread, task, action, and activity will be examined in Chapter 5.

Classic functional modeling, not object oriented, have their adepts and a lot of blueprints still populate industrial drawers. Even, a good object modeler must understand what a "bubble" is because it is not excluded that some days; he will have to read, maintain or reengineer an old project made with a functional dataflow methodology (Yourdon, Ward-Mellor, Yourdon De-Marco, Gane-Sarson, Jackson, etc.). The DFD (Data Flow Diagram) can be found currently in modern CASE tools or general drawing tools as Visio. DFD comes from one (among the oldest methodologies) proposed by Yourdon which represented a C function as a circle (bubble) with inputs (parameters passed within the function) and outputs (parameters returned or passed by reference then modified when the function executes). A form other than a circle can be used to represent function, for instance, a rounded-edge Gane–Sarson rectangle.

2.3.2 How Process is Converted to UML

As mentioned earlier, control and process modeling have not lost their importance in UML since processes in object technology are viewed as operations defined in classes (structural part of a project) or actions/activities (behavioral part of a project). Control flows in the UML activity diagram can convey signals and be used to coordinate the execution sequence of tasks. Therefore, the "go" signal in Figure 2.3.1.1 is converted into a control flow in the UML. A control flow cannot pass an object or data but only "token" to activate the next process. A control flow is notated as an arrowed line connecting two actions/activities. Data flows are replaced by object flows. Datastores in the Gane–Sarson diagram will

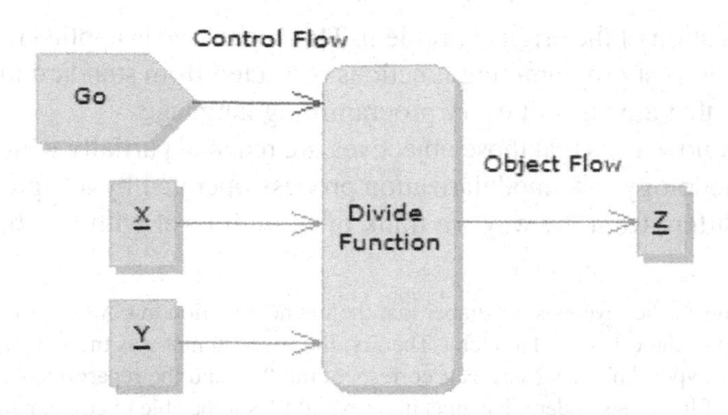

Fig. 2.3.2.1 Same Process as Figure 2.3.1.1 with an UML activity diagram. X, Y, and Z are objects. The arrows representing a data flow is now replaced by object flows (arrows connecting X, Y, and Z to the Divide function). Control is issued from the "Send" element which generates a "Go" signal. The bubble is replaced by an action or activity in the UML

be depicted as stereotyped objects. So, at low level, the UML has elementary equivalents to achieve conversion of all processes.

2.3.3 The Limit of Functional Decomposition: Half a Century of Modular Programming

With the resurgence of business processes for modeling business applications, we find it interesting to discuss what we have learnt from the past in order to avoid spending time in the wrong direction and find the best way to integrate various applications. One of the well-known mechanisms to handle complexity proposed in the early days of programming (Parnas, 1972) was *modularization* or the process of chopping a big program into smaller chunks, which is more easily understandable and manageable. *Structured design* (Yourdon and Constantine, 1979) addressed this problem from the functional design standpoint. The primary goal of the decomposition into modules is the reduction of overall software cost. The rationale of modules was multiple. Modules can be designed, developed with easily understood structure, and can be revised separately without affecting the behavior of other modules.

Good module interfaces, complete module descriptions, data flow analysis, task decomposition, information hiding (local variables), etc. were the buzzwords of modularizing process of functional generation. Fundamentally, *structured design* and *modular programming* consider every program as a huge process to be discovered and decomposed assuming that, at any level inside the hierarchy of decomposition, the subproblem stands alone as the original problem. People suppose that if all of the subproblems in the decomposition are solvable, then, solutions of these subproblems can be combined together to build the solution of the original problem. This technique is applied recursively towards the lowest programming functions extracted from standard libraries or instructions that are parts of every programming language.

We know nowadays that those objectives are reached partially with the functional methodology. The modularization process operated by a C programmer generally differs from the way we think of at high level with an object/agent model.

> Students in the seventies remember that the main() function in C was always at the top of the call stack hierarchy. The first intuitive attempt was the separation of prototypes ".h" files from their corresponding ".c" and the repartition of the grape of functions under the main() into several files to be able to compile them separately, both to save compilation time and help organizing a tremendous number of functions spawned within medium scale program.
> The criteria adopted by C programmers to decide on what function must go into what files varied substantially from individuals. If main() calls A1(), B1(), C1(), those subfunctions and their dependencies can go into 3 separate files (vertical grouping) but programmers can decide that A1(), B1(), C1() form the first calling layer (horizontal layering). Besides those "geometrical" methods, some programmers group functions according to their usage, their domains, for instance I/O, low-level access, mathematical, etc. *Criteria are variable from one person to another.* Earlier, some rare birds, born with object paradigm in the head, grouped functions naturally as they must be in the object paradigm.

Functional decomposition does not fit well with modern software engineering since this idea seems to work only at low level for small local control program.

> When working with a function, the first called function, let us say A(), is at the top level of the call stack, so all called processes at any level below A() are hierarchically under A(). It looks like A() is the master and it "holds" all other processes just by the fact that it can wake them up. If other hierarchies B() or C() share a common function L(), the latter will be separated and stored in a library to allow B() and C() to reach its compiled code of L(). A() does not hold L() anymore. An enhancement notion of code reentrance allows B() and C() to call L() and creates a separate copies of independent state variables. In this sense, reentrancy was the faraway ancestor of object instance and code reuse.

Nowadays, we know with MDA that a system could be developed by exploring successively various models. Models are tied to domains and ontologies. Even if there is a one-to-one mapping of the classical development phases into models, the reality of models differs substantially from the classical way of viewing development steps. Therefore, it would be very difficult (not unfeasible) for any developer to operate a functional decomposition without any risk of mixing domains and ontologies through an inextricable hierarchy. On a single blueprint, it is frequent to see logical concepts mixed up with platform-dependent nodes.

Moreover, modern software is not based only on hierarchical structuring process but relies more heavily on a middle structural model mixing both hierarchical processes to collaborative tasks in which intelligent and autonomous agents participate with their specific roles in a universe crowded with objects. If we simply consider a server in client–server architecture, the server must serve several clients in various contexts and when a client makes a call to a server for servicing, the calling side or the client does not "own" the server just for its own use by the fact that it can call the server for servicing. In a company, even if a president instantiates a production line, he does not own all his employees or other servicing companies, he only instantiates a collaborative process to reach a specific goal. We do not need to nest functions into functions as programming languages Modula and Pascal did and in this sense, modern software is "democratic," mimics human organization and is built on a *collaborative model* as each computing block identified as agent/object/component is more and more intelligent and versatile.

Let us now compare programming languages as Modula, Pascal, and C relative to the way they supported modularization. Pascal and Modula were built on the philosophy of modular programming as they support nested functions directly in the language. Nested functions are not allowed in C so the structure of C does not fit obviously to this way of thinking and the call stack of C is treelike with each node in the tree being a procedure or function. But a program is more than a bunch of processes. Niklaus Wirth, the Pascal inventor, my neighbor in the same university, stated that program consists of process and data with his well-known formula

$$\text{Algorithms} + \text{Data Structures} = \text{Program}$$

In fact, his equation fits only to a class of problems in which the system can be structured, not awkwardly, into processes. Moreover, algorithms and data can be found and fitted into this structure. The call stack constitutes itself a control structure but data that circulate inside this control structure belong to another different data flow network as data repositories must be added and data paths are superposed to the network of control flows. If we do not have appropriate data flows for a function, we must rip out a large branch of the tree and replace it with one that has better data coupling.

> A similar example is found in architecture. When we build a full residential district, every house must have water, electricity, phone lines, internet connections, etc. Several networks are therefore superposed to give a full residential service package. If the ground relief is complicated, it can impact on the way houses must be built to accommodate all servicing networks.

This way of programming considers that the nature of every solution to a problem is initially a process or a set of actions. Later, people try to fit data inside this structure as an overlaid structure. This is essentially a development process including two nearly independent steps which can give rise to structural mismatch between the tree control structure and the data flow network.

To share data between processes, *global data* were a programming buoy that provides a way for processes at different levels to communicate together. But when programmers are asked on what theoretical basis, they created such global variables, answers are rather hazy. Moreover, global data could easily get corrupted or incoherent, as they are accessible to all functions, even to those which do not have any right to access them.

However, functional modeling is still of topical interest as few of us are computer scientists. Financial analysts in business branch always think of plans, course of action, means, directives, goals, results, etc. Their domain vocabulary is mainly function-oriented, with actions, inputs, and outputs. In fact, computer scientists and business analysts are in two different spheres (domains). To conciliate things, it would be interesting to start talking in the language of business where the application is. Curiously, we can reason with functions and processes at a very high and abstracted level owing to the over simplification of things and due to the absence of complex control structure and data circulation network, still not identified at this stage of project development.

In the uniform methodology explained later, we still make use of processes at two levels, the highest and the lowest. In the requirement engineering phase, business process modeling is used to communicate with clients to identify tasks, plans, required inputs, and desired outputs and to understand their vision of the system. At the lowest level, in the implementation phase, when dealing with object operations packed inside classes, we need inputs and outputs. In the dynamic view, the activity diagram of UML version 2 reestablishes and legitimates the data flow concept. So, functional methodology can be used as an auxiliary tool for complexity reduction at the highest and at the lowest ends of the development process but not as the main concept for structuring the whole system.

2.4 How Objects Technology Handles Complexity

2.4.1 Object Paradigm

From a programming viewpoint, in a typical nonobject-oriented development, we have two separate things: code that is a bit of logic to perform task and data manipulated by code. Data and code are bundled together inside a new entity called *object* in object technology. Data can be everything necessary to model classical systems (data of a database system, state variables of a real-time system or function constants for a mathematical formula). Data receive the specific name of "attributes" at the early days of object programming in the implementation model and codes are organized around "methods," which are merely object operations.

Objects are both physical and conceptual things found in the universe. Hardware assemblies, software components, electrical devices, mechanical machines, documents, human beings, and even concepts are examples of objects. To model a company, a designer views buildings, divisions, departments, employees, documents, and manufactured products as objects. An automotive engineer will see motor, carburetor, tires, doors, etc. as objects. Atoms, molecules are object candidates for a chemist in an OOS of a chemical reaction. Transistor, logic circuits, and op-amps are objects for electronics engineers. Memories, stacks, queues, windows, and edit boxes are objects for software programmers. The whole earth is an object when seen from the space.

Objects have states while working. States of a car are its "running, parking, accelerating, decelerating" states. A state of a bank account object is its current balance, the state of a clock object is its current time, and the state of an electric contact is "on" or "off." Humans or agents have more complex states, the number of state variables is tremendously large but, fortunately, while modeling, we typically restrict the possible states to only those that are relevant to our models.

Simplest objects are *passive objects*. An unanimated mechanical piece, for instance, the plastic cap of your keyboard, a character sent to the screen, or an electronic resistance or capacitor are passive objects. Their state may change when exposed to actions performed by active objects. Active objects can instantiate changes to the universe and to themselves. A clock is a simple *active object* in this sense as it can change its state every second. If we put a computer inside a robot and give it some energy, the robot can work on a production line. If sensors and some dose of AI are added, this dumb robot can become intelligent and its behavior mimics that of a human. Humans are biological, intelligent, and active objects as they can instantiate changes, though not always intelligent, to the universe. They are called agents in social or business contexts. The term "component" is used currently to account for passive and/or active and/or intelligent aggregate of objects that can be deployed repeatedly in many applications and contexts. To alleviate our description, according to the context, we consider the term "object" in its largest sense (as common vocabulary) including agents and components. An agent can be an entire and complex hierarchical organization and a component may also be a very complex system, but part of a larger system.

In object/agent/component technology, every task is performed by an object alone or a collaboration of objects. Each object contains some data of the system and objects are created with procedures and operations for acting on internal data. Once again, data are understood with their largest sense. They can be some data records of the system, data describing object states, or data manipulated by objects. Data can be considered as objects if necessary, but not all the time. For simplicity, attributes can contain values and operations can have values as inputs and outputs.

In the programming world, the access to data is made under the control of the object that holds these data (It is not the universal model of real world. In real situations, objects can modify directly properties of other objects without their consent or approval). In other words, data can be accessed only by calling to an available service or operation of this object. So doing, the system is partitioned into entities that isolate changes. To accomplish their tasks, objects communicate with each other through messages. Messages can request a data or a change of data through encapsulated operations. Depending on how the change procedure is implemented, an object can control the identity of the object that requests the access to its data, either in read or write mode. Some degree of security is obtained by encapsulating the process and imposing systematic control to data access.

2.4.2 Information Hiding, Encapsulation

How object technology copes with complexity?

First, an object is a black box and it hides the underlying implementation. In the object world, only the creator or the company who builds the object knows the implementation details. Providing access to an object only through its messages, while keeping the details private is called *information hiding*. An equivalent buzzword is *encapsulation*. The consumer who uses this object has three possibilities:

1. Use directly the object through its public interface. "Public" means "visible to everybody."
2. Derive and create a specialized version of the original object if he wants to add or change slightly the object behavior (inheritance concept).
3. In the presence of parameterized class (a class is a mold for creating identical objects), supply parameters to create an instance.

So, complexity is partly cancelled by the information hiding mechanism. We can drive a very complex and high-tech car knowing nothing about multivalve ignition, antilock brakes, and vehicle stabilization system as the high-tech car offers nearly the same interface (not the same comfort) for driving as all other regular cars. A complex object is often monolithic, meaning that its structure is indiscernible from the outside (black box concept).

2.4.3 Composite Objects are Built from Aggregates

The perception of complexity comes from two aspects, the complexity at the interface (the object is difficult to use) and the complexity of a composite (objects are made from a concatenation of a large number of smaller components). If the structure is indiscernible from the outside (monolithic), we can reduce the second factor of complexity feeling but not the first one.

Composite objects are built by aggregating simpler objects and the numbers of objects and levels are not limited. If the object is not a finished product (e.g. a vision system), must be integrated into other systems, the term *component* is used to describe it. If the object is a human or implies a complex mixture of humans, business, and social objects, then terms such as "agent," "organization," "process," or equivalents prevail.

Aggregation/composition is the most important mechanism (at run-time, in real world) to deal with complexity in object technology.

> A human body is composed of a respiratory system, a circulatory system, a nervous system, a skeletal system . . . An aircraft contains a propulsion system, a navigation system . . . A program is composed of a user interface, a main core, local and distributed components. This relation is been called the *part-of, part-whole* relationship in the era of semantic networks.

Objects may also exhibit a complex structure involving both collaborative objects and aggregates. A computer is composed of a microprocessor, memory, interfaces, communication structure that is in turn built from two chip sets, etc. A computer is an aggregate of these objects but at the next level of decomposition, memory, microprocessor, and interfaces are collaborative objects. As long as we do not need to know "composition details" to make use of aggregated objects or monolithic objects, the process of building monolithic objects from aggregates is the main technique of complexity reduction. The interface with the user must be simplified to enforce this complexity reduction process.

2.4.4 Class Factorization as Complexity Reduction Mechanism

Another complexity reduction mechanism was the identification of identical objects and the class factorization mechanism. Class acts as category. Objects are individual instances of classes. For example, objects like Beethoven and Rantanplan can be instantiated from a same class Dog. The Dog class embodies all descriptive (attributes) and behavioral (methods) characteristics of a Dog object, and all the messages a Dog object can receive or send. A class was compared to some sort of factory which "manufactures" objects, a kind of mold. Each object created is an instance from its class or molded with the same class. Objects work through their defined operations (dog can watch over a house, bite foreigners, etc.). Operations can be activated by the object itself or by another object through activation messages. Messages are the only means used by objects in the object world to communicate with each other and evolve.

The act of building a class in programming environment allows n instantiated objects $X_1 \ldots X_n$ from class C to share the same code snippet defined in object languages without having to replicate a code. This sharing mechanism is limited to operations, not attributes. Each instance has memory allocated for its own set of instance variables, which store values peculiar to each instance.

Class factorization and classification step is early at the design level of a product. Composition/aggregation hides complexity at run-time, at user level. While making a composite object, a manufacturer must assemble all pieces together so he or she can master this complexity dimension.

2.4.5 Unique Messaging Mechanism for Object Interaction

In the object world, all communications are reduced to a unique and universal messaging mechanism that simplifies singularly the way objects interact. For instance, when writing the procedural sequence:

int a, b, c; a = 1; b = 2; c = a + b;

We can reason, in object technology, as having a thread object T responsible for executing the following thread:

1. int a, b, c

   T instantiates three "variable" objects named a, b, c

2. a = 1

   T sends a message to object a *and urges* a *to initialize its data property to* 1

3. b = 2

   T sends *a message to object* b *and urges* b *to initialize its data property to* 2

4. c = a + b

   T sends *messages to* a *and* b *to get their values, performs the addition* (3 = 1 + 2) *then sends a message to object* c *and urges* c *to initialize its data property to* 3

Message is a useful metaphor. An object behaves like a social actor in this respect. It has a particular role to play within the overall design of the program, and all objects act independently, each inside its own task script. The "actor view" changes the way we reason with objects. Instead of calling a function a function in conventional procedural methodology, we send a message to an object requesting it to perform one of its operations.

For instance, in an object-oriented programming language, some methods are fairly standard: a *constructor()* method builds the instance, a *destructor()* method kills the object, a *draw()* method produces an image, *get()* and *set()* methods are called *accessors* as they allow to access to read/write internal variables.

All objects issued from a same class do not have the same state at a same moment; they share only the same behavior.

2.4.6 Inheritance as an Economy of New Class Description

The easiest way to explain something new is to start with something old. This is an economy of concept description by avoiding redundancy, an element of complexity reduction. Object technology allows making a new class from an existent class. The *base class* is called a *superclass* and the new class is its *subclass*. The subclass adds *differential properties* to the base class but keeps all of the original definition.

From an implementation viewpoint in object language programming, nothing is copied from the base class to its subclass. Instead, the two classes are connected so that the subclass inherits all the methods and instance variables of its base class. In semantic networks, the connection between the base class and its subclass is named "is-a" relationship.

A subclass modifies slightly the behavior of the original class as it adds new methods and instance variables. We actually get a newer version with more possibilities through this inheritance process. The new version may simply override some old methods, and change or update the way of doing things. One of the great features of inheritance is the possible coexistence of both old and new versions inside a same system. If we have done some work with an old version of a development platform P6.0 and we do not want to develop newer applications with this old version, then with a newer P7.0, we can still let the two versions coexist inside our system. Such functionality is current in object technology. Versions coexist in the same system without any risk of inconsistencies that can jeopardize older works.

2.5 Object Paradigm Applied to Control Applications: Reactive Systems, Control Arrangement

A real-time system is first a reactive system (Harel and Pnueli, 1985). A reactive system has the following organization:

1. *Presence of an environment*. A reactive system has at least implicitly three subsystems: the system to be controlled, the controller, and its environment.

2. *Concurrency*. All subsystems evolve in parallel.

3. *Presence of inputs and outputs*. A reactive system reads information from their sensor inputs and acts on the environment through their actuator outputs.

4. *Closed-loop*. A reactive system is a closed-loop system that uses feedback information to control its outputs dynamically.

5. *Continuous operation*. A reactive system is a continuous process without a defined end point. It is called a never ending process.

6. *Synchronization burden*. A reactive system observes variations of its own settings, scrutinizes the behavior of the environment, and affords responses or corrections very quickly. A reactive system has the burden to synchronize itself to its environment. It differs therefore from an interactive system by the fact that, in an interactive arrangement, the two partners communicate to each other but there is no constraint or burden put neither on the synchronizing aspect nor the preponderance of any partner for controlling the situation.

A real-time system is therefore a reactive system whose timing are clearly specified.

Control (Curtis et al., 1992) is a very old engineering notion that traces its roots to the industrial revolution of the last century. Control applies to feedback and reactive systems to bind the output to control variables or parameters.

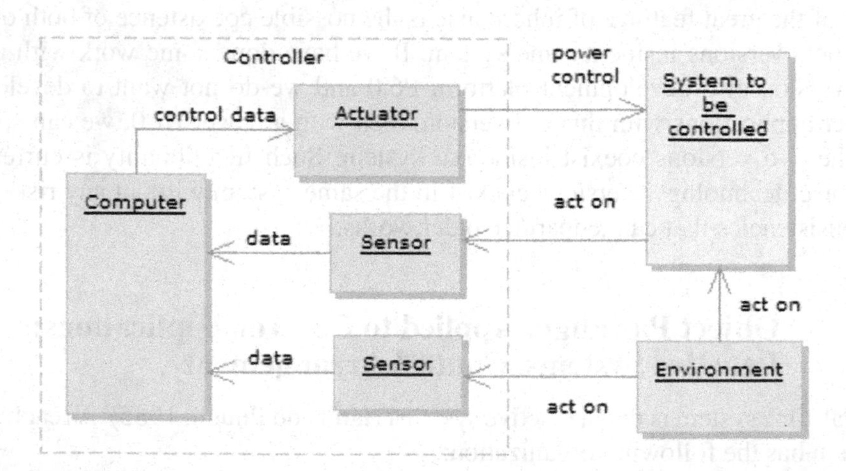

Fig. 2.5.0.1 Typical arrangement of a reactive system. Control includes sensing, actuation, and computation, mixed together to produce a working system. Actuators act on the system to be controlled. Output sensors sense system outputs. The environment acts on the system but the system perceives environmental disturbances via Environment Sensors. All these information are fed as inputs to the Computer that calculates in real time the necessary correction to feed corrective data to the actuator to maintain correct outputs

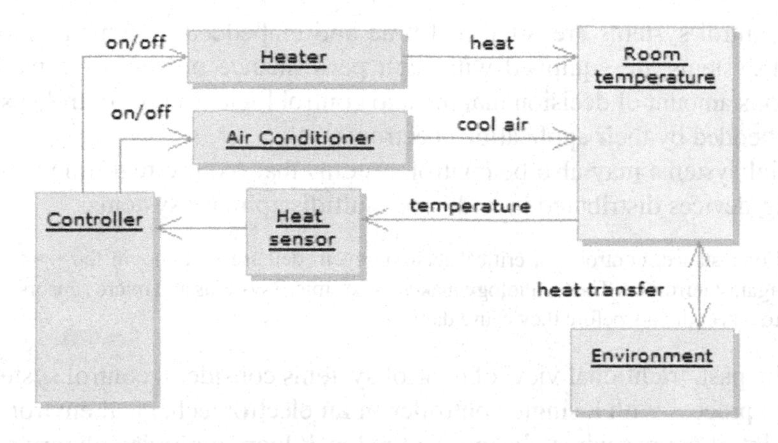

Fig. 2.5.0.2 Thermostatic control built on the principle of a reactive system

Figure 2.5.0.1 represents a typical control arrangement. If control was assimilated in the past to transfer functions (Bode or Nyquist plots, gain phase margin problem), modern control uses algorithms and computerized feedback. Control plays an essential part in the development of domains like electrical and mechanical engineering, and computer sciences. Examples of control arrangements can be found in applications such as power consumption and production, communication, Internet message routing, transportation, manufacturing, factory automation, aerospace, and military applications.

Many tasks performed by a manipulator arm in a manufacturing environment as arc welding, spray painting, continuous soldering or deburring, require that the end effector follows a defined trajectory. The typical arrangement described in Figure 2.5.0.1 could be adapted to a fast real-time system, for instance a welding robot which follows a welding path on a metal sheet. The "seam tracking" can be described as follows.

> When the robot is welding, the camera (Environment Sensors) scans the joint ahead of the welding torch. The camera feeds constant information containing either camera co-ordinates or path correction information to the robot (Actuators) to guide the welding torch. As this is happening while the robot is welding, there is no delay time between the scan and the weld execution (real-time aspect). If the joint changes along the path of the weld, the robot is able to accommodate for the variation (tightly coupled control). The arc welding system of this example is composed of a robot arm, the welding torch, everything just necessary to perform the initial function (welding). Others components, such as sensors, computer, control algorithms and actuators, guarantee that the work can be done efficiently, correctly. The Computer controls the operation of the system by taking information from sensors, compares the result against the desired behavior, computes corrective actions based on a feedback model and actuates the system to effect the desired change.

Tight control systems are often real-time and embedded systems as modern control systems are equipped with high performance processors capable of enormous amount of decision making and control logic. Many control systems are embedded by their application spectrum.

Social systems may also be control systems that require real time and embedding devices distributed in wide and multidisciplinary systems.

> For instance, control is a critical technology in defense systems, in the fight against terrorism. The technology makes use of micro systems and micro sensors to detect threats before they cause damage.

If, in the past, traditional view of control systems consider a control system as a single process with a single controller in an electromechanical environment with a lot of copper wires closing the feedback loop, multidisciplinary trends and modern problems recognize control systems as an heterogeneous collection of physical and biological information systems with intricate interconnections and interactions, not necessarily "copper wired" together. The media used to close the loop can be Hertzian waves and the components of a system can be distributed all around the planet including space.

2.6 Object Paradigm Applied to Database Applications. Data Modeling: Handling Database Complexity

Data are the life-blood in modern economy. We collect data every day, use it to decide, to guide our behavior. The tremendous amount of data available dictates efficient processes to classify, store, retrieve, and use them. If physical drawers and files are sufficient at home, business or governmental organizations need to automate tedious manual business processes and secure data repositories. Database business has been the most flourishing and profitable activity in the second half of the last century (about 13.8 billion in 2005 according to Gartner [available at: www.gartner.com]). To manage data, we need database which is more than a simple repository. Data are organized, indexed, and classified to allow quick access via queries. Databases need a good data model and high performance software motor called the DBMS, either relational (R type giving RDBMS), object types (O type giving ODBMS), or combined type as ORDBMS (Object Relational DBMS). Prerelational databases like hierarchical and network models are still in use.

A data model is a conceptual representation of data as structures. Data are modeled as entities/objects with their attributes, relationships. Data models focus on what data are required and how they should be organized rather than what operations could be performed on the data, relayed to the role of DBMS. Data designers, in this sense, have a better role than control and process designers: they must orchestrate only the *structural view* of systems and delay all the tricks of dynamic table joining at the query phase to the DBMS.

There are two major methodologies used to create a data model: the Entity–Relationship (E-R) (Chen, 1976) approach, its derivatives, and the Object Model leading to object databases. Relational database, with theoretical foundations as Codd algebra (Codd, 1990) and his well-known rules of normalization, is a mature and well-proven technology. For constructing, maintaining, querying, and updating data, it uses a long-established SQL (structured query language) standard, which has been adopted by the International Organization for Standardization (ISO) and the American National Standards Institute (ANSI).

To stress at this introductory stage the differences between the relational and the object view of data, consider a banking application. If we build a teller machine to service a person, the object view is perfect for that application because the Person object is identified by its name and its account number. Other information about this person includes his address, a list of his accounts with all their individual attributes, a list of recent transactions, and so on. All this information can be organized as a hierarchical view of data related to the business relationship that this person holds with the bank. That is the perfect view for the teller which corresponds to the object description.

However, at the end of the day, for the clerk who needs to count how many transactions happened that day, what is the total amount of deposits and withdrawals, how much money he still has in his strong safe, he has a totally different view of the data. Data are transaction oriented, not person oriented. In this case, a relational, "table", or "spreadsheet" view is more appropriate. He needs an account table that is joined to a transaction table, a person table and so on. Tables are more appropriate to answer to questions like: how many, how much . . . assorted with conditions because the tabular form is easier to handle while counting, summing.

Relational database includes a collection of data organized as tables, with columns representing data categories and rows representing the data records, for example, a list of product orders containing columns of product type, customer code, date of sale, and price charged to customers. Users can enter data rows in chronological order and view data from multiple perspectives by reindexing the table, for instance from the lowest to highest price.

Relational database is a good candidate for managing a huge amount of alphanumerical data of nearly the same size and type, but its rigidness comes from the unique implementation structure such as *tables* storing records of values (that is why relational is called "value-based"). Record identities are fabricated by adding keys (primary, candidate, and surrogate). When applying this tabular structure for storing semantic data structured as an ontology forest of trees or multimedia data, we see the limit of the relational model, and naturally look for other alternatives. Moreover, in the way relational tables are designed, information is distributed among several tables, and a time consuming and costly operation of joining tables must be triggered to get the answer from queries.

A database expert said that if we model all the pieces of a Boeing or an Airbus plane with relational tables of screws, molded parts, plastic pieces, we must decompose the aircraft into tables and then join tables to reconstruct the aircraft.

This metaphor explains that not all data are best modeled with relational tables. In engineering, multimedia data are more efficiently stored inside object databases which binds objects through a richer and complex set of relationships. The resulting storage structure bears a resemblance to natural structure, something like a big grape of objects linked together inextricably like a complex semantic network, but surely not a spreadsheet. Each solution tends to occupy a niche in the market and for the moment the niche of RDBMS is quite huge compared to that of ODBMS that has not got through the step of reaching a critical mass in the marketplace. The success of relational models comes from many facts:

1. *Nature of applications*. Most database applications are transaction based and this situation will not change in the future. A bank, an insurance company, or a stock market is not concerned with the hierarchical and network structures of data.
2. *Rigorous mathematical formalism.* Relational model is supported by Codd algebra that allows some preliminary verification to be made.
3. *Easy model.* The relational model is implemented with tables. Spreadsheet and tabular forms are intuitive for everybody, particularly for financial analysts or business developers. The object model is powerful but is really too complicated and needs special training.
4. *Large operational base and maturity.* RDBMS has occupied the market for a long time and currently supports nearly 80% of database applications. Even if object or another paradigm seems to be promising, generally, people are reticent to change if they do not see any business opportunity. Companies involved in database loathe scrapping their systems for a new technology unless it offers a compelling business advantage.
5. *Dynamics of the market.* ODBMS manufacturers slow to propose an equivalent of RDBMS solution at the end of the last century with full OQL (object query language) support, high reliable environment. Furthermore, ODBMS needs a huge amount of memory to cache data in the client application's memory to eliminate extra call to ODBMS back end. Some technical problems still remain in the research domain.
   - There was a lack of object specialists. Object databases appeared as black art.
   - The mapping of relational schemas directly to ODBMS is not straightforward as inheritance is not natural with tables.
   - Object paradigm itself penetrated the market very slowly. Older versions of UML itself needed important enhancements that arrived finally only in 2003.

- RDBMS manufacturers offered attractive extensions to support data such as audio, video and images.
- RDBMS manufacturers proposed alternate solutions with ORDBMS.

The solution must be adapted to the nature of the problem to be solved. So, object databases will have their niche when ODBMS finally reaches its maturity and when researchers or software companies arrive to solve main technical problems.

Data modeling produce schemas. Large database projects may contain tens of thousands of entities/classes. Database modeling is still a job of an artisan and data schema grows with daily needs by patches and often without any evolution plan. Team members are often not permanent, so huge database is often a nightmare to maintain and diagrams frequently unreadable. Sometimes, reengineering is the only way to recover a badly maintained database becoming unmanageable though it could be perceived initially as a good design. If schema redesign is already an important step, the recovering process of data between two inconsistent sets of schemas is often a job of computer artists.

The solution of managing complexity in huge database passes through complexity reduction techniques of layering, hierarchy-zing and separation of concerns, either in a centralized or decentralized implementation model. Frequently, some management cultures assert that centralized control is better control. From a technology standpoint, this is justified by data integrity and from an economic standpoint by unnecessary replication of redundant systems. But, redundant systems may enhance data security and cache replication accelerates data access. Decentralized systems suffer from problems of linking multiple systems. Moreover, once decentralized, data sharing may never occur; technical concerns and local human culture are closely related. Decentralization puts human behavior and culture inside the initial problem. Independence, controllability, communication, database coherence, data security, etc. are parameters that must go inside constraint analysis phase. From a technical standpoint, Internet federation is a good example of a working decentralized implementation.

2.7 Objects and Databases: Bridging or Mingling?

At the implementation level, object software and relational databases are built for different purposes, and probably it would be easier to find a bridge between the two worlds than unifying them under the same development tool. At the conceptual level, more interesting things can be done and will be discussed in the uniform methodology. The incompatibility comes first from the application concern.

A car is described by the engineering staff as a hierarchical system with many levels of aggregation/composition. When an engineer asks his system "what are all pieces that constitute a motor?", an object design allows the system to reach automatically all components by navigating from the first level of composition. Object technology does not deal only with structural aspect of a system. It can handle functional, dynamic aspects at the same time. So, the state of the motor may be determined dynamically by computing the states of all components that make up this motor and this principle may be conducted recursively through all levels of the composition hierarchy. The object view appears in this case as the most attractive natural design.

Opposite to the engineer who studies only one motor in real time to optimize his motor, the director of the storehouse of spare parts of the same company may need a table view of the same data since he has for instance 20,000 spare parts of such a motor in his stock and he must compute how many pieces he must manufacture to support the spare part market. The table view will look more natural in his case.

So, inside the same environment there exist two specific views of apparently the same reality *Car*. In fact, there is very little connection between the two problems. The semantic structure build by the engineering staff describes how a Car object is connected to all of its parts and the maximum number of different objects that we can instantiate through this structure is often 4 or 5 (4 wheels, 5 passengers, etc.). The storehouse of spare parts is concerned with a large inventory of pieces. In fact, a person who works in this store may ignore completely that a mechanical with reference X must go with another with reference Y three levels higher, only their ID numbers, the number of pieces in stock, their costs are relevant to his concern. If we take the engineering model and instantiate tens of thousands of pieces out of this model, we must come back to the tabular structure to store all the piece descriptions and when we want to find out information, table jointure made through a hierarchical structure of tables are more difficult than having a model of tables laid out with classic relational model.

Accessorily, if a mechanical piece must change its dimensions or its description, there is a need for an update that must be propagated to maintain the coherence of the whole system.

A logical approach would be "respecting the nature of things." Each problem has its optimal solution and solutions of problems are in different domains. Bridging between domains by creating a "bridging domain" is sometimes a better approach then mingling them. Moreover, the "unification'" or "uniform" view of everything sound, always well, and, as a first reaction, people suppose that these magic terms are equivalent to process simplification, cost saving, and better business. In this sense, some terms are "political" or "managerial" and we must be very careful with them. When looking at the title of this book, we are concerned also with a "uniform" methodology. It is up to the reader to make the same critical analysis before approving it.

2.8 Object Paradigm and Component Reuse

Reuse attacks the complexity reduction at the design level. Complex systems can be built on complex components that have already made their proof in the past and validated by the market. Reuse is a very sensitive and passionate research subject. Naysayer and fan camps are crowded and arguments on both sides are solid. We do not really want to waste the reader's time by entering this large debate. We think that a technology or a way of doing things must be mature and validated with time, so intrepid researchers need naysayers to alleviate their passion and these two opposite forces are both needed to get closer to the truth, that is, as said in the foreword, always questionable.

Reuse in the small level was already a known issue. Modular programming already promoted code reuse and libraries encouraged compiled code reuse. In a large context, software products could be built entirely from reused components. The virtues of that kind of thinking are obvious since a lot of software no longer would need to be written from scratch; and as components have already passed numerous quality and reliability tests, people expect a significant enhancement of productivity and quality.

Large reuse inside a company is already a reality. Many pieces of a car are standardized inside car companies to give customers the illusion of diversity. Reuse outside a company is a difficult issue because it would be hard to find a component that fits perfectly to a targeted application unless people ask the component vendor to make an OEM version especially for their company. Some experience of external reuse is already current on the Internet (OpenDoc, OLE, ActiveX controls, COM and CORBA objects, JavaBeans, etc.). On the design side, *design patterns* (Gamma et al., 1995), reusing design commonalities is an emerging idea, though not yet mature. Reuse may be at a compiled code level, source code level, design level, documentation level, data level (database replication), or cultural level inside or outside of a company. Reuse is also a cultural issue because if we organize our job so that newcomers can be productive immediately, we are practicing reuse.

Reuse problems (Biggerstaff, 1992; Krueger, 1992; Frakes and Fox, 1995; Sonnemann, 1996) may come from the fact that a component editor does not adhere to good object-oriented design practices and makes poor ad hoc implementation. In this case, reuse is not the issue but problems must be found elsewhere. Reuse success needs management commitment, investment strategy, organizational structure, and staff experienced with reuse concepts. It is also an attitude coopted by developers, system architects, and project managers. Managers who typically want to see progress earlier should plan and gratify reuse effort. External reuse (outside a company) needs standards, so organizations like OMG are very important in the process. Reuse must be planned, preferably, early in the development process, at the requirement analysis phase as an important requisite. Very often, its usefulness is not perceived as mandatory by developers. The following questions (some marked with * are inspired from the work of Frakes and Fox [1995]) stress the most important factors influencing reuse success.

Table 2.8.0.1 Survey questions to test the commitment to reuse concept in an organization

Questions: in your organization	*Comment or remedy*
Are developers masters of core object technology?	Employee training. Project manager must be a good object designer
Are reusable assets actually used and how are they found valuable?*	The past is the future of the present (Japanese proverb)
Do developers make use of programming languages which provide reuse support (e.g. by supporting abstractions, inheritance, strong typing, etc.)?	Watch your development environment. Adhere to good object-oriented design practice. Employee training
Does your organization make use of CASE tools that support reuse?	Buy the right CASE tool
Do developers prefer to build from scratch or do they make efforts to find available components?	Change organization culture
Do recognition rewards increase or promote reuse?*	Gratify reuse effort. Develop incentive
Do legal problems inhibit reuse?*	Consult good lawyers
Is there any repository for code libraries and components?	Create database and document repositories for this purpose
Are you interested in Model Driven Architecture for reuse?	Complete reading this book

*Inspired from the work of Frakes and Fox [1995]

We need experienced project managers who know how to properly evaluate risks and opportunities, to bring their teams through the constantly changing technological and business world. Reuse is also a cultural issue and can be practiced at any level in the organization. Avoid generalities, platitudes, or evident development patterns; developers must learn through experience how to design, implement, maintain, and reuse software components and frameworks. To practice reuse, we must first have available and good materials designed for reuse. So, it would be time to review our best projects and give them a second life.

2.9 Mastering Development Complexity: A Roadmap Towards Model Driven Architecture

The MDA is an industry-standard architecture developed by the OMG in late 2001 (Available at: www.omg.org). The MDA focuses primarily on the functionality and behavior of the application or system, and separate the technology in which it will be implemented. Thus, it is not necessary to repeat the process of modeling the application's functionality and behavior each time a new technology comes along.

The MDA unifies every step of the development of an application. It separates clearly *Platform Independent Models* (PIM) from one or more *Platform Specific*

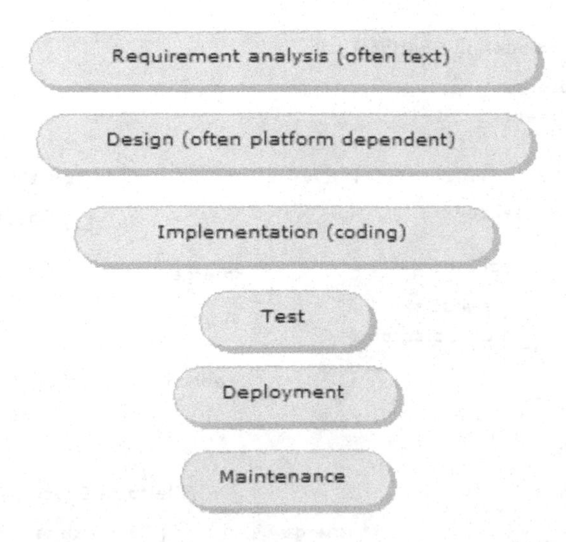

Fig. 2.9.0.1 Classic development roadmap. The design is often platform dependent. It takes into account all the requirements specified at the previous phase. The implementation is often assimilated directly to the coding phase

Models (PSM). Portability and interoperability are supposed to be built into the architecture. The PIM remains stable as technology evolves, extending and thereby maximizing software return of investment. More precisely, the goal of the MDA is to separate business and application logic from its underlying execution platform technology so that changes in the underlying platform do not affect existing applications. The evolution from one model to another normally needs a *model transformation* that can be done automatically by a tool assisted eventually by a human operator if needed. The benefit of this approach is that it raises the level of abstraction in software development at least in the early stages. Instead of going directly to platform-specific code, software developers focus on developing models that are specific to the application domain but independent of the implementation platform. MDA, contrary to its name, is only a conceptual framework. It does not define any particular software architecture or any architectural style.

Figure 2.9.0.1 displays a classic development roadmap. Figure 2.9.0.2 shows a development scenario compliant to the MDA concept.

In Figure 2.9.0.2, the PIM package is derived partly from the domain model that stores the whole assets of development knowledge, classes, logical architectures, schemas, and patterns. The new application under development will create new domain objects that enrich, in return, the domain model. The PIM is then transformed into several PSM models. Theoretically, if a common PIM is used as a basis to generate applications for two different platforms, these two applications will share the common PIM and will therefore

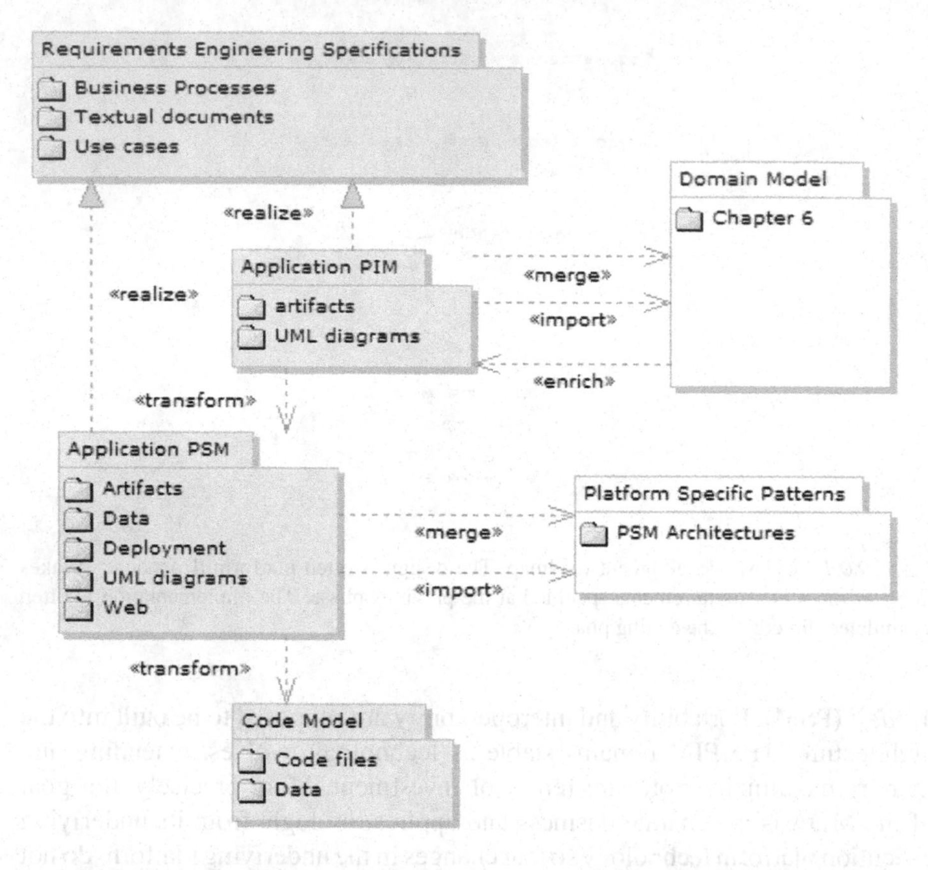

Fig. 2.9.0.2 Instance of development scenario compliant to the MDA (to be compared with the classic development roadmap) that enacts high reuse. As specifications contain both platform independent and platform specific, specifications must be taken into account by the PIM and PSM. Reuse assets are materialized by Domain Model and Platform Specific Patterns. Packages are shown with some of their contents. More than one PSM can be built from one PIM

have increased chances of interoperability at the logical level. To enforce this interoperability, bridges can be created to finalize the process as shown in Figure 2.9.0.3.

The MDA also addresses the problem of middleware proliferation. Middleware is a general term for any software that serves to glue together or to mediate between two separate and existing programs. Often found in a distributed environment, middleware is a layer above the operating system, above the application programming interface (API of Windows) but below the application program. Middleware masks some heterogeneity that programmers of distributed systems must deal with. They also mask networks and hardware intricacies. If an operating system hides hardware details and provides a homogenous programming model to programmers, we can say that middleware masks complexity details in a distributed environment. Examples of famous

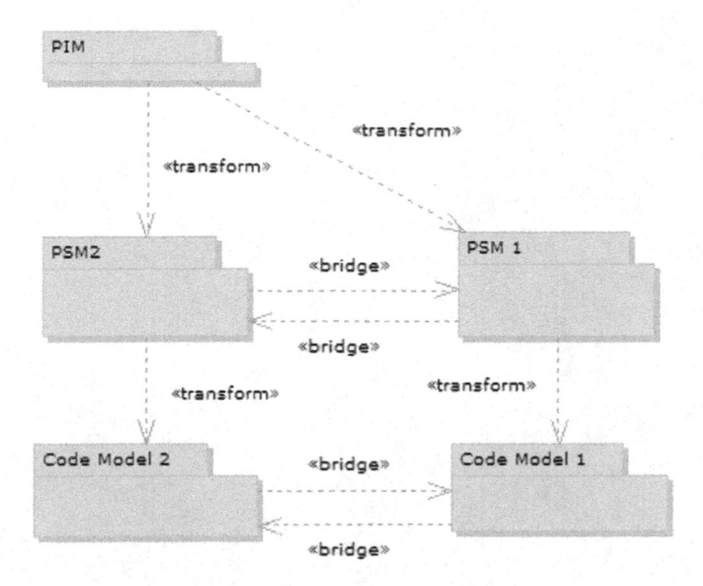

Fig. 2.9.0.3 Transformations and bridges between models. From one platform independent model (PIM), we can make several platform specific models (PSMs). Bridges at platform specific level and at code level can be created for interoperability

middleware are CORBA, Enterprise Java Beans, XML/SOAP, COM+, and .NET.

MDA-based development involves large initial investment in configuration, transformation mechanisms and transformation rule sets, assisted by humans at the early design phase. The transformation could be more easily automated at coding phase (PSM) as programming languages, grammar, and platforms are formal and known issues. If development time should be longer for a first application or for a company without any development assets, development time should decrease for subsequent applications developed inside known domains and platforms.

To sum up, the MDA makes a distinction between the logic of business, the logic of the application, and the platform deployed for this application or this business. The MDA proposes a framework to develop and create models and possibly later, create code directly from these models.

Chapter 3

UML Infrastructure: Core Concepts

The definition of the second major release of UML 2 has been a long-awaited process. In 2000, the OMG issued four Requests for Proposals (RFP) for UML 2.0, Infrastructure, XML Metadata Interchange (XMI), and Superstructure. Most users are concerned only with the last Superstructure that contains modeling concepts having a direct relationship with their application blueprints. As this textbook is partly addressed to researchers, when necessary, we must expose the Infrastructure that contains basic constructs merged by the Superstructure. Some figures are reproduced as is from the standard. Others are synthesized, to explain this complex standard in an understandable way. Comments in quotes (") or in italic are extracted from the UML. For those who are interested in details about metamodeling, please consult Atkinson and Kühne (2002), Kobryn (2004), and Alanen and Porres (2005); the following description gives only a snapshot and an overview of the current UML version contents necessary to understand how the UML is built from its first stone. The rationale behind metamodeling proposed by the OMG is to create a family of modeling languages (UML is one of them) based upon a common core of concepts and to avoid a collection of unrelated specialized languages that could impair interoperability.

To explain the metaconcepts in the Infrastructure and the Superstructure, the UML makes use of graphical notations of itself. This kind of "bootstrapping" supposes that the reader who reads the Infrastructure Book is already familiar with the UML notation (class, association, aggregation/composition, inheritance/derivation, constraints, roles, multiplicities at the association endpoints, etc.) or mathematics like subsetting or derived union (Rundensteiner and Bic, 1992) to understand diagrams at the metalevel. That supposition does not stand for readers who want to learn UML for the first time. So, normally, the neophyte must start with the Superstructure Book (Chapter 4) without paying any attention to the abstract and the way modeling packages are tied together and tied to packages defined in the Infrastructure Book. Once the UML is fully understood, readers can reiterate through Infrastructure (Chapter 3) and Superstructure to strive to understand why he cannot draw, for instance, an association between the two object states ("not UML compliant"), but only dependencies.

D. M. Bui, Real Time Object Uniform Design Methodology with UML, 47–75.

Researchers interested in ontology engineering must read everything because UML is itself an example of ontology.

(Note: If there is some difficulty in interpreting the diagrams of this Infrastructure part, do not be frustrated, have a quick look at the kind of diagrams and concepts developed in the Infrastructure, then skip this arid, hermetic, and highly technical description because some symbols used are in fact described later when studying UML specification. The diagrams in this chapter will appear clearer when you finish reading this book and may then serve as a reference for metamodeling (some tools in the market let you make metamodeling and define your proper syntax extension to UML for your own need, not profiles).)

3.1 Core Concepts Compliant to MOF: Modeling Levels

In the real work, we have only objects that interact together to make a working system, no classes. User objects and user data, real or simulated, are found at the "User Object Level" called M0 level. They are images of real objects that we manipulate in everyday life and M0 starts as our reference abstraction level.

As many objects can share common properties, to economize description and make models clearer and manageable, we can factorize properties of similar objects. For instance, the four tires of our car need only a unique description even if all these four objects are distinct and will be worn out differently with time (the four tires have four different states but a same structural description). If there are 20 cereal boxes on the shelf of a store, they may be categorized

Table 3.1.0.1 A four-layer architecture of the UML, which is defined at M2 (specification level or metamodel viewed by users). Users build software at level M1 (user model) to create a running system at M0 (objects and data). The meta-metamodel at M3 level is the basic concept level used for managing interoperability of object standards

Layers viewed by users	*Levels used in this book*	*Scope*	*Defines*	*Constituents*
Meta-metamodel	Basic concept *M3 level*	metamodeling	Language for specifying metamodels	Metaclass, MetaAttribute, MetaOperation
Metamodel	UML specification *M2 level*	UML specification	Language for creating a user model	Class, Attribute, Operation
User model	User model *M1 level*	Project using UML	Language to describe your project	Client, Company, Order_Status, Edit_Order()
User objects and data	User objects *M0 level*	Run-time system	A working system	Order_1234, Client_Paul, Order_Status = "Paid"

into a unique description. The main categorizing mechanism of object technology is *classification*. For the car description, we need to create a *Tire* class with *attributes* to describe tire size, load index, speed rating, etc., and *methods* (operations) to show that tires and wheels must rotate to move car.

The *Tire* class belongs to the User model depicted at M1 level. We need to create at the same time the class *Car* and thousands of classes to describe all other car components. As a car is a complex object, a flat list of thousands of classes is an unstructured set and an unmanageable project. We must reorganize classes into packages, connect classes together via a set of relationships, and describe them with various semantics to account for the way objects inside a car are connected together. So, the user model contains, besides classes, all relationships and artifacts to make models more understandable.

To make user models, we need an object-modeling tool like the UML. CASE toolmakers in the market propose a software to model and develop systems. They adhere to the current standard UML defined and maintained by the OMG. The UML is an object modeling language and is, at the time of this writing, the most comprehensive and object-oriented standard. If our "Car" model is called the "model" (user model at M1 level), then, the UML itself is called a *metamodel* (a model to create model).

Metamodels are developed at M2 level. The UML is defined with graphical guidelines, naming conventions, organizational artifacts like packages, association/relationship connection rules, etc. To build correct and understandable models at M1 level, modelers must fully understand UML conventions and rules defined at M2 level. The UML or the metamodel, once understood, allows us to produce application models at level 1 and instantiate a working system at level 0.

As the UML is known as four-layer model architecture, in fact, it has a fourth level called meta-metamodel or M3 level. This "Basic Concept Level" of the UML comprises a set of low-level modeling concepts and patterns that are in most cases too rudimentary or too abstract to be used directly in modeling software applications. The rationale behind this fourth level is multifold. First, the UML is a language among many other object languages developed and maintained by the OMG. This organization needs a common base for all object languages to fully reuse common object concepts, ease data interchange, and automate model transformations. The Infrastructure of the UML defines base classes that form the foundation not only for the UML but also for other objects standards, for instance, MOF (Meta Object Facility, a common object model metalevel for most object standards promoted by the OMG).

Classes, attributes, operations, and associations at M3 level are sometimes referenced in the literature as MetaClass, MetaAttribute, and MetaOperation. As the prefix "meta" could be confusing as the "meta" applied to model is one "meta" higher than applied to elements they describe (see 1st and 3rd columns of Table 3.1.0.1), the understanding of the term "meta" in our text is contextual and we let the readers interpret these concepts at their proper level according to the discussion context. The OMG is currently upgrading all of UML chunks to version 2. It is a large specification, made up of four parts:

- **Infrastructure**. This package is reused at several metalevels in various OMG specifications to deal with general modeling process.
- **Superstructure**. This package is used to define specifically UML.
- **Object Constraint Language**. This package defines a formal language used to describe string expressions on UML models. UML diagrams are not refined enough to provide all the relevant aspects of a complete specification. Modelers need to describe constraints and requirements in natural language. However, natural language carries ambiguities incompatible with programming, so a formal language OCL (Object Constraint Language) has been developed. Typically, the OCL is used to specify queries, pre/postconditions, operations, types, etc.
- **Diagram interchange**. The goal of this package is to define a smooth and seamless exchange of documents compliant to the UML standard. For instance, in previous version of UML 1.x, XMI has been used to exchange diagram information. If objects can be easily transported from model to model, all their layout information is lost. Layout information locates modeling objects on a diagram, specify their sizes, etc. This limitation is not due to XMI but comes from the fact that UML metamodel has not yet defined a standard way of doing things. Moreover, to assure the exchange of tools that do not understand model elements, but only lines, text, and graphics, bridging from XMI to Scalable Vector Graphics (SVG, an XML-based format that has been adopted as a W3C recommendation) if necessary.

3.2 Infrastructure: Core Package

Hereafter, we make an overview of the M3 level to show how UML is built. Diagrams are inspired from original and public documentation available at the end of 2005 on the OMG web site (Infrastructure Book). For research purposes, please consult the OMG web site. If there is misinterpretation by us, please take the interpretation on the OMG web site as reference. We try to synthesize the information for learning purposes, apologize for such an eventual incident, and would be grateful that somebody reports the error. We prune off details and focus only on the way (philosophy) the UML has been built. As graphical notations are examined in detail within the UML syntax of Superstructure Book, they are not explained in this chapter. As said before, the reader can bypass this section if necessary and come back to it later if needed.

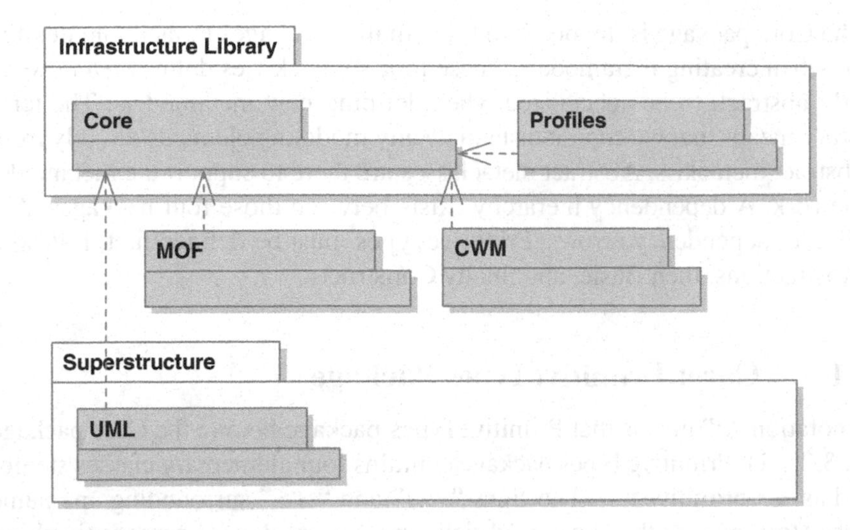

Fig. 3.2.0.1 Reuse of Common Core package by the UML and MOF. The dashed arrow says that the UML is importing and merging modeling elements from Core package. The import and merge notions are detailed in the Superstructure exposé

The Infrastructure of the UML is defined by the *Infrastructure Library* package which is a metalanguage core used in various metamodels, including the MOF, CWM (common warehouse metamodel) and of course, the UML. The MOF is viewed as a language at M3 level, the UML and CWM at M2 level. In Figure 3.2.0.1, the profiles package contains an extension mechanism that allows "second class" customization of UML. "First class" extension could be made via the Infrastructure level. So, it would be interesting for researchers to have an idea of the Infrastructure organization. For the Superstructure, the extension can be made only via profiles, a ruled process defined in UML itself. So, "first class customization" is more powerful but can potentially diverge from UML standard.

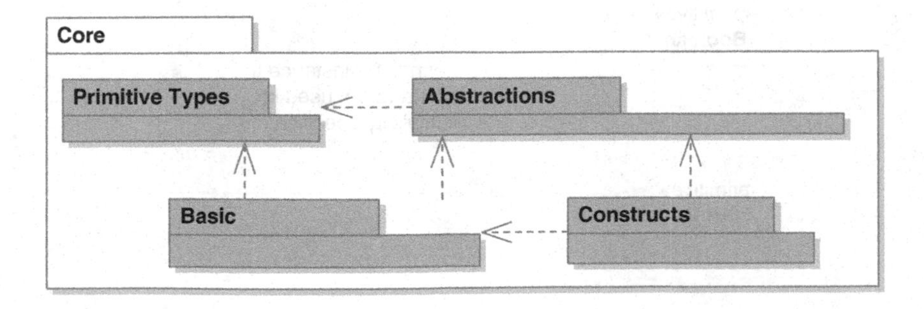

Fig. 3.2.0.2 Four Core packages defined in the Infrastructure. Arrows show their dependencies "merge" or "import" (Fig. 3, Infrastructure Book)

The Core package is divided into four smaller packages to facilitate flexible reuse when creating metamodels. These four subpackages define *metaclasses*, mostly abstract, to be specialized when defining new metamodels. The term *abstract* means that we cannot instantiate any modeling elements directly from an abstract metaclass. Abstract metaclasses are there to support the metamodel framework. A dependency hierarchy exists between those four packages. According to dependency arrows, PrimitiveTypes must be defined first, followed by Abstractions, then Basic, and finally Constructs.

3.2.1 Core::PrimitiveTypes Package

The notation "::" means that PrimitiveTypes package belong the Core package (Fig. 3.2.1.1). PrimitiveTypes package contains four elementary classes stereotyped as <<primitive>>. The signs "<<" and ">>" surrounding any name are used to announce that the named element is a *stereotype* (a preconceived and oversimplified idea of the characteristics which typify a person or thing). These predefined types are commonly used in metamodeling. The <<primitive>> classes are intended for very restrictive graphical uses, specifically to define elementary means to name modeling elements (String class), to count occurrences or multiplicity in a diagram (*Integer*, *Unlimited Natural* classes), and to lay out constraints and evaluate conditions (*Boolean* class). For more advanced context where we need to define ways of writing method specifications, functions, parameters, actions, events, etc., OCL, another important component of

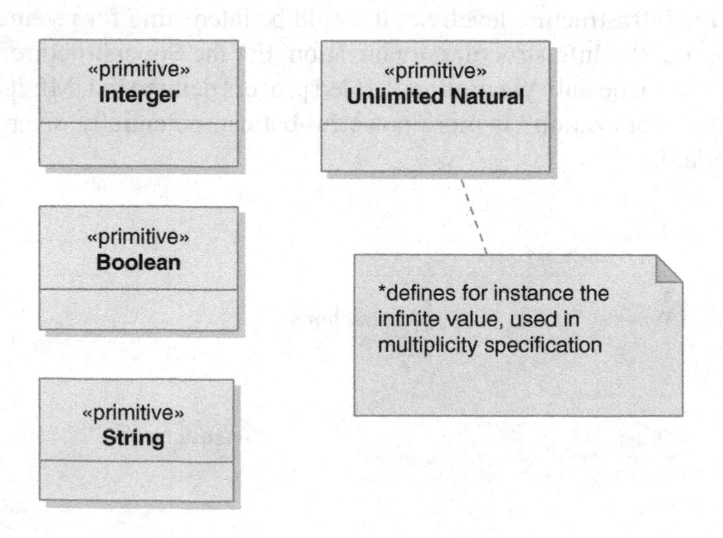

Fig. 3.2.1.1 Core::PrimitiveTypes package contains four classes

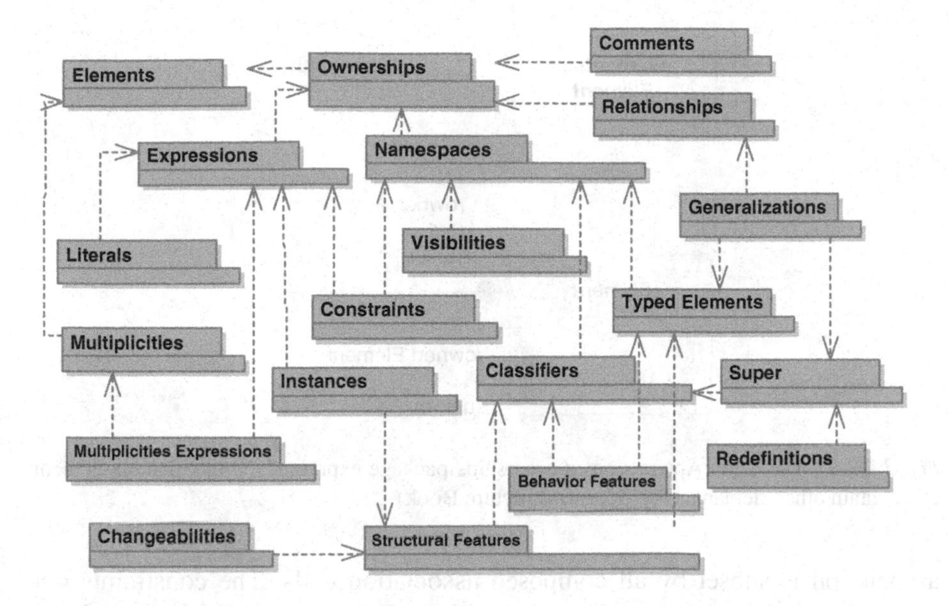

Fig. 3.2.2.1 Contents of Core::Abstractions. This package is composed of 20 elementary packages connected together via a network structure (Fig. 12, Infrastructure Book)

the UML, is more appropriate to unambiguously describe textual constraints in natural language. Please refer to OCL specifications on the OMG site.

3.2.2 Core::Abstractions Package

At the root of Core::Abstractions, we found a unique modeling *Element* which is an abstract metaclass with no superclass. An *Element* is an abstract constituent of a model. The package Core::Abstractions::Elements contains just the class *Element* (Figs 3.2.2.1–3.2.2.3).

An abstract element has the capability of owning other abstract elements. The relationship drawn with a diamond is an aggregation/composition, the diamond is the "owing side" and the side without adornment is the "owned side." The slash ("/ownedElement") is a role specification ("/" means "derived") and, with the association end constraint "{union}," it states that the derived owned Element

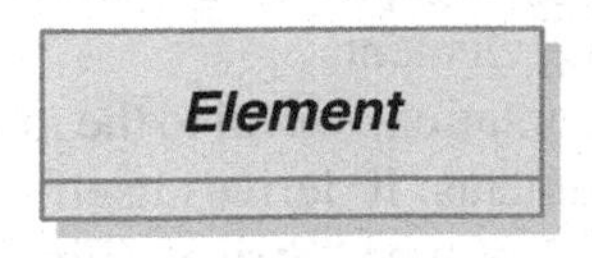

Fig. 3.2.2.2 The Core::Abstractions::Elements package contains the metaclass Element (Figs 27 and 28, Infrastructure Book)

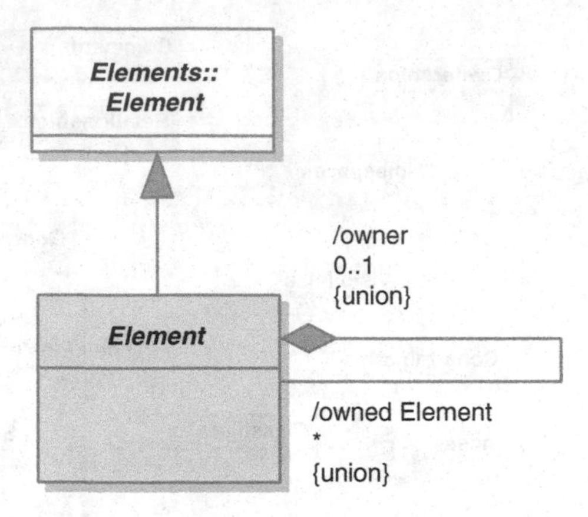

Fig. 3.2.2.3 The Core::Abstractions::Ownerships package expresses the fact that an element can contain other elements (Fig. 50, Infrastructure Book)

association is subset by all composed association ends. The constraints put on an association end are surrounded with curly braces "{ and }" and give a complementary information (constraint) to interpret the role end of an association. "*" or "0..1" are specifications of multiplicities.

> The derived symbol "/ " allows marking any model element as derived from other element or elements and thus serving redundant data. The symbol is applicable mainly to attributes and association ends, to indicate that their contents can be computed from other data.

Actually, we cannot find all the information by analyzing only the graphical notation of a diagram. For instance, "an element cannot own itself" or "an element that must be owned must have an owner" are constraints added as textual specifications in the standard. They may also appear through a set of available "queries" accompanying the specification of this package. Together, *constraints* and *queries* must normally complement the graphical notation and specify the package unequivocally. All this information is available in the Infrastructure Book. For this overview, we retain only that a modeling element can own many other elements and the hierarchy "owner/owned" is not limited when applied to the tree structure that this association may generate. This very simple aggregation, together with the defined multiplicity, can in fact generate an infinite tree structure of abstract modeling elements.

A *Comment* is a textual annotation, useful to the modeler that can be attached to an *Element* or a set of Elements. It derives itself from an *Element* class and has an attribute named "body" of type "String" (Fig. 3.2.2.4) that is a primitive type defined in Core::PrimitiveTypes. When deriving from *Ownerships*, any modeling element inherits properties defined in Ownerships and may contain any other elements, in particular, one or several comments. The association at

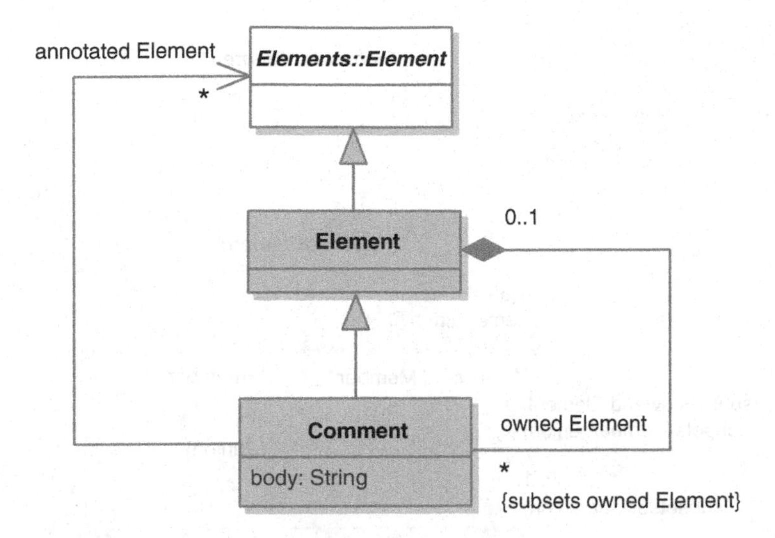

Fig. 3.2.2.4 Core::Abstractions::Comments package (Fig. 20, Infrastructure Book)

the left references the element, which has been commented. The asterisk or star sign (*) means 0, 1, or N elements.

A *Namespace* derives from a *NamedElement* that in turn derives from an *Element* with ownership property. A namespace provides a container for named elements. A *NamedElement* has two attributes. A multiplicity [0..1] applied on an attribute specifies that a *NamedElement* may have no name (which is different from an empty name) or one name. A *NamedElement* has a usual name (that may be ambiguous) and a qualified name (of the form name1 :: name2 :: . . . :: nameN) that allows it to be unambiguously identified within a hierarchy of nested namespaces. It is constructed from the names of the containing namespaces starting at the root of the hierarchy and ending with the name of the *NamedElement* itself. "/" before *qualifiedName* means that it is a derived attribute. The aggregation–composition relationship generates a tree structure. A collection of *NamedElements* are identifiable within the *Namespace*, either by being owned or by being a member, i.e. by importing or inheritance (Fig. 3.2.2.5).

In Figure 3.2.2.6, a *Visibility* provides a mean to constrain the usage of a named element in different namespaces within a model. The diagram shows that a *Visibility* is a *NamedElement* with a *visibility* attribute that takes values from *VisibilityKind* that is of enumeration type. Although enumeration type is defined later in the Core::Basic, *VisibilityKind* is represented temporarily as a stereotype. A *public* element is visible to all elements that can access the contents of the namespace that owns it. A *private* element is only visible inside a namespace that owns it. When a named element ends up with multiple visibilities, *public* visibility overrides *private* visibility.

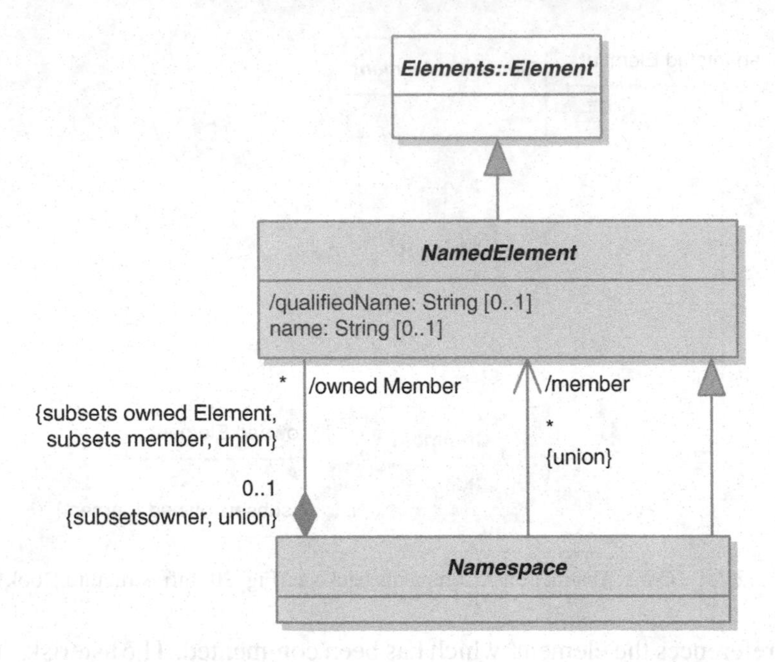

Fig. 3.2.2.5 Core::Abstractions::Namespaces package. A namespace provides a container for named elements (Fig. 48, Infrastructure Book)

Expressions support the specification of values, along with specializations for supporting structured expression trees and opaque expressions (opaque: not interpreted). An *Expression* represents a node in an expression tree. If there are no operands, it represents a terminal node. If there are operands, it represents an operator applied to those operands. Various UML constructs use expressions, which are linguistic formulas, that yields value when evaluated in a context.

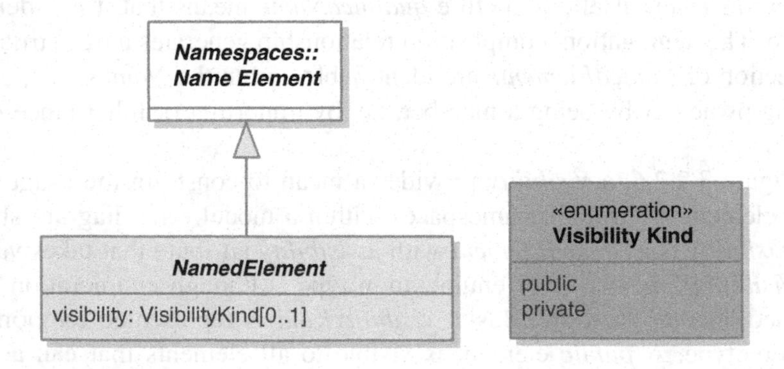

Fig. 3.2.2.6 Core::Abstractions::Visibilities package. This package defines how a element can be visible to other modeling elements (Fig. 62 and 63, Infrastructure Book)

Combined with properties defined in Ownerships package, a complex expression may be generated as a structured tree of symbols. An opaque expression contains language-specific text strings used to describe a value or values, and an optional specification of the corresponding language.

A *Constraint* is a condition or restriction expressed in natural language text attached to the constrained elements. A constraint may or may not have a name; generally it is unnamed. It contains a value specification indicating a condition that must be evaluated to *true* by a correct design. A constraint is defined within a context that is a namespace so a namespace can own constraints. It does not necessarily apply to the namespace itself, but may apply to elements in the namespace. Owned rules are well-formed rules for the constrained elements. To be well formed, an expression must verify the type of conformance rules of the language (e.g. we cannot compare an Integer with a String) (Figs 3.2.2.7–3.3.3.9).

A *Classifier* is defined at its simplest form in a Core::Abstractions package (Fig. 3.2.2.10). It is an abstract namespace whose members can refer to features. A classifier describes a set of instances that have features in common. A *Feature* can be of multiple classifiers.

The Super package (Fig. 3.2.2.11) provides mechanisms for specifying later generalization relationships between classifiers. Any classifier may reference more general classifiers in the reference hierarchy. As a consequence, an instance of a specific classifier is also an (indirect) instance of each of the general classifiers. Therefore, features specified for instances of the general classifier

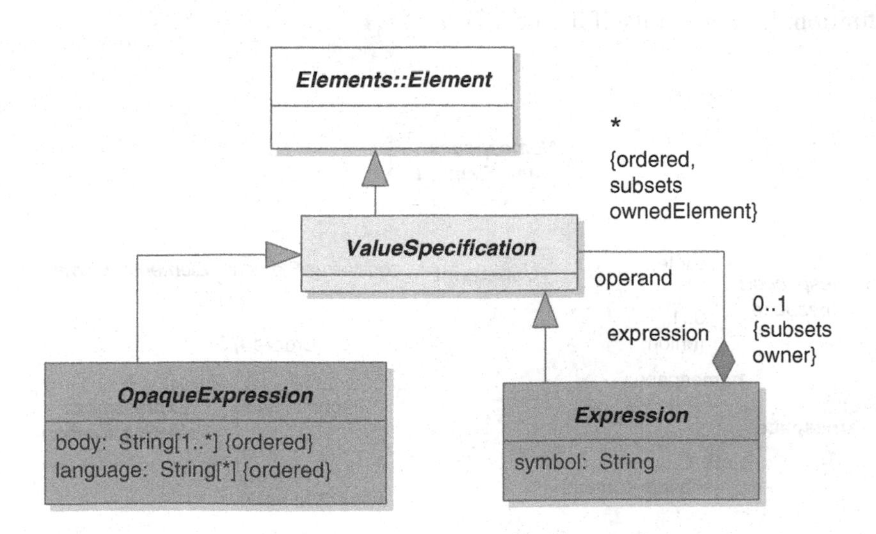

Fig. 3.2.2.7 Core::Abstractions::Expressions package. Expressions are used to support structured expression trees (Fig. 30, Infrastructure Book)

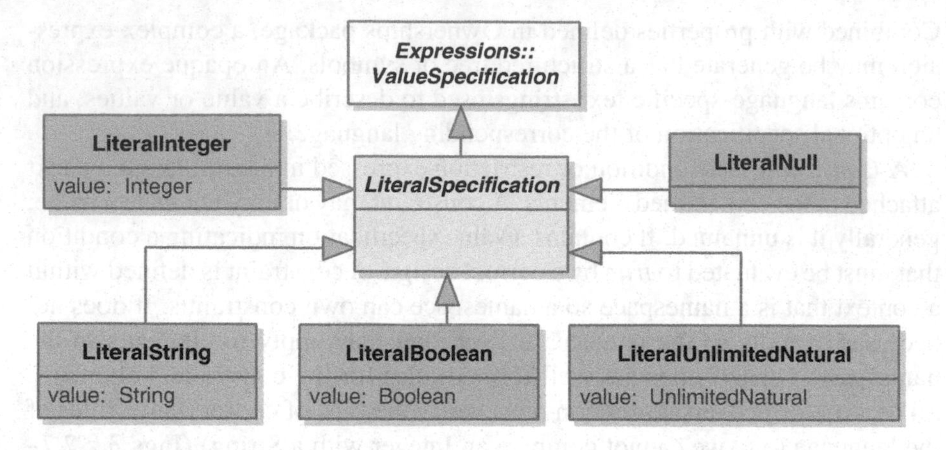

Fig. 3.2.2.8 The Core::Abstractions::Literals package specifies metaclasses for specifying literal values. A literal Boolean is a specification of Boolean value. A literal integer contains an integer-valued attribute, etc. (Fig. 40, Infrastructure Book)

are implicitly specified for instances of the specific classifier. Any constraint applying to instances of the general classifier also applies to instances of the specific classifier.

TypedElement and *Type* (Fig. 3.2.2.12) are abstract classes derived from *NamedElement*. The type constrains the range of values represented by a typed element. Simply defined at this stage, a type represents a set of values. According to this definition, a *TypedElement* requires at least a *Type* [0..1] as part of its definition. It does not itself define a *Type*.

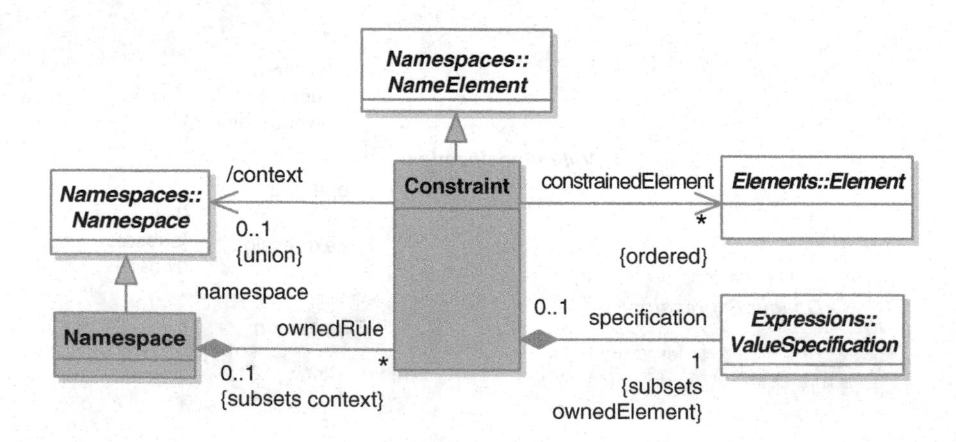

Fig. 3.2.2.9 Core::Abstractions::Constraints package. Constraints use Expressions to textually define conditions that must be satisfied (expressions evaluated to true) in a model (Fig. 23, Infrastructure Book)

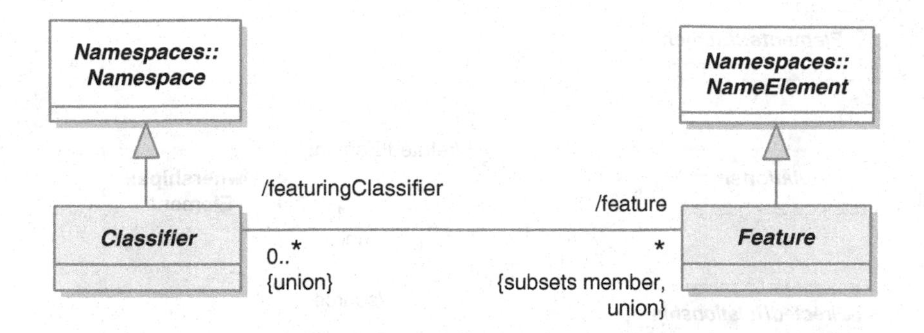

Fig. 3.2.2.10 Core::Abstractions::Classifiers package (Fig. 18, Infrastructure Book). Classifier in the Core package is in its simplest form. It is a namespace and refers to (not contains) features that are namespaces also

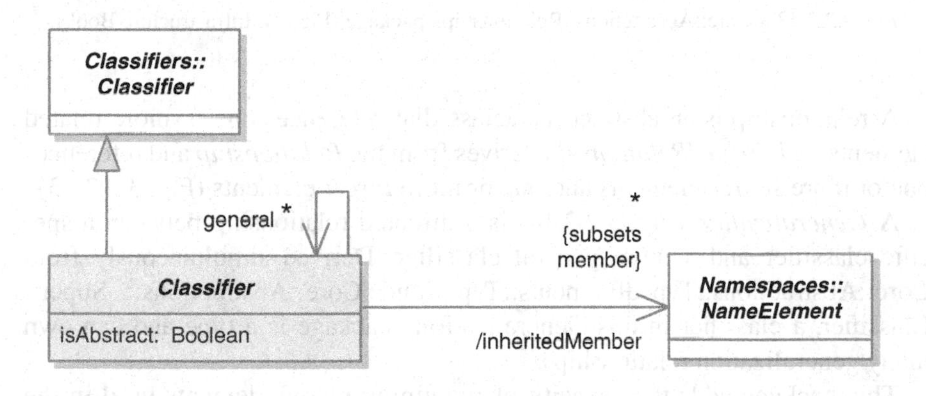

Fig. 3.2.2.11 Core::Abstractions::Super package (Fig. 58, Infrastructure Book)

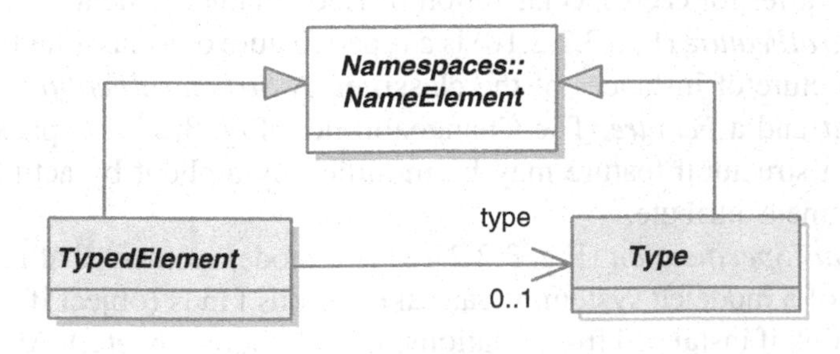

Fig. 3.2.2.12 Core::Abstractions::TypeElements package (Fig. 61, Infrastructure Book)

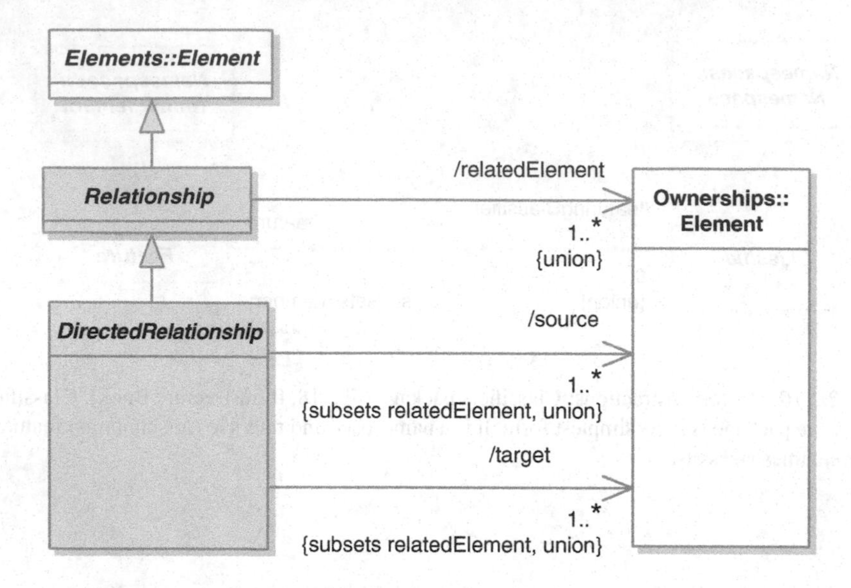

Fig. 3.2.2.13 Core::Abstractions::Relationships package (Fig. 54, Infrastructure Book)

A relationship is an abstract metaclass that references one or more related elements. A *DirectedRelationship* derives from the *Relationship* and references one or more *source* elements and one or more *target* elements (Fig. 3.2.2.13).

A *Generalization* (Fig. 3.2.2.14) is a directed relationship between a specific classifier and a more general classifier. Derived simultaneously from Core::Abstractions::TypedElements::Type and Core::Abstractions:: Super:: Classifier, a classifier in this Generalizations package is a type and can own many generalization relationships.

This package adds the capacity of redefining model elements used in the context of a generalization hierarchy. The statement in the standard is "A redefinable element is an element that, when defined in a context of a classifier, can be redefined more specifically or differently in the context of another classifier that specializes the context classifier" (Fig. 3.2.2.15) The detailed semantics of redefinition varies for each specialization of RedefinableElement.

A *StructuralFeature* (Fig. 3.2.2.16) is a typed feature of a classifier that specifies the structure of instances of the classifier. A *StructuralFeature* is both a *TypeElement* and a *Feature*. The Changeabilities (Fig. 3.2.2.17) package defines when a structural feature may be modified by a client by acting on the *isReadOnly* meta attribute.

An *InstanceSpecification* (Fig. 3.2.2.18) is a model element that represents an instance of a modeled system. It can take various kinds (object if instanced from class, link if instanced from relationship, collaboration, etc.). An instance specification reveals the existence of an entity in the modeled system, and

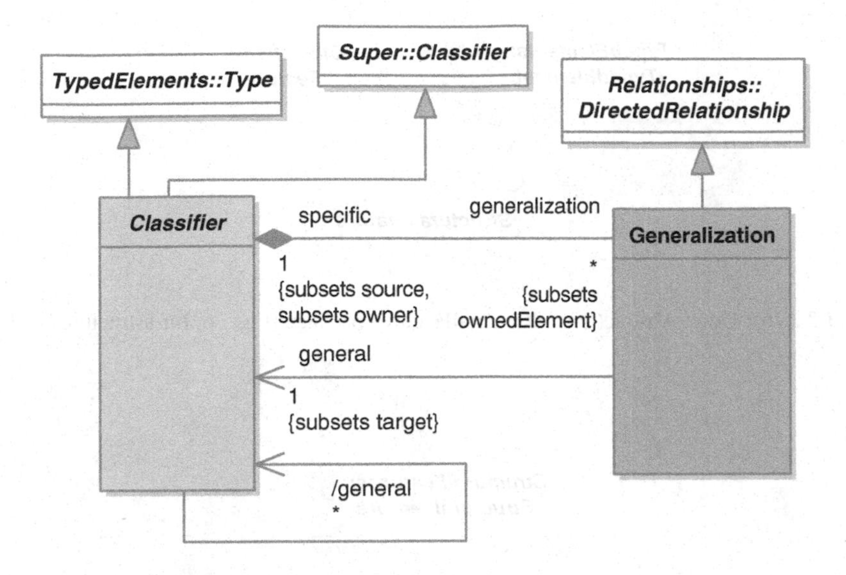

Fig. 3.2.2.14 Core::Abstractions::Generalizations package (Fig. 32, Infrastructure Book)

completely or partially describes this entity. This description embraces: classification of the entity with one or more classifiers, precision on the kind of instance, and specification of values for defining structural features. Details can be incomplete since the purpose of an instance specification is to show what is of interest about an entity in a modeled system. From this metalevel diagram, *InstanceSpecification* is a concrete class derived from *NamedElement*. It references one or many classifiers, has possibly *ValueSpecification* to build the instance, and has many slots, each described by a *StructuralFeature*. The

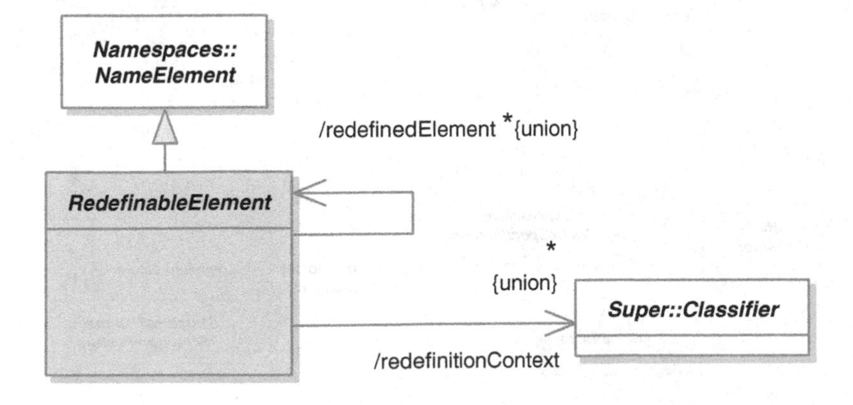

Fig. 3.2.2.15 Core::Abstractions::Redefinitions package (Fig. 52, Infrastructure Book)

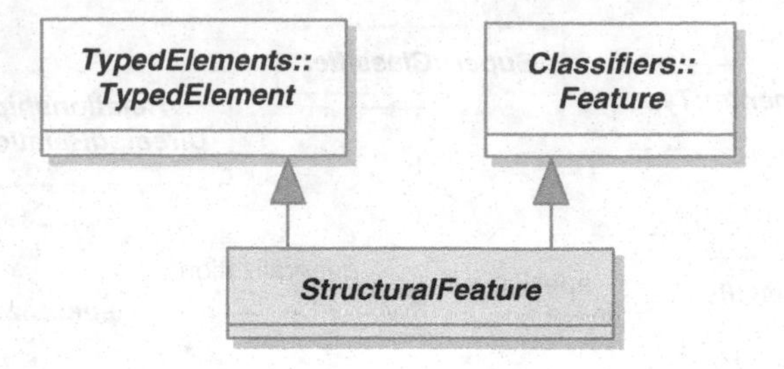

Fig. 3.2.2.16 Core::Abstractions::StructuralFeatures package (Fig. 56, Infrastructure Book)

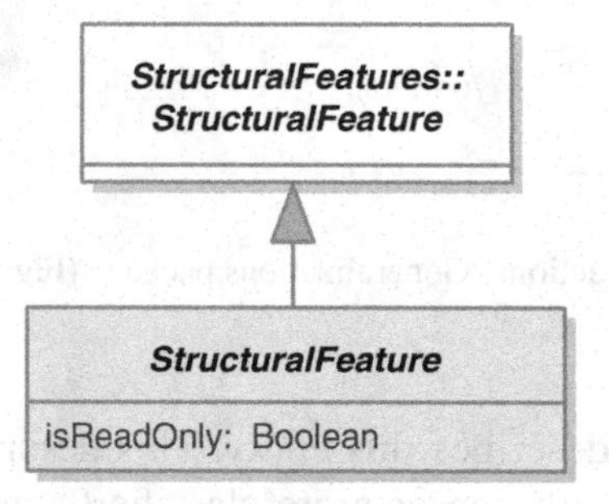

Fig. 3.2.2.17 Core::Abstractions::Changeabilities package (Fig. 16, Infrastructure Book)

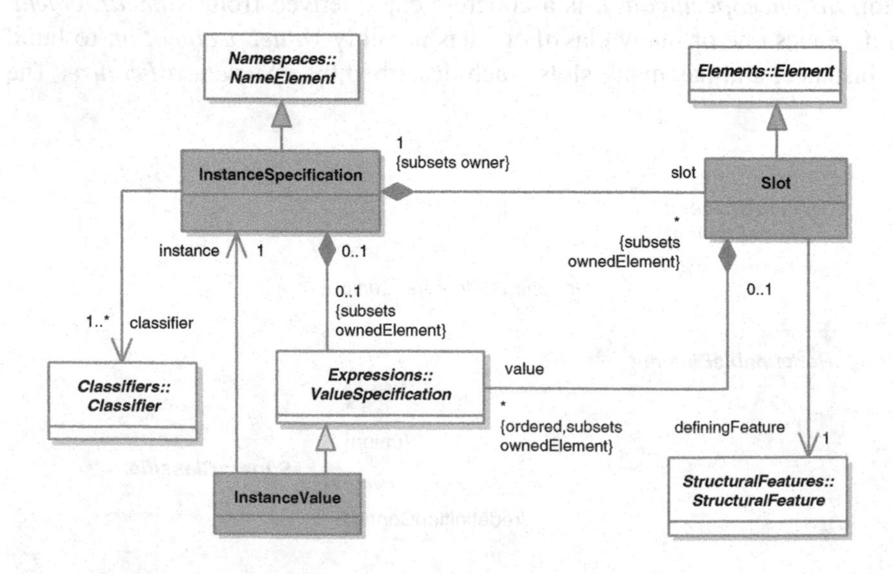

Fig. 3.2.2.18 Core::Abstractions::Instances package (Fig. 35, Infrastructure Book)

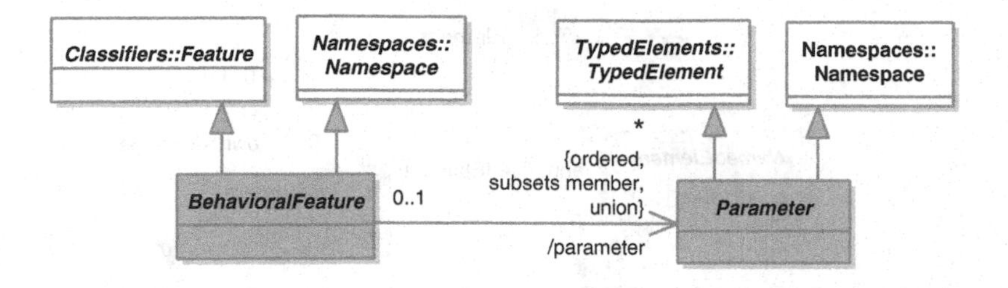

Fig. 3.2.2.19 Core::Abstractions::BehavioralFeatures package (Fig. 14, Infrastructure Book)

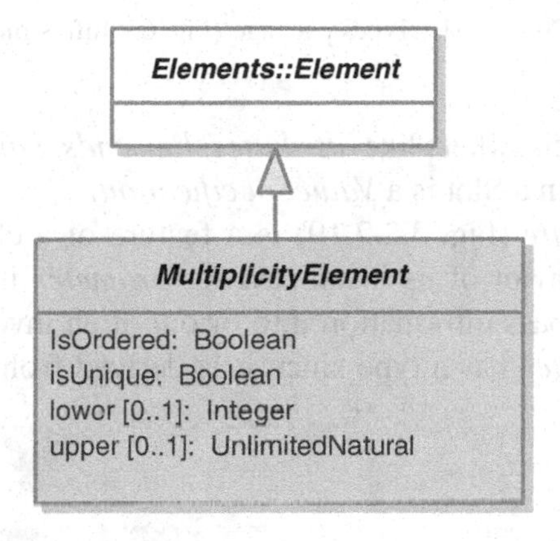

Fig. 3.2.2.20 Core::Abstractions::Multiplicities package (Fig. 42, Infrastructure Book)

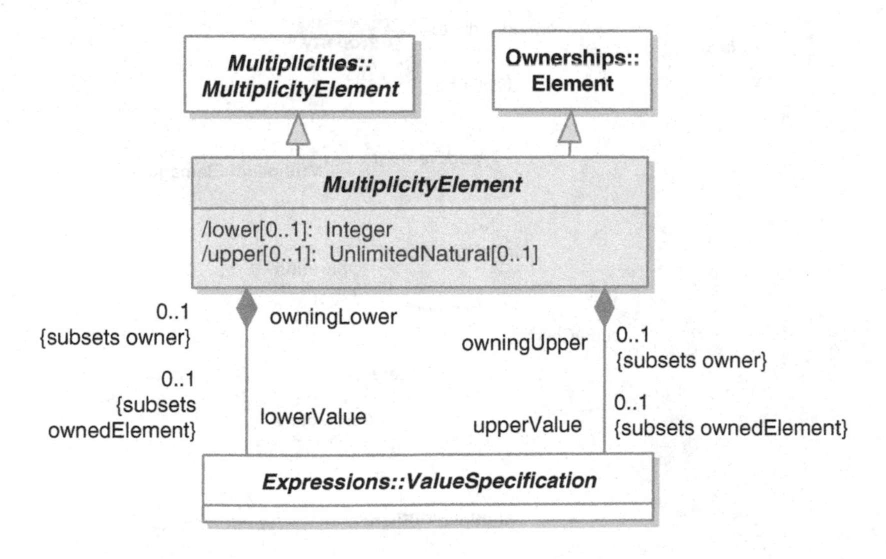

Fig. 3.2.2.21 Core::Abstractions::MultiplicityExpressions package (Fig. 46, Infrastructure Book)

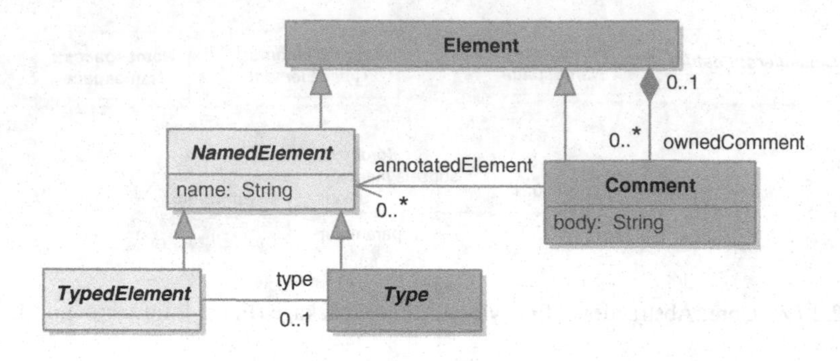

Fig. 3.2.3.1 Core::Basic::Types package (Fig. 65, Infrastructure Book)

structural features of classifiers like *attributes, link ends, parts*, etc. in UML are called *Slots*. A value in a Slot is a *ValueSpecification.*

A *BehavioralFeature* (Fig. 3.2.2.19) is a feature of a classifier that specifies an aspect of behavior of its instances. A *Parameter* is a specification of an argument used to pass information into or out of an invocation of a behavioral feature. *Parameter* has a type since it is derived from *TypeElement* and *Namespace*.

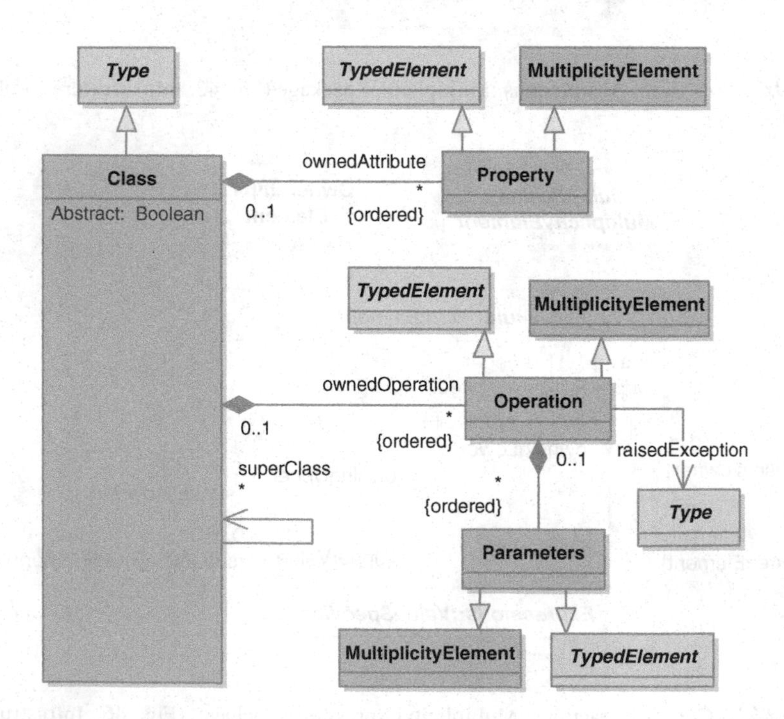

Fig. 3.2.3.2 Core::Basic::Classes package (reduced version of Fig. 66, Infrastructure Book)

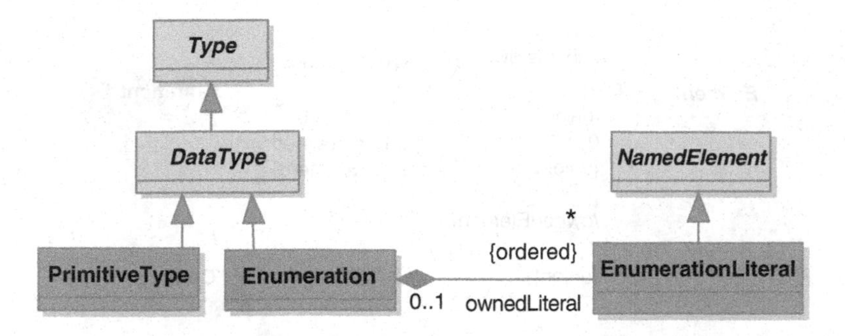

Fig. 3.2.3.3 Data types in the Core::Basic::DataTypes package (Fig. 67, Infrastructure Book)

A *Multiplicity* (Fig. 3.2.2.20) is an inclusive interval of nonnegative integers beginning with a *lower* bound and ending with an *upper* bound, which can be infinite. Multiplicities are mostly used to specify later association ends. A multiplicity specifies the allowable cardinalities of an element and regulates the number of instances we can generate from this element.

This MultiplicityExpressions package extends the multiplicity capabilities to support the use of expressions. *MultiplicityElement* is an abstract metaclass, which includes optional attributes for defining the bounds.

3.2.3 Core::Basic Package

Core::Basic package provides a minimal class-based modeling language on top of which more complex languages can be built. Core::Basic was effectively reused in MOF and contains four diagrams: Types, Classes, DataTypes, and Packages. Core::Basic imports model elements from PrimitiveTypes package and contains metaclasses derived from Core:: Abstractions.

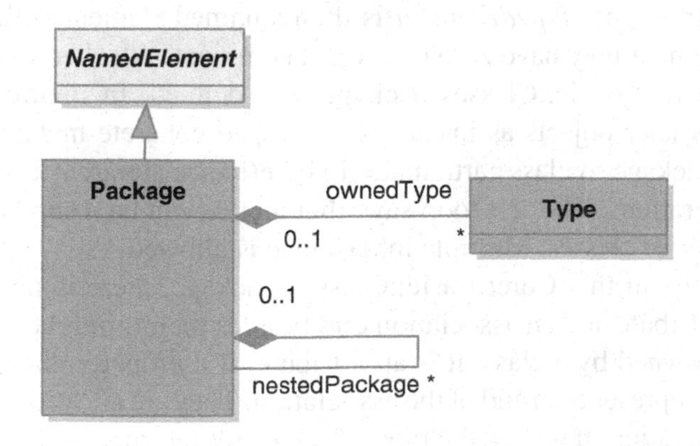

Fig. 3.2.3.4 Core::Basic::Packages package (Fig. 68, Infrastructure Book)

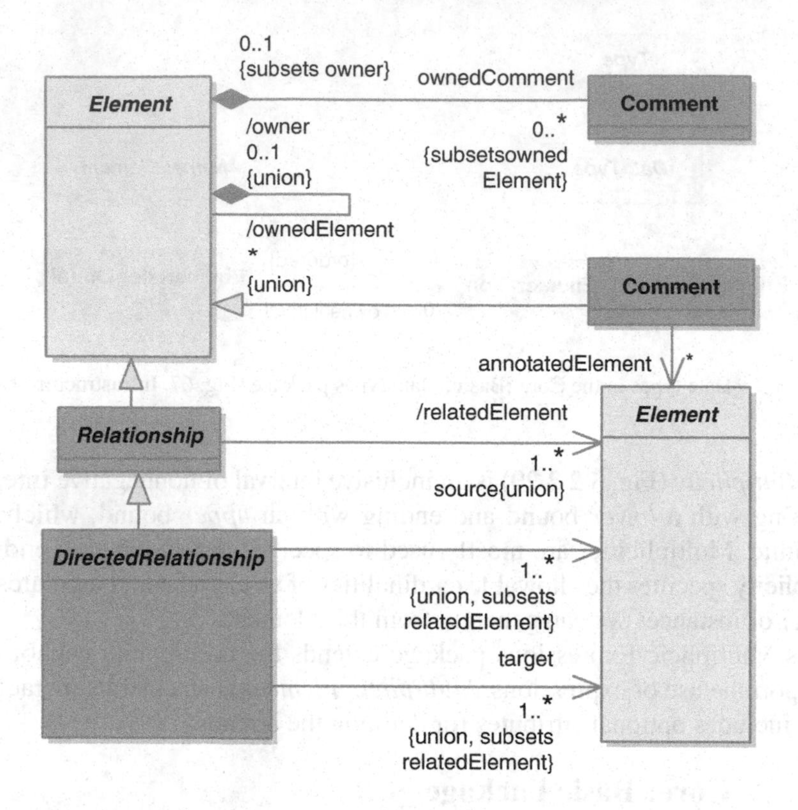

Fig. 3.2.4.1 Core::Constructs::Root package (Fig. 71, Infrastructure Book)

Figure 3.2.3.1 defines abstract metaclasses that deal with naming and typing of elements. *Element* is the root element defined in Core::Abstractions. *NamedElement* represents "element with a name." A *Type* is a named element used to specify a type. *TypedElement* is then a named element with a defined type. Each element may have zero or several comments attached to it.

Next, the Core::Basic::Classes package was defined. In Figure 3.2.3.2, a *Class* that has later objects as instances is a typed concrete metaclass in this Core::Basic package. A class participates in inheritance hierarchy and has properties and operations. The self loop says that a class can be a superclass of an infinite number of classes. Multiple inheritance is allowed.

Conceptually, in this Core::Basic::Classes package, there is no difference between an attribute and an association end besides their different notations. If a property is owned by a class, it is an attribute. If a property is owned by an association, it represents an end of the association. They are all "properties." The type of the operation, if any, is the type of the result returned by the operation. The multiplicity of the operation is the multiplicity of its result. An operation

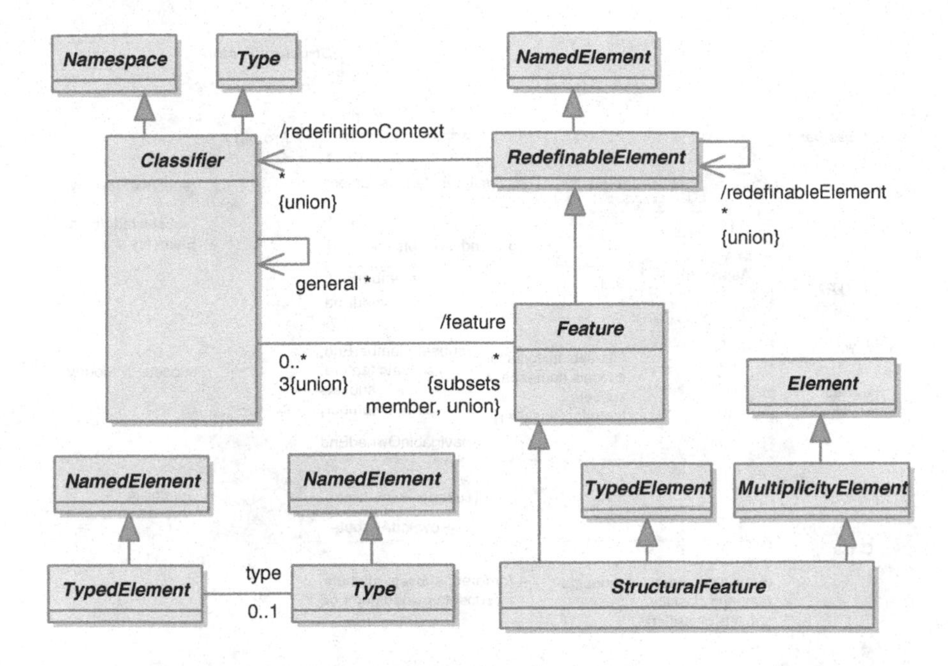

Fig. 3.2.4.2 Core::Constructs::Classifiers package (Fig. 84, Infrastructure Book)

can be associated with a set of types that represent exceptions raised by the operation. Each parameter of the operation also has a type and multiplicity.

The next package is the Core::Basic::Datatypes. In Figure 3.2.3.3, *DataType* is an abstract class that acts as a common superclass for different kind of data types. *PrimitiveType* is a data type. Primitive types used at the Core level are *Integer, Boolean, String*, and *UnlimitedNatural*. An *Enumeration* is composed of a set of literals used as its values.

The last package is Core::Basic::Packages (Fig. 3.2.3.4). Packages provide a way of grouping types and packages for managing a model. In this figure, a package cannot contain itself; but it can include other packages, their contents, and various types. This definition of *Package* allows us to put practically everything in a package, named or unnamed.

3.2.4 Core::Constructs Package

The last package of the Core level is the Constructs package that is dependent on all other packages of the Core.

It merges practically all constructs of other packages. It imports model elements from Core::PrimitiveTypes, it also contains metaclasses from Core::Basic and shared metaclasses from Core::Abstractions obtained by copy. It defines nine new packages: *Root, Expressions, Classes, Classifiers, Constraints, DataTypes, Namespaces, Operations, and Packages*. Except

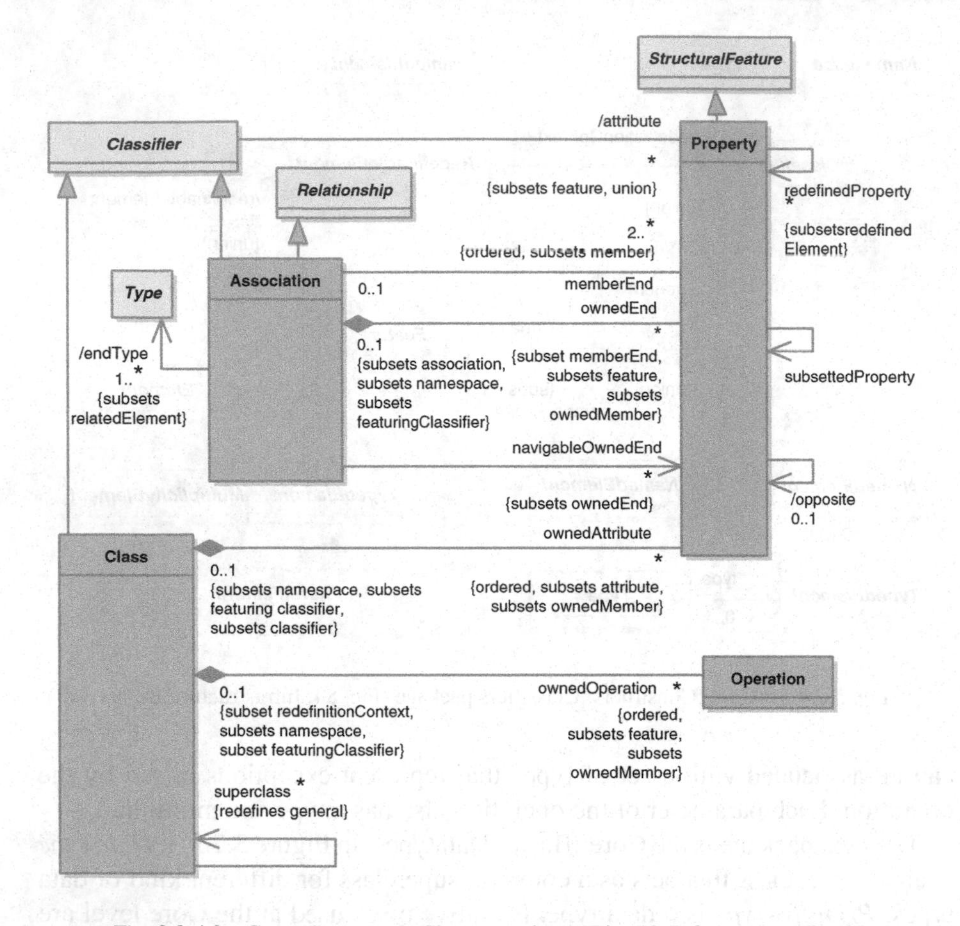

Fig. 3.2.4.3 Core::Constructs::Classes package (Fig. 73, Infrastructure Book)

Operations package, all other names are already present in more elementary packages like Core::Abstractions or Core::Basic. In fact, Classes coming from Core::Basic::Classes or Core::Constructs::Classes are viewed in this namespace organization as two independent packages. We can say that the Core package has two definitions of "Classes." A modeling language specification at M2 level will then have the choice of reusing Classes from Core::Basic or from Core::Constructs from level M3.

The Root diagram of Figure 3.2.4.1 specifies the *Element*, *Relationships*, *DirectedRelationship*, and *Comment* constructs. All these concepts are already defined in Core::Abstractions; they are simply merged together inside one "Root" package.

This package specifies the concepts *Classifier, TypedElement, Multiplicity Element, RedefinableElement, Feature*, and *StructuralFeature*, merging heavily concepts from Core::Abstractions and Core::Basic.

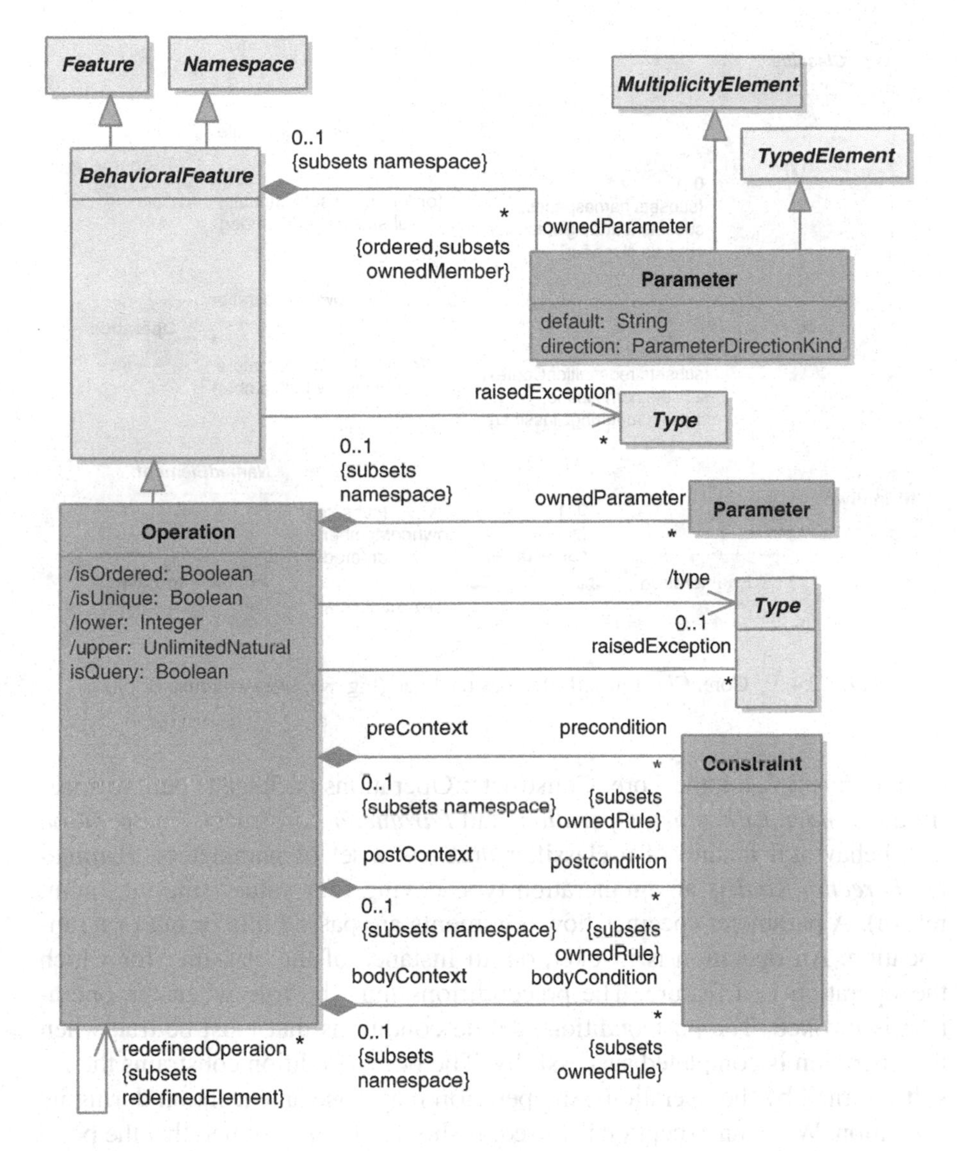

Fig. 3.2.4.4 Core::Constructs::Operations package (Fig. 93, Infrastructure Book)

In the Figure 3.2.4.3, Core::Constructs::Classes package is defined. From this Core::Constructs::Classes package, we learn that a class derives directly from abstract *Classifier*. More precise than the definition of class in Core::Basic::Classes package, a new *Association* class is now derived simultaneously from the Core::Abstract::Classifiers and Core::Abstract:: Relationships packages, so "Property" is now split into two distinct entities, *attribute* and *association end*, through the creation of an *Association* class. Classes have properties; operations and are now connected through associations.

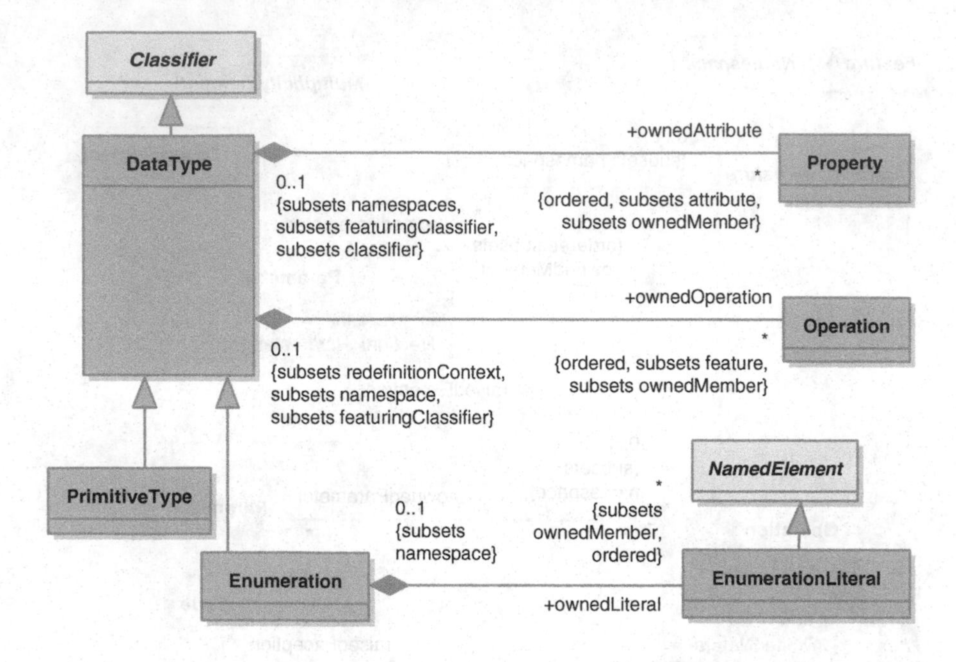

Fig. 3.2.4.5 Core::Constructs::Datatypes package (Fig. 86, Infrastructure Book)

In Figure 3.2.4.4, the Core::Constructs::Operations package is built to specify the *BehavioralFeature, Operation*, and *Parameter* constructs. An operation is a behavioral feature of a classifier and has a set of parameters. *Parameter DirectionKind* is an enumeration type having four values (in, out, inout, return). A parameter specifies how arguments are passed into or out of an invocation. An operation is invoked on an instance of the classifier for which the operation is a feature. The preconditions must be true when the operation is invoked. The postconditions define conditions that must be true when the operation is completed successfully. The body condition constrains the result returned by the operation. An operation may raise an exception during its execution. When an exception is raised, it should not be assumed that the postconditions or body conditions are satisfied. An operation may be redefined in a specialization of the classifier. In this case, it can refine the specification of the operation.

In Figure 3.2.4.5, Core::Constructs::Datatypes package complements the Core::Basic:: Datatypes package by adding a data type to Classifier. A data type in the package Core::Constructs contains attributes to support the modeling of structured data types. There exists a similitude in the way DataType and Class are defined at this stage. Both Class and DataType are derived from Classifier. DataType differs from Class in that instances of DataType are identified only by their values. All copies of an instance of a data type and any instance

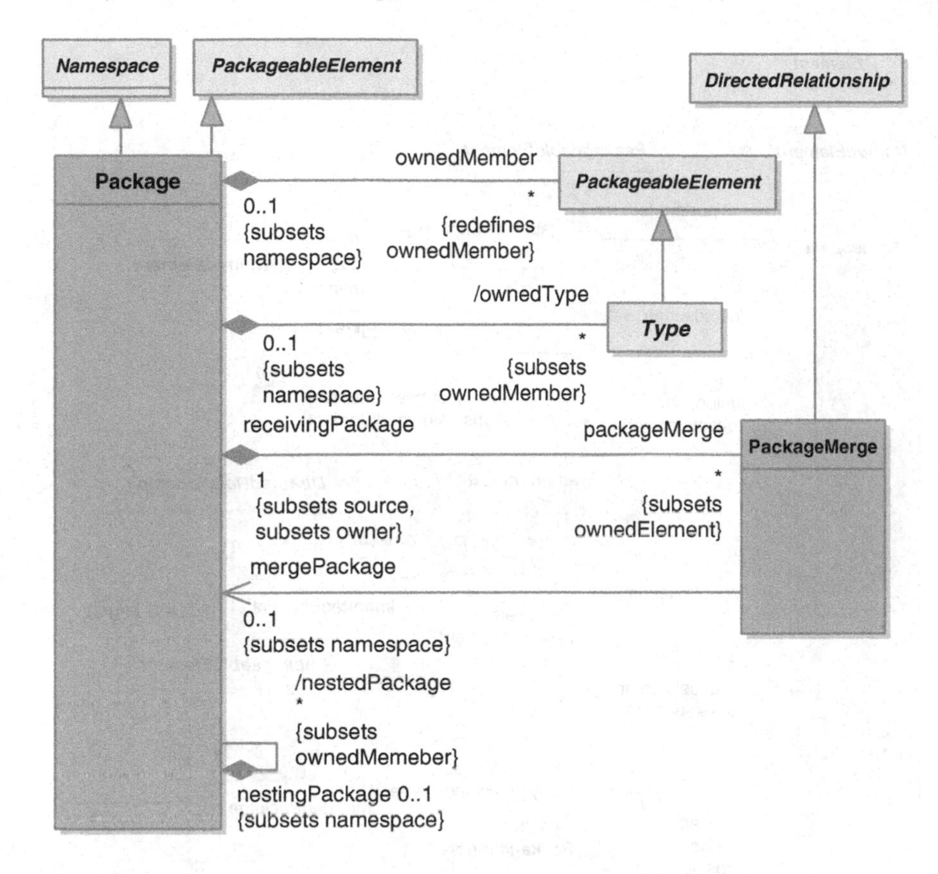

Fig. 3.2.4.6 Core::Constructs::Packages package (Fig. 94, Infrastructure Book)

of that data type with the same value are considered to be the same instance. They are value based.

Core::Constructs::Packages package of Figure 3.2.4.6 specifies a new *Package* and *PackageMerge* constructs. A package is a namespace for its members and may contain other packages. Only packageable elements can be owned members of a package. Being a namespace, a package can import either individual members of other packages or all their members. The principal mechanism governing the construction of package is package merge.

The Core::Constructs::Namespaces package (Fig. 3.2.4.7) specifies the general capacity for any namespace to import all or individual members of packages. An *ElementImport* is a *DirectedRelationship* between an importing namespace and a packageable element or its alias to be added to the namespace of the importing namespace. A *PackageImport* is a *DirectedRelationship* that identifies a *Package* whose members are to be imported by a namespace. The notion of package import differs from the package merge and will be explained later.

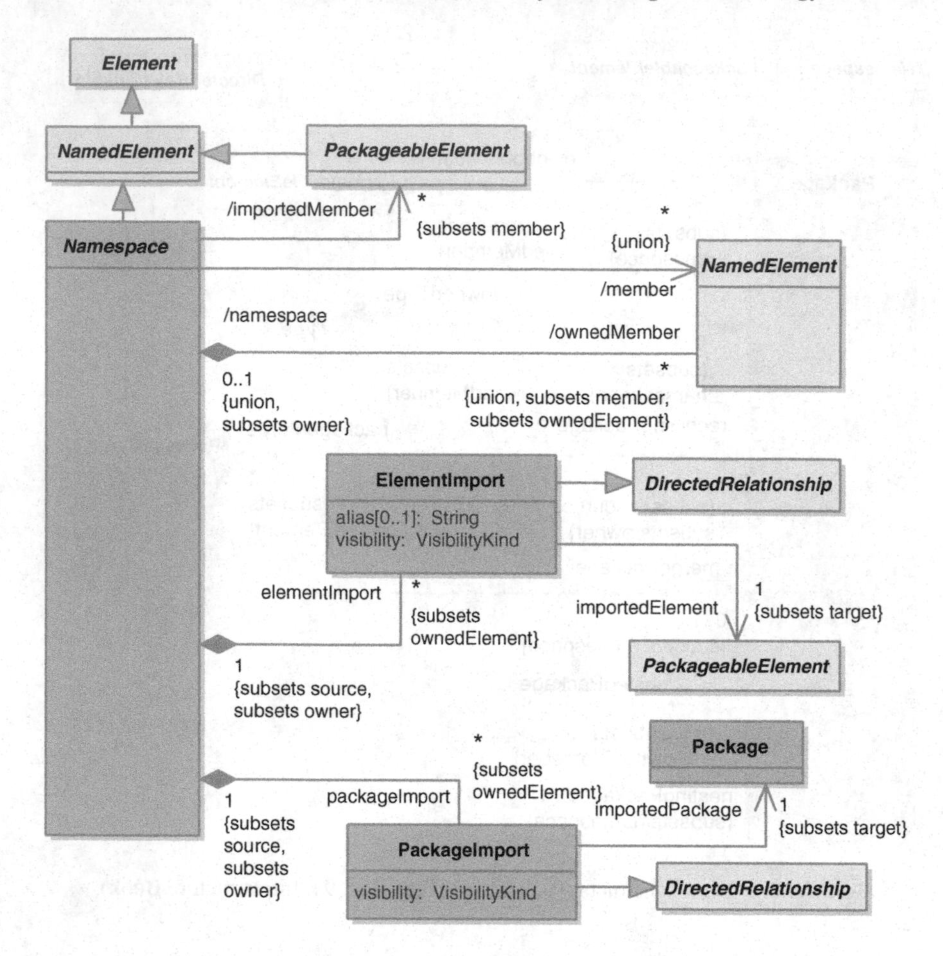

Fig. 3.2.4.7 Core::Constructs::Namespaces package (Fig. 89, Infrastructure Book)

The Core::Constructs::Expressions package is very close to the Core::Abstractions::Expressions. Instead of the Core::Abstractions::Ownerships::Element, we now derive an element from *PackageableElement* (named element that may be owned directly by a package) and a *TypedElement* (please consult Core::Abstractions::Expressions for comparison).

The Core::Constructs::Constraints package differs only from Core:: Abstractions::Constraints by the fact that Constraint derives now from a *PackageableElement*, and not from a *NamedElement*.

3.2.5 Core::Profiles Package

In the Core::Constructs::Profiles package, *Profile* is a package issued by itself from a namespace. A *ProfileApplication* indicates which *Profile* has been

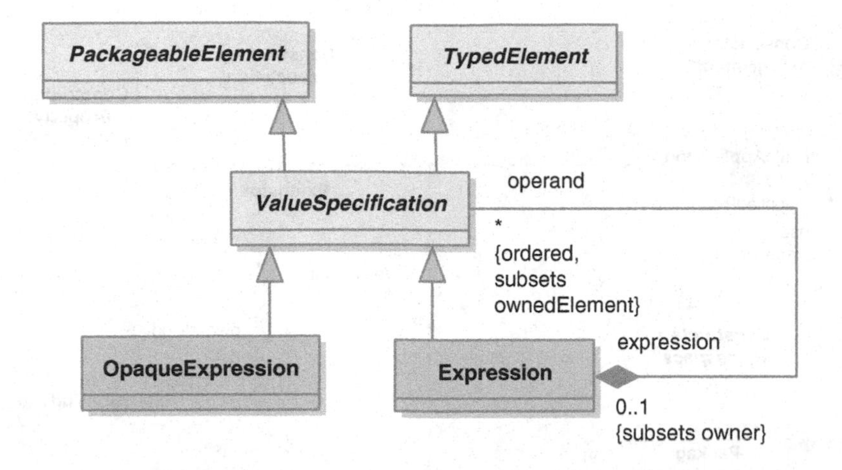

Fig. 3.2.4.8 Core::Constructs::Expressions package (Fig. 72, Infrastructure Book)

applied to the package. It comes from *PackageImport* which is a *DirectedRelationship.* A stereotype defines how an existing metaclass can be extended. A class may be extended by one or more stereotypes. *Extension* is a kind of *Association*. An *ExtensionEnd* is used to tie an extension to a stereotype when extending a metaclass. A stereotype can change the graphical appearance of the extended model by using an attached icon. Finally, metaclass can be individually extended, and so can the complete package.

The Profiles package provides mechanisms for adding new semantics to a metamodel without creating contradictions with existing packages (*profiling* versus *metamodeling*). Rules could be perceived as constraining but the result is worth the exercise. Moreover, Profiles can extend by restricting. For instance, generalization of classes should be able to be restricted to single inheritance (multiple inheritance not allowed), without explicitly assigning any stereotype.

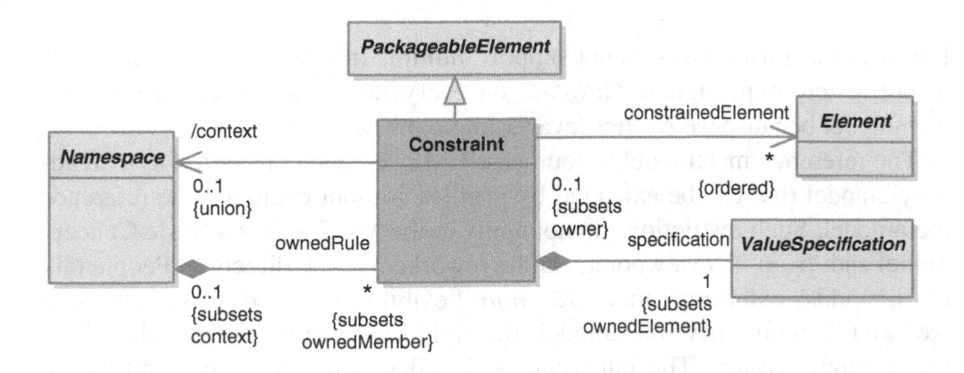

Fig. 3.2.4.9 Core::Constructs::Constraints package (Fig. 85, Infrastructure Book)

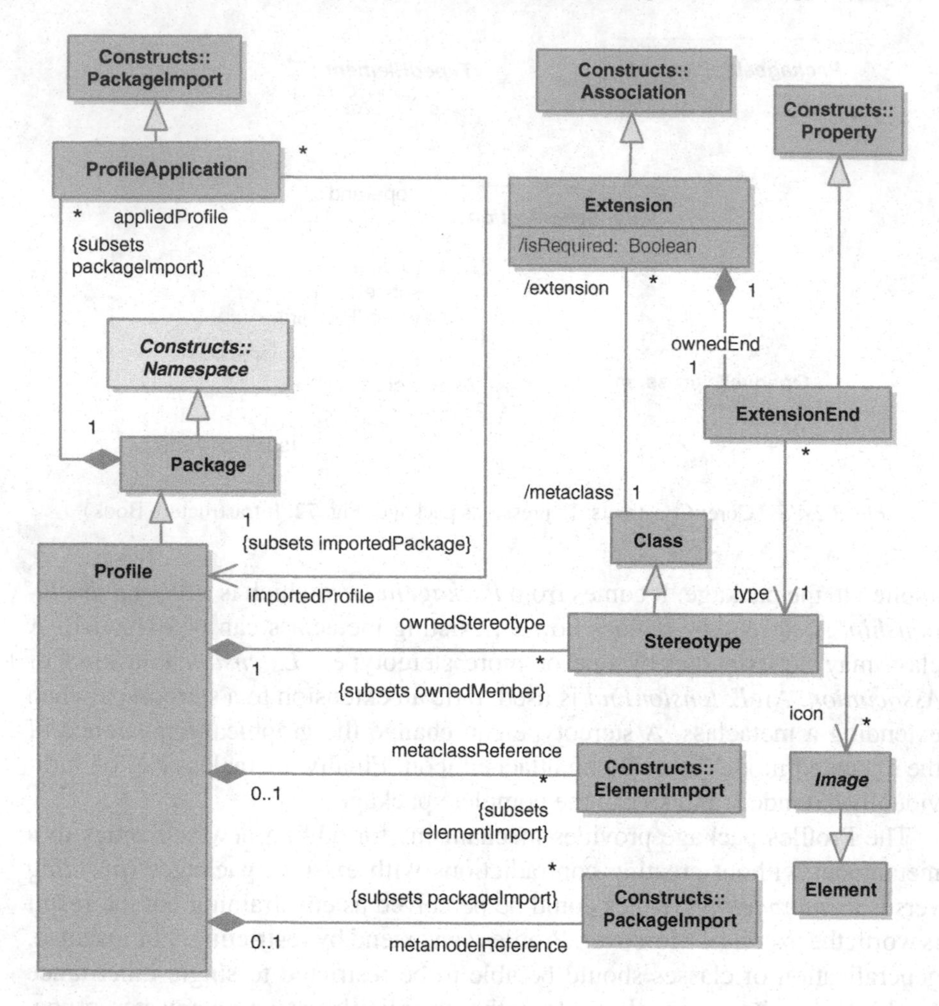

Fig. 3.2.5.1 Core::Profiles package (Fig. 109, Infrastructure Book)

For instance, Java classes do not support multiple inheritance. We do not have to put a stereotype sign <<Java>> on every Java class instances from the metamodel because we restrict Java to single inheritance.

The reference metamodel (in our case UML) is then considered as a "read-only" model that can be extended by profiles without changing the reference metamodel. Such restrictions do not apply to the MOF as it is a Basic Concept model and, from this viewpoint, can be reworked in any direction. People talk of "first class extension" that offer more flexibility in regard to semantics. If we want to define new metamodel, the right way would be from the MOF, not through Profiles. The intentions of Profiles is to give a straightforward mechanism for adapting an existing metamodel with constructs that are specific to a particular domain, platform, or method (second class extension). In other words, the profiling approach does not compromise the existing metamodel and shields user extensions from each other.

From a research point of view, the first class extension mechanism is more interesting to extend the object paradigm towards unexpected frontiers. The Superstructure is tied to the current version of the UML. The Infrastructure of the UML, just exposed, is a good example of that could be inspired as part of ontology engineering, future object concept building.

Chapter 4

UML Superstructure: Language Definition and Diagrams

4.1 Making Our Own Development Methodology with UML

As a general graphical language, the UML is not just targeted for modeling object-oriented software. Being very flexible and customizable, it enables us to create our own in-house methodology that can describe real-time workflow, system activities, state evolution, object, relational databases, etc. The "U" in the UML means "Unified" since the first version of the UML came from the effort of three persons, Rumbaugh who developed the OMT methodology (Rumbaugh et al., 1995), Booch (1993), who developed the Booch methodology, and Jacobson who developed the Objectory methodology (Jacobson et al., 1992). In 1997, they submitted the version 1.0 of the UML to the OMG, an independent standards organization which took over the UML development and released subsequent UML versions. In 2003, the latest version 2.0 was a quantum leap from older 1.x releases with the arrival of a true object diagram, a very long-awaited instantiation diagram necessary to study real-time systems, a more precise semantic framework, four supplementary diagrams, and a deep change in the role of some diagrams (e.g. the Component Diagram is not considered as an implementation diagram but a design diagram). The UML is an open modeling standard designed and supported by software companies, consultants, large corporations, and governments. The UML has a built-in extension and customization capabilities, like stereotypes and profiles that enable users to customize the UML to their needs.

Why should we model with the UML? For the moment, the UML is the most complete modeling standard in the software market and is a very dynamic standard as it tries to adjust rapidly, at the time of this writing, to changing needs of the fast moving software industry. To answer more fundamental aspects of this question, let us draw a parallel between software engineering and mechanical or electrical engineering. Mechanical engineers draw mechanical plans of parts and put geometrical dimensions before molding and feeding data to the CNC (Computer Numerical Control) machines. Before implementation,

D. M. Bui, Real Time Object Uniform Design Methodology with UML, 77–190.

electrical engineers draw schematics containing electrical devices in a standard way using common graphical standards so that, everywhere in the world, the schematics are always interpreted in the same way. There is no reason why software engineers tend to jump directly into code without correctly analyzing and designing the software.

Models are convergence points for an organization. We make a requirement model to ensure we build what customers want. Requirement models help us take good business decisions. Design models should provide understanding, communication, and discussion between multidisciplinary engineering staff to discover the optimal solution. Models are created in a logical way: implementation models are elaborated from design models that are in turn elaborated from requirement models. If testers are not getting involved directly in model building, understanding previous models is useful to create significant and systematic tests. As the business of modern software industry relies heavily on maintenance, it is imperative to have an in-house process for dealing with model maintenance and evolution over time. As human labors are a changing resource, models and domain knowledge warehoused in a database CASE tools are invaluable assets for an organization to react to market trends, evolution, platform changes, interoperability, and human resource mobility.

Moreover, the UML can be enriched with Profiles (user extensions) or complementary modeling tools, all maintained by the OMG, for instance Business Process (BP) Model. The Business Process Management Initiative (BPMI; available at www.BPMI.org) and the OMG have announced the merger of their Business Process Management (BPM) activities to provide thorough leadership and industry standards for this vital and growing industry. The combined group has named itself the Business Modeling and Integration (BMI) Domain Task Force (DTF) (for more information about the BMI DTF's activities, please visit their homepage at http://bmi.omg.org). In this book, we make use of Business Process diagram as the 14th diagram at the requirement analysis phase (UML 2.0 already offers 13 diagrams). Good standards are evolving standards; UML 2 is very different from its first 1.0 version. With all the welcomed corrections in version 2, with the MDA as a supervisor framework and a synergistic support of all complementary object standards developed and centralized by the same OMG organization, the UML now becomes an interesting standard and has really taken off since the beginning of 2004.

In this chapter, we start studying the UML contents from the authors' viewpoint and try to follow their interpretation and semantics. In the following chapters, before getting into the uniform methodology, we expose fundamental modeling concepts and try to map UML to fundamental concepts used when modeling and give, at a suitable time, our comments. The UML is a visual language with a visual notation but it is supported in the background by a set of theoretical and abstract constructs. On the superficial layer viewed by users, it offers two main views of all systems: Structure and Behavior. We decompose further the behavioral view into two distinct views: functional and dynamic. We choose to separate functions in order to connect them to use cases and business processes. According to the context, we will make use of Structure/Behavior or Structure/Function/Dynamics.

4.2 Structure and Behavior: 13 UML Diagrams Classified Along Structure and Behavior Axes

To structure the abstract concepts, the UML divides them into two families: Structure and Behavior.

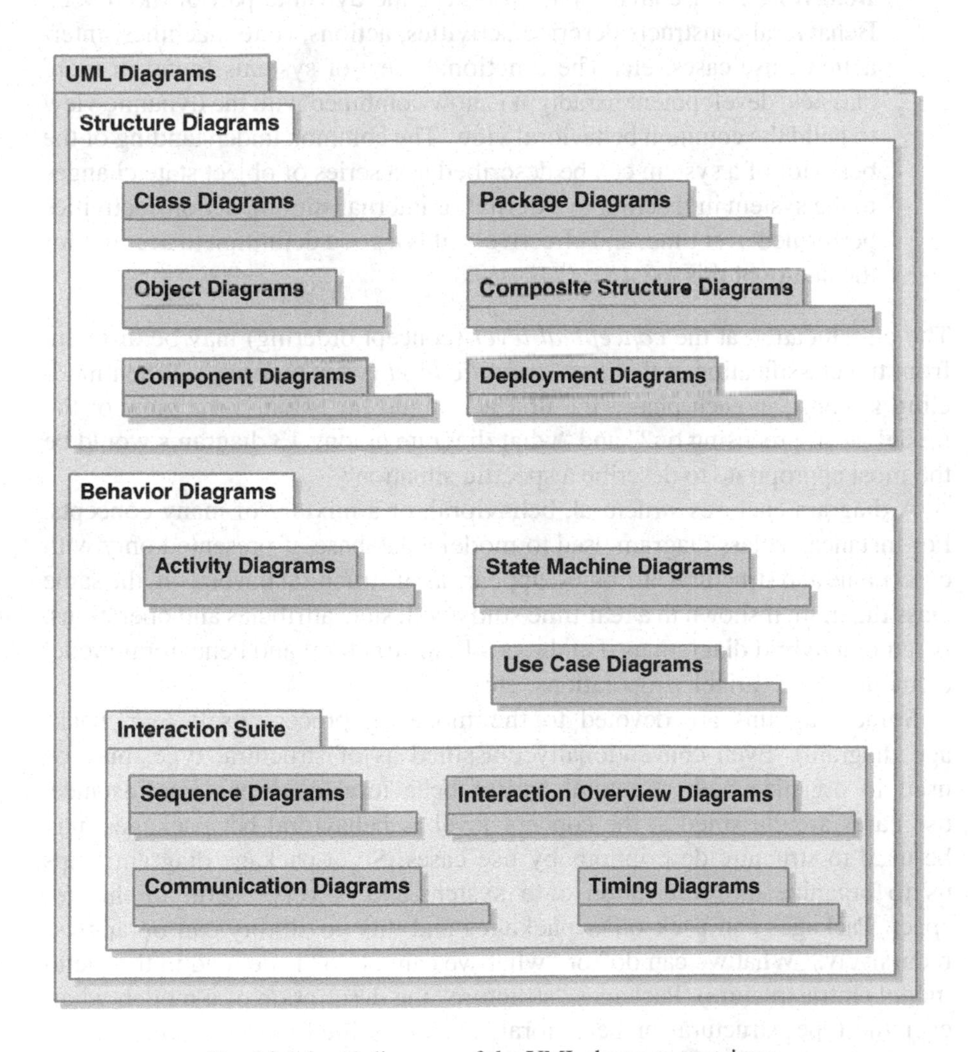

Fig. 4.2.0.1 13 diagrams of the UML shown as containers

- **Structure**. The Structure axis embraces the static part of the model. Structural constructs describe classes, instances, interfaces, packages, composite structures, relationships, nodes, components, deployments, etc. With regard to the development process, this part covers all the development phases, starting from the requirement analysis toward to testing phase, going through multiple design and implementation models. This part corresponds to the classical structural view of all systems. Structure is mostly static, irrespective of time (so sometimes, people assimilate structural view to static view). A structure may be real or abstract.

- **Behavior.** The Behavior axis embraces the dynamic part of the model. Behavioral constructs describe activities, actions, state machines, interactions, use cases, etc. The functional view of systems found in more classical development paradigm is now combined with the dynamic view to build the common behavioral view. The common understanding of the behavior of a system can be described as a series of object state changes to the system in reaction to external or internal stimuli, actions/activities performed over time, and effectively, it is a good definition to account for the nature of this axis.

The classification at the *conceptual level* (concept ordering) may be different from the classification at the *diagrammatic level* (view ordering). When modeling systems, in each phase, the first act would be "what is the view of the model we are focusing on?" and "what diagram among 13 diagrams would be the most appropriate to describe a specific situation?"

A diagram encloses structural, behavioral, or a mixture of many concepts. For instance, a class diagram used to model a database, if presented only with class name and structural attributes, appears as of structural flavor. But, the same class diagram, if shown in a real-time study with state attributes and operations, is rather a hybrid diagram as it embraces both structural and behavioral model elements (state variables, operations, etc.).

Some diagrams are devoted to the modeling process itself (e.g. package diagram). Even conventionally classified as of structural type, may be used to organize both structural and/or behavioral concepts. For instance, use cases are classified at the concept level as behavioral but packages may be used to structure descriptions by use cases. So, a package diagram helps us to organize the description of a system and is very useful in this respect. Packages can pack other packages and this possibility can be applied recursively. "What we can do" or "what we cannot do" is defined in the metamodel (Infrastructure). Packages "structure" the description of a model, whatever the type, structural or behavioral, even classified as "structural" in the standard.

Another example is the Component Diagram. Structure and Behavior are packed inside a unique concept "component." So, the component diagram is not uniquely "structural" as it contains a whole working subsystem.

Opposite to the standard that chooses a description by concepts, most tool vendors opt for the description based on diagrams (not on concepts) in their user's manual. A diagram collects inside a graphical representation of well-chosen concepts necessary to explain a specific functionality, view, model, or aspect of a system. Before going into a more detailed description of each diagram and their associated concepts, let us summarize their common use.

Table 4.2.0.1 Six structural diagrams of the UML standard. The table explains their nature and mentions when they are mainly created in the development cycle. Hybrid means "structural and behavioral." Organizational means "needed to organize the modeling process." "Drawn mainly" column states when these diagrams are drawn for the first time but they can be used in subsequent phases and refined if needed

Structural diagrams	*Drawn mainly*	*Comments*
Package Diagram Organizational	Anytime	Mostly used at top-level description, this diagram helps organizing a complex description into packages. A Package of packages is allowed to create a multilevel hierarchy. A package groups several model elements in a single unit and provides a namespace. Package may contain almost everything.
Class Diagram *Structural or Hybrid*	Early in design and later	Shows a collection of static model elements such as classes, interfaces, instances, packages and their relationships. In a database development, a class replaces an entity of the classical Entity–Relationship Diagram. In real-time system, this diagram factorizes common properties of objects into real-time classes.
Object Diagram *Structural or Hybrid*	Early in design and later	Shows objects and message flows between them in a given context (a given class may have many instances in an object diagram). This diagram is essential in real-time systems where separate object diagrams are drawn to describe various scenarios of object collaboration.
Composite Structure Diagram *Structural or Hybrid*	Early in design and later	Shows internal collaborations of classes, interfaces, or components to describe functionality. A Composite Structure Diagram is used to express run-time architectures, usage patterns, and relationships among participating elements. The object diagram is closed to the Composite Structure Diagram but is targeted to describe more elementary tasks.
Component Diagram *Hybrid*	During design and later	Describes components that compose a system, their public interfaces, and their relationships. Components are a set of connected objects and the whole structure is ready for executing high-level tasks. In the past, components often constitute physical pieces of software but their use has been extended in UML 2 towards abstract components. Class Diagram is more elementary than Component Diagram. Components pack inside high-level entities both structure and behavior of many objects.

(*cont*).

Table 4.2.0.1 (Continued)

Structural diagrams	*Drawn mainly*	*Comments*
Deployment Diagram *Hybrid*	Design and in PSM (Platform-Specific Model)	Used mostly at the end of the implementation phase, the Deployment Diagram shows the execution architecture of a system and the relationships between hardware and software. Main entities of this diagram are "nodcs" connected together through communication paths. Nodes may be nested. Deployment is often specific to a predefined or a typical environment. "Artifacts" represent concrete elements in the physical world being included in this diagram.

Table 4.2.0.2 Seven behavioral diagrams of the UML 2 standard. Behavioral diagram includes both functional and dynamic aspects

Behavioral diagrams	*Drawn mainly*	*Comments*
Sequence Diagram *Behavioral*	Early in design and later	First diagram of the Interaction Diagram Suite selects and shows one execution path through a program. It identifies all collaborating objects, messages, calls, and data exchanged between objects in a structured way, along a time axis. Objects are represented with their vertical lifelines. All possible sequences, merged together, make up a program or an algorithm.
Communication Diagram Behavioral	Design and later	Second diagram of the Interaction Diagram Suite can be generated automatically from the Sequence Diagram and vice versa. It shows interactions between objects at run-time. Objects are not arranged "from left to right" as in Sequence Diagram but are distributed uniformly in the graphical area. The focus is put on the inter-object relationships and the way messages are exchanged over time is visualized with message numbering. As its sequence counterpart, communication illustrates only one sequence in a program. Communication Diagram can be seen as Object Diagram with an ordered set communication messages.
Interaction Overview Diagram *Behavioral and Organizational*	Design and later	Third diagram of the Interaction Diagram Suite is a variant of the Activity Diagram. The focus is put on the flow of control where nodes are Interaction (Unit of observable behavior) or InteractionUse (a reference to Interaction). This high-level diagram is targeted to structure elementary sequences, communication chunks, or pieces of timing inside a control structure identical to that found in an activity diagram. It is a kind of "Composite Behavior" to make a parallel with "Composite Structure".

Table 4.2.0.2 (Continued)

Behavioral diagrams	*Drawn mainly*	*Comments*
Timing Diagram *Behavioral*	Design and later	Fourth diagram of the Interaction Diagram Suite derives from the well-known electronic "timing diagram" often used to specify hardware timing of processors or memory chips. Time is represented as a linear axis with conditions. This diagram is precious for real-time specifications of time and delays.
State Machine Diagram *Behavioral*	Design and later	A state is an interesting snapshot on the evolution of a system. This diagram models the behavior of a single object, specifying the sequence of states that this object goes through, during its lifetime in response to events. With regions, concurrent systems may be represented on a single diagram to show their synchronization. From a theoretical viewpoint, a State Diagram packs many sequences and the control flow is represented by "state transitions." This diagram can be used to merge sequences or elaborate software algorithms.
Activity Diagram *Behavioral and Organizational*	Design and later	This diagram put the focus on actions and activities. As explained later in the uniform methodology, action, and activity, state, event and condition come from the same continuum and the activity diagram is just another view of the State Diagram. Conditions on "transitions" are transferred to "Decision Nodes." Join and fork nodes allow the representation of parallel processes. This diagram is closest to the Petri network and the way we write algorithms. Used to elaborate software algorithms, it is an alternative to the state machine diagram.
Use Case Diagram *Functional*	Requirement analysis	Of functional flavor, this diagram is used mostly for specifying required usages of a system. The key concepts of the use case diagram are "actors" and "use cases". Known as a diagram that is not "object oriented" at all, this diagram is much useful for the Requirement Specification step. Used in conjunction with the Package Diagram, a complex system may be specified hierarchically with use cases.

4.3 Hierarchy of UML Metaclasses

4.3.1 UML and "L" as Language

By establishing a parallelism with natural language, we can compare the process of building a modeling language (L as language in UML) to the inception of the natural language. First, we need to define the atom or elementary particle of all languages. A "character" has its counterpart "*model element*" in the UML. The first character of the alphabet has the name "A" and an infinite number of ways to write it, each way corresponds to a glyph or font, so does the model element.

Table 4.3.1.1 Parallel between the natural language and the UML

Aspects	*UML*	*Natural language*
Nature	Iconic and pictorial	Textual or character based
Element of language	Model element with its graphic notation (e.g. class, association, state, etc.)	Character with their glyphs and fonts
Full set	All elementary model elements (a model element is considered as an "elementary" concept)	Extended alphabet (characters, numbers, signs, special characters)
Element at communication level	Modeling concept (chunk of model information, e.g. a message between two lifelines, a transition between two states, a class with attributes and operations, etc.)	Word
A communication message	Any diagram combining several modeling concepts	Sentence
Dictionaries	All modeling concepts (to be discovered, practically infinite)	Vocabulary (can be enriched but finite)
Rules	UML rules defined at the metalevel	Grammar
Element at project level	Models composed of diagrams and artifacts generated by a CASE tool	Textual document

If the model element represents a class, it is drawn as a rectangular box. An association or a relationship is drawn with a line, directed or not, connecting one, two or several classes together. The "glyph" of a model element is its graphical "notation," term used in the UML standard.

Next, characters are combined together to build words (natural language contains roughly one million if we count all scientific words). The corresponding "alphabet" of UML is composed of primitive types as classes, instantiated objects, associations, inheritance–derivation relationships, aggregation–composition relationships, nodes, use cases, object states, packages, etc. "Words" in the UML are elementary chunks of communication elements built from application, for instance a class Car with their attributes and operations are equivalent of a word in natural language, an Acceleration state of this Car is also a dynamic element of information.

The way words are combined together in natural language to form sentences is ruled by English grammar. The way modeling elements in the UML are tied together with relationships composing an understandable message (diagram) is ruled by current "grammatical" rules defined in Superstructure Book.

UML rules are defined in Superstructure Book (708 pages). To define the way that expressions are written to specify constraints, conditions, queries, operations/actions, OCL (Object Constraint Language) book adds another 185 pages. Two complementary books (Infrastructure Book, 226 pages) and Diagram Interchange (82 pages), complete the UML version 2 documentation.

If the current semantics of the UML is not as precise as mathematicians expect it to be (defining precise mappings from symbols to abstract entities and sets, and thus giving meanings to symbols by virtue of the logical properties of the corresponding abstract entities and sets), from a practical viewpoint; this fact must not be a sufficient reason for not deploying a methodology of development inside an organization. We can see current UML as an informally defined, popular set of graphical conventions and semantics, which can be put to use in whatever ways appropriate. If the UML cannot fit our needs, we can extend it via Profiles or even define our unique and personal standard with meta CASE tools. To resume, the UML must be a tool, not a constraint.

4.3.2 Package Import and Merge in Metamodeling

The Superstructure of the UML makes wide use of two relationships between packages: Package Import and Package Merge, to build the language. There exists semantically an important difference behind these two concepts.

Import. Importing process identifies an element in another package and allows the element to be referenced using its name. The name of the imported element or its alias is to be added to the namespace of the importing namespace. An element import works mainly by reference. An imported element can be further imported by another namespace. A package import is a directed relationship that allows the use of a name to refer to package members from other namespaces. As notation, a package import is represented by a dashed arrow with an open arrowhead oriented from the importing package to the imported package. The predefined stereotype notation is ≪import≫. The package that imports is named the Source package. The Source package's namespace will gain access to the Target classes; the Target namespace is not affected. In other words, when using an ≪import≫, it appears as if the Source package has the model element of the Target package inside its own package by means of the a mechanism of naming and referencing.

Merge. Package merge allows modeling concepts defined at one level to be extended with new features while keeping the same original namespace. Merging process enables interchange. A package merge between two packages implies a set of transformations, whereby the contents of the package to be merged are combined with the contents of the receiving package. Deriving a new class from an existing class declared in another package is an example of package merge. Merge implies generally an import (to localize the merged package) followed by a copy/duplicate (the copy/duplicate process is necessary to preserve the integrity of the original packages) and then a transformation of the original package to allow its reuse in a new context. A merge is a more complex process if compared to import and generally needs package reworks. For instance, if we merge two packages PA and PB into a new package PC, possibly, there could be some inconsistencies between elements of PA and PB that must be resolved (Pottinger and Berstein, 2003). The merged package retains mostly all nonduplicated information in PA and PB, collapses redundant information, and solves conflicts if necessary (automatically or assisted). From our abstract level, criteria of the merge cannot be defined yet, since they could be very diversified and contextual (type preservation, relationship preservation, conflict resolution, etc.).

4.3.3 Comment on Inheritance Relationship at the Metalevel of UML

Before digging into the details of 13 UML diagrams, let us examine the inheritance relationship that the UML make extensive use of to build up its whole syntax and semantics. One of the most powerful notions in object paradigm is inheritance, considered as one of its most representative features. Modeling quality and reusability of designs are largely accredited to inheritance at the conceptual phase. Code sharing at the software implementation step is one of the most powerful direct consequences of inheritance.

The basic idea of inheritance is simple. It allows new definitions to be based upon existing ones. When a child class is derived from a parent class, only differential properties between the child and his parent must be specified in the child. Properties defined in the parent class are automatically shared by the child class. Inheritance is transitive, so if C derives from B that in turn derives from A, then C derives from A. This implies that a derived class will have all the properties, not only properties of its parents, but also those of their grandparents, and so on. The UML allows inheritance from several parents at the same time in its metalanguage. This fact is called *multiple inheritance*. Despite the fact that multiple inheritance introduces many conceptual and technical intricacies, it is supported by most object systems, at least at the conceptual level.

4.4 Superstructure and Compliance Levels

The Superstructure Book defines the UML constructs and four compliance levels to support model interoperability, graphic representation, and model exchange. New constructs are added at each level. Existing constructs in Core packages have been merged or imported into new UML package. Developers and toolmakers can therefore choose to adhere to the desired level of compliance. Hereafter is the four-level compliance proposed by the UML; they characterize the degree of conformance to the current standard.

Table 4.4.0.1 Four compliance levels of UML L0 to L3 defined with regard to the abstract syntax dimension. From level to level, packages are enriched. (For details, see Tables 2.3–2.5, Superstructure Book.)

Compliance level	*Language units*
L0 Foundation level	Class-based structures encountered in most popular object languages. Basis for interoperability
L1 Basic level	Extends L0 constructs to introduce Usecases, Structures, Actions, Activities, Interactions, GeneralBehavior
	Complement L0 constructs for Classes
L2 Intermediate level	Extends L1 constructs to introduce Components, Deployments, StateMachines, and Profiles
	Complement L1 constructs for Actions, Activities, GeneralBehavior, Interactions, Profiles, and Structures
L3 Complete level	Extends L2 constructs to introduce InformationFlows, Templates, and Models (model packaging)
	Complement L2 level constructs for Actions, Activities, Classes, Components, Deployments, StateMachines, and Structures

At each level, the compliance is separated into:

1. *Abstract syntax compliance*: A compliance with metaclasses, their relationships, constraints, and queries
2. *Concrete syntax compliance*: A compliance with the graphical notation specification and diagrams
3. *Diagram interchange compliance*: For tool implementer, it entails the ability to output models and to read in models based on XMI schema corresponding to that compliance level

In other words, a compliance has two dimensions and for a given level i, can be expressed as "L_i," "L_i with concrete," and "L_i with concrete and interchange."

Figure 4.4.0.1 displays the topmost packages of the UML. Classes package, complemented by AuxilliaryConstructs are at the root of all levels. CompositeStructures contains both static and behavioral elements. Components are merged from CompositeStructures and Deployments are merged from Components. Deployments are hybrid structures that combine both structural and behavioral constructs. CommonBehaviors are originally issued from Classes and serves as basic modeling elements for Activities, Interactions, StateMachines, and UseCases. Actions are merged from Activities. All behavioral constructs borrow model elements from Composite Structures.

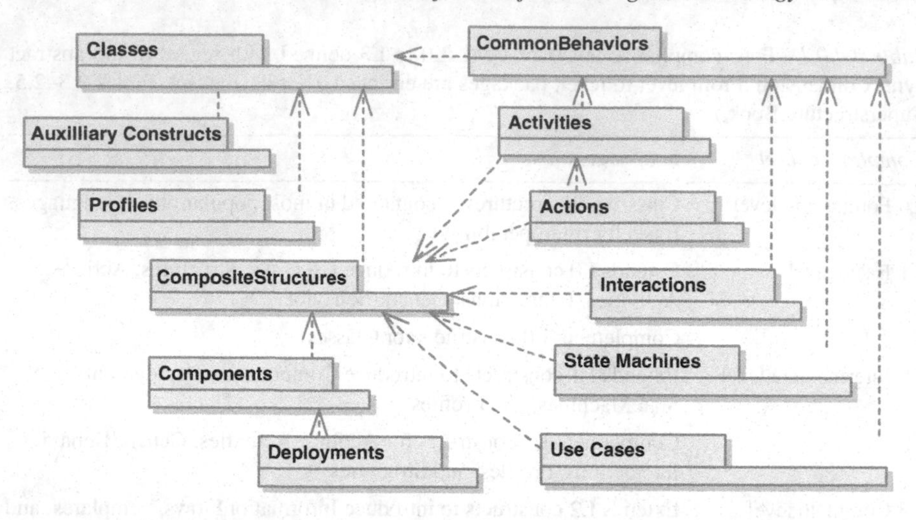

Fig. 4.4.0.1 Top-level package dependencies. Relationships are "import" and/or "merge"

4.5 Reuse of The Infrastructure: Generalization Hierarchy of Metaclasses in The Structure Part

The Kernel package represents the core modeling concepts of UML that include basic definitions of classes, associations, and packages. The reuse of Core is accomplished in the Kernel by merging Core::Constructs with relevant subpackages of Core::Abstractions.

After the merging process of Figure 4.5.0.1, as a result, the Kernel of the UML contains 12 diagrams not organized as subpackages but as a flat hierarchy of 16 diagrams in 16 packages, which borrow heavily from the Infrastructure.

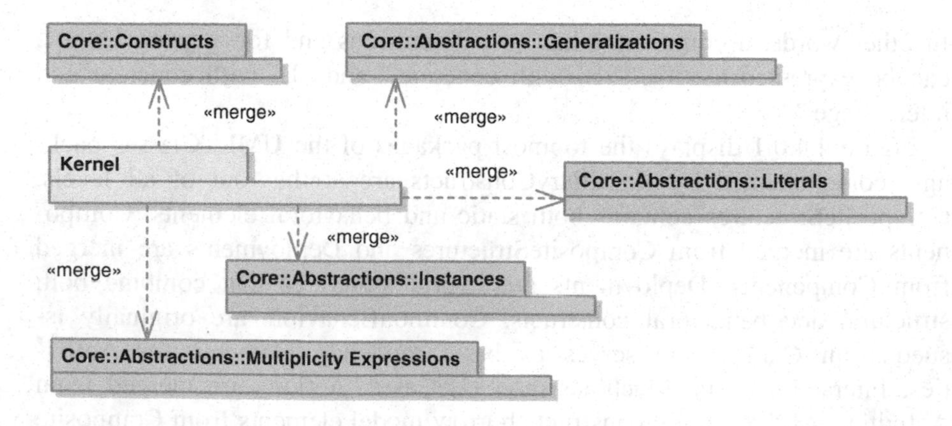

Fig. 4.5.0.1 Infrastructure packages merged into the Kernel of the UML

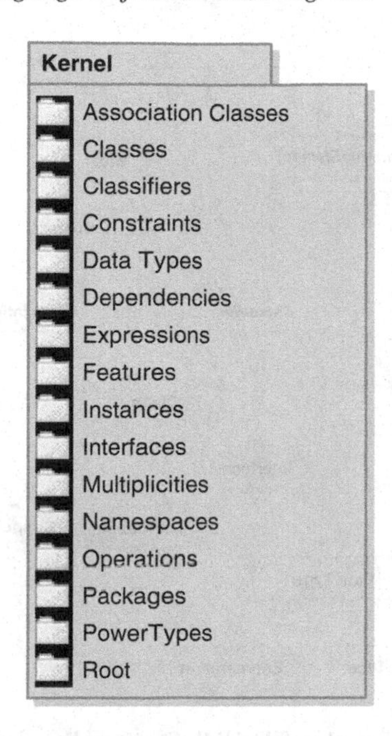

Fig. 4.5.0.2 Flat 16 packages of the Kernel merged from the Infrastructure

The detailed description of the 16 packages of the Kernel can be found from Figures 7.3 to 7.18 of the Superstructure Book (Fig. 4.5.0.2). When we have difficulty explaining some intricacies of concepts, we will reproduce some of them to find out explanations. The UML standard is explained in more than 1,000 pages, so the reader must consider this chapter as guidelines to understand the overall structure of the UML. For detailed information, he must dive again in the alphabetically ordered Superstructure Book of the UML. Toolmakers must read the Standard documentation, not practitioners.

The Kernel package contains the core modeling concepts of the UML. Constructs and Abstractions packages of the Infrastructure contribute heavily to the Kernel definition. During this reuse process, the reused classes (metaclasses) are extended in the Kernel with additional features.

If we extract information from the Kernel packages, the derivation hierarchy of the UML model elements can be summarized by the three Figures 4.5.0.3, 4.5.0.4, and 4.5.0.5.

Figure 4.5.0.3 shows in broad outlines how the UML is built for deriving complex model elements. Elements shown in italic are *abstract* and cannot be instantiated to give rise to modeling objects. For instance, we can model a *Class* with UML but not a *Classifier*. Inheritance is represented by an arrow oriented from the child class towards the parent class. For instance, *Comment* class is a child class of *Element*.

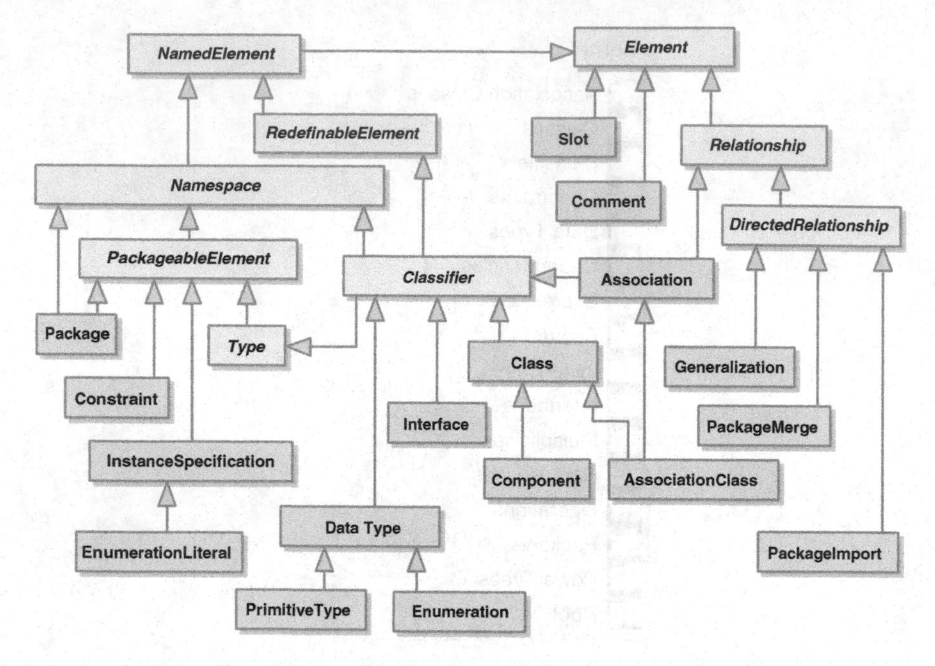

Fig. 4.5.0.3 Generalization hierarchy of the UML Structure Part showing principal metaclasses and their structure in the metalanguage (Part1)

Figure 4.5.0.4 shows the definition of three concrete classes *Property*, *Operation* and *Parameter*. In UML, *Classifier* has only Properties but *Class* has both *Properties* and *Operation*s. The definition of *Operation* needs *Parameters*. Notice that *Operation* derives from a *BehavioralFeature* even if it is defined in the Structure part of the UML.

The diagram of Figure 4.5.0.5 shows all model elements used to specify the multiplicities accompanying each relationship or any expression written on the UML diagram. Opaque expression is a non interpreted textual statement.

Starting from the first diagram of the series (Part1), an abstract model *Element* is defined at the root of the language as an abstract metaclass with no superclass. An *Element* can own other elements (not explicit). An abstract metaclass *NamedElement* is then derived from *Element* to add a name to a model element. Now, a model element may have a name and may own other named elements.

Namespace is an abstract metaclass derived from *NamedElement*. *Namespace* is thus a model element that contains a set of named elements. From the property of its *Element* superclass, a *Namespace* can own other named elements. A namespace then provide a container for named elements. It provides a means for resolving composite names such as *name1::name2::*

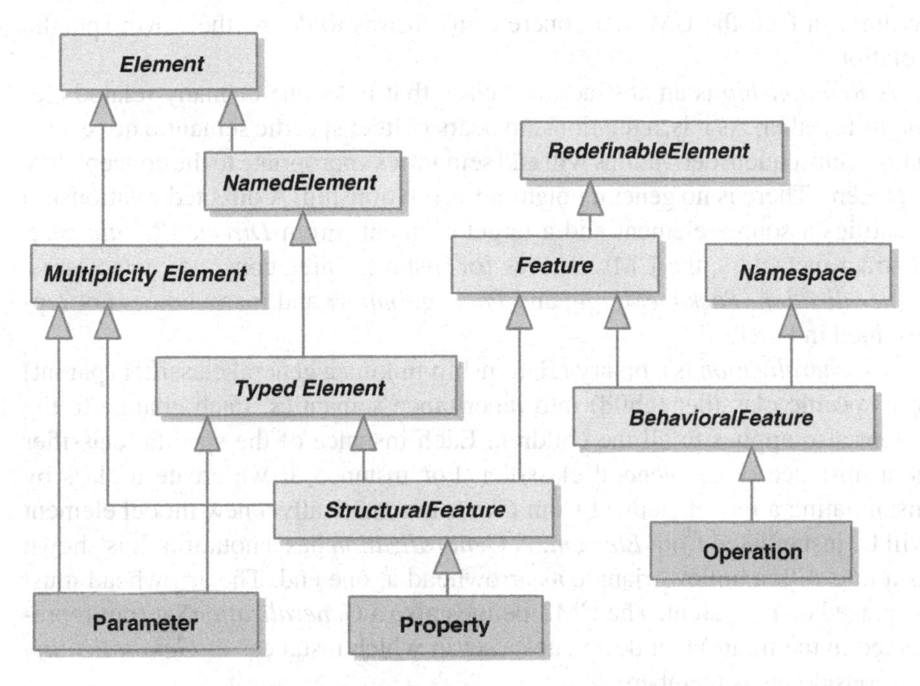

Fig. 4.5.0.4 Generalization hierarchy of the UML Structure Part (Part2). Operation is a BehavioralFeature and Property is a StructuralFeature

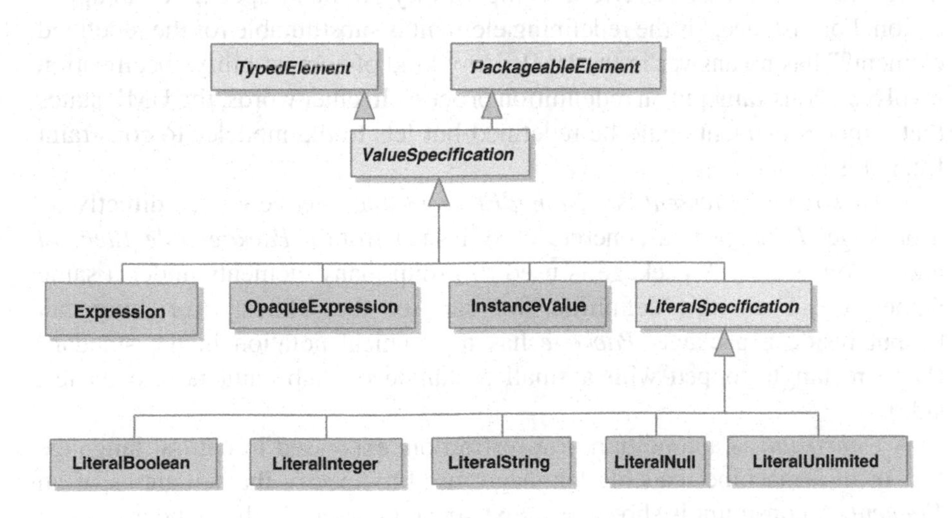

Fig. 4.5.0.5 Generalization hierarchy of the UML Structure Part (Part3)

name3. There is no explicit notation defined for the way namespaces must be written. In fact, the UML let concrete subclasses to define their own specific notation.

A *Relationship* is an abstract metaclass that links one or many related elements together. As it is, a relationship bears neither specific semantic nor even a name. Subsequent derivations will add semantics appropriate to the concept they represent. There is no general notation for relationship. A directed relationship identifies a source element and a target element. From *DirectedRelationship* abstract metaclass, the UML derives for instance three concrete metaclasses, *Generalization, PackageMerge*, and *PackageImport* and many others not represented in Part1.

A *Generalization* is a binary relationship linking a general classifier (parent) to a specific classifier (child) into inheritance semantics. Each change to the parent also applies to all the children. Each instance of the specific classifier is an instance of the general classifier. For instance, if we create a class by instantiating a model element from *Class*, automatically a new model element will be instantiated from *Element*. A *Generalization* has a notation. It is shown as a line with a hollow triangle as arrowhead at one end. The arrowhead must be placed on the parent. The UML defines also a *GeneralizationSet* (not represented in the figure) that designates a set in which instances of *Generalization* are considered as members.

From *NamedElement*, a *RedefinableElement* is derived to offer to a model element the ability to be redefined in a context of a *Generalization* relationship. The detailed semantics of redefinition depends upon the context and pertains mostly to structural or behavioral compatibility problems specific to an application. For instance, "is the redefining element is substitutable for the redefined element?" has no answer in the UML. Any kind of compatibility specification involves a constraint put on redefinition process. In other words, the UML states that a model element could be redefined but lets to the modeler to constraint freely his model.

A *PackageableElement* is a *NamedElement* that may be owned directly by a package. *Package* is a concrete class issued from a *Packageable Element* and a *Namespace*. A package is used to group many elements under a same namespace. From this definition, we can see that almost everything can be put inside a package. *Package* has a graphical notation in the standard (large rectangle topped with a small rectangle or "tab" attached to its left side).

A *Constraint* is a condition or a restriction, expressed in natural language text or in a machine readable language, used to specify the semantics of an *Element*. A constraint is shown as a text string in braces in the standard. It can be attached to any model element. A *Type* constrains the values represented by a *TypedElement*.

A *Classifier* is then an abstract metaclass that derives from a *Type*, a *Namespace*, and a *RedefinableElement*. We can say that a *Classifier* has a *Type*. We can search some analogy with the notion of "type" used in programming language to constraint a variable to be an integer, a real, or a string but at this conceptual stage, the notion of type is very general and the semantics of the *Type* is not yet defined. All we know is that a *Type* restricts or constraints values taken by a *Classifier*.

A concrete metaclass called *Class* is derived from Classifier and allow UML users to create classes of their system. A class may be stuffed with *Properties* (attributes) and *Operations*. Properties are *StructuralFeatures* if read from the derivation hierarchy. There is also behavioral properties that came from the Behavior part of the UML.

4.6 Structure: Class Diagram

Classes are defined in the Structure Part, in the Kernel package of the UML as the first important modeling concept. When the UML defines its Class, the Class is a metamodel element of the UML. When a user defines his Class "Car" by using the UML as a modeling tool, he is instantiating a modeling element "Car" Class from the metaclass Class defined in the UML. Later, when he extracts a "Ferrari" from his "Car" class, he is instantiating a real world object car from his "Car" class that belongs to his User Model named for instance "Car project." Good developers are not concerned only with "Car" class and "Ferrari," but some elementary knowledge about the metadefinition of Class is mandatory if they do not want to fail later when checking his model against UML rules. These rules constrain for instance a class to have an identity (for instance *Car*) to name it. The class can have only two sets of properties named *Attributes* and *Operations*, it may be connected to other classes of the user's model through a predefined set of relationships offered in the current version of UML plus stereotyped relationships.

As suspected, the modeling class concept is hosted mainly by a class diagram but this fact is not exclusive. We can find the class concept elsewhere in other UML diagrams if needed. So, by adopting a description by diagram, we group the most representative concepts bound to this kind of diagram in one place but the reader can find them eventually in other diagrams. The reader can refer to synthetic tables in this chapter that summarize UML modeling concepts, their graphical notation, short examples, and comments to understand their usage proposed by the UML.

Let us start with the most important model element of the object paradigm: *Class* model element. A class is a kind of classifier whose features are attributes and operations. Before defining the metamodeling element *Class*, a metamodeling abstract *Classifier* was defined in Figure 4.6.0.1 and afterwards

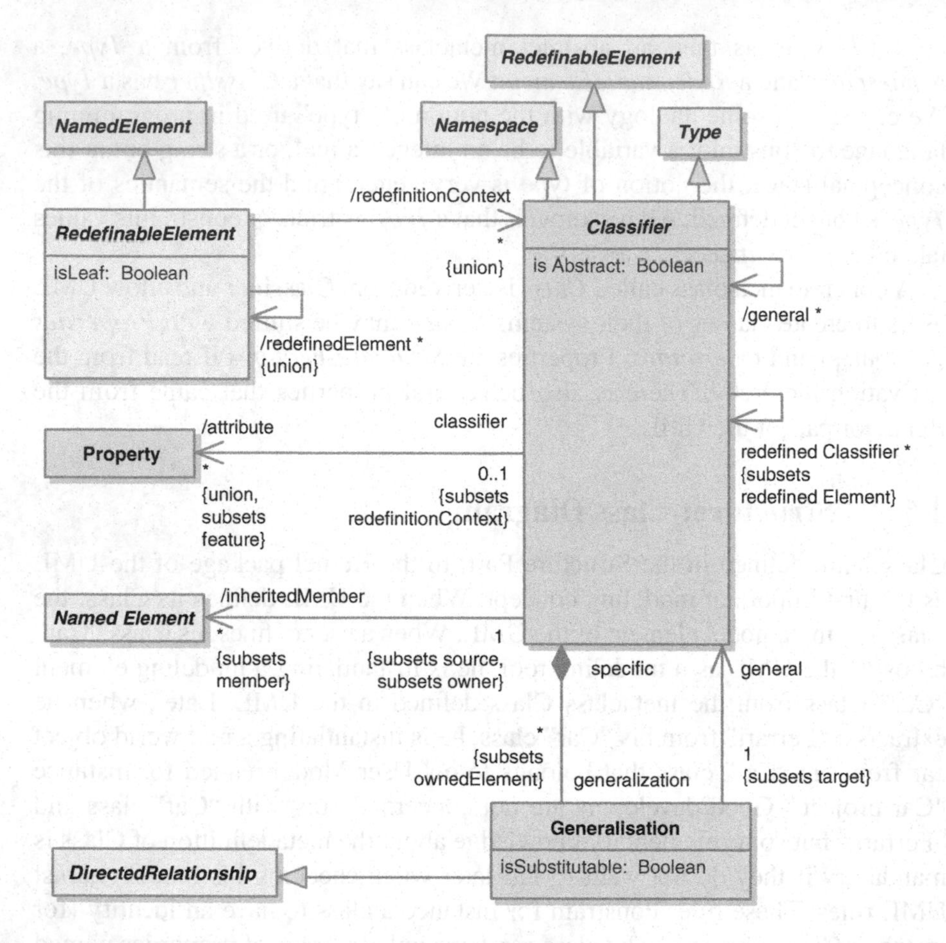

Fig. 4.6.0.1 Classifier concept of the Kernel package. (Fig. 7.9, Superstructure Book)

Class in Figure 4.6.0.2. If we compare the current definition of Classifier in Figure 4.6.0.1 with its equivalent in the Infrastructure (Core :: Abstractions :: Classifiers package), *Classifier* in the Infrastructure was simpler than the current definition of *Classifier*. In Infrastructure, there is no definition of *Class*.

In Figure 4.6.0.2, a concrete *Class* is built from the abstract *Classifier*, so *Class* inherits all *Classifier* properties, in particular, a class may be implied in an inheritance hierarchy. *Class* owns *Properties* and *Operations*. *Property* derives from *StructuralFeature* and can represent an attribute of a class or an end of an association (*association end*). So, we can find the usual definition of a Class in Object programming with *attributes* and *methods*. At UML design level, there is no *Method* metaclass, only *Operation*; there is no *Attribute* metaclass, only *Property* that can be mapped indifferently as *class*

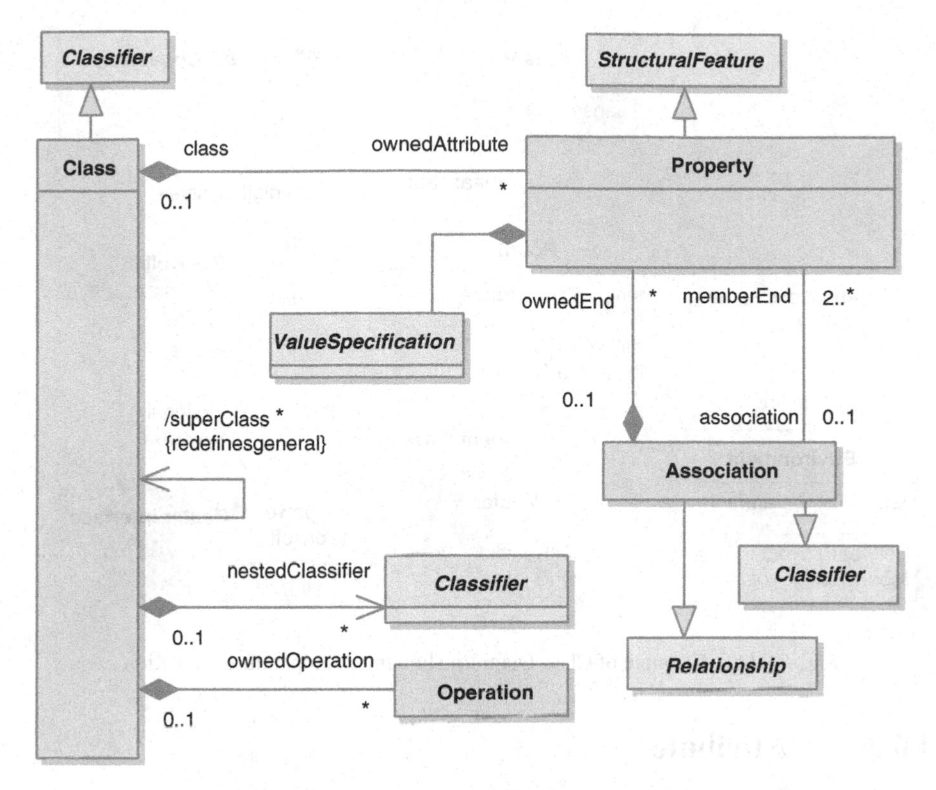

Fig. 4.6.0.2 Class and Property concepts of the Kernel package. (simplified version of Fig. 7.12, Superstructure Book)

attribute in programming language and *association end* at the design phase. The term "attribute" or "ownedAttribute" appear timidly as *role name* of *Property* metaclass.

4.6.1 Example of User Class Diagram

Figure 4.6.1.1 illustrates the use of Class at the M1 level. The *Heater* transmits energy to the *Room* whose temperature must be controlled. The *Room* loses or gets heat from both the *Heater* and the *Environment*. The *Sensor* captures the temperature of the *Room*, then feeds a small voltage to the *Analog Digital Converter* that supplies digital information to the *Controller*. The *Heater* is connected to the *Controller* through a *Heater Interface*. All connections are stereotyped with "high level interpretations" in the context of the *Heater* application. Each class has three slots: the upper one stores the name of the class, the middle contains declared attributes, and the lower is used for declared operations.

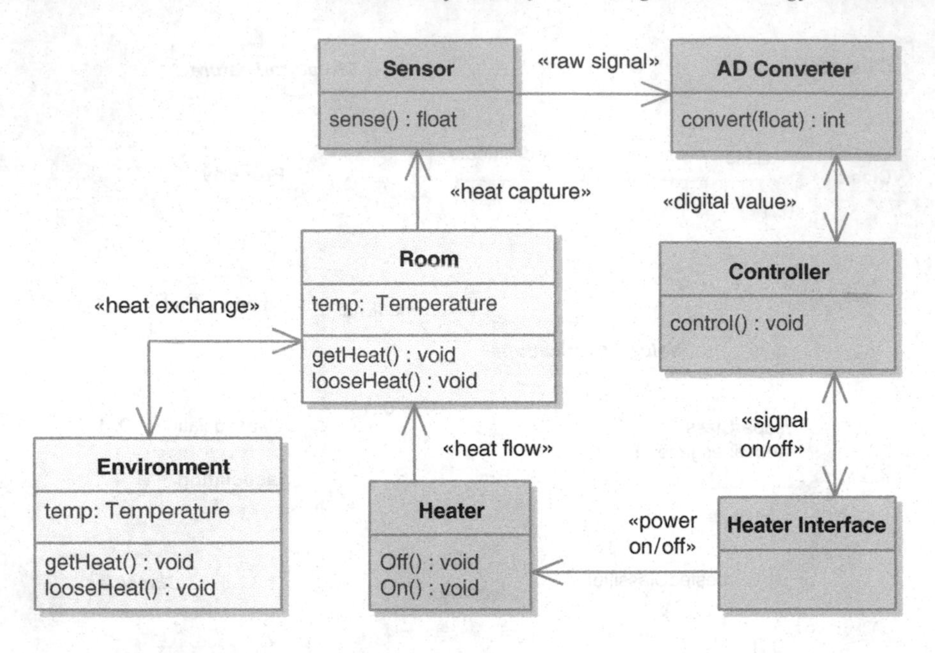

Fig. 4.6.1.1 Example of Class Diagram showing a temperature regulation

4.6.2 Attribute

Concrete *Property* metaclass is used to generate user class attributes. Attributes of child *Class* can redefine attributes of parent *Class*. *Property* inherits the *isUnique* and *isOrdered* meta-attributes from *MultiplicityElement* (*StructuralFeature* derives from *MultiplicityElement*, *Feature*, and *Typed Element* in the Kernel). There is a difference between attributes instantiated from *Property* and a set of meta-attributes that control the process of building the user model. Meta-attributes are not transported to the user model. When the meta-attribute *isUnique* is true, the collection of values cannot contain duplicates. When *isOrdered* is true, the collection of values is ordered. Tool vendors can offer these settings as options to create mathematical collections composed of mathematical elements named *set*, *ordered set*, *bag*, and *sequence*.

> In a *set*, every element appears in the collection only once. A *set* is *ordered* if elements are ordered. There is an index number for each element. Ordered does not mean "sorted." In a *bag*, an element can appear in a collection more than once. A *sequence* is a bag in which all elements are ordered.

An attribute may have an initial value. Each attribute has its own visibility and when redefined, this visibility may change. When a property is derived, its value or values can be computed from other information (for instance, age can be computed from date of birth, so age is a derived attribute). Very often, derived properties are read-only and clients of the class cannot modify them. To summarize, a full attribute declaration consists of:

1. A *visibility* specification
2. A *name* starting a lower case letter ("lowercase" is guideline, not mandatory)
3. A *type* separated from the name by a colon ":"
4. A *multiplicity*, e.g. array, set, bag, or sequence. Multiplicity by default is 1. Multiplicities are enclosed in square brackets "[" and "]"
5. An *initial value* separated from the type or multiplicity by an equal sign "="
6. A *constraint* that is a textual string enclosed in a pair of "{" and "}"

Each feature (attribute or operation) can be marked as *public* (+), *protected* (#), or *private* (−), meaning that the feature can be used by other classes, respectively, by any outside class, by any descendant of the current class, or only by the class itself.

Everything except the name may be optional. For instance, when specifying the languages that the class *Candidate* knows in a curriculum vitae,

$$+\text{programmingLanguage} : \text{String}[3]$$

is a specification of a "public" attribute of type "String." The attribute *programmingLanguage* is multivalued (for instance, the candidate John knows C, C++, and C#). Notice that the multiplicity is attached to the type in conceptual modeling, not to the attribute (an erroneous interpretation would be "the programming language is a unique attribute written with an array of three characters").

Another example,

$$-\text{account} : \text{Account}[0..5]\ \{\text{unique}\}$$

shows that a customer may have zero to a maximum of five private accounts. All accounts are different from each other. The last fact is a constraint and is enclosed in a pair of curly braces "{and}." In programming language, we implement the multiplicity as an array of five account numbers, but at this conceptual stage, we do not have to mention "array," which is only one possible implementation artifact among others.

4.6.3 Association End, Navigability, Role, and Qualifier

The concrete *Association* class in the UML derives from the abstract class *Relationship*. An association is used to define the semantics that exist between classifiers. An association, once instantiated from *Association* metaclass, is composed of two parts: an association object (with an optional name) and a collection of association ends. Concrete *Property* metaclass is used to characterize an association end, which may be part of a binary or an N-ary association. An association end may have several elementary properties:

1. A *role* name
2. A *multiplicity*
3. A *navigability*
4. An *attached qualifier*

An association has roles. Each role specifies a type that an object bound to that role must satisfy. Role names are textual specifications that indicate the roles played by objects linked by an instance of an association that is called a *link* when it connects objects together. The use of role and qualifier came from database modeling in the past E-R era. E-R diagrams are used mainly to implement relational databases that are considered as value-based. The object world is very different in that two different objects may have all values identical (true clones) but different identities. Roles and qualifiers are still conserved to deserve the adaptation of the class diagram to the E-R community. In object technology, their use is somewhat questionable as they can blur the underlying structure (a same class may have multiple roles according to the set of operations currently in use).

In Figure 4.6.3.1, *Company* has "employer" as role name and *Person* "employee" as role name. When role names are explicit, it is not necessary to name the association. Without roles, we must put a stereotype ≪employ≫ to name the association. 1 and * are multiplicities. A company has 0, 1, or any arbitrary

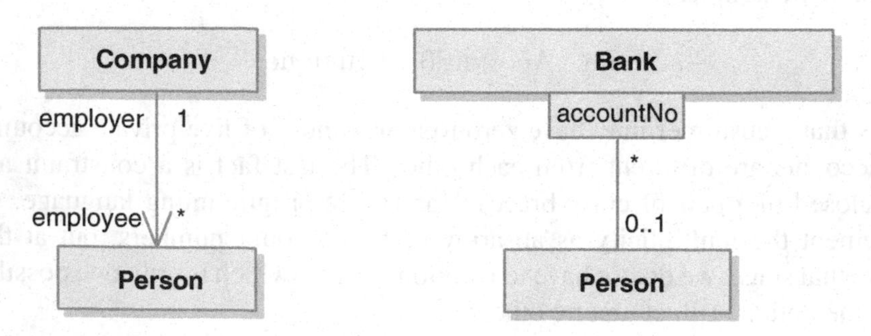

Fig. 4.6.3.1 Role, multiplicity, navigability, qualifier attached to an association

number of employees (*). The arrow put on Person is the navigability of the association end connected to Person.

The second example, drawn from Superstructure Book, shows that the qualifier "accountNumber" identifies an internal object "account" of the composite Bank object. Bank is linked to Person through accountNumber. A person may have any arbitrary number of accounts. Zero means that the bank can create an account for their own use without having to attach it to any identified individual. The semantics of qualifier is difficult to grasp as it hides composite objects. As composition may have several levels, it would be better to express this E-R modeling feature more explicitly. Moreover, a qualifier partitions the set of objects that could be instantiated from a class into subsets. The multiplicity on the end opposite to the qualifier end may be influenced by the presence of the qualifier. The Bank in this example is a composite object built up from a multitude of smaller objects. Only account numbers are involved in the relationship maintained with customers. The object Bank may have a lot of other "qualifiers" devoted to other kinds of association. The semantics of qualifier is potentially confusing.

The navigability of an association end indicates whether a relationship (*Association* derives from *Relationship* abstract metaclass) can be traversed in the direction of the arrow. For a binary relationship, we can have no navigability at all (no arrowhead), bidirectional navigability (two arrowheads), and two distinct directed navigabilities (left and right arrowhead). The semantics of "navigability" is not really clearly mentioned in the standard (Génova et al., 2003), but, if we correlate navigability to "graph traversal," at the class level, it may indicate a possibility of information retrieval, for instance, if class C1 is connected to class C2 with a directed arrow pointing to C2, then, in a database implementation, from an instantiated object of type C1, we can retrieve all instantiated objects of type C2 but the navigability in the reverse direction will need a bidirectional navigability to force the implementer to plan for a reverse procedure. To alleviate the notation, "No navigability" is often interpreted as "full navigability" and can be specified as a general constraint for the whole project. "No navigability" can be understood as "at this time, we do not worry" so the decision is let to the implementer. In fact, no navigability, if not interpreted as bidirectional, is a "dead relationship" if we can do nothing with the association. Navigability, visibility, and how the association must be interpreted and how it must be related to the instantiated "link between objects" are sometimes confusing and general interpretation rules must be written down to avoid misunderstanding.

In Figure 4.6.3.2, we discuss one of the E-R diagram replacement issues. Attributes (properties) of a class are bound to this class through a particular relationship named aggregation. In the UML, attributes go inside the middle slot of a class. If we envision each attribute as an object instantiated from a separate class *Property*, *Property* is thus connected to the metaclass *Class* through a relationship. As the old E-R diagram often represents the entity as a rectangle

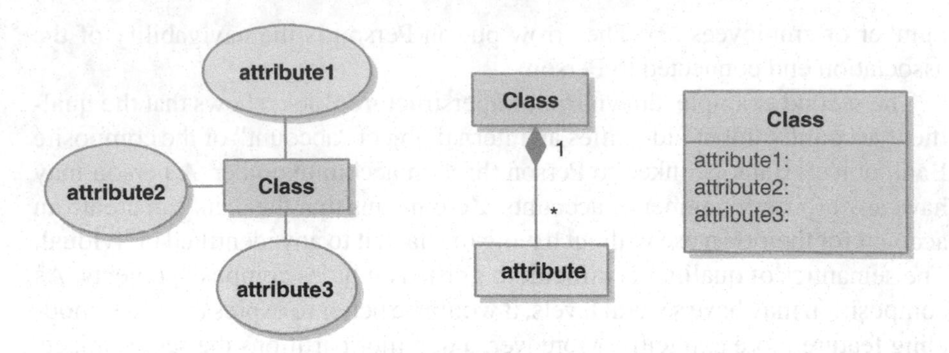

Fig. 4.6.3.2 E-R diagram replacement. The class diagram can replace the E-R diagram to model database applications. Attributes are instantiated from the UML Property metaclass associated to Class through aggregation/composition relationships. The graphical notation of the class diagram offers a very compact graphical representation that save space if compared to graphical bulbs in the E-R diagram. There is full compatibility even at the metalevel (see text)

connected to circles/ovals containing attribute names, it is interesting to notice that the E-R representation complies with the UML even at the metalevel because, attribute and class are involved in a relationship too. So, there is a real continuity in the way the UML is conceived to handle database applications. Class diagram can now definitely replace its "parent" E-R diagram with a more compact graphical notation.

4.6.4 Operation, Parameter List, Return Type, Signature

> An operation is a behavioral feature of a classifier that specifies the name, type, parameters, and constraints for invoking an associated behavior. (Superstructure Book)

A *method* or *function* in object programming languages (C++, C#, Java, etc.) implements the modeling concept *Operation*. A complex operation can be implemented by many contributing or collaborative methods. The opposite case is possible too: a method may implement many operations; each operation may be invoked with a particular parameter. In other words, an operation could be implemented by several methods and a parameterized method may implement several conceptual operations. A full operation specification consists of, in order:

1. A *visibility* specification.
2. A *name* starting with a lower case letter (name is mandatory, lower case is a guideline, not mandatory).
3. A comma separated *parameter list* enclosed with a pair of signs "(" and ")." In the absence of parameters, the pair of parentheses is mandatory to distinguish an operation from an attribute.

4. Each parameter can be preceded with a *direction* "in," "out," or "inout." If the operation makes use of the parameter, it will be "in." If the operation writes to the parameter, it will be "out." "inout" represents a read-then-write parameter.
5. The parameter type is separated from the parameter name by a colon ":".
6. A *multiplicity* can be specified for each parameter to indicate that the parameter may be multivalued. Multiplicities are enclosed in square brackets "[" and "]."
7. If a *default value* for the parameter type is mandatory, this initial value must be separated from the parameter type by an equal sign "=."
8. Once the parameter list is terminated, the *return type* of the operation may be specified. If it is mandatory, a colon ":" is used to separate the preceding parameter list from the return type. Return type may be null (no result returned). Otherwise, it is commonly admitted that the type of an operation is assimilated to the type of its return value (conventional even questionable as we can return a pointer to a structure of unrelated elements of various types, known as heterogeneous structures).
9. Finally, an *operation property list* assimilated to a *constraint list* enclosed in a pair of curly braces "{" and "}." The operation property list contains comma-separated textual strings that may take the following values: *redefines* if the operation redefines an inherited operation; *query* means that the operation does not change the state of the hosting object; other user defined constraints, for instance *constructor* or *destructor* to indicate operations performed at the creation or the destruction of objects.

Hereafter are some valid samples of operation specifications:

$$+\,\mathrm{computeTotal}\ (\mathrm{itemPrice} : \mathrm{Float}\ [1..10], \mathrm{includeExtras} : \mathrm{Boolean}, \mathrm{Extras} : \mathrm{Float}\ [0..3]) : \mathrm{Float}\ \{\mathrm{redefines}\}$$

$$+\,\mathrm{createWindow}\ (\mathrm{location} : \mathrm{Coordinates}, \mathrm{container} : \mathrm{Container}\ [0..1]) : \mathrm{Windowdisplay}()$$

$$+\,\mathrm{toString}\ () : \mathrm{String}$$

The operation *computeTotal()* may contain prices of 1–10 items. Up to three "Extras" amounts could be added to the total (if the Boolean "includeExtras" is true). The result returned is of type "Float", and so is the *computeTotal()* operation. This operation redefines another *computeTotal()* that exists in a parent class. The three last examples are extracted from the Superstructure Book : *display()* has no parameter list, no return type; *toString()* has just a return type and *createWindow()* has user defined types.

In programming languages, the parameter list of a function is referred to as its *signature*. A function in object technology is characterized by three elements:

1. Its *name* that cannot be unique (for instance, a name as "get" or "set" is very common. In this case, the namespace helps to determine the right function)
2. Its *signature* that embraces the parameter list and their respective type
3. Its *return type* that is considered as the type of the function. With "void" or "null" as return value, the function is considered as "not typed". As a function may return values in its parameter list in case of "out" or "inout" directions, the return value can be considered as the (N + 1)th element of the parameter list of N parameters. In fact, we can return systematically a "void" return type and put the return parameter in the parameter list. So, the type of the function based on the return type is somewhat artificial and is not really useful in most situations. The signature of the function including the return type is its *extended signature*.

The name and extended signature of a function uniquely identify it. As we can see later, we can make several functions with the same name but different signatures or extended signatures. For instance:

display (int) : void
display (String) : void

are two operations with the same name but different signatures. This fact allows us to model *function invocation at run-time* and the function that will really execute must match the signature or the extended signatures.

Several satellite notions accompany a classical definition of operation:

1. The *preconditions for an operation* define conditions that must be true when the operation is invoked or allow the function invocation
2. The *postconditions for an operation* define conditions that will be true when the invocation of the operation completes successfully, assuming all preconditions are already satisfied
3. *An operation may raise an exception* during its invocation. When an exception is raised, postconditions may not be satisfied.

In a nonobject language, these notions are in fact very imprecise as we do not know what objects are concerned by preconditions and postconditions. If they seem to be very important in more classical functional development, with object technology, the alternative is replacing pre- and postconditions by a complete dynamic diagram that describes step by step the evolution of all concerned objects. As for exception, the event generated by the exception will be assimilated to a message directed by the object assigned to handle this exception (that could be the same object). So, things can be modeled differently without having to call for low level concepts at the logical level.

4.6.5 Interface

As defined in the standard:

> An interface is a kind of classifier that represents a declaration of a set of coherent public features and obligations. An interface specifies a contract; any instance of a classifier that realizes the interface must fulfill this contract.

So defined, an interface is a collection of operation signatures and attribute definitions that support a cohesive set of behaviors. Interfaces are "realized" by one class or many classes or components that implement the operations and attributes defined in the interfaces.

> If an interface declares an attribute, this does not necessarily mean that the realizing class will necessarily have such an attribute in its implementation only that it will appear so to external observers.

To explain the usefulness of the interface modeling concept, let us deal with the Car analogy. If we know how to drive a car, then we must be able to drive any car: American, European, or Japanese. The Car interface must be standardized: steering wheel, left/right turn signals, gas pedal, gear lever, speed indicator, brake etc. So designed, no matter what is implemented on the backend of the interface, we do not really care because this car interface subscribes to the basic Car *contract*.

In the "code world," the concept of interface allows us to keep a relative constant interface offered to the users and isolate the implementation from what they can see. We do not have to reprint, for instance, the manual of a product each time we make a change to the underlying structure. This separation or shield allows developers to build reusable, maintainable classes at their development level that could bear any significance to high-level users. Users only see what they used to see, what they want to see and just enough information for them to make efficient use of the software without being embarrassed with implementation details and evolution constraints.

Provided interfaces are those offered to clients and *required* interfaces are those needed for an underlying structure to fulfill its own obligations to its clients. They are represented with different graphical notations in the Table 4.6.7.1 summarizing all concepts used in a class diagram. If there are few interfaces between two classes, small graphical objects are added to existing classes (small circles or balls are used to represent the provided interfaces, and small half circles or sockets are used for required interfaces). If the interface contains several declarations that must be grouped together, a rectangular shape identical to the class notation will be used with the stereotype ≪interface≫

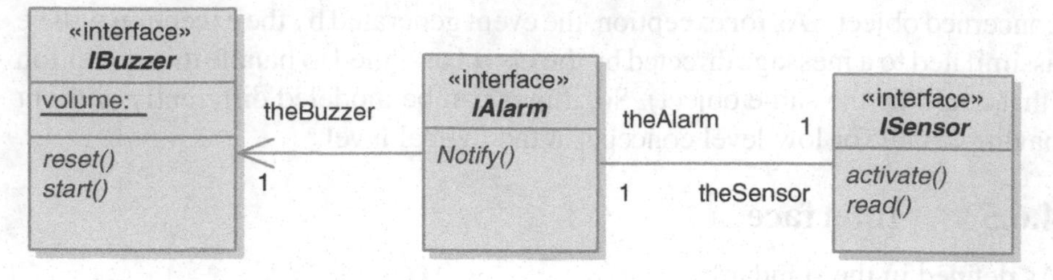

Fig. 4.6.5.1 Using interfaces to describe systems (example 7.57 of Superstructure Book). This example describes the operation of three interfaces involved in a "collaboration" to detect thief intrusion. Interfaces are represented with their attributes and operations signatures. IAlarm and ISensor are involved in a bidirectional association and the communication between IAlarm and IBuzzer is one-way navigable. This diagram needs another class diagram to show classes that "realize" these interfaces

above the name of the interface. Slots can be created to store extended signatures of operations and attributes of the interface.

Interfaces provide a high level of abstraction that makes programs easier to understand. As interfaces allow developers to omit details when defining what the user must see, some developers make extensive use of them to describe all systems. The example of Figure 4.6.5.1 illustrates such an attitude. However, remember that interfaces are only declarations, signatures of operations and after all, they are only a superficial description of system targeted to facilitate its use. A system completely described with interfaces is insufficient to develop a system because background classes and detailed dynamics are still ignored.

Interfaces may be compared as a *view* of an object or several objects that collaborate to execute/receive a service. In this sense, they are not real objects but only views. On the other hand, interfaces may give a false impression to create objects based only on their functionalities and in so doing, there could be a temptation to backtrack towards functional methodology or the creation of very artificial objects if the mapping is somewhat direct between interfaces and classes. But, when used sparingly and with a correct mapping, essentially not direct, interfaces are powerful artifacts to create the interface between clean reusable objects system and whatsoever we can create at any level to adapt arid and impenetrable professional software engineering models to user perceptions.

4.6.6 Association Class

AssociationClass is a concrete metaclass that derives both from *Class* and *Association* in the Kernel. An instance of *Association* class is a model element that has both association and class properties. It holds a set of classifiers properties and a set of features that belong to the relationship itself. Semantically, properties in an association class cannot be affected uniquely to either sides of the association without distorting the interpretation, so it must remain in the association to enrich its semantics. Moreover, each time we instantiate the association, we need a parallel instantiation of the association class. If those conditions are not satisfied, may be the concept being modeled does not really need an association class. For instance, a contract that is part of a job must be instantiated each time a person decides to work for a company. A man and a woman involved in a relationship do not necessarily need a marriage.

The association class comes from the old notion of attribute of a relationship in the E-R modeling, for instance, the attribute "Salary" is often modeled as the attribute of an relationship ≪employ≫ between two entities Employee and Company. In Object Technology, as attributes cannot be a dangling notion and therefore must be attached to a class, attributes of a relationship metamorphose into an association class. Some attempts (Gogolla and Richters, 2004) transform the association class to a ternary relationship between three classes. In fact, there is a subtle difference in the fact that the association class has implicitly a multiple of 1 for each association instantiation. A class put at the same place of the ternary relationship may have 0, 1, or * as multiplicity, so the semantics is slightly different.

4.6.7 Graphical Notations

Table 4.6.7.1 summarizes modeling concepts mostly (these modeling concepts may be found in other diagrams) used in a class diagram. It gives the name of the concept, its graphical representation, and a short summary.

Table 4.6.7.1 Modeling concepts mostly used in a class diagram

Name and graphical notation	*Comments*
ClassName privateAttr: Boolean protectedAttribute: String publicAttribute: int privateOperation() : void protectedOperation(String, int) : Boolean publicOperation() : int	**Class** A *class* is represented with a rectangle. A class has by default three compartments. The top compartment holds the class name. An *italicized name* specifies an abstract class that can not be instantiated. The middle compartment holds a list of attributes and the bottom compartment a list of operations. Nothing prevents a user to create a user-defined named compartment.

*Attribute*s and *operations* can be displayed with their visibility or constraint. Attributes have a type and may have an initial value. Operations may have a return type different from "void" and parameters of different types (see text for complete discussion on attribute and operation formats).

(*cont.*)

Table 4.6.7.1 (Continued)

Name and graphical notation	*Comments*

Interface with ball-and-socket notations

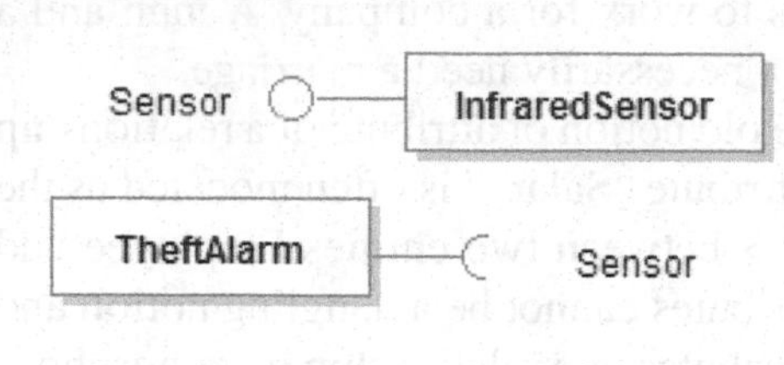

The InfraredSensor class has a *provided interface* (circle or ball). The TheftAlarm has a *required interface* (half circle or socket). Balls and sockets can be attached to classes and components. Interfaces can be connected to parts of a Component (see Component Diagram)

Package

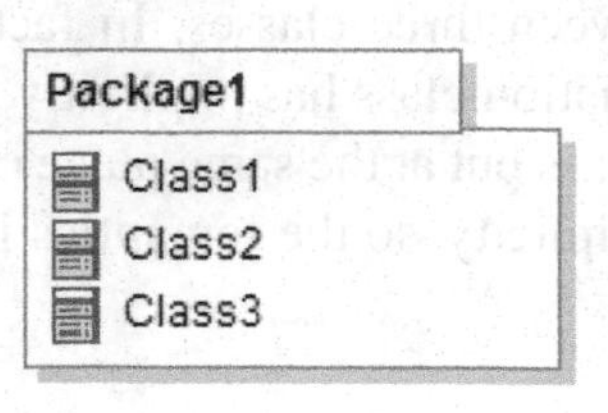

See "Package Diagram". Package diagram may be used to structure classes of the domain or classes of an application. A package may also include other packages.

Interface with class notation

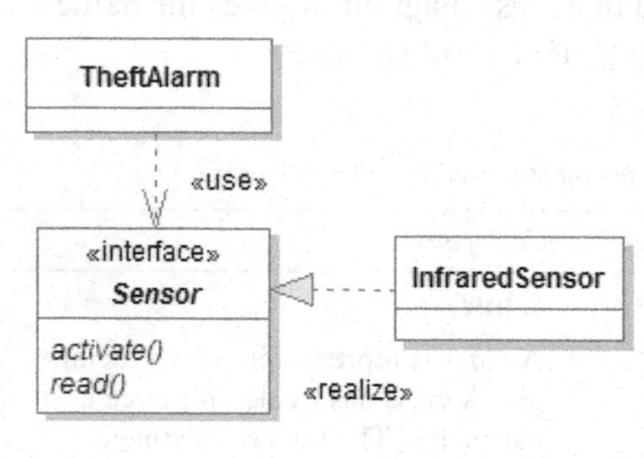

The same classes InfraredSensor and TheftAlarm are now connected through an interface represented with a class notation. Within the class notation, interface attributes and operation signatures may be added. Interface is an abstract metaclass and object cannot be instantiated from interface. The InfraredSensor "realize" the interface. The TheftAlarm is connected to the Interface through a <<use>> dependency.

Aggregation/Composition

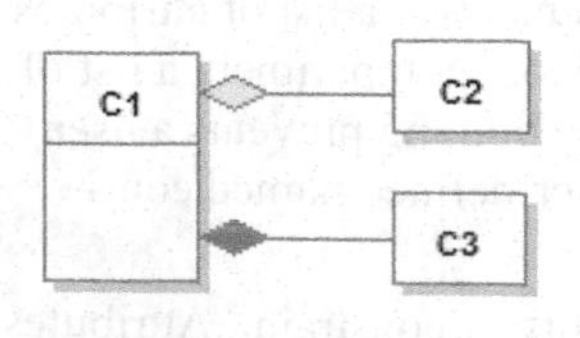

Aggregation/Composition is known as "part/whole," "part-of," or "containment" relationship. An aggregate or a composite object contains other objects. In the standard, the composition is a stronger form of aggregation since parts are created and destroyed with the composite object. Aggregation is represented with a hollow diamond on the container side, composition with a black diamond.

Table 4.6.7.1 (Continued)

Name and graphical notation	*Comments*
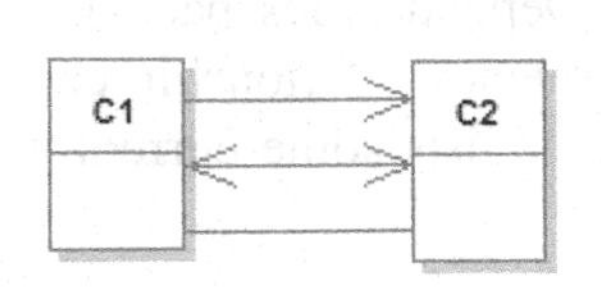	**Binary Association** Three associations between classes C1 and C2 are shown: directed association, bidirectional association, unspecified navigability (see text for detailed comments).
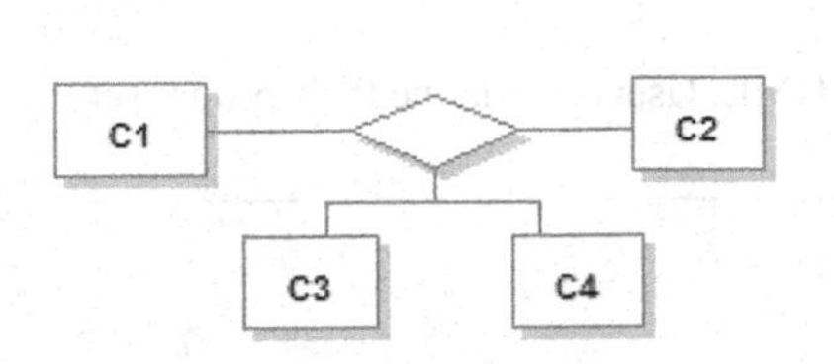	**N-ary association** An *N-ary association* between classes is represented with a diamond shape with an optional name. The lower multiplicity is typically zero. A lower multiplicity of 1 or more implies that one link or more must exist for every possible combination of values for all other ends.

Generalization: dependency and realization relationships

The *Generalization* relationship is represented with a full line with a triangular arrowhead lying on the parent class.

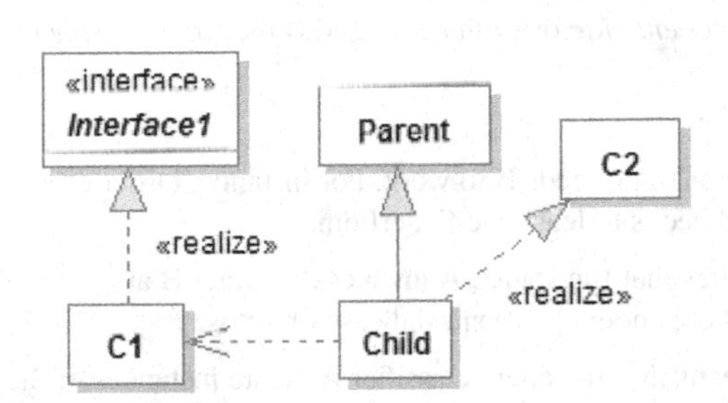

A *dependency* is a directed relationship (dashed line with simple arrow) that signifies that the child class (Client) requires C1 (Supplier) for its specification or implementation. Dependency may receive a stereotype, for instance <<usage>> if the modeler wants to point out the nature of the dependency.

The *realization* relationship is shown as a dashed line with a triangular arrowhead lying at the realized side. Classes may realize other classes or interfaces. At the conceptual stage, the realization meaning is very large and embraces refinement, optimization, composition, synthesis, implementation, etc.

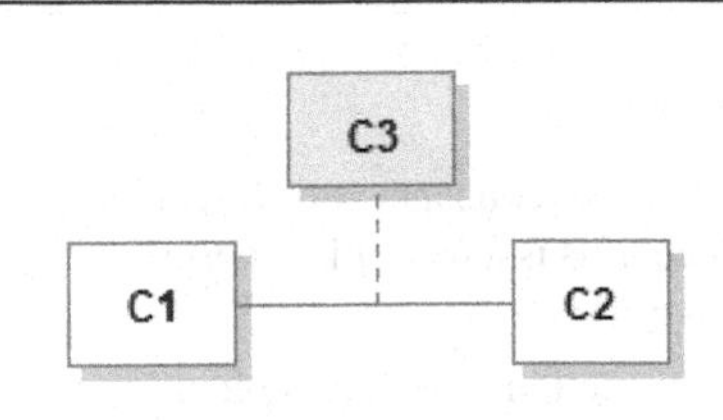	**Association Class** An association class C3 is shown as a class symbol attached to the association path by a dashed line. A single name for the class covers the semantic of the association and could be placed on the association or on the class or on both but the name must be the same. Be careful to distinguish an *association class* from a *ternary association* between three classes.

4.6.8 Dependencies

Dependencies are used to assert some degree of dependence between modeling elements. Some standard stereotypes are defined in Dependencies package. New stereotyped dependencies can be added. A dependency relationship can be freely drawn between classifiers. The following table lists some currently used dependencies.

Table 4.6.8.1 Standard dependencies defined in the UML. Users can define their proper set of dependency stereotypes freely

Name of dependency and stereotypes	*Comments*
Dependency A —–> B (read “A depends on B”)	In a basic dependency with no other specification, if A depends on B (dashed arrow with arrow head on B) means that changes on B may affect A (not always) from a structural or behavioral viewpoints. Some dependencies must be navigable in both sides semantically, in this case, suppress the arrowhead. A is called the *client side* or *source side* and B the *supplier side* or *target side*.
Dependency of Usage type	
Usage ≪use≫	If A make use of B, A needs B to work. For instance, Order uses OrderItem and needs at least one OrderItem.
Call ≪call≫	A calls B implies that the source A invokes the target B at run-time. This dependency is a special case of ≪use≫.
Create ≪create≫	A create B means that the client classifier A create instances of the supplier classifier B. This dependency is a special form of Usage. For instance, if A calls a creation method of B, if no ≪use≫ stereotype is attached to the same link, only the creation method can be called, so ≪create≫ restricts ≪use≫.
Instantiation ≪instantiate≫	The client creates instances of the supplier. Its semantics is identical to ≪create≫.
Dependency of Abstraction type	
Abstraction ≪abstraction≫	“A is an abstraction of B” puts the two concepts at different viewpoints or levels. For instance, if B is a real object described by a class A, so A is an abstraction of B.
Manifestation ≪manifest≫	Manifestation is a subtype of Abstraction. It replaces older ≪implement≫ stereotype in previous version, so its usefulness is found more in the deployment phase. We can say that any platform dependent implementation is a “manifestation” of all

Table 4.6.8.1 (Continued)

Name of dependency and stereotypes	*Comments*
	artifacts used to describe a system. At more elementary level, an object is a manifestation of an abstraction.
Realization ≪realize≫	If A realizes B, the client A implements one fraction (or all) of the functionality of B.
Substitution ≪substitution≫	Substitution denotes run-time substitutability that is not based on specialization or any inheritance of structure. If a classifier offers a clean interface contract, then any classifier that complies with this contract can be a client for substitution.
Trace ≪trace≫	Traces are mainly used for tracking requirements and changes across models. As changes are bidirectional, traces are better used without any directional navigation.

4.7 Structure: Object Diagram

Objects are defined in the Structure Part, in the Kernel package of the UML as instance specifications. An object diagram shows how a specific set of instances from a given subset of classes in a system are linked together to perform a given task, to constitute a real world system. When modeling an elevator, for instance, we have an Elevator class, a Door class, a Button class, and a Controller class, but a real elevator instantiates one elevator, two gliding doors (one left, one right), and several command buttons connected to a controller. So the class diagram, which is very useful in database models, falls short in describing a real-time system. Moreover, in complex situations, designers will instantiate only specific objects taking particular roles in a given situation to describe their interactions. An instance specification is defined at the metalevel as follows:

From the definition of *InstanceSpecification* concrete metaclass (Fig. 4.7.0.1), we can instantiate an entity that we call "object" if the classifier is a class, and a "link" (term used in UML) if the classifier is an association. An object can refer to more than one classifier. An object or a link holds many "slots" composed of values that describe their structural features. An instance specification provides an illustration or an example of a classifier or a link. It is not necessary to have a complete description of the classifier, i.e. it is not necessary to give a value to each feature. Sometimes, an object with just a name would be sufficient. If all slots are not defined, it does not mean that the entity does not have all the features but merely that the missing features are not relevant in the current context.

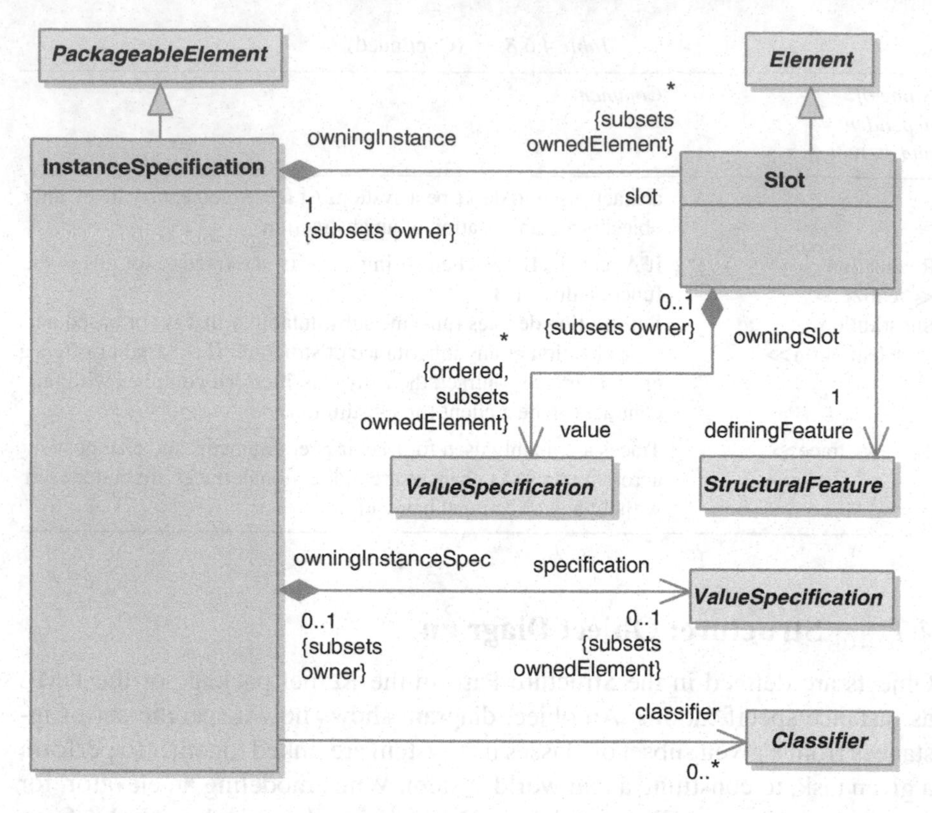

Fig. 4.7.0.1 InstanceSpecification concept of the Kernel package (Fig. 7.14, Superstructure Book) used to instantiate both objects ands links

4.7.1 Object, Instance of A Class

For most people, instances usually have more meaning than do classifiers (a red Ferrari is more suggestive than an abstract class "Car"), because classifiers represent at least the first level of abstraction. People often gather several instances under the same classifier to help people understand what classifiers are. But class is not a collection of instances (group of objects sharing the same description). Saying that "a class is a set of objects" is simply incorrect. Class is just a mold defining all the properties of instances.

Normally, at the conceptual stage, it is not necessary to define the mold before creating an object in your model, especially when this object is unique. Modeling tools must let us define object without having to decide on its classifier immediately. In Figure 4.7.0.1, *InstanceSpecification* refers to *Classifier* with a multiplicity starting from zero, so the definition of classifier is not mandatory or can be postponed.

4.7.2 Link, Instance of An Association

A link is defined in the UML as an instance of an association. According to multiplicities, many objects could be instantiated from each side of a binary association. All the instantiated links connecting objects together will constitute a collection of all instances of an association. The size of the collection depends upon multiplicities. If the end is marked with an "ordered" constraint, then the collection will be ordered. If the end of the collection is marked with a "unique" constraint, then the collection is a set, otherwise, elements could be duplicated in a bag.

The semantics of aggregation/composition are fully transmitted from the association form (in a class diagram) towards its instantiated link form (in an object diagram), except that nothing is mandatory in an object diagram. Objects with well-chosen values and connected links are instantiated to explain a run-time context. An object belonging to Class "SportCar" can be instantiated as

Ferrari : SportCar (object name : name of its classifier) or only
: SportCar (unnamed object, meaning a generic element from class "Car")

We can optionally give values to some attributes (any combination from zero value to all values) to the object "Ferrari : SportCar" by using the list of attributes hold by the class "Car," for instance *year*="2002" and *engine*="*Engine: 6.0 V12*" letting all other attributes undefined. If we instantiate now four wheels *front-left : Wheel, front-right: Wheel, rear-left : Wheel, rear-right : Wheel* from the class *Wheel*, they must be now connected to the object *Ferrari : SportCar* with four composition links.

Generalization relates child and parent classifiers. Each time, a child object is instantiated, in the semantics of UML, implicitly; a parent must be instantiated to give to the child all parent features and the child must be instantiated too, to fill in differential features. Normally, when instantiating a link, objects located at the association ends must be represented. This rule is relaxed when applying to a directed relationship like inheritance or composition. We do not have to represent an instance of class Car when instantiating *Ferrari : SportCar*.

Generalization is defined at the class level as an association. At the instantiated level, generalization link between objects is not represented as in the way programming language is implemented to save code. This fact could be problematic for some real situations where an object can effectively be a child of another object and mimic all their properties at run time. In the current standard, we must create two classes and represent this inheritance in a

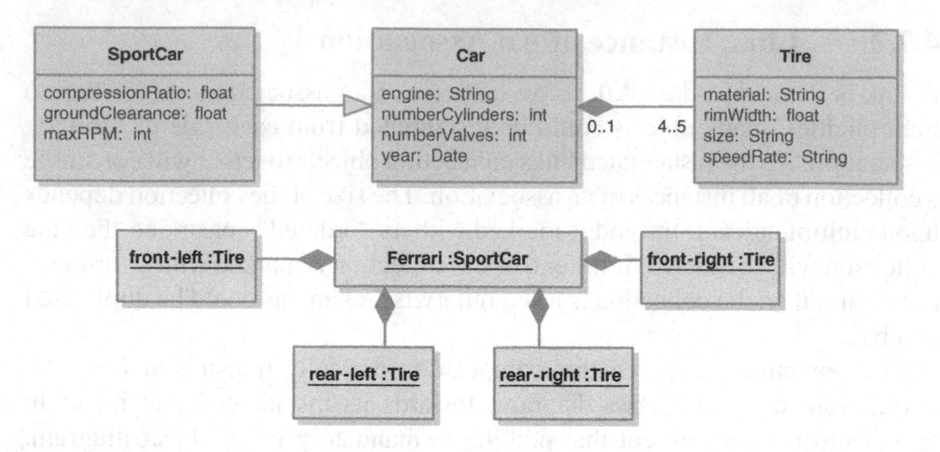

Fig. 4.7.2.1 Class diagram and Object diagram

class diagram. In an object diagram, we must create a stereotyped relationship ≪child-of≫ with two instantiated objects (parent and child) by using the extension mechanism offered by the UML. Probably, in programming language, we will never have to handle this case but this reverse interference of the low-level programming and implementation model to higher conceptual level could be harmful as it reduces the expressiveness of the conceptual space.

In Figure 4.7.2.1, the class SportCar derives from the class Car with Generalization association. The class Car is connected to the class Tire with a Composition association. When instantiating an object of type SportCar, four objects tires are instantiated and four Composition links are instantiated from the Composition association to connect the object Ferrari to four objects front-left, front-right, rear-left, and rear-right. Ferrari inherits all the properties of the Car class but this fact can only be found in the Class diagram. Many composite classes may exist between the class Car and the class Tire, as the Composite relationship is transitive, we cut them short to simplify the example.

If all links between objects must come from associations in the class diagram, this fact will subjugate all object diagrams to the semantics of the class diagram and negate, from the metamodeling viewpoint, the full existence of the object diagram as a standalone diagram. Fortunately, the notions of "connector" are introduced in version 2.0 of the UML (Connector metaclass is not defined in version 1.4 of the UML) to repair this semantic breach of version 1.4. Connectors is a standalone modeling concept and do not require an equivalent association in the class diagram. So the object diagram can now have an independent existence. The object diagram is one of the most important tools for real-time applications.

4.7.3 Graphical Notations

Table 4.7.3.1 Modeling concepts found in an object diagram

Name and graphical notation	*Comments*

Object, instance of a Class

Object, instance of a Class

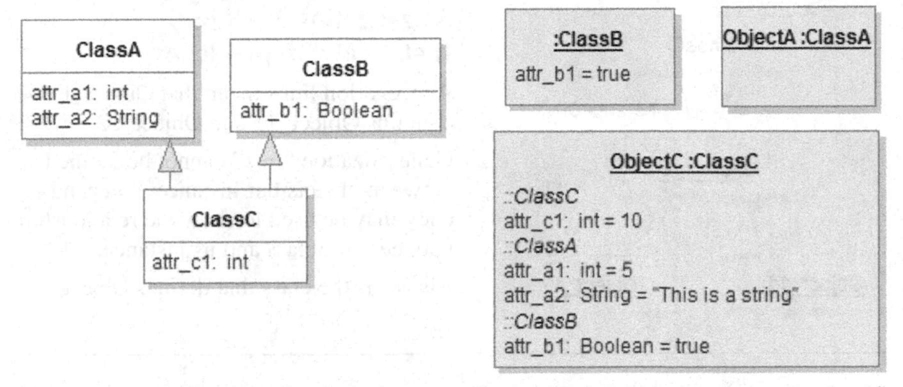

An instance specification of a class shares nearly the same graphic notation as its classifier, except that the object name is always underlined.

Objects may have a specific name if there is more than one issued from a same class. A colon (":") separates the object name from its classifier name. If the object is generic, the object name can be omitted. In this case, the name of the classifier appears after the colon ":". Unnamed object and unnamed classifier (unnamed object that has not yet received any assigned class) has just an underlined colon.

Attributes of object may be completely or partially showed in the attribute compartment, each slot is composed of the feature name followed by an equal sign ("=") and a value corresponding to the type of the feature. If the type itself must be shown, it will be inserted between the name and the value. A colon (":") is used in this case to separate the name from the type. In the example, ClassA, ClassB and ClassC are classes. ClassC derives from ClassA and ClassB. ClassC has differential property attr_c1.

The object : ClassB instantiated from the ClassB is not named in the context. It takes the name of its class. The object ObjectA : ClassA instantiated directly from class ClassA has a name but its attributes are not instantiated in the context shown. The object ObjectC : ClassC is complete with all the attributes instantiated. As the ClassC derives from ClassA and ClassB, attributes of each parent class are grouped together and instantiated separately, so does attr_c1 that belongs specifically to ClassC.

Link, instance of an association or Connector (see Composite Structure Diagram)

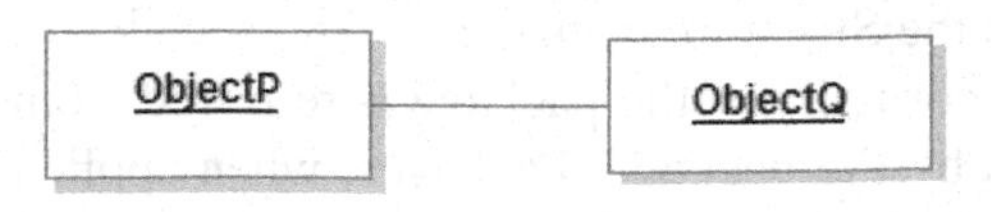

If a link, instance of an association, is instantiated, it must be preceded by the instance of objects coming from classes involved in the association (except for Generalization).

(cont).

Table 4.7.3.1 Continued

Name and graphical notation	*Comments*
Some UML tools consider association and link as the same, so objects are connected together with "association" but from a theoretical view, link is an instance of association. Moreover, there is no graphical distinction between a link and a connector (Connector will be defined later in Composite Structure diagram. Subtleties are discussed in that diagram).	
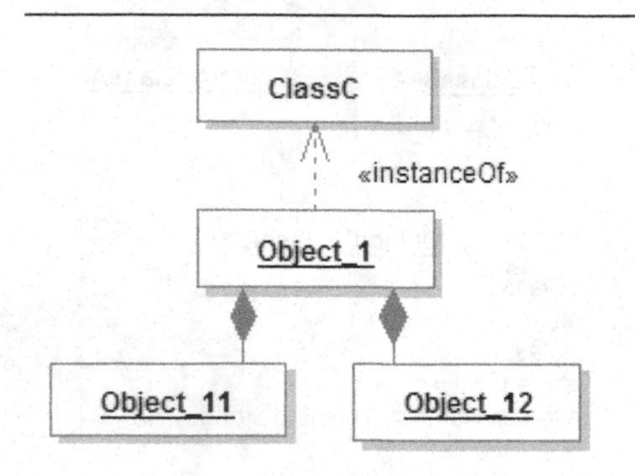	**Aggregation Link and InstanceOf dependency** Aggregation links mean that Object_1 contains Object_11 and Object_12. Generalization "link" cannot be defined between objects but InstanceOf dependency may be used to show the relationship between one class and its instance. ClassC is the class that defines Object_1.

4.8 Structure: Package Diagram

Package metaclass is defined in the Structure Part, in the Kernel package of the UML as an important modeling concept for structuring the modeling process itself. Figure 4.8.0.1 shows an example of the package diagram that gives an overview of this modeling concept before discussing its usage. *Requirement Analysis, Design, Structure,* and *Behavior* act as namespaces to group smaller real packages. Package *Design* is connected to *Structure* and *Behavior* packages through nesting relationships (+ sign circumscribed by a circle). The dependency ≪realize≫ connects *Design* package to the contour of *Requirement Analysis*. As Behavior and Structure packages are nested inside the Design namespace, any package in the Structure and Behavior may realize what is defined in the Requirement Analysis. Properties of contours were explained by Harel (1988), with our "selective" interpretation, roughly, "some" packages taken in the Structure and Behavior packages are used to realize "some" packages defined in the Requirement Analysis. The dependency ≪bind selectively≫ connects the Behavior package to the Structure package to indicate that some dynamic packages are used to explain the dynamic properties of some structural diagrams in the Structure package.

A package owns its members, so normally, if the package is removed from a model, so are the elements owned by the package. Packages, when applied effectively, structure the whole project organization. Packages organize models and do not have any semantics at run-time. At the time of modeling, a package

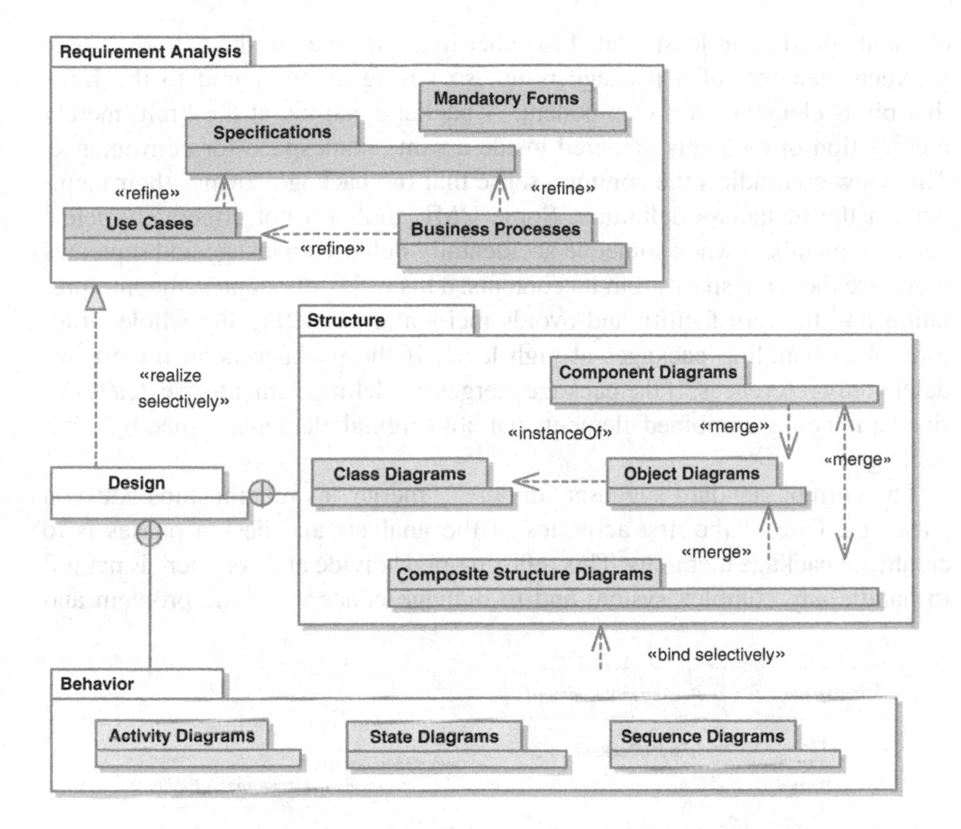

Fig. 4.8.0.1 Example of package diagram and relationships defined with packages. Nesting relationships between packages (equivalent to aggregation/composition) are represented with a + sign circumscribed by a circle

may accidentally correspond to a component or a composite class but fundamentally, they are only units of organization. The name given to a package is merely a namespace to reference the package.

A package *import relationship* may import modeling elements from other packages and make them usable in the importing packages as they are defined in the importing package. A *PackageMerge* acts first as the import relationship but may create copies, rename them, modify, add features and combines modeling elements from different imported packages to create new modeling elements that will be made available for enriching the arsenal of modeling concepts. Technically, a package merge is a directed relationship between two packages and the contents of the source package must be reworked, contrarily to package import that references only elements of the source package.

Packages are usually organized to maximize internal coherence within each package and to minimize external coupling among packages. Modeling elements must be at least related together in a package but the cohesive force between elements of a package is not so strong as compared to the force that binds elements of a Component. A package can be, at the limit, merely a collection of elements gathered inside a same namespace for convenience. This view contradicts the common sense that the package "owns" their members in the metaclass definition. Some UML tools do not effectively delete package members when someone accidentally deletes a package and separates therefore the namespace from its contents. This reflex dissociates the interpretation and the tool facility and avoids incidentally deleting the whole structure when handling packages at high level. If the package is an import, we delete only references. If the package merges modeling elements, we delete derived and newly combined elements but not original elements owned by other packages.

The current standard suggests "import," "merge" as relationships between packages. One of the first activities of the analysis and design phases is to create the package hierarchy. This reflex named "divide and conquer" is natural to handle any complex system and to manage concepts of the problem and

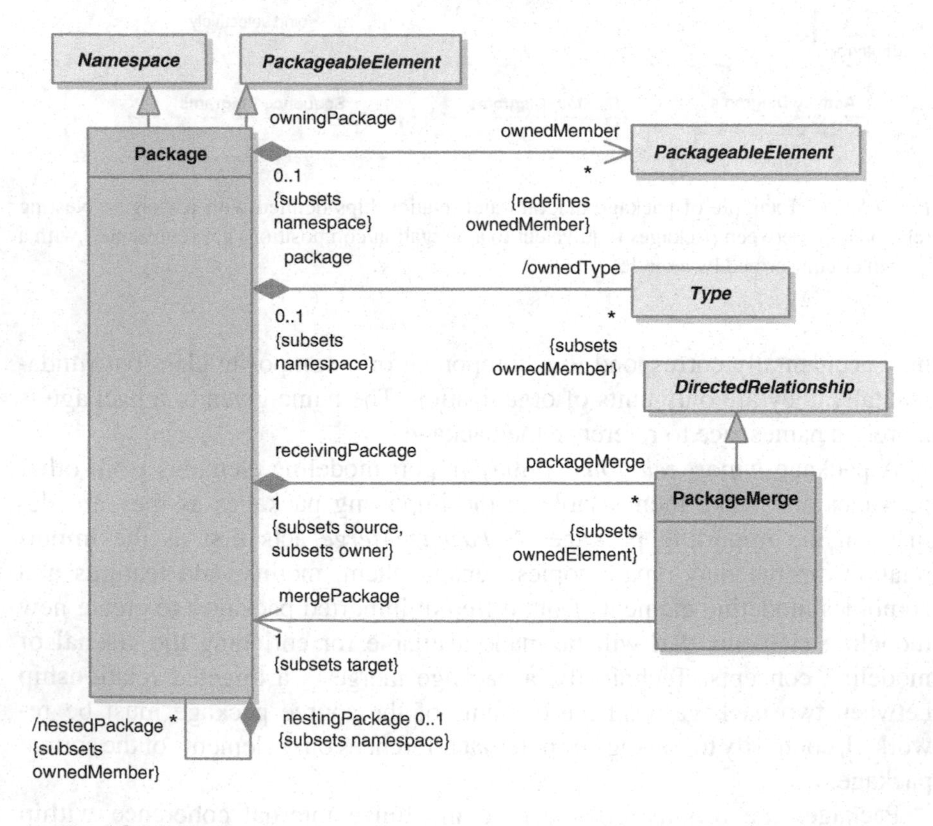

Fig. 4.8.0.2 Package concept of the Kernel package (Fig. 7.14, Superstructure Book)

solution space in a comprehensible way. A package does not add any new concept to the solution space; it acts only as a way of organizing them and dealing with complexity. Packages may contain every "packageable" element of the model: diagrams, components, interfaces, classes, use case, and so on. As a package may pack diagrams, it can pack, not only structural diagrams, but also dynamic diagrams and concepts (Fig. 4.8.0.2).

If every element in the model is owned by exactly one package, this fact does not limit this element to appear in several diagrams (belonging to other

Table 4.8.0.1 Modeling concepts found in a package diagram

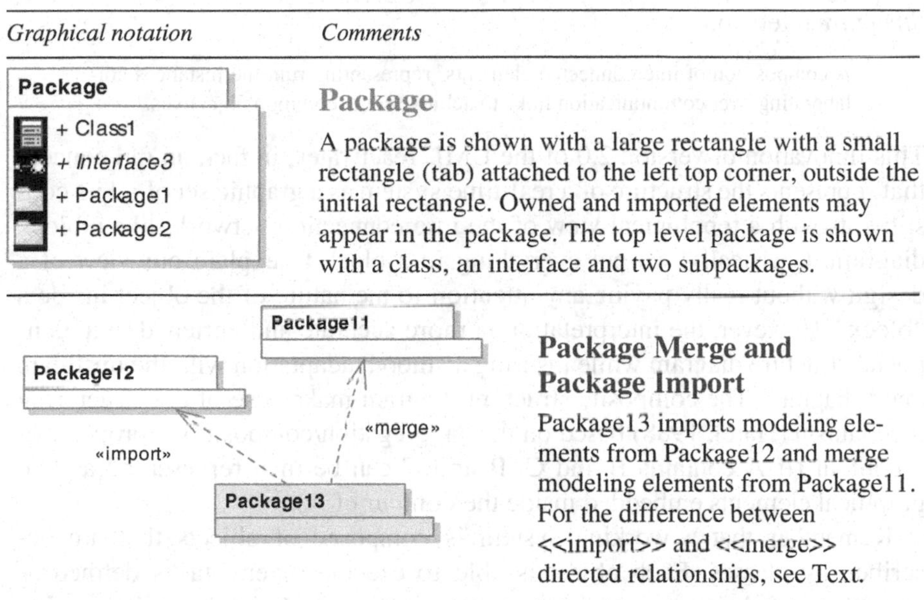

Graphical notation	*Comments*
	Package A package is shown with a large rectangle with a small rectangle (tab) attached to the left top corner, outside the initial rectangle. Owned and imported elements may appear in the package. The top level package is shown with a class, an interface and two subpackages.
	Package Merge and Package Import Package13 imports modeling elements from Package12 and merge modeling elements from Package11. For the difference between <<import>> and <<merge>> directed relationships, see Text.

<<import>> and <<merge>> are supplier-client relationships shown as dashed arrows between two model elements. At the tail of the arrow, is the Client that depends on the Supplier at the arrowhead.

Any stereotyped or user-defined dependencies

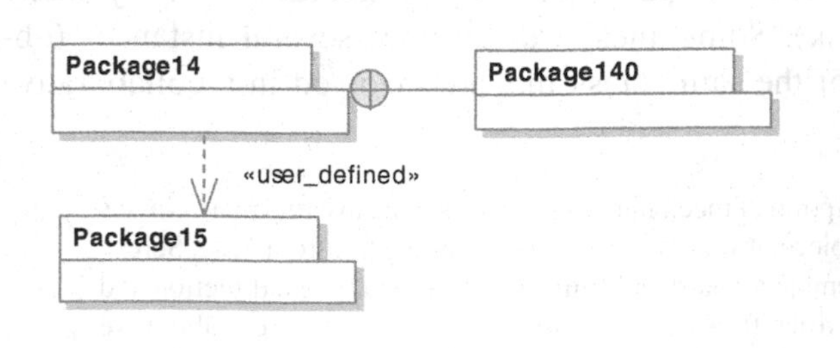

Nesting relationships is a kind of Aggregation/ Composition between packages. Package140 is part of Package14 and Package14 depends on Package15 through a user defined stereotyped relationship.

packages) and participate inside complex relationships with modeling elements in other packages. Classes in a package can be made public or private. A public class in a package is visible to and can be used by elements outside of the package.

4.9 Structure: Composite Structure Diagram

UML 2 provides a new way to represent the internal structure of classifiers, a new decomposition mechanism mixing functionality and structure, a notion of "instance of an application context from a general design" that give more accuracy and expressiveness to the modeling process. The term *Composite Structure* refers to:

> A composition of interconnected elements, representing runtime instances collaborating over communication links to achieve some common objectives.

This innovation of version 2.0 of the UML reactivates, in fact, an old concept that represents the structure of a real-time system as a graphic set of all its constituents with a topological view of the interconnection network, like a block diagram. Classically, we put everything in a block to explain our view of a design without really paying any attention to the nature of the object inside a "block." However, the interpretation is more accurate and enriched in a composite structure diagram while insuring a smooth adaptation with the upstream class diagram. The composite structure diagram makes use of the structuring mechanism (Harel, 1988) based on the "aggregation/composition" property of a contour (if A contains B and C, B and C can be then represented as two graphical elements embedded inside the contour of A).

Remember that a working system is composed of objects that are described by classes. Each object is able to execute several tasks defined as operations of their respective classes. A complex task need a collaboration of many objects, each participates with a specific role in collaboration. To understand working mechanisms of a design, it would be interesting to describe a system with a series of identified tasks, each task requires a set of instantiated objects that collaborate together to perform the global task. So, a task can be executed by a single object (no collaboration) or by many objects (collaborative task). Sometimes, we can have several instances (objects of the same type) of the same class that are involved in a collaborative task.

> Let us consider a description of mechanical assemblies. We have two main classes Roller and Board (flat piece of wood) and a lot of mounting fixtures (Axle, Screw, Nut, etc.). We can assemble a board and four rollers aligned in one direction and use the assembly to move a refrigerator. "Move_furniture()" is then a collaborative task executed by an instance of the class Board, four instances of class Roller and

some instances of fixture classes. By changing the way the rollers are mounted, we can make a rotating base that needs collaboration with another set of instances. A board itself can be used just as a shelf, but this fact is not described as collaboration when the board is alone with its "shelf" task. The participating "role" of an object changes with the way it is used in a system.

To describe this complex reality, the UML defines, in its metamodel, a series of new concepts as *StructuredClassifier, ConnectableElement, Property, parts, ports, connectors, Collaboration, CollaborationUse*, and three new stereotype dependencies ≪role binding≫, ≪occurrence≫, and ≪represents≫. Let us start with *StructuredClassifier* and *Property* modeling concepts, necessary to define the notion of *parts* and its containing classifier.

A *StructuredClassifier* is an abstract metaclass whose behavior can be fully or partly described by a collaboration of owned or referenced instances. Each *StructuredClassifier* contains *parts* (roles of *Property*) and *connectors*.

StructuredClassifier (Fig. 4.9.0.1) defined in InternalStructures package is derived from Kernel::Classifiers::Classifier metaclass. *StructuredClassifier* references *ConnectableElement* that represents an instance playing a specific role. In other words, a "role" is a *ConnectableElement* referenced by a

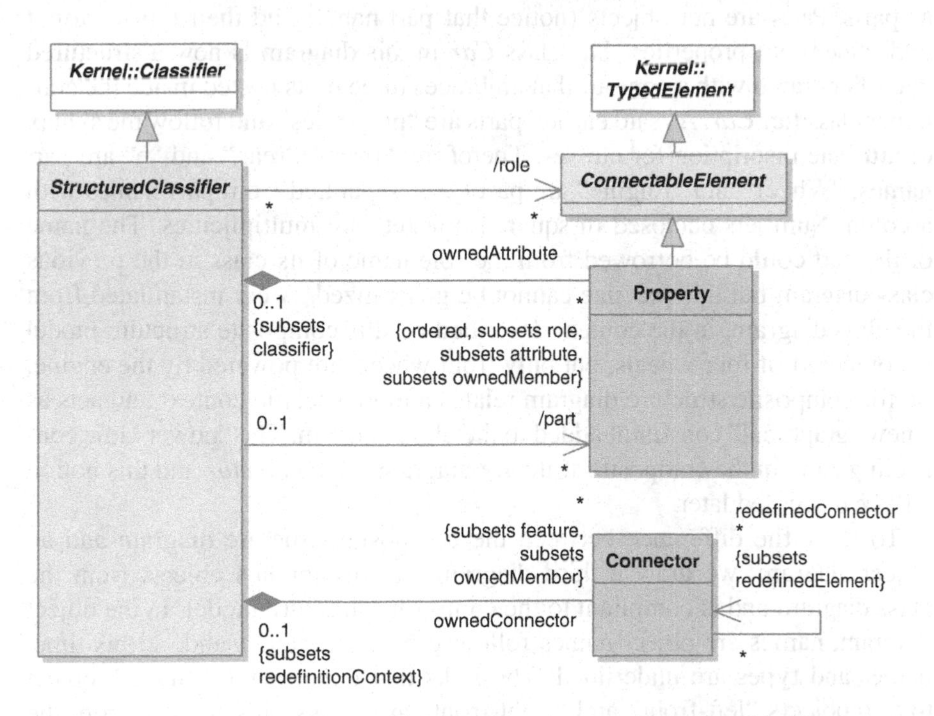

Fig. 4.9.0.1 StructuredClassifier of the InternalStructures package (Fig. 9.2, Superstructure Book)

StructuredClassifier. The real semantics of *ConnectableElement* is given by its concrete subtype. As said, a *StructuredClassifier* contains attributes called "parts." A part is a property specifying instances that a StructuredClassifier owns by composition. Parts are not objects but properties, so think of part as "part in a play" instead of more common sense of part as a "component of a whole." All parts are roles but not every role is a part. As the term "part" and "role" are intensively used in common description language, take care of the context when using these terms.

4.9.1 Part

A part declares that an instance of a containing classifier (StructuredClassifier) contains a set of instances by composition. Figure 4.9.1.1 represents first a Class diagram showing three classes Car, Wheel, and Engine.

The first class diagram declares that, generally, a *Car* is composed of four wheels, an engine playing a given role "e." Any instance of class *Engine* may power any arbitrary number of instances of class *Wheel* (to move the car). The association named "power" connects *Engine* class to *Wheel* class.

The second diagram is a composite structure diagram that represents classes as parts. Parts are not objects (notice that part names and their types are not underlined) but properties. The class *Car* in this diagram is now a structured classifier drawn with a contour that embraces three parts nested inside the container classifier *Car*. As said earlier, parts are "properties" and follow the syntax of attribute inscription for classes. Therefore, "front," "rear," and "e" are part names, "Wheel" and "Engine" are part types separated from part names with a colon. Numbers enclosed in squared brackets are multiplicities. The name of the part could be borrowed from the role name of its class in the previous class diagram but this practice cannot be generalized. A car instantiated from the class diagram, in the context described by this composite structure model is composed of four wheels, but only front wheels are powered by the engine. So the composite structure diagram relates a more specific context and acts as a new "graphical" constraint added to the class diagram. The "power" line connecting parts in the composite structure diagram is a *connector* and this notion will be discussed later.

To show the difference between the composite structure diagram and an object diagram, we draw a third diagram that instantiates objects from the class diagram and is compliant to the composite structure model. In the object diagram, names are object names followed by class names and, at this time, names and types are underlined. The links that connect the "engine" object to two objects "left-front" and "right-front" are "links" instantiated from the "power" association in the class diagram.

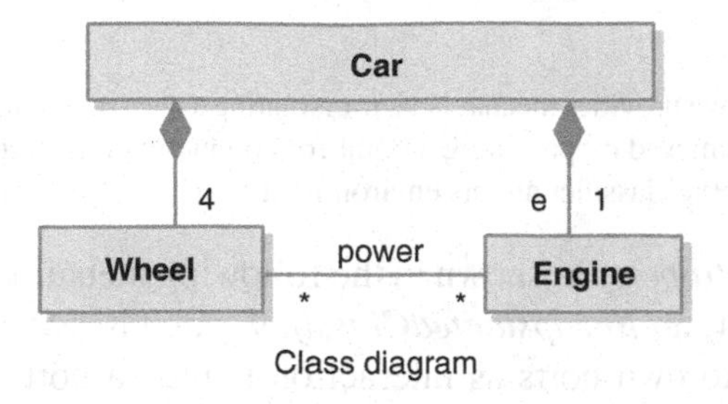

Class diagram

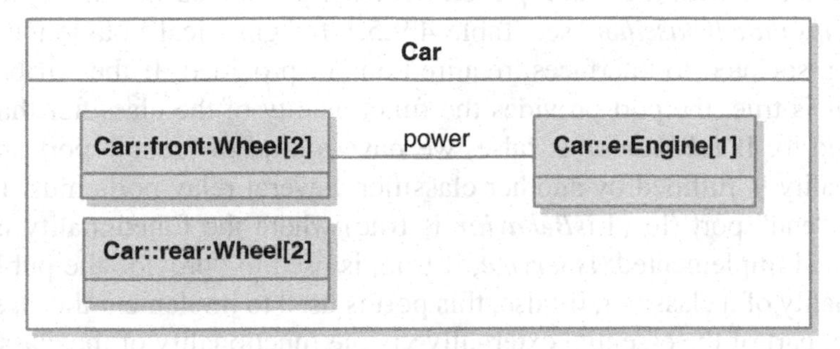

Composite Structure Diagram

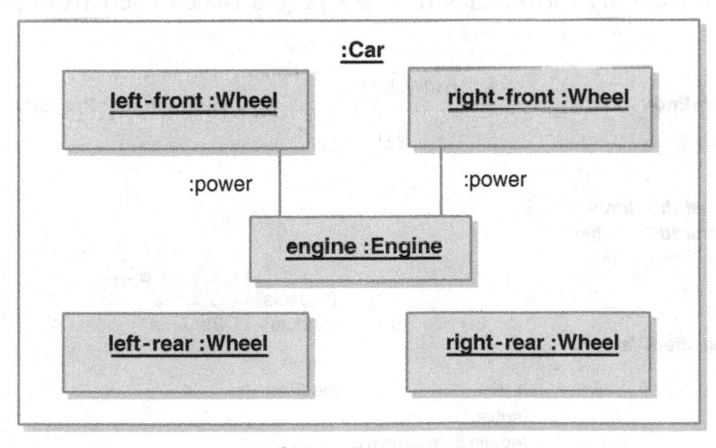

Object diagram

Fig. 4.9.1.1 Differences between a class diagram, a composite structure diagram and an object diagram. The first diagram is a class diagram. The second is a composite structure diagram that makes use of the Harel contour property to embed "parts" in a specific context. The third diagram is an object diagram that conforms to both the class diagram and the composite structure specification. The contour of :Car object replace five composition links between the Car objects and all of its constituents

4.9.2 Port

> The Ports package provides mechanisms for isolating a classifier from its environment. This is achieved by providing a point for conducting interactions between the internals of the classifier and its environment.

Port derives from *Property* as shown in the following metamodel (Fig. 4.9.2.1).

In Figure 4.9.2.1, an *EncapsulatedClassifier* extends a *StructuredClassifier* to add the ability to own ports as interaction points (a port is represented as a small square, with the name placed near by, positioned on the contour of the *EncapsulatedClassifier*, see Table 4.9.5.1 for graphical notations). Ports may be associated to interfaces, required and/or provided. If the attribute *isBehavior* is true, the port provides the functionality of the classifier that it is connected to. If *isBehavior* is false, we have a kind of "relay" port and the functionality is fulfilled by another classifier. Several relay ports must finally have an "end" port (local *isBehavior* is true) where the functionality is performed and implemented. *IsService*, if true, is used to "provide the published functionality of a classifier, if false, this port is used to implement the classifier but is not part of the essential externally visible functionality of the classifier." In other words, the meta-attribute *IsService*, if false, declares that the port is for internal use only for the classifier. As port is issued itself from property,

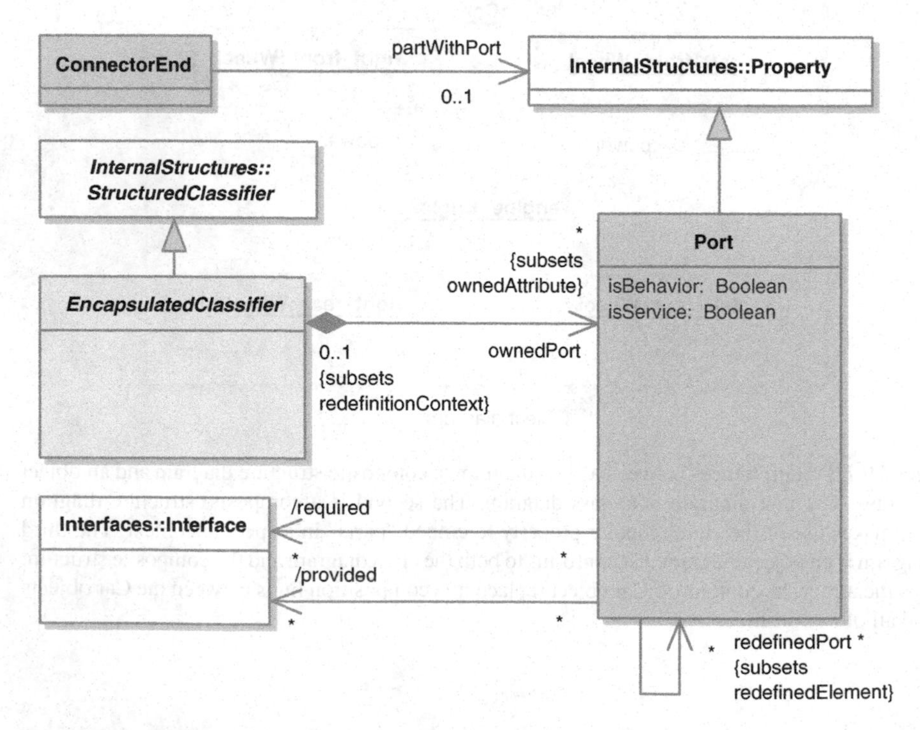

Fig. 4.9.2.1 Port metaclass defined in CompositeStructures::Ports package (Fig. 9.4, Superstructure Book)

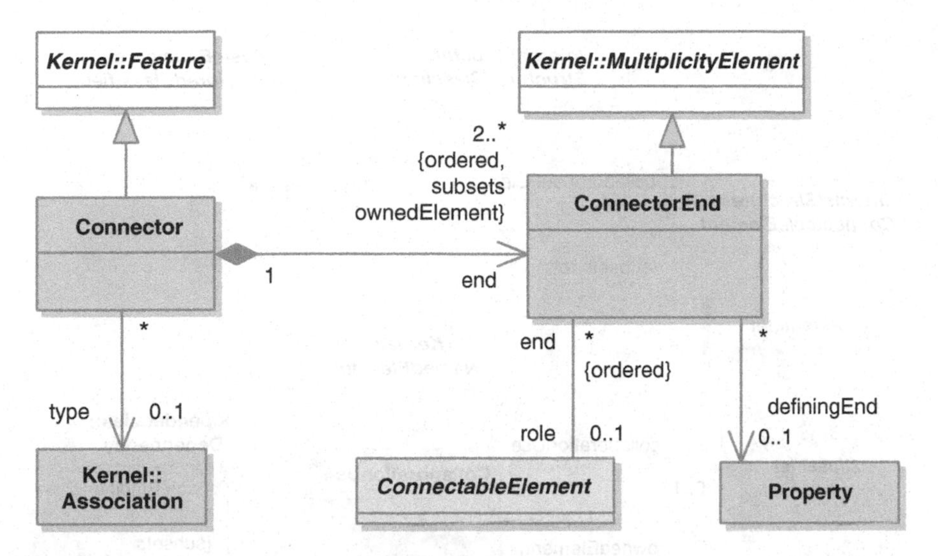

Fig. 4.9.3.1 Connector and ConnectorEnd metaclasses as defined in CompositeStructures:: InternalStructures package (Fig. 9.3, Superstructure Book)

there is some resemblance in the way "parts" and "ports" are defined at the metalevel.

4.9.3 Connector

A connector is a link that enables communication between two or more instances. We can connect two ports (attached to classes/interfaces/ components) together. In this case, classes are interpreted as an instance set from classifiers and interfaces are interpreted as their contextual instances. Generally, connectors may connect ports, parts, and objects directly. When connecting two objects, a connector may be interpreted as a link typed by an association, which exists between their corresponding classifiers, but this duality between "class–association–class"/"instance–link–instance" is not always mandatory as connectors can be defined exclusively between instances, independently from the fact that a corresponding association may or may not exist between classes (Fig. 4.9.3.1).

4.9.4 Collaboration and Collaboration Use

> Collaboration describes a structure of collaborating elements (roles), each performing a specialized function, which, collectively accomplish some desired functionality. It primary purpose is to explain how a system works and, therefore, it typically incorporates only those aspects of reality that are deemed relevant to the explanation. Thus, details, such as the identity or precise class of the actual participating instances are suppressed.

The *Collaboration* metaclass derives from both a *StructuredClassifier* and a *BehavioredClassifier*. A *CollaborationUse* (Fig. 4.9.4.1) represents one particular

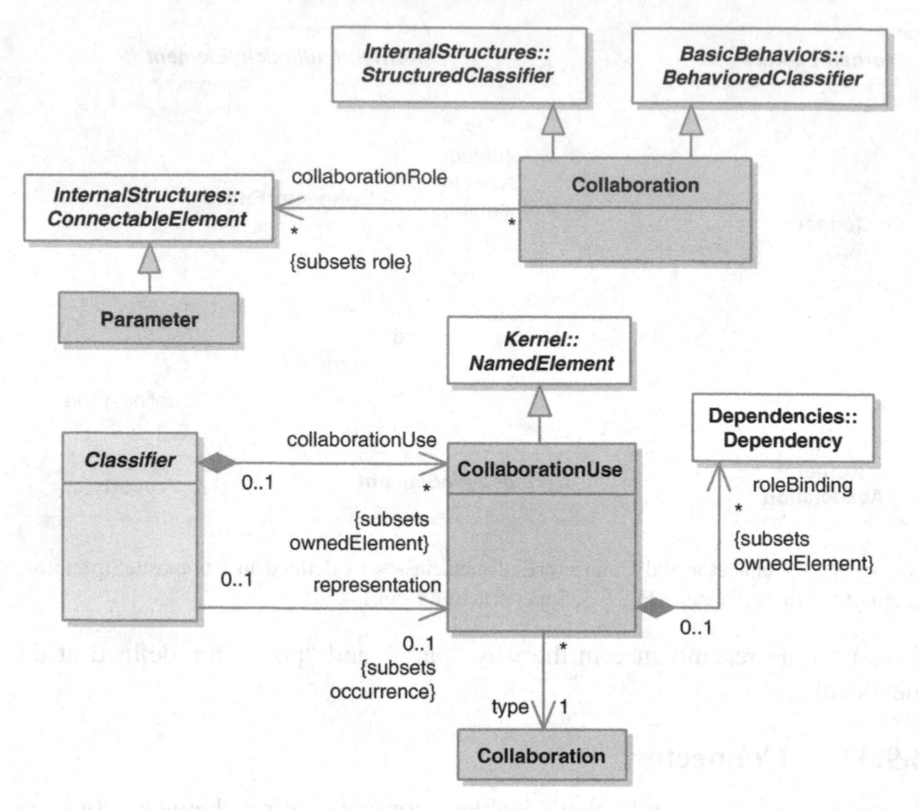

Fig. 4.9.4.1 Collaboration and CollaborationUse metaclasses in CompositeStructures:: Collaborations package (Figs 9.6 and 9.7, Superstructure Book)

use of Collaboration in a given context. There may be multiple occurrences of a given collaboration within a classifier, each involving a different set of roles and connectors.

The following example taken in the mechanical domain explains the use of *Collaboration, CollaborationUse*, and three dependencies used in conjunction with these modeling concepts. Notice for the moment that everything put in an oval form represents functionality (like the graphic notation used for the use cases).

In Figure 4.9.4.2, the collaboration "Create_Movement" is represented with three parts. "Create_Linear_Movement is a collaboration use that is a specific occurrence of the collaboration "Create_Movement." The three parts ":Board," ":Screw," and ":Roller" constitute the solution for a particular context of the collaboration. These parts play roles defined in a more general collaboration. All classes used to affect a type to "parts" are listed at the bottom of the figure.

The composite structure diagram used with parts, *Collaboration* and *CollaborationUse*, together with three new stereotyped dependencies ≪role binding≫, ≪represents≫ and ≪occurrence≫ allow the modeler

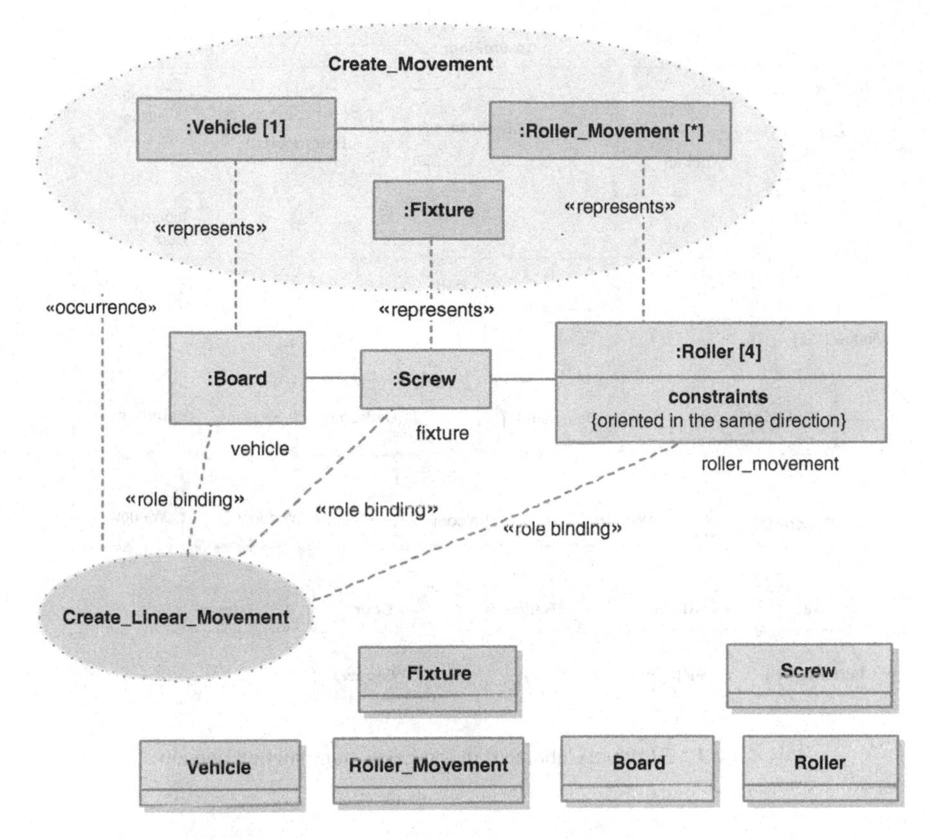

Fig. 4.9.4.2 Example in the mechanical domain that illustrates the use of "Collaboration" and "CollaborationUse" modeling concepts and three associated kinds of dependency

to describe a functionality (e.g. macroscopic task), show a general solution and sketch out plans to propose solutions for particular contexts. This diagram is very versatile as we can mix almost everything together to explain our model.

Hereafter is another sample of the UML applied to the domain of architecture. We can transform a textual description into a composite structure diagram easily.

In Figure 4.9.4.3, a composite structure diagram with parts (rooms and main spaces), ports (windows, external doors, and balcony), assemblies (internal doors), and connectors (when adjacent pieces are not separated by doors) is used to store the description of a specific instantiated ground floor. Classes are listed at the bottom and they must store attributes and operations (not displayed in the figure). Every object instantiated on this layout receives a full description (dimensions, orientation, material, etc.) in the development database. According to that information, a real architect's plan could be drawn automatically. Operations defined in classes could then be used as procedures to assist the automatic design process.

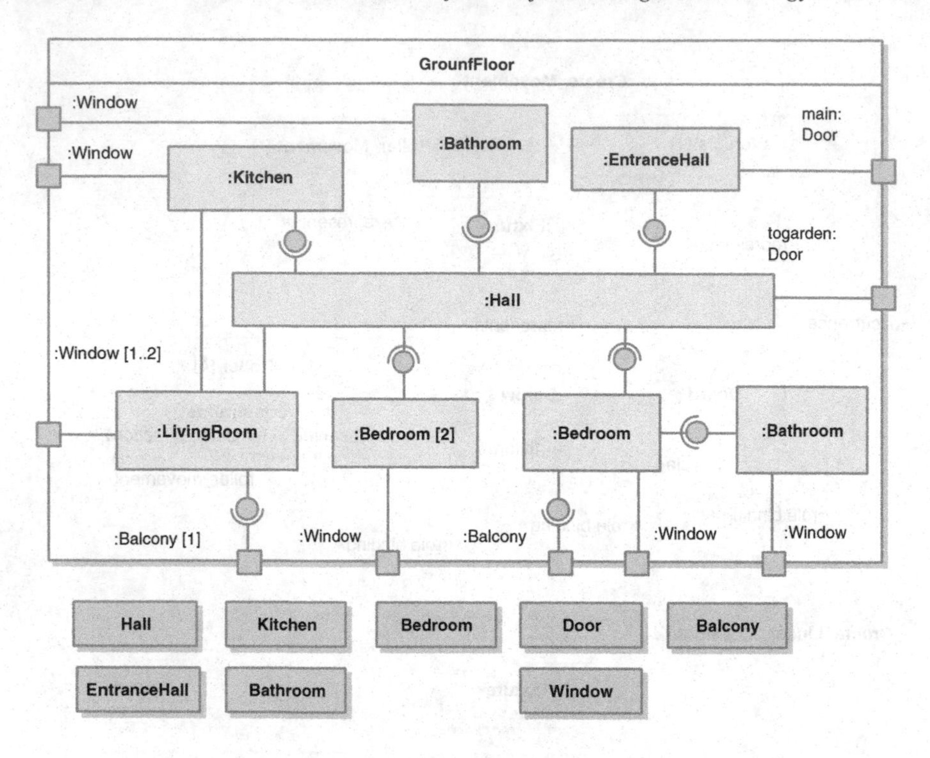

Fig. 4.9.4.3 Example elaborated in the house construction domain

4.9.5 Graphical Notations

Table 4.9.5.1 Modeling concepts in a composite structure diagram

Name and graphical notation	*Comments*
Part partname partname :PartClass [1..*]	A part is a property and its notation is identical to the syntax of class attribute. A part may be created with just a name or with a name, a type and a multiplicity. Parts represent classes in the context of a composite structure diagram. For the exact use of part and its subtleties, see Text.
Port and Connector Ports (small squares attached to the peripheral of class, interface, component, part, and object) are interaction points between a class, an interface, an instance and their environment. Connectors (no visual difference with an association in a class diagram) are links that enables communication between two or more instances.	

(*cont.*)

Table 4.9.5.1 (Continued)

Name and graphical notation	*Comments*

Connectors can be drawn directly between objects and parts with or without ports. Classes, interfaces, and components can be connected by connectors, but not directly, through

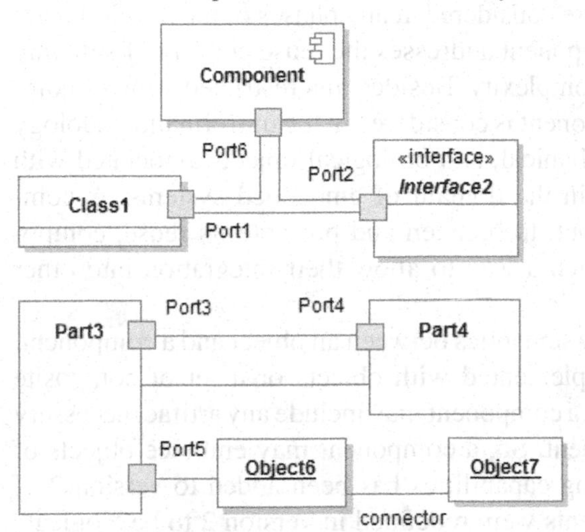

ports only.

Port can have a name, a type, and a multiplicity. The syntax is, like attribute, defined in class:

<port name> : <port type> [<multiplicities>]

A connector may be named and typed by an association. It has at least two ends. The graphical syntax for the connector's ends is similar to the syntax of an association end for a class diagram. The type of the connector end and the type of the instance it is connected to must be compatible.

Port with interfaces

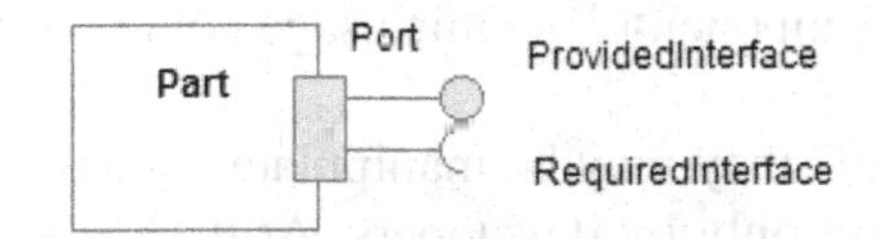

Ports can be represented with provided interface and required interface. In this case, the small square becomes a small rectangle if there are many interfaces connected to the same port.

Collaboration, CollaborationUse and stereotyped dependencies <<role binding>>,<<occurrence>>, and << represents>>

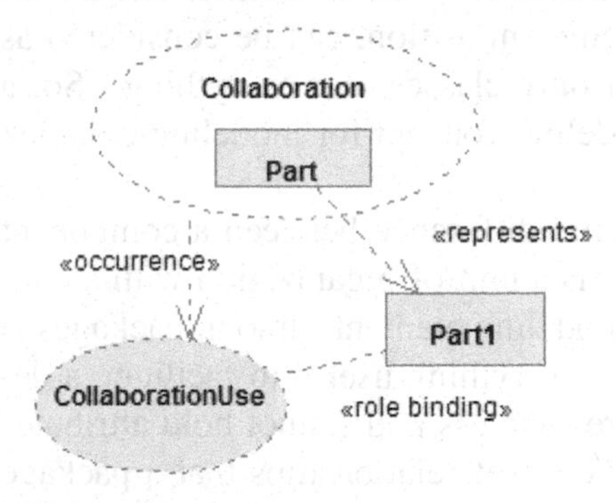

A Collaboration is a functionality, a macroscopic task that is executed by calling for collaboration of more elementary instances. A CollaborationUse is a collaboration applied to a specific context.

All dependencies are drawn with a dotted line (dependency type) with their respective stereotype.

<< occurrence>> shows the adaptation of collaboration to a specific context (collaboration use).

<<role binding>> is the mapping between a collaboration occurrence and parts needed to implement a specific situation.

<<represents>> connects elements in a Collaboration to a CollaborationUse to show the adaptation from a general case to a specific context.

4.10 Structure: Component Diagram

A component is a modular unit with well-defined functionalities and interfaces. In previous versions of the UML, components are seen as an implementation concept but in the current version, its use has been widened considerably and components are considered at any phases of the development, particularly in design. The component addresses the reuse concept of software systems of arbitrary size and complexity. Besides this restricted view of component as software unit, a component is considered in the uniform methodology as a mixture of electrical, mechanical, and biological objects associated with intelligent pieces of software in the domain of embedded systems. A component could be autonomous but, to broaden and promote its reuse, components should be designed in such a way to allow their integration into other systems.

Traditionally, there is a foggy semantics between an object and a component. In fact all components are implemented with objects or a set of composite and/or collaborative objects and a component may include any artifact necessary to support its use and deployment. So, a component may embrace objects of very dissimilar types (packaging capabilities has been added to version 2 of the UML). Moreover, components were redefined in version 2 to be "logical" instead of only "physical," so it can be used early in the design phase to describe business components, process components, and abstract concepts. There is no limitation in the UML to define "abstract components," so this usage could be anticipated.

As components are becoming "high level" they may be manipulated at any level (requirement or commercial level), not only by developers. At the metalevel, a Component metaclass derives directly from a StructuredClasses::Class, so it is only a special kind of classifier that contains more general packageable elements. So, a component acts simultaneously as a classifier and a container. Notice that a class, by aggregation/composition, can be considered as a container too, but classes can contain only classes, not everything). So, a component is a very handy and rich modeling concept for modeling complex systems.

Another possible question is the obscure difference between a component and a package. The answer is: a package is a unit of organization with a convenient namespace. The assignment of modeling elements among packages is a mental mechanism of arranging almost everything useful to facilitate a description in a given context. Packages are stateless and cannot hold attributes or operations as defined in the UML. The set of relationships that a package maintains with other packages are very limited (≪merge≫, ≪import≫, or stereotyped dependencies). Conversely, a component is a product, a piece of software, and/or hardware, a unit of exchange and reuse.

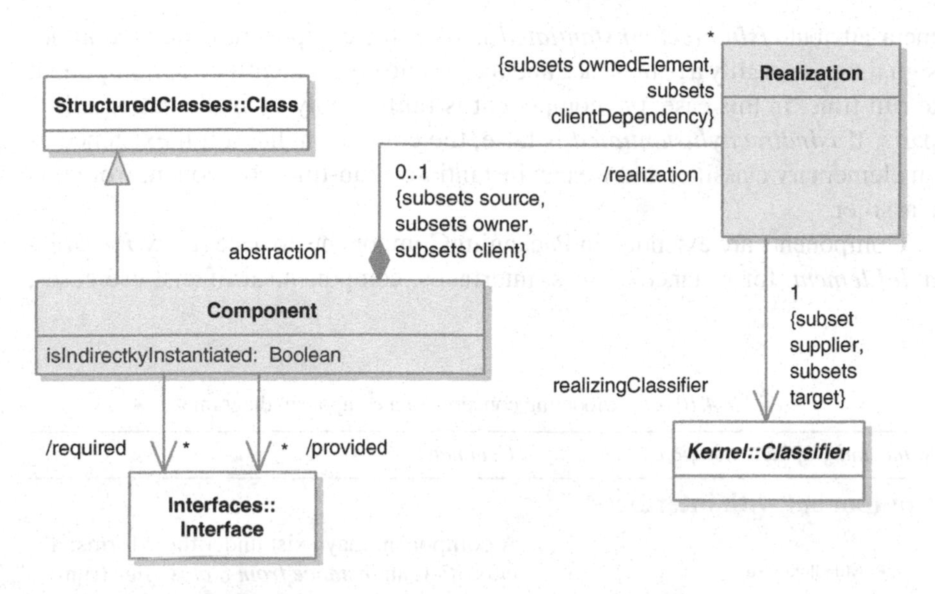

Fig. 4.10.0.1 Component concept as defined in BasicComponents package (Fig. 8.2, Superstructure Book)

Constructions that go inside a component may be merged from many packages (the UML itself may be considered as a "graphical modeling component" inside the long process of designing a system). Components may contain attributes and operations. A component may communicate with other components through signals and data at run-time. Occasionally, if the frontier of a package matches the boundaries of a component, it is only a matter of coincidence. Their use and the way these two concepts are defined in the UML are very different. We do concede that current spoken language make few difference between these two words.

In Figure 4.10.0.1, a *Component* is a subtype of *Class*. It has attributes and operations, and participates in associations or generalizations. As *Class* is taken from *StructuredClasses* package instead of *Class* from Kernel package, the components may have an internal structure, own *ports* as interaction points, and hold *required/provided interfaces* to communicate with instances. If the

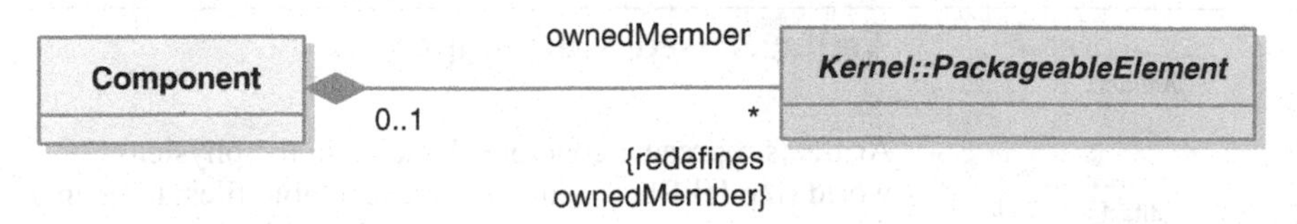

Fig. 4.10.0.2 Extension of the Component in PackagingComponents package (Fig. 8.4, Superstructure Book)

meta-attribute *isIndirectlyInstantiated* is true, the component is defined at design time as roughly a namespace and there is no object identified as component at run-time. In this case, the component is built simply by instantiating all its parts. If *isIndirectlyInstantiated* is false, the component has a full existence as an elementary classifier and we can instantiate a run-time component from this classifier.

Components are extended in PackagingComponents to accept any *PackageableElement*, for instance classes, interfaces, components,artifacts, use cases,

Table 4.10.0.1 Modeling concepts of a component diagram

Name and graphical notation	*Comments*

Component with interfaces

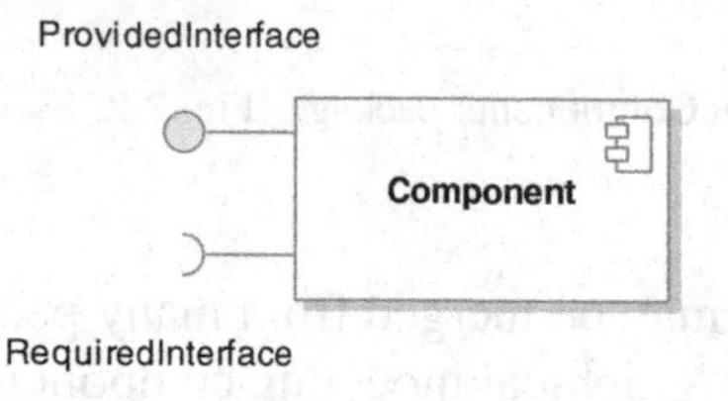

A component may exist under three forms: a *classifie*r, an *instance from a classifier* (run-time component) and finally an *entity built up by instantiating all of its constituents*. Instantiated forms, direct or indirect, contain an underlined name.

Its graphic notation contains a component icon in the right hand corner. Provided and required interfaces can be added to the contour of the component.

Component with attributes and operations

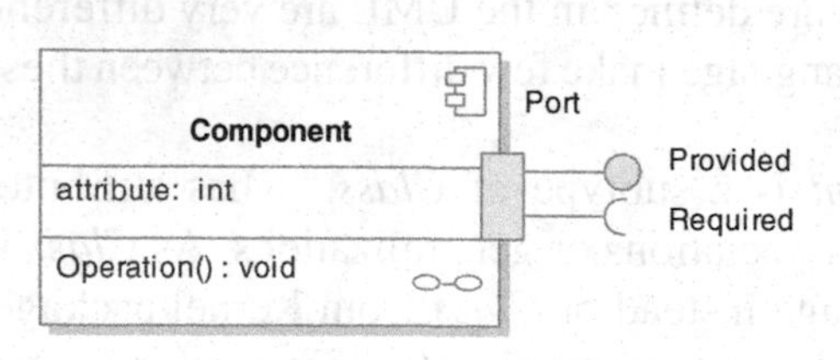

A component is a structured class, so it may contain attributes and operations.

The syntax for writing attributes and operation is the same than that was described previously for a class (class type component) or an object (runtime component).

Artifacts used with components

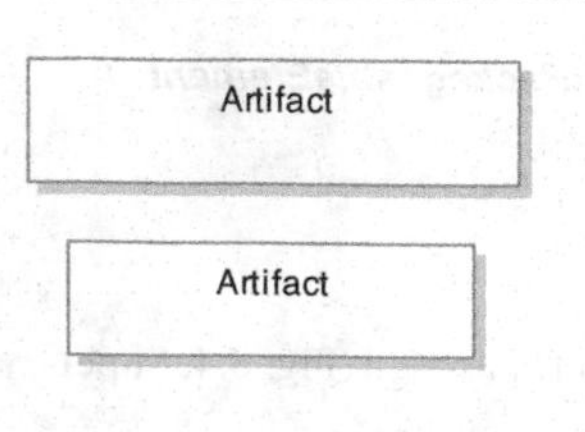

Artifacts represent concrete elements in the physical world (model files, source files, executable files, table in a database system, a development deliverable, word processing document, mail message, etc.). Artifact metaclass is defined in Artifacts Package.

Artifacts can be packed with components.

Table 4.10.0.1 (Continued)

Name and graphical notation	*Comments*

Association between components. Connectors between instances

All the associations (regular, Generalization, Composition) may be extended from classes to components. The first diagram shows Component2 that derives from Component1 (Generalization). Component2 and Component3 are involved in an undefined association.

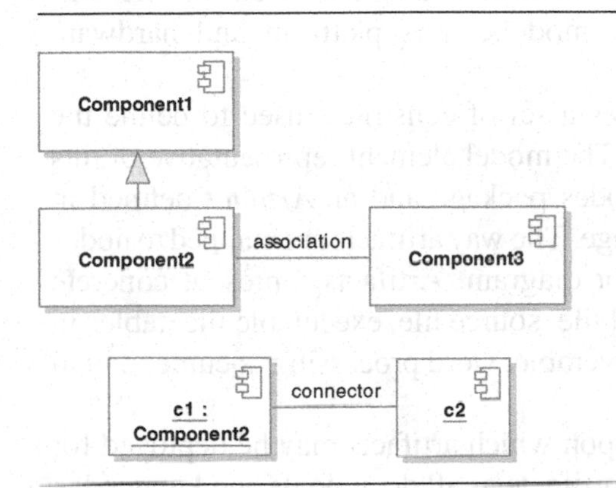

c1 and c2 are instances of components (their names are underlined). Like objects, connectors can be drawn between instances.

Connectors, like links for objects, may be instantiated from their corresponding associations defined at the classifier level but this fact is not mandatory as components may exist as a "runtime" entity only.

White box view of a component. Ports Assembly and Delegate connectors

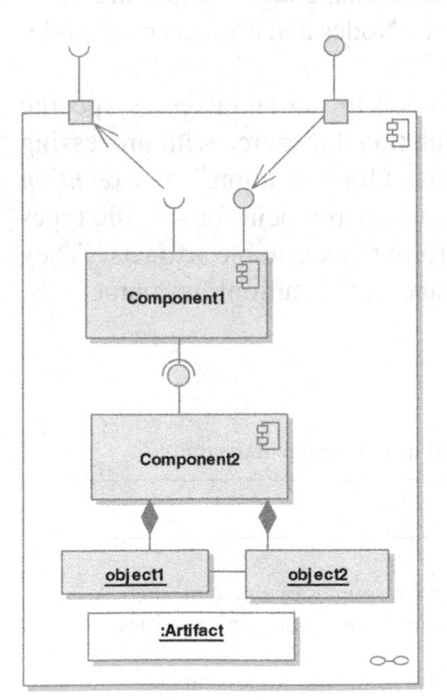

Components are composed of "packageable elements". As said in the text, a component may pack almost everything.

A black box view of a component shows only its interface with ports (see composite structure diagram for "Ports") and hides its internal structure.

A white box view of a component may be drawn with a large contour delimiting the component and containing all its constituents. The figure shows two new elements that could be used in a component diagram: an assembly connector between Component1 and Component2, delegate connectors connecting ports and interfaces.

An *assembly connector* is a connector between two components; one provides the services that another component requires.

A *stereotyped delegate connector* can connect interfaces and ports. A port may be connected to many components with "delegates". The functionality available at the interface of the component is accomplished internally by Component1.

dependencies, packages, etc. This property comes from the “namespace property” of a Component that can own almost everything.

4.11 Structure: Deployment Diagram

Deployment diagrams show how different parts of a system (objects, components, artifacts, etc.) are deployed onto the hardware environment. The deployment diagram is typically a structural diagram used in the implementation phase and deployment models. It is platform and hardware dependent.

The Deployments package specifies a set of constructs used to define the “execution architecture” of a system. The model element representative of this diagram is a *Node* defined in the Nodes package and an *Artifact* defined in Artifacts package of UML metalanguage. The way artifacts are mapped to nodes is typically assigned to a deployment diagram. Artifacts represent concrete elements in the physical world (model file, source file, executable file, tables in a database system, a development deliverable, word processing document, mail message, etc.).

“A node is computation resource upon which artifacts may be deployed for execution”. We can consider roughly artifacts as all documents and knowledge injected and developed before the first deployment diagram is established and nodes are devices, execution environment, networks, geographical, and hardware resources needed to deploy those artifacts. Nodes and artifacts exist under two forms; classes and instances.

Device and ExecutionEnvironment are added in the language as specific nodes. A *device* designates a “specific computational resource with processing capability upon which artifacts may be deployed for execution.” An *execution environment* is a “node that offers an execution environment for specific types of components that are deployed on it in the form of executable artifacts.” They are marked with stereotypes ≪device≫ and ≪executionEnvironment≫, respectively.

Table 4.11.0.1 Modeling concepts used in a deployment diagram

Name and graphical notation	*Comments*
Node 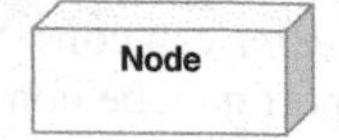	A node represents computational resources of a system, such as servers, personal computers, experimental components of a real-time application system. A node has a name that describes its nature. Nodes can be nested and have a composite internal structure. A node is displayed as a 3D box.

Table 4.11.0.1 (Continued)

Name and graphical notation	*Comments*
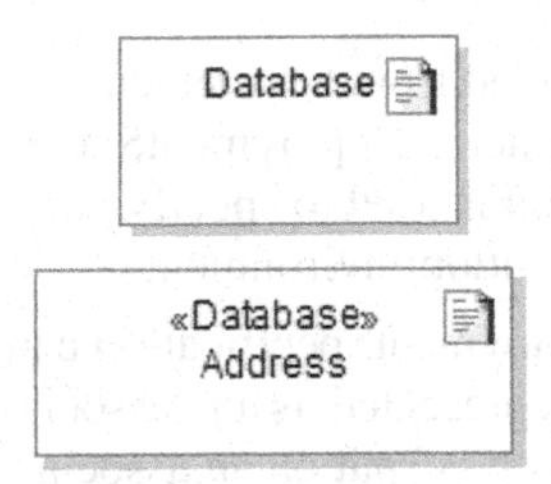	**Artifacts** Artifacts are used both in deployment diagrams and component diagrams as well. They are documents, artifacts and knowledge injected and developed within the design process: executables, libraries, software components, databases, textual documents, etc. Stereotypes can be used with artifacts
Device, executionEnvironment and other stereotyped nodes 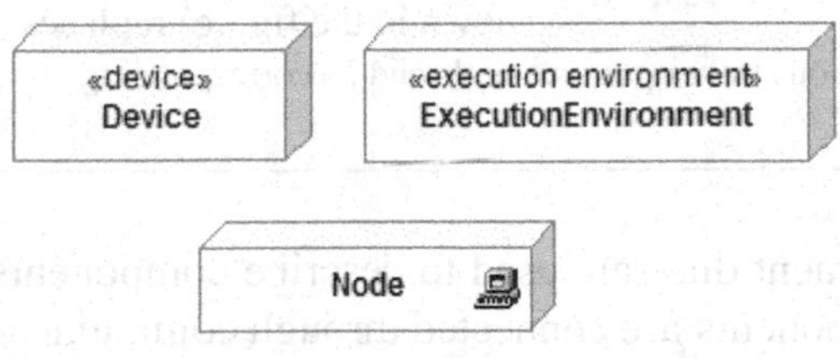	A *device* is used to specifically name a computing resource. An *executionEnvironment* specifies a specific execution platform, such as an operating system or a database management system. Stereotyped nodes represent specific hardware or environment. Their shapes could be replaced by personalized icon to make them more suggestive.
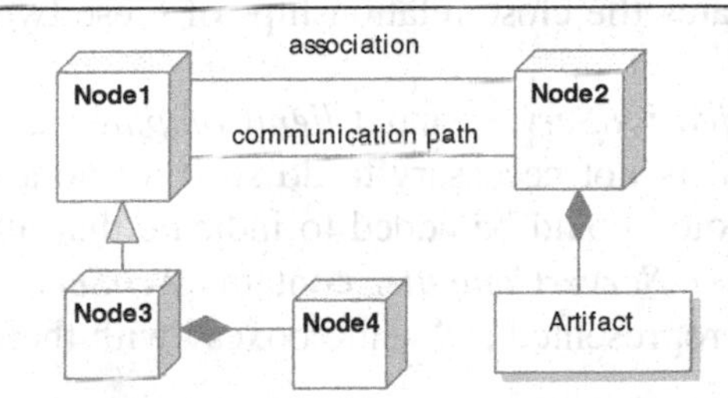	**Generalization, Aggregation/Composition, CommunicationPath and other associations** As node and artifacts are classifiers, all associations defined for classes can be used with nodes and artifacts. A communication path is a special form of association devoted to the exchange of signals and messages. Association ends are typed by Deployment targets.
Deployment, Manifest relationships, and other dependencies 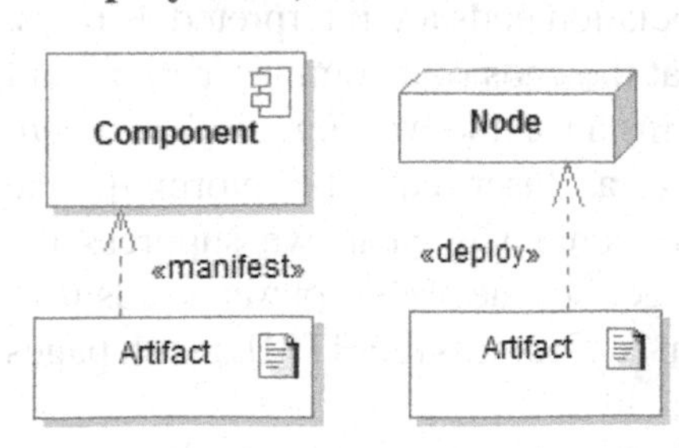	For <<manifest>>, see "Manifestation" concept that replaces "Implementation" concept in "Structure: Class Diagram" chapter. The component is a "manifestation" of the artifact used to describe a system. An artifact holds many possible manifestations. <<deploy>> is a kind of <<manifest>> used at more elementary level. A node deploys an artifact. Any stereotyped dependency can be added to this phase.

(*cont.*)

Table 4.11.0.1 (Continued)

Name and graphical notation	*Comments*
Deployment Specification 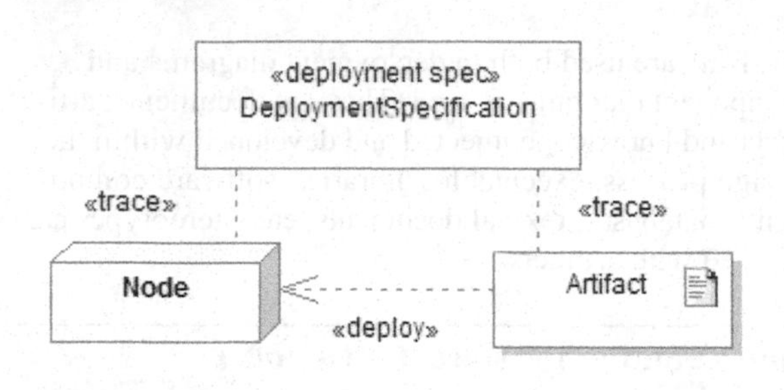	Derived from the metaclass Artifact, a DeploymentSpecification is used to specify an ExecutionEnvironment. DeploymentSpecification can be represented as an "association class" put on an association stereotyped as <<deploy>>. An alternative (as shown in the figure) replaces the association class by two simultaneous traces put on a node and its corresponding artifact.

Figure 4.11.0.1 shows a deployment diagram used to describe components and nodes of a web database. Components are connected through communication paths. Inscriptions at each end of the communication path represent message types. This example shows a combined version of a component diagram and a deployment diagram, and demonstrates the close relationships of these two diagrams.

ServerComputer, WebServer, ApplicationServer, and *ClientComputer* are nodes. "Devices" could be used but it is not necessary to do so if all nodes are "devices" by default (A "UML note" could be added to indicate that all nodes are stereotyped by ≪device>>). *ServerComputer* contains *WebServer* and *ApplicationServer*. All nodes are represented as "white boxes" with their contents exposed.

Components are connected through communication paths. A communication path is a special form of association devoted to the exchange of signals and messages. Normally, association ends are typed by the deployment targets. For an association, inscriptions at the association ends are interpreted as roles, for the deployment diagram, inscriptions at the ends of a communication path could be interpreted as message types. For instance, the *Web Server Component* interprets everything sent by a *Web browser* as "user requests." Normally, the communication is bidirectional, but if the context is clear, we suppress the double arrowheads to avoid visual surcharge. So, the *Web browser* sends user requests to *Web Server Component* and then later returns the HTML result pages to the client.

Communication paths between components are interpreted in this example by users as "high level" messages. Communication paths between nodes are interpreted in this example as "low level" TCP/IP (Transmission Control

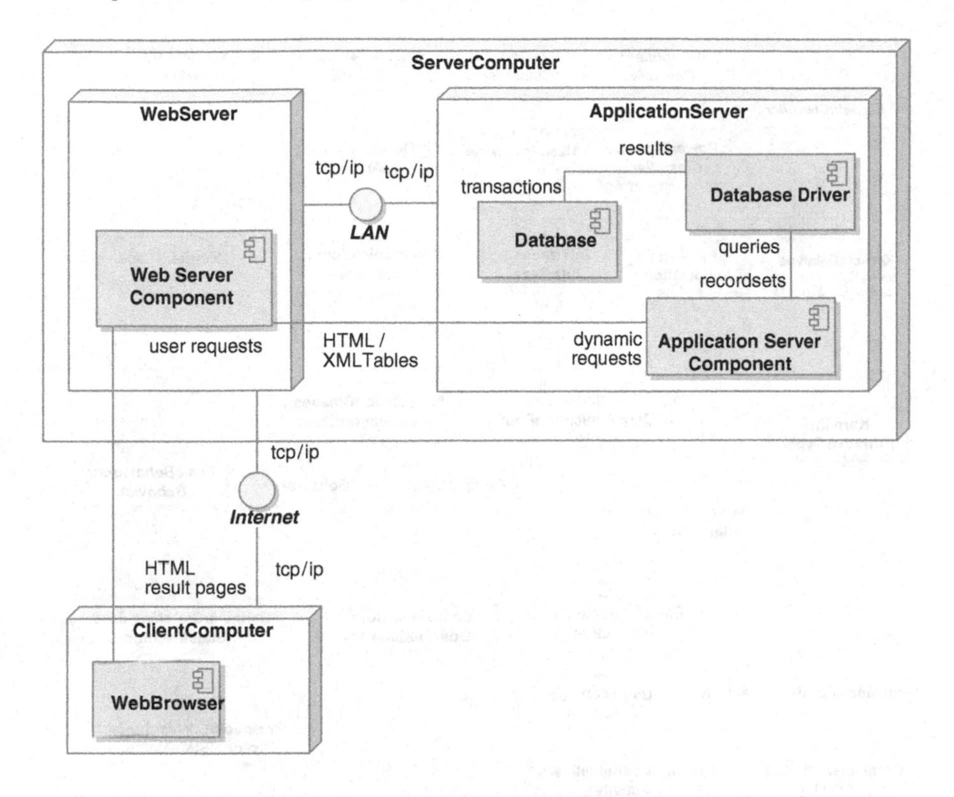

Fig. 4.11.0.1 Deployment diagram showing components and nodes of a web database

Protocol/Internet Protocol) packages exchanged between computers. Generally, components are more abstract and "higher level" than nodes. Small circles, named "Internet" and "LAN," are interfaces. The communication path between Internet and the *Web Server Component* crosses the contour of *ServerComputer* node to reach the *Web Server Component* node. This fact means that the *ApplicationServer* is not connected to the Internet (Harel's interpretation of contour properties). To allow the *ApplicationServer* and the *Web Server Component* to be connected both to the Internet; we must stop the link on the contour of *ServerComputer*.

4.12 Hierarchy of Metaclasses Defined in the Behavior Part: Abstract Metaclass Behavior

The Behavior part of the UML is more difficult to grasp than the Structure part. Modeling concepts with the same name are present in several packages and are successively adapted (or patched). For instance, the metaclass *Activity* is defined four times in four different packages. We supposed that this state results from the adaptation, evolution, and reengineering process of the UML through its multiple versions. The hierarchy of main modeling

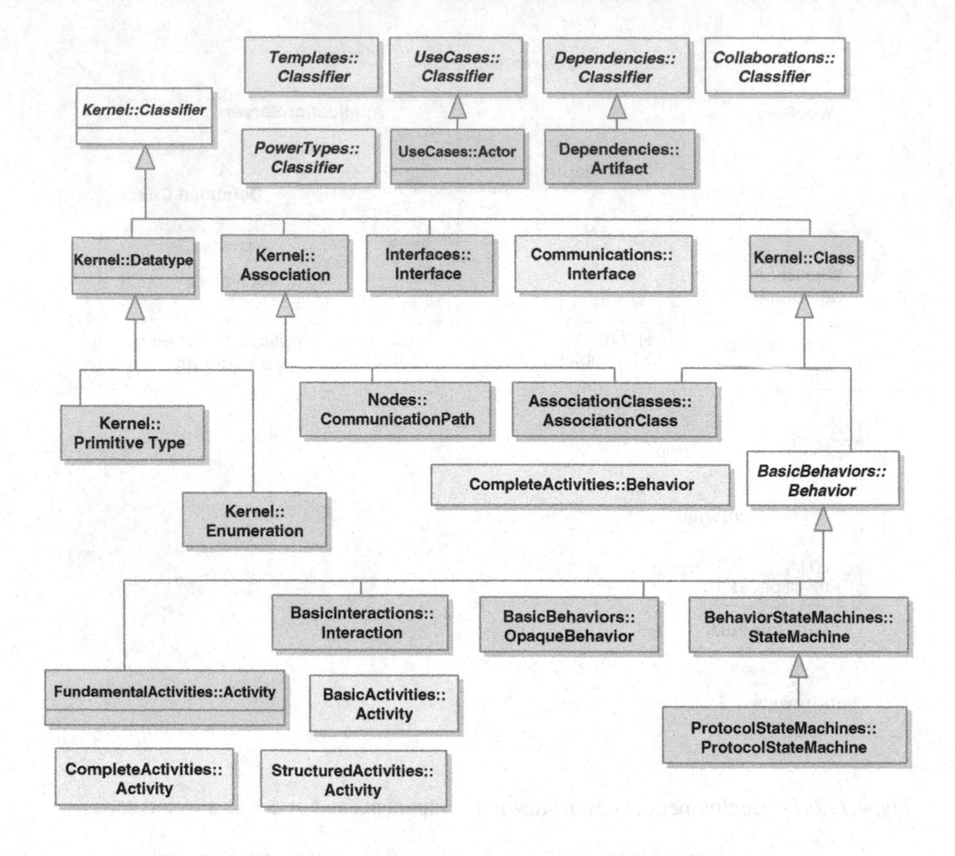

Fig. 4.12.0.1 Generalization hierarchy of the UML Behavior (Part1)

concepts in the Behavior Part of UML 2 is presented in Figures 4.12.0.1 and 4.12.0.2.

These diagrams show that everything still derives from *Classifier* in the Kernel package. They show "roughly" how the UML is built to derive complex model elements in the Behavior Part. Elements shown in italic and not colored (or grayed) are abstract and cannot be instantiated. Elements that are colored (or grayed) differently represent different versions of the original class with the same name, for instance *Interface* is defined twice, the first time in Interfaces package and a second time in the Communications package. The Communications package merges the definition of *Interface* from the Interfaces package.

Possibly, readers not exposed to the UML may have some difficulty in understanding this dynamic part of the UML for lack of knowledge about dynamic concepts. In this case, the solution would be browsing quickly the rest of this chapter then reading Chapter 5 before coming back to this chapter. This iterative process is advisable and will help us in keeping a good structure for a reference

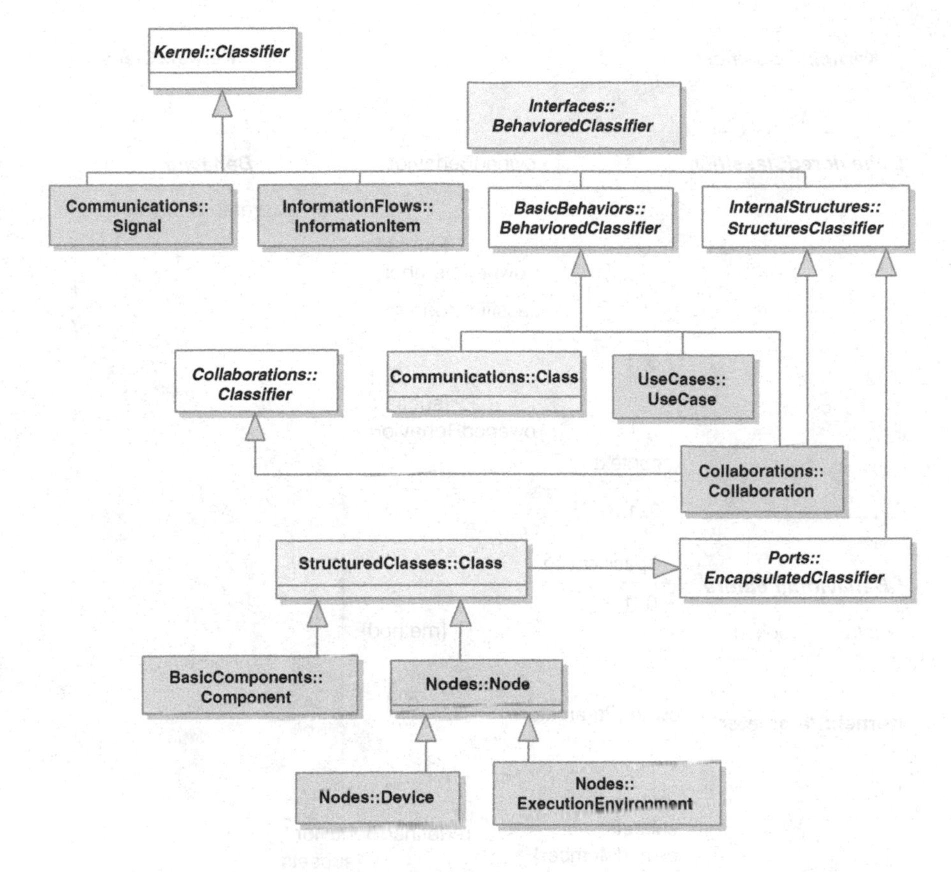

Fig. 4.12.0.2 Generalization hierarchy of the UML Behavior (Part2)

book. It is illusory to think that we can remember all complex conventions and interpretations of the UML by reading a book or the standard just once. Unless we are exposed to all modeling subtleties of the UML everyday, it would be hard to remember all details.

A "piece of behavior" is first defined in BasicBehaviors package that merges directly model elements from the Kernel package of the Superstructure. In Figure 4.12.0.3, the metaclasses *BehavioredClassifier* and *Behavior* in BasicBehaviors package are all abstract and provide reference classes for building of specific dynamic model elements. Derived from a Kernel::Classifier metaclass (Fig. 7.9 Superstructure Book), *BehavioredClassifier* holds properties and generalization relationships.

A *BehavioredClassifier* has many *Behavior* elements. *Behavior* derives from Kernel::Class metaclass (Fig. 7.12 Superstructure Book). As a classifier, a "Behavior" object can be instantiated (so an activity, a piece of dynamic behavior, is a "conceptual" object in the syntax of the UML). Instantiating a behavior is referred as "invoking" the behavior. When a behavior is invoked, it receives a

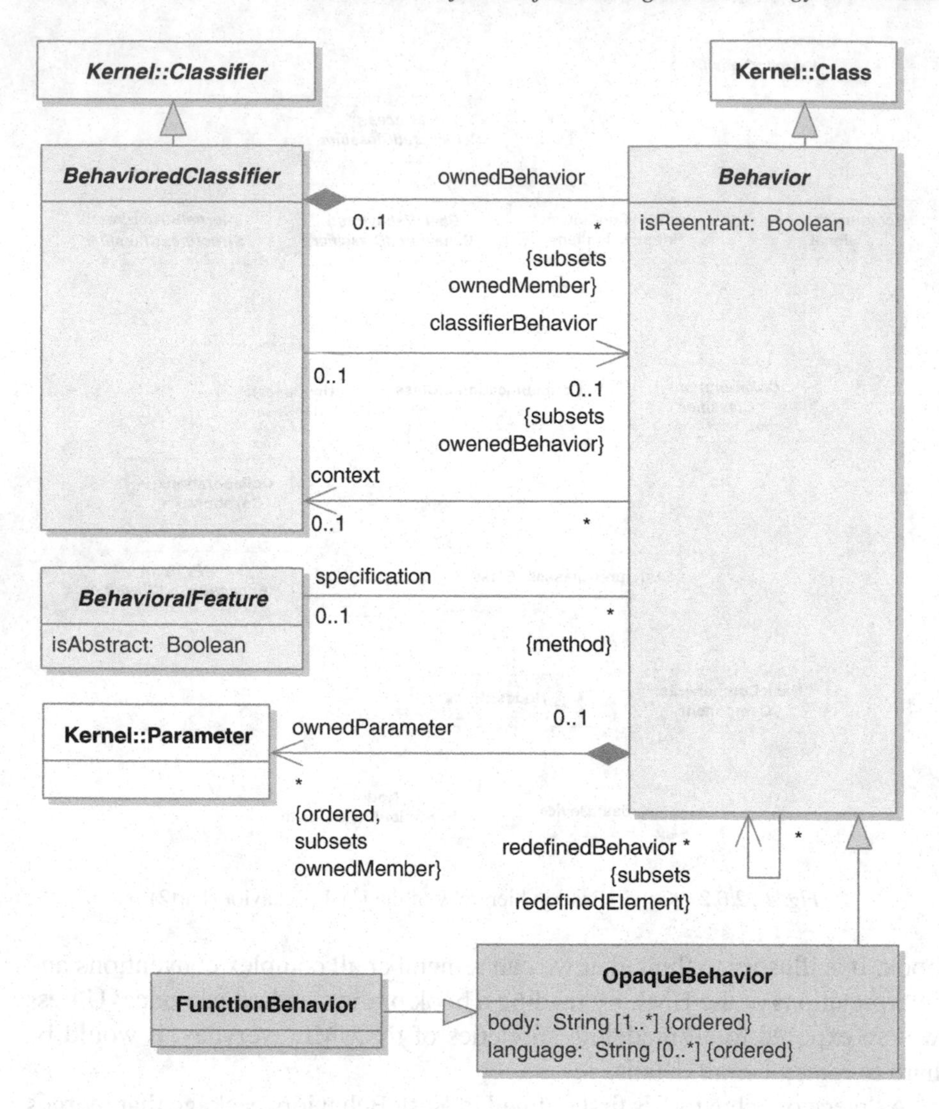

Fig. 4.12.0.3 Behavior metaclass as defined in BasicBehaviors package

set of input values and its "execution" produces a set of outputs, as specified by its parameters (Kernel::Parameter metaclass). The meta-attribute *isReentrant* tells whether the behavior can be invoked while it is still executing from a previous invocation. A *Behavior* may be redefined. If a *Behavior* is owned by a *BehavioredClassifier*, that classifier acts as its context. A *BehavioralFeature* (according to Fig. 7.10 of the Superstructure Book, a BehavioralFeature is a simple Feature that can own a Parameter set) is realized by a *Behavior*. An *OpaqueBehavior* acts as a placeholder for implementing specific behaviors.

4.13 Behavior: State Machine Diagram

State Machine diagrams specify the life cycle of objects, integrating many sequences of states, actions, activities, events, conditions, and state transitions inside a diagram with a particular emphasis on states. State machine diagrams in the UML pack several state diagrams inside one unique representation showing their concurrency. They do not describe functionalities of bare finite state machines at the early days but are variants of statecharts (Harel and Gery, 1997). Harel (1987) added hierarchy (also known as depth) and orthogonality (also known as concurrency).

The UML defines two kinds of state machines: behavioral state machine and protocol state machine. The first type specifies general behavior of a model element and the second type expresses usage protocols. A protocol expresses legal transitions that a classifier can trigger during a life cycle.

A hair dryer is represented by a composite state made of two orthogonal regions named Heater and Motor (Fig. 4.13.0.1). "Heater States" contained in "Heater" region is called a substate. When entering a composite state of type "orthogonal" (having more than one region), each region must have a "starting state." As there is no starting state indicated by a pseudo "initial" state, any state in each region could be a candidate for a starting state. So, when plugging this hair dryer, the dryer can start immediately, for instance with "Motor LOW" and "Heater HOT."

The states of these two objects Heater and Motor of the hair dryer are independent and are represented by a composite state "Hair Dryer States" composed of two concurrent composite substates embedded in two regions "[heater]" and "[Motor]." The motor has three states: OFF, LOW, and HIGH; LOW and HIGH are "Motor ON states." The "Motor OFF" state can switch to "Motor LOW." The "Motor HIGH" state can be reached only through "Motor LOW" state but not directly from "Motor OFF."

The initial state named "When Plugging" is a pseudostate indicating that the state diagram must be read from this point. All arrows are "transitions" or "state changes." The pseudostate "When Plugging" state is connected to the contour the "Hair Dryer States," so, when entering this contour, the system can reach any pair of states, one in the motor region and another in the heater region. For instance, when plugging the power cord, the hair dryer can be at "Motor LOW" and "Heater HOT." Later, the heater may be switched to the "Heater OFF" state, but it must pass through the "Heater WARM" state.

States of the heater and those of the motor are independent but "concurrent." There is no connection between their states, so from one state of the motor we cannot go to another state of the heater. However, objects necessary to implement this hair dryer (special switches, connections, etc.) may communicate together to produce such a result described by this state machine. An object

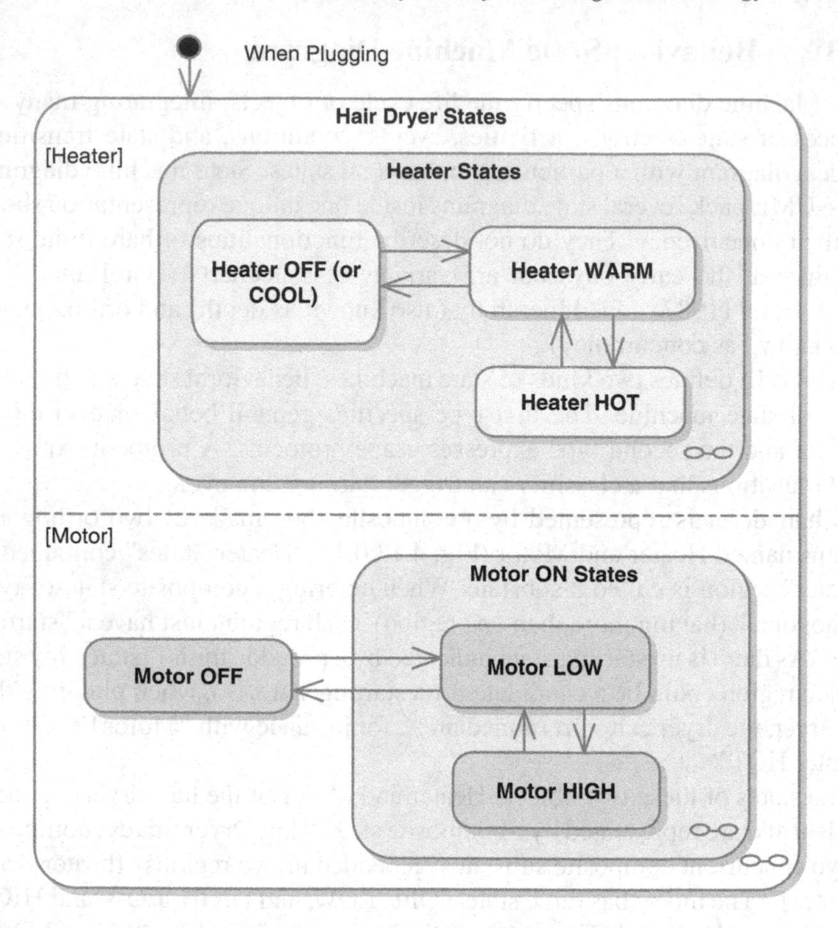

Fig. 4.13.0.1 Hair Dryer State Machine

diagram is more appropriate to describe such a communication view, not a state machine.

To illustrate this communication, when switching from "Motor OFF" state to "Motor LOW" state, the button used to create this state transition must send a message to enable the heater, which can take any on the three heater states: COOL, WARM, or HOT. When switching from "Motor LOW" state to "Motor OFF" state, the same button will send another message to the heater to disable all states; hence the user can never start the heater when the motor is off. We can name/comment/constraint "Motor ON states" as "Heater Enabled State" and "Motor OFF" as "Heater Disabled state" but a name, a comment or a constraint must be translated into real messages or commands in other diagram.

To understand the modeling concepts of State Machines, it would be interesting to analyze the metaclasses of a state machine defined in the standard (Fig. 4.13.0.2). A state machine is a *Behavior*, as *Behavior* derives from

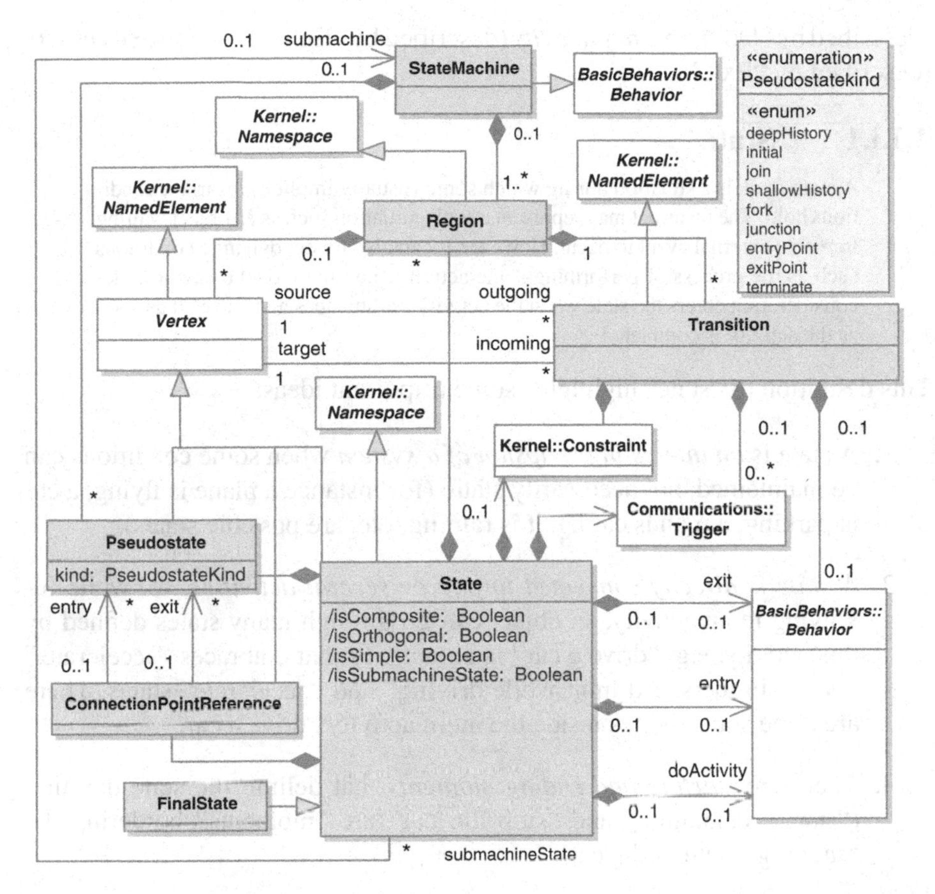

Fig. 4.13.0.2 Metaclasses StateMachine, State, Region, Transition, Pseudostate, FinalState defined in BehaviorStateMachines package

Kernel::Class, so instances of a state machine are objects. A state machine or a state may contain one or more regions. Regions describe concurrent states of many objects simultaneously.

A state machine contains edges (transitions) and vertices (pseudostates, states, and connection point references). *Pseudostates* are vertices (thus named elements, not states. They have various graphical notations in a diagram. It would be interesting to notice that "Initial" is a pseudostate but *FinalState* derives from *State*, so "Initial" helps us where to start interpreting but *Final* is a sink state where an activity arrives to its end.

A Transition is an edge that links two vertices, one named "source" and the other "target." It holds triggers, may be constrained with guards and may have an optional effect (Behavior) when the transition fires. A State has several meta-attributes (isSimple, isComposite, isOrthogonal, isSubmachineState) that describes different forms of a state. A state may contain an *internal activity* (described by "Do"), an *entry activity* (described by "Entry"), and an exit activity (described by "Exit").

4.13.1 State

> A state models a situation during which some (usually implicit) invariant conditions hold. The invariant may represent a static situation such as an object waiting for some external event to occur. However, it can also model dynamic conditions such as the process of performing some activity (i.e., the model element under consideration enters the state when the activity commences and leaves it as soon as the activity is completed).

This definition of "state" highlights some important ideas:

1. A state is *an interesting snapshot of a system* when some conditions can be maintained, not necessarily static (for instance a plane is flying, a car is parking, a man is eating, it is raining, etc. are possible states).
2. *A state is directly connected to one or several activities*, so, while involving in an activity, an object can go through many states defined by this activity, e.g. "drive a car" is an activity that embraces "accelerate," "maintain the speed limit while driving" and "decelerate" states. There are three subactivities inside the main activity "drive a car."
3. There are *starting and ending moments* that delimit the state duration ("start accelerating" and "stop the car" are "moments" bordering the "running" state of the car).

A state is a model element included in a more important concept known as a "state machine" that packs states, transitions, and other model elements as stated by the standard:

> A state machine owns one or more regions, which in turn own vertices and transitions.
>
> The behaviored classifier context owning a state machine defines which signal and call triggers are defined for the state machine, and which attributes and operations are available in activities of the state machine. Signal triggers and triggers for the state machine are defined according to the receptions and operations of this classifier.
>
> As a kind of behavior, a state machine may have an associated behavioral feature (specification) and be the method of this behavioral feature. In this case the state machine specifies the behavior of this behavioral feature. The parameters of the state machine in this case match the parameters of the behavioral feature and provide the means for accessing (within the state machine) the behavioral feature parameters.
>
> A state machine without a context classifier may use triggers that are independent of receptions or operations of a classifier, i.e. either just signal triggers or call

> triggers based upon operation template parameters of the (parameterized) state machine.

From this textual definition of state machine, we can understand roughly at this time (details will be exposed later when studying the State Machine diagram) the semantics of *State* and *StateMachine* as follows:

1. A state machine *can be owned by a behaviored classifier* so we can attach a state machine to any classifier (class, component, interface, etc.) to describe the classifier
2. A state machine *may contain many regions*, so it can represent the state machines of many objects at the same time. A region contains vertices (states, pseudostates) and transitions (links that connect states and pseudostates)
3. A state machine *holds signal triggers and call triggers* (a trigger is something that may fire a state transition). Triggers may come from the outside (receptions)
4. A state machine *is related to some specific attributes and/or operations of a classifier*. Some attributes and/or operations defined in classifiers are parameters of the state machines
5. A state machine *shows many activities*
6. A state machine *is a kind of Behavior*. As *Behavior* derives from *Class* metaclass, so an instantiated state machine is an object
7. A state machine *is associated to a behavioral feature* and it *is the method (operation)* of this behavioral feature
8. We can have *a state machine without a context classifier*. In this case, triggers are not from outside and operations of a classifier are not concerned. This particular case allows a state machine to be established for templates

4.13.2 State Transition

Now, let us focus on details of states and transitions.

> A state models a situation during which some (usually implicit) invariant condition holds. The invariant may represent a static situation such as an object waiting for some external event to occur . . .
>
> State machine can be used to express the behavior of part of a system. Behavior is modeled as a traversal of a graph of state nodes interconnected by one or more joined transition arcs that are triggered by the dispatching of series of event occurrences. During this traversal, the state machine executes a series of activities associated with various elements of the state machine.

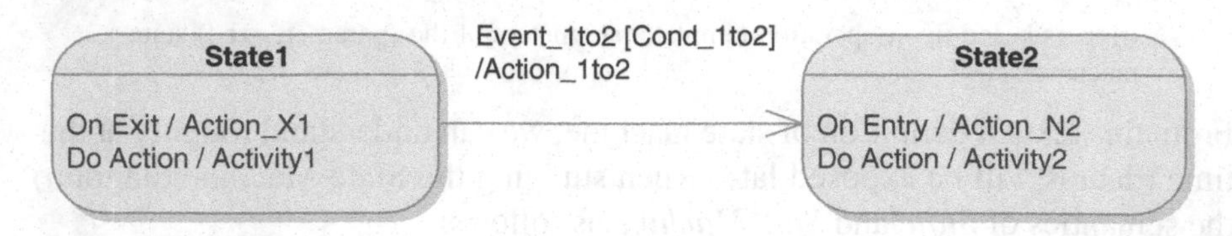

Fig. 4.13.2.1 Abstractions defined in a state and on a transition. Each State has a "Do Action" activity. The state may have an "On Entry" action and/or an "On Exit" action. A transition specification contains an event, a guard followed by an activity expression that can be reduced to a simple action

According to this definition of the UML, a state machine diagram focuses on the event-ordered behavior of an object. It shows all possible states that an object can go during its lifetime, by reacting to discrete events. Actions, activities that accompany any state transition can be represented inside the states or put on the transition if their meaning is bound or closer to the transition than to the states linked by this transition. Moreover, a state transition can be effective only if a guard or guarded condition attached to the transition is verified to be true. Figure 4.13.2.1 explains these subtleties.

In this figure, states are characterized by their corresponding activities. "Do Action" (DoActivity in the standard) points out the main activity executed while in the state. When the object under focus receives an event Event_1to2, if the guard Cond_1to2 is verified to be true, then the object switches from State1 to State2. The object undergoes a state transition and executes Action_1to2 when switching. The object may execute Action_X1 before leaving the State1. When reaching the State2, an action Action_N2 may be required on entry. The Action_1to2 is semantically attached to the transition itself and must not be muddled with the Action_X1 performed On Exit of State1 or Action_N2 performed On Entry of the State2. There is no default timing constraints that specify that Action_X1 must be executed before Action_1to2 or Action_1to2 before Action_N2.

The notion of data is not present in this type of diagram that highlights only states and how an object can change (or be forced to change) its state dynamically. Normally, we must be able to rebuild all possible sequences from a state machine diagram by exploring all possible paths in the graph. The pseudostate "initial" (black and small circle in the Figure 4.13.0.1) can be arbitrarily considered as the root node of the graph. So, a state diagram gives us a "state" view of an algorithm.

If we detail all the actions, activities, events, and conditions, we can build a program with this diagram. As a program is also based on processes or actions, the state view does not give a direct "mapping" to procedural view of a program although we can establish a perfect correspondence. Moreover, this diagram packs several model elements inside two unique concepts, state and transition, so the expressiveness of a state machine diagram for modeling is somewhat limited as it is very compact (the default of the quality).

4.13.3 Trigger and Event

"A trigger specifies an event that may cause the execution of an associated behavior. An event is often ultimately caused by the execution of an action, but need not be." "A trigger is used to define an unnamed event. Syntax for the event is defined by different subclasses of "Event" (call-event, signal-event, any-receive-event, time-event, and change-event) (Fig. 4.13.3.1).

Figure 4.13.3.2 shows event metaclasses.

"A message event specifies the receipt by an object of either a call or a signal." *AnyReceiveEvent* specifies that the system will trigger for all applicable message receive events.

"A signal event represents the receipt of an asynchronous signal." This asynchronous signal comes from some external asynchronous process.

> A call event represents the reception of a request to invoke a specific operation... A change event occurs when a Boolean-valued expression becomes true... A change event is raised implicitly and is not the result of an explicit action.

This change could be related to some change of an attribute value.

"A time event specifies a point in time by an expression. The expression might be absolute or might be relative to some other point in time".

4.13.4 Simple, Composite, and Submachine States

The UML defines three subdivisions of state: simple, composite, and submachine:

1. A *simple state* is a nondecomposable state, does not contain any substate, region, or submachine state machine

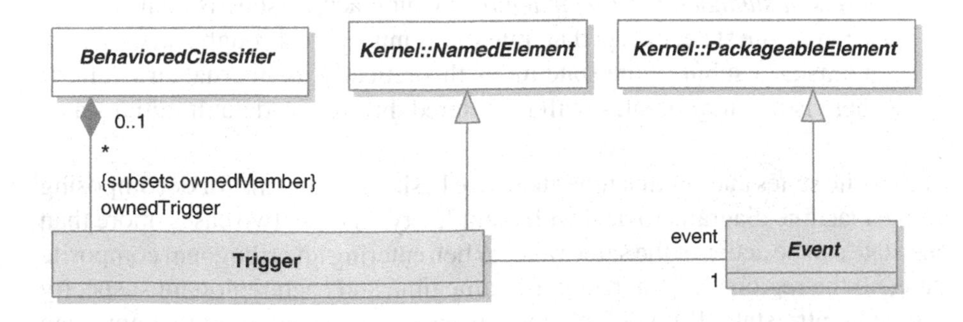

Fig. 4.13.3.1 Trigger and Event metaclasses defines in Communications package

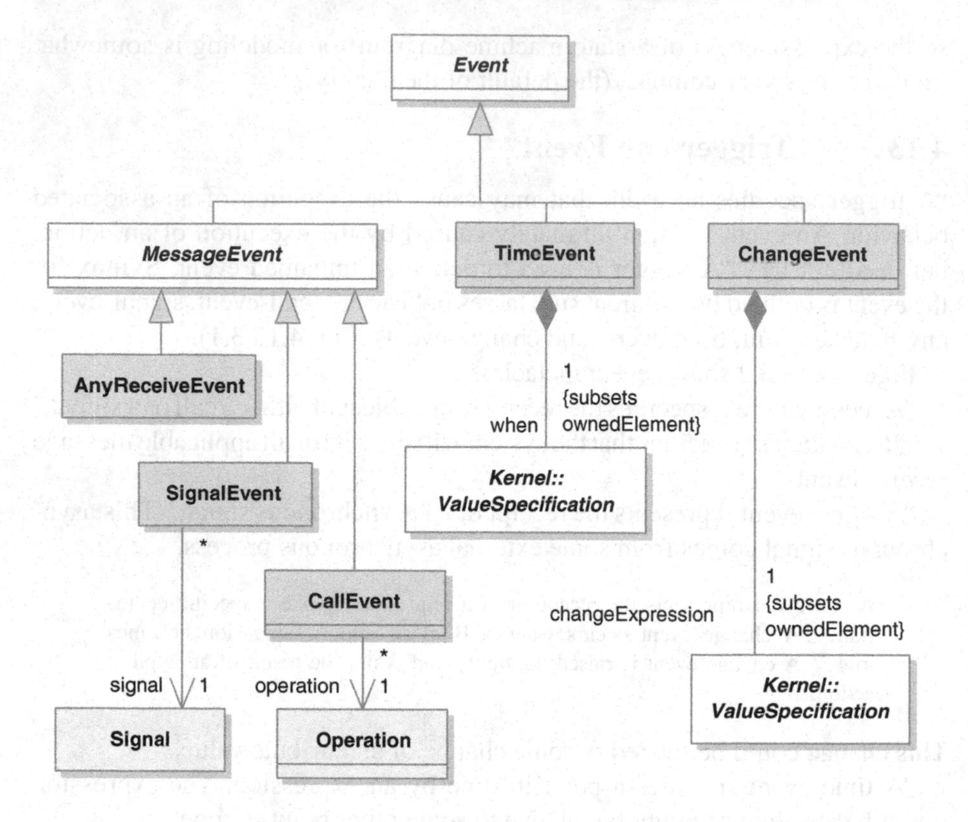

Fig. 4.13.3.2 Subclasses of Event metaclass defined in Communications package

2. A *composite state* is either a simple composite state with just one region or an orthogonal state (with more than one region). States enclosed in a region are called *substates*

3. A *submachine state* specifies the insertion of the specification of a submachine state machine. The containing state machine may have *several levels of submachine state machine*. A submachine state is equivalent to a composite state, except that transitions must pass through entry and exit points as a submachine state hides theoretically its internal structure. If there is no entry point, it will be entered through its default initial state.

Composite states and submachine states are basic mechanisms for decomposing a state machine diagram, to deal with complexity. In these two cases, more than one state can be active at the same time. When entering an orthogonal composite state, all the regions are entered at the same time and each region must specify a default entry state. If no default entry is specified, we suppose that any state could be a candidate as a starting state.

4.13.5 Graphical Notations

Table 4.13.5.1 Modeling concepts used in a State Machine diagram

Name and graphical notation	*Comments*
Simple state	A state is shown as a rectangle with rounded corners. The state name is inside the rectangle or can be on a tag. A state is defined by observing the main activity performed by the object at this moment. The term "Do" is used to identify this main activity. The number of activities or actions is not limited.
State On Entry / Action_N On Exit / Action_X Do Action / Activity State On Entry / Action_N On Exit / Action_X Do Action / Activity	When entering the state, the object may execute an action identified by "Entry". The action performed by the object when exiting the state is identified by "Exit".
State list 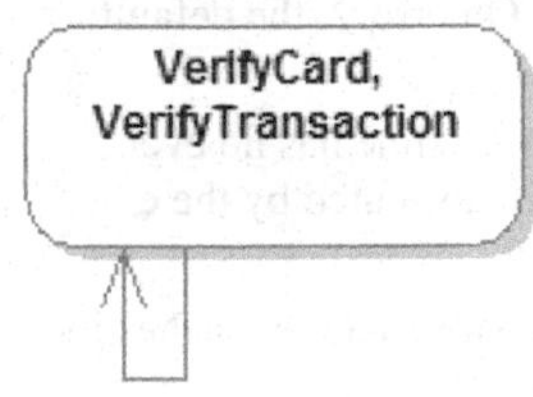	A state may be redefined. A simple state can be redefined or extended to become a composite state by adding regions, vertices, states, entry/exit/do activities, or transitions to inherited regions. A state list (of names) is then used to enclose redefined states into the name. This state is halfway between a black-box representation and a white-box representation and must be made clear as soon as possible.
Composite State 	In some cases, it is convenient to hide the decomposition of a composite state. There may be a large number of states, nested or not. Notice the composite icon in the lower right-hand corner. The "hiding" is purely a matter of graphical convenience and does not bear any semantics for access restriction if compared to the syntax for a class (information hiding or encapsulation).

(*cont.*)

Table 4.13.5.1 (Continued)

Name and graphical notation	*Comments*

Orthogonal Composite State with regions, Initial State, Final State

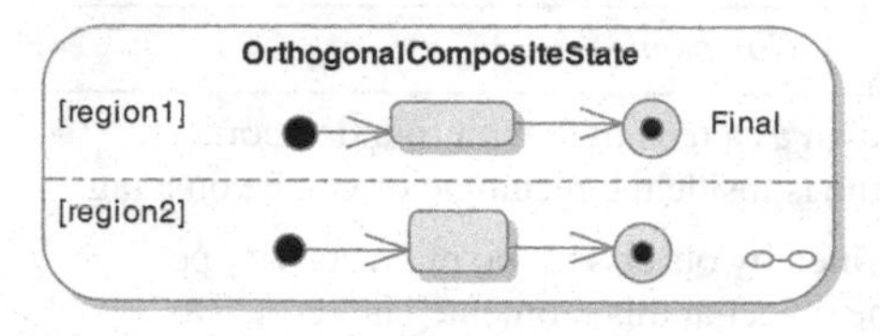

An orthogonal composite state has at least two regions. In the case of one unique region, the state is only "composite" . In each region, separate state machine. If orthogonal (*AND-states* configuration), each region has one active state and all states of all regions contribute to the overall states of the system. In an *Or-states* configuration, only one state of one region is active at any given moment. In this case, state is not orthogonal.

State transitions in different regions can be synchronized but this feature cannot be represented.

An *Initial* state (pseudostate) identifies the first state of a state machine. A *Final* state (not a pseudo state but a state derived from State metaclass) identifies the last state. It is not necessary to name them.

Submachine state, Entry Point, Exit Point

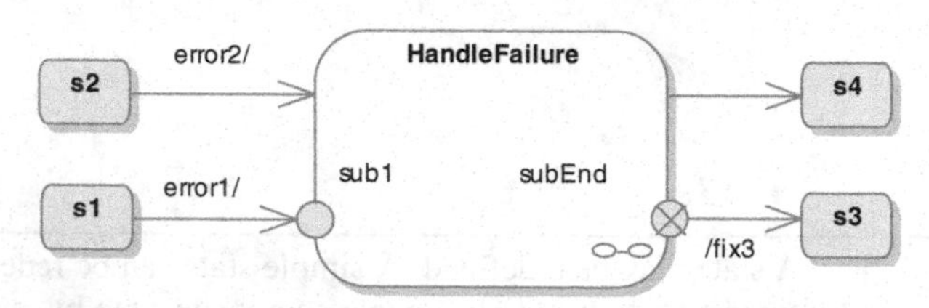

The transition triggered by event "error1" will terminate on entry point "sub1". An Entry Point is a pseudostate attached to the peripheral of a state.

On error2, the default transition inside the HandleFailure submachine will be invoked.

The transition emanating from the subEnd exit point of the submachine has no event, so the machine will execute the fix1 action in addition to what is executed by the exit point subEnd. An Exit Point is a pseudostate.

The unnamed transition leads to s4, so when the HandleFailure submachine reaches its final state, the system exits to s4.

Transition, Trigger, Guard, Effect

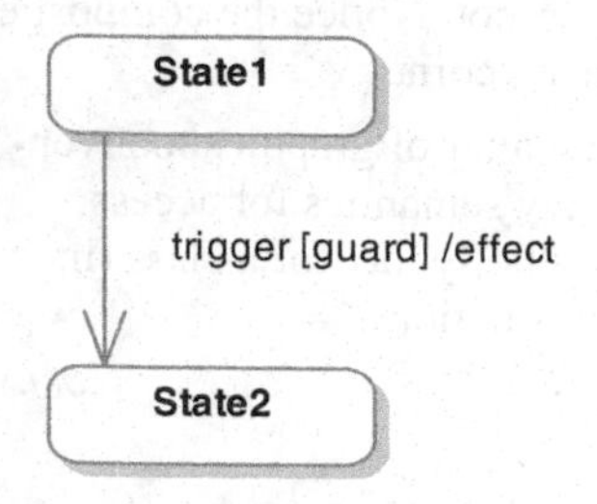

A *transition* is a directed relationship between a source vertex and a target vertex.

A transition contains a *trigger* that may fire the transition. A *guard* is a constraint that provided a way to control the firing of the transition.

The *effect* is a behavior that can be performed when the transition fires. The effect is an action or an activity.

Table 4.13.5.1 (Continued)

Name and graphical notation	*Comments*

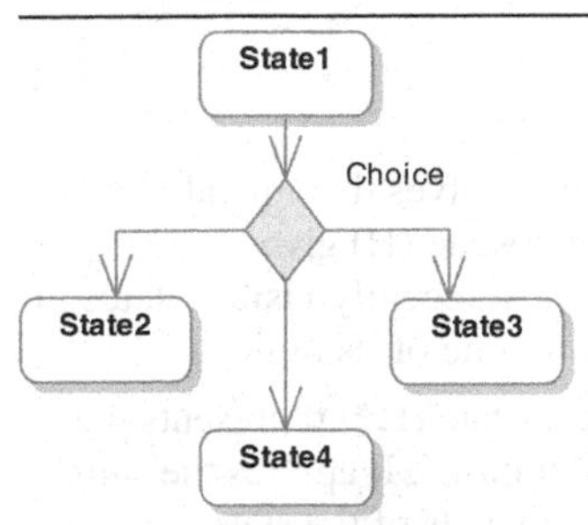

Choice pseudostate

The choice pseudostate is used to compose an IF or CASE transitional paths, where the outgoing transition path is determined by dynamic, run-time conditions. The action performed by the state machine on the path leading to the choice generates the test value. The choice pseudostate is considered as a transient state that will decide on what transition to fire.

Junction pseudostate

A Junction pseudostate is used to design complex transitional paths. A junction can combine or merge multiple paths into a shared transition path like a join pseudostate. Alternatively, a junction can split an incoming path into multiple paths, similar to a fork pseudostate. Each incoming or outgoing transition can be tested against its guard so, if the guard expression is false, the transition is disabled.

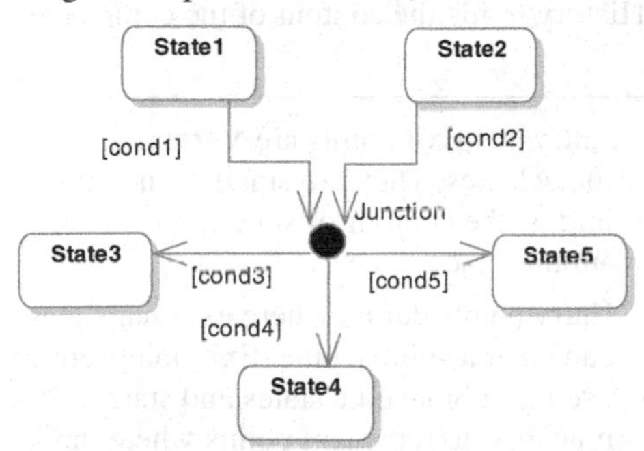

If in the example, cond1, cond2, and cond4 are evaluated to be true, and cond3 and cond5 evaluated to be false and if the system is in State2, it will switch to State4.

Fork and Join pseudostates

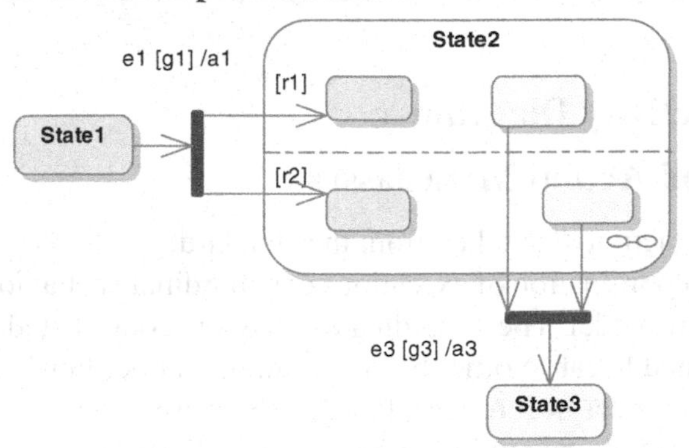

A *fork pseudostate* splits an incoming transition into two or more transitions terminating on orthogonal target vertices (i.e. vertices in different regions of a composite state). The segment outgoing from a fork vertex must not have guard or event (actions/activities are accepted instead).

A *join pseudostate* merges several transitions emanating from source vertices in different orthogonal regions. The transitions entering a join vertex cannot have guards or events.

(*cont.*)

Table 4.13.5.1 (Continued)

Name and graphical notation	*Comments*
shallowHistory and deepHistory pseudostates	
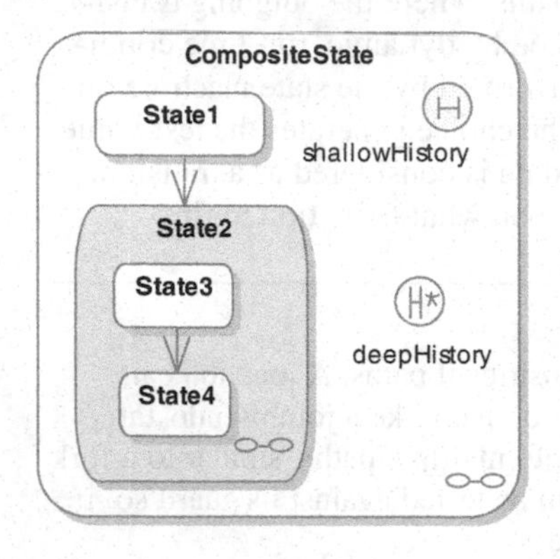	As the State Machine evolves dynamically, a shallowHistory pseudostate (H) gives information on the most recently visited states in a composite state or in one of its region. A deepHistory pseudostate (H*) represents the most recent "configuration". It applies the same rule recursively to all levels of the state hierarchy of the composite state. If the composite state has only one level of nesting, shallowHistory and deepHistory are equivalent. In the example, if State4 is active, shallowHistory maps to State2 and deepHistory maps to "State4 in State2". If we define a deepHistory inside the contour of the composite State2, it will map to State4.
Entry, Exit Points	
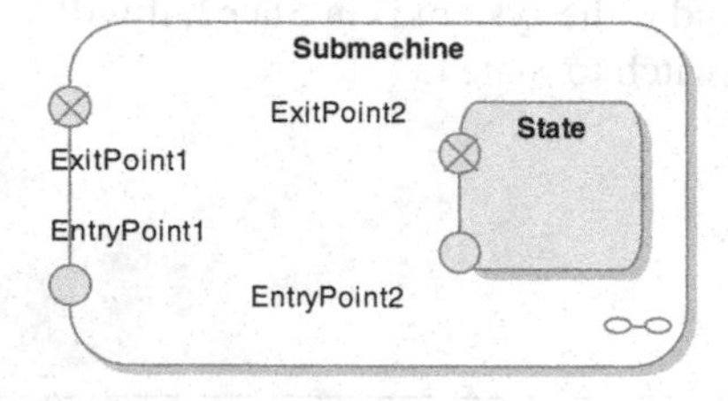	Entry and Exit points are Vertex metaclasses. They are small round circles laid on the contour of states and submachines. Entry points define where external states can enter a submachine. Exit points are used in submachine states and state machines to represent points where the machines will be exited.

4.14 Behavior: Activity Diagram

4.14.1 Activity and Action Metaclasses

Hereafter the definition of "activity" is taken from the standard.

An activity specifies the coordination of executions of subordinate behaviors, using a control and data flow model. The subordinate behaviors coordinated by these models may be initiated because other behaviors in the model finish executing, because objects and data become available, or because events occur external to the flow. The flow of execution is modeled as activity nodes connected by activity edges. A node can be the execution of a subordinate behavior, such as an arithmetic computation, a call to an operation, or manipulation of object contents. Activity nodes also include flow-of-control constructs, such as synchronization, decision, and concurrency control. Activities may form invocation hierarchies invoking other activities, ultimately resolving to individual actions. In an object-oriented model, activities are usually invoked indirectly as methods bound to operations that are directly invoked.

From this definition, rather "impenetrable" for many of us, we can extract some main ideas:

1. An activity involves one or many objects
2. The main reasoning methodology to understand/describe an activity would be a diagram that can highlight the flow of control and data
3. The description of an activity includes that of activity nodes (central centers where data manipulation, computing, world transformation, synchronization, decision control, concurrency control, etc. take place) and that of activity edges (connections between activity nodes, communication links, data flows)
4. An activity is often event driven
5. An activity may invoke many other activities so an activity can be composite and decomposable
6. Activities are finally object methods

There are four versions of *Activity* metaclass defined in four different packages (FundamentalActivities, BasicActivities, StructuredActivities, and CompleteActivities). The package merging hierarchy is shown in Figure 4.14.1.1.

The *Activity* metaclass defined in FundamentalActivities package derives from BasicBehaviors::Behavior. As *Behavior* is a class, an activity is also a class. So, "invoking an activity" is, at the metalevel, as "instantiating an object." In this package, *Activity* may contain activity nodes that are actions. *Action* is not a metaclass *Class* but only a *NamedElement* in the FundamentalActivities package. An activity, refined in BasicActivities package, may have control flows and object flows attached to it. Many incoming and outgoing flows may be connected to an activity.

An action is a named element and a fundamental unit of executable functionality. To take into account the fact that actions need inputs and outputs, the UML defines "pins". A *pin* represents an input to an action (*input pin*) or an output from an action (*output pin*).

"Pin multiplicity controls action execution, not the number of tokens in the pin". To compare an action with a C function, all the input parameters must be present and all control conditions must be satisfied before firing the function. When the function terminates, all the output repositories must be made available to avoid run-time errors.

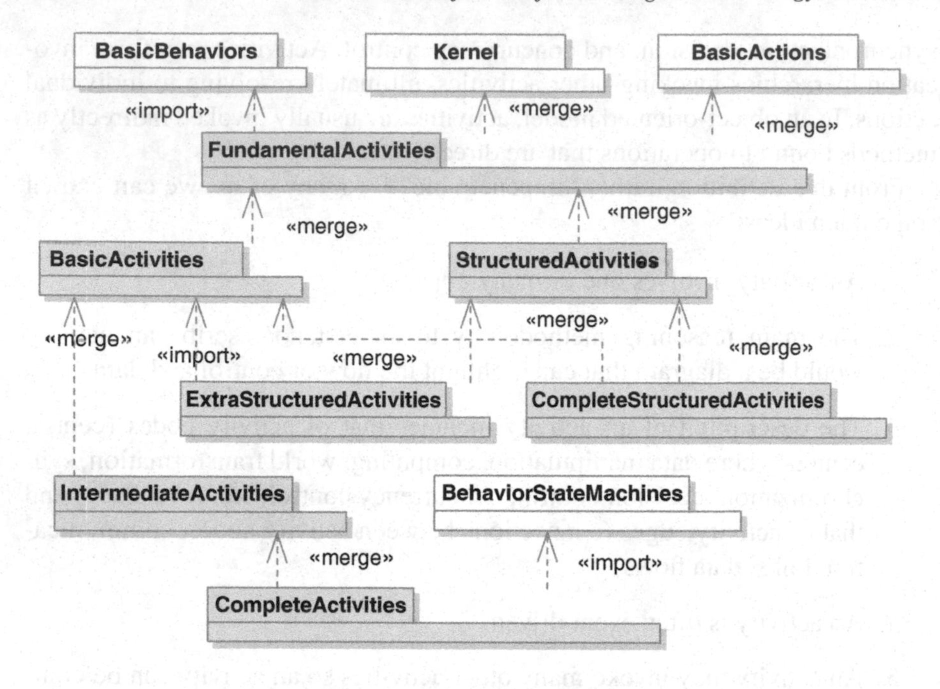

Fig. 4.14.1.1 Activity package dependencies

The UML defines a very long list of nearly 37 metaclasses representing specific actions available by default in the standard, for instance, the UML has actions to create objects, read attributes, set attributes, invoke functions or call procedure, reply to an incoming call, raise exception, etc. Besides those that add new modeling concepts, some predefined actions of this long list are somewhat of minor importance as they make use of the elementary action

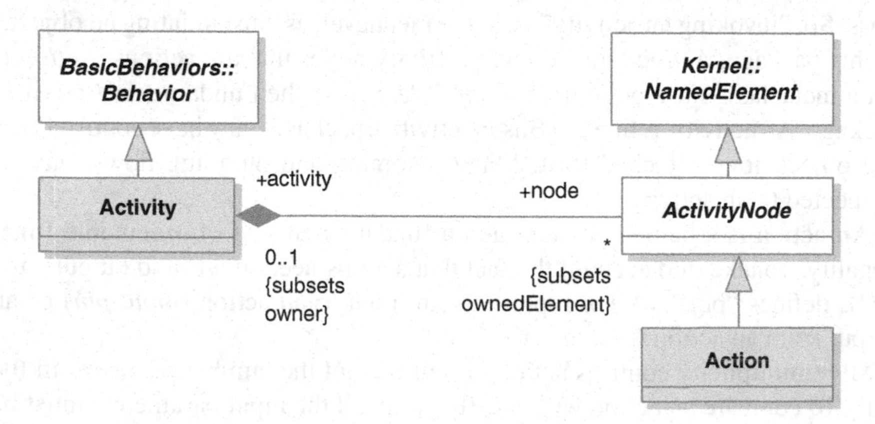

Fig. 4.14.1.2 Activity metaclass defined in FundamentalActivities (Fig. 12.3 Superstructure, Book). An activity, derived from BasicBehaviors::Behavior metaclass may contain nodes that are actions

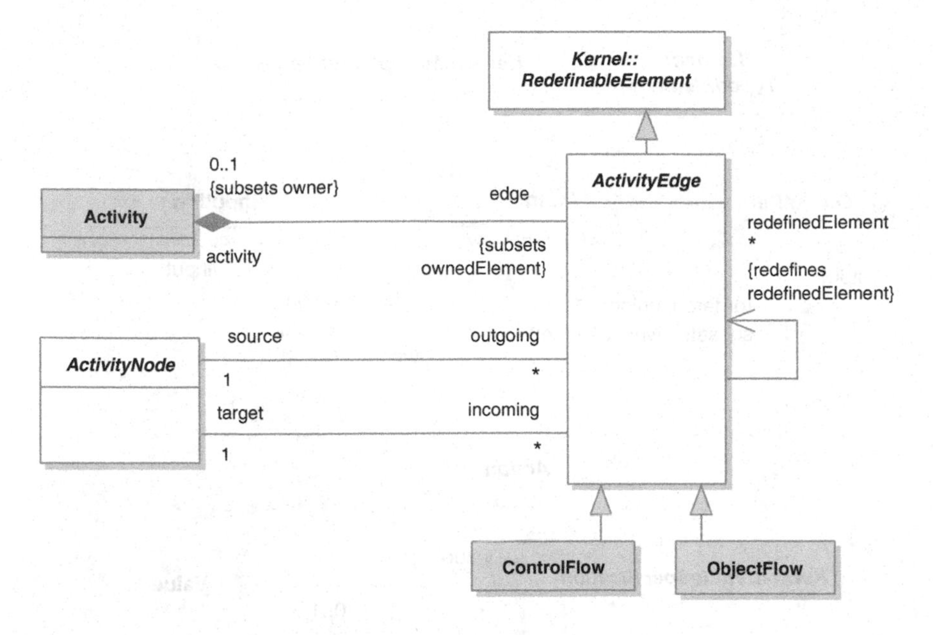

Fig. 4.14.1.3 Activity metaclass refined in BasicActivities (Fig. 12.5 Superstructure, Book)

defined in BasicActions package, delivered with semantic variations, particular to computer programming or specific application domains.

Activity and action are not similar in the metalanguage as *Activity* is a *Class* and *Action* is only a *NamedElement*. For instance, if we said that "an action is an elementary and non decomposable activity," the UML disagrees; it says

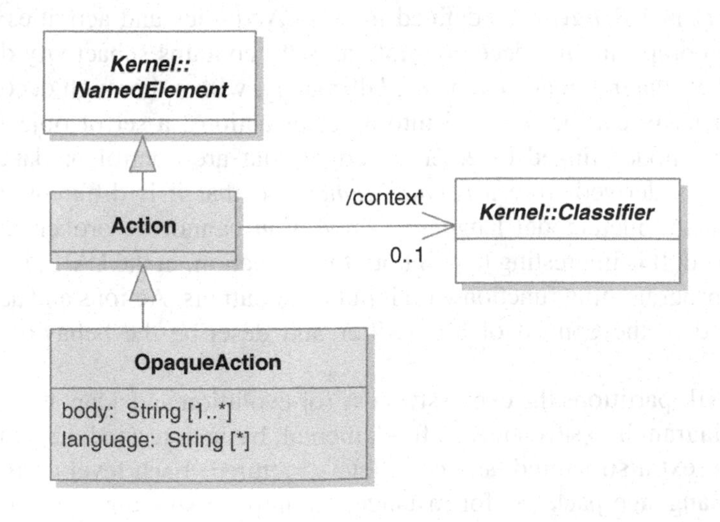

Fig. 4.14.1.4 Action metaclass as defined in BasicActions

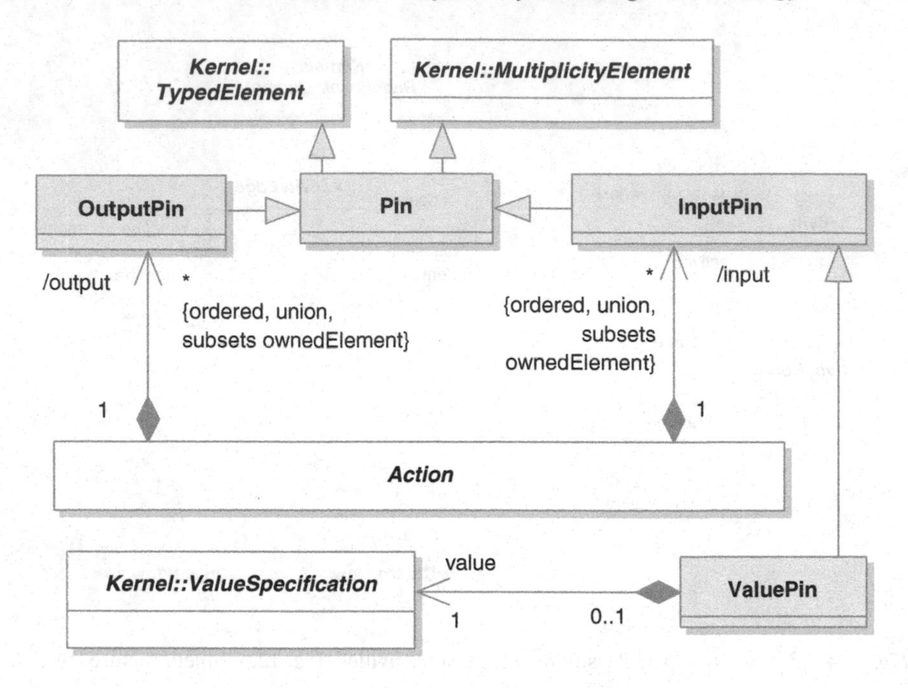

Fig. 4.14.1.5 Pin metaclass defined in BasicActions package

"find a smallest and non decomposable activity and wrap this action into this activity."

To summarize, an elementary activity diagram is a graph containing nodes and edges. Nodes are activities, actions, objects, and control nodes (fork, join, initial, final, etc.) that are constructs for controlling the data and control flow. An *Activity* is a *Behavior* as defined in BasicActivities and activities can be nested. A composite and decomposable activity contains subactivity diagram with all their diagrammatic elements. Ultimately, with a thorough decomposition, an activity can be resolved into a set of actions, a set of object nodes and control nodes linked together by edges that are control or data flows. An *Action* is derived from a *NamedElement* so that it is different from an *Activity* in the metamodel language. An *Action* cannot be broken down or decomposed. It is interesting to point out that an action, in the UML, is considered as a programming function with inputs and outputs. Actions and activities are defined in the context of a classifier and describe the behavior of this classifier.

The UML partitions the expressiveness (or evolution and adaptation) of the activity diagram into seven levels: fundamental, basic, intermediate, complete, structured, extra structured, and complete structured. Each level corresponds to a metalanguage package, for instance, a complete structure gives rise to a CompleteStructuredActivities package (Table 4.14.1.1).

Table 4.14.1.1 Seven levels of Activities definitions. This enumeration shows the complexity and the adaptation of the standard to new definitions without jeopardizing older ones

Level	*Contents*
FundamentalActivities	This level defines only Activity and Action metaclasses. An ActivityGroup is defined as a container for Action
BasicActivities	This level adds ActivityEdge abstract metaclass that defines control flow and data flow. ActivityGroup can now refer to ActivityEdge
	Adds ObjectNode, instance of a particular classifier, available in a particular point of an activity diagram. Abstract metaclass ControlNode is defined at this level to implement ActivityFinalNode and InitialNode
IntermediateActivities	CentralBufferNode metaclass is defined at this level to collect or distribute data and flow tokens
	Specific control nodes (ForkNode, JoinNode, FlowFinalNode, DecisionNode, and MergeNode) enrich those already defined in BasicActivities package
CompleteActivities	Adds extra specification (weight) to edges, constraints to actions. Defines DatastoreNode, ParameterSet, and InterruptibleActivityRegion
	Allows an activity to instantiate separate executions (multiple tokens in Petri net network)
StructuredActivities	Adds classical traditional structured program constructs (sequence, loop, if) to activity nodes
CompleteStructuredActivities	Adds support for "data flow output pins of sequences," ConditionalNode and LoopNode metaclasses
ExtraStructuredActivities	Adds exception handling and invocation of behaviors on sets of values

4.14.2 Activity Diagram

We are now ready to draw the first activity diagram. An UML activity diagram (Fig. 4.14.2.1) emphasizes activities and flows of data/control. Activities are high level and complex behaviors containing algorithms while actions are more elementary, low level units of transformation, signal, or data processing. At first sight, an activity diagram seems to be a special form of state diagram where states are interpreted as activities. If a state diagram seems to be event-triggered, an activity diagram undergoes state changes upon completion of activity executions. The notion "event-triggered" may let an ambiguous impression that if there is no event, then a system cannot progress.

An activity diagram could also be recognized as a classical flowchart as this oldest computer diagram highlights program calls and actions. The main difference arises from the presence of the new concept of object flows (data

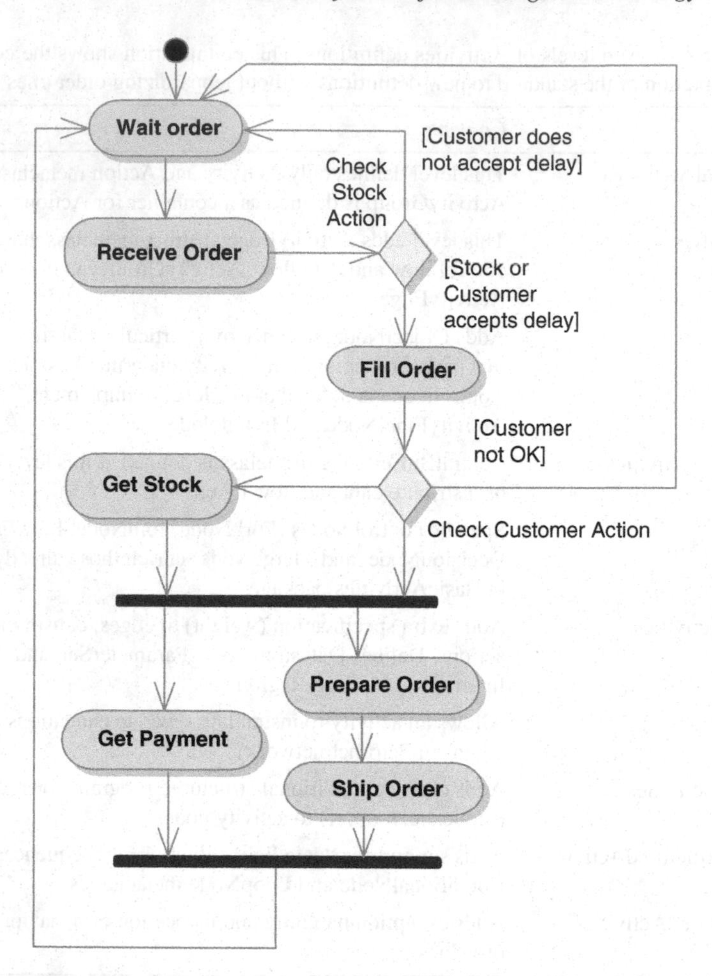

Fig. 4.14.2.1 Example of activity diagram with activities, decision nodes with conditions, fork and join nodes, initial and final activity nodes. Nodes are activities, actions, special control nodes. Nodes are connected by edges that are control flows. This diagram presents several problems if interpreted with Petri net logic (see text for discussion)

circulation) and the apparition of "partitions" that identify all objects responsible of the system dynamics. Therefore, an activity diagram is very versatile as it is simultaneously a flowchart, a state diagram, and a DFD with a support for object orientation.

In Figure 4.14.2.1, a priori correct if considered as a flowchart, has in fact, several problems for someone familiar with Petri net reasoning:

1. As "Get Payment" and "Ship order" need to be synchronized. A token (a Petri marker that traces the evolution of an activity diagram) is not returned to "Wait order" so the system cannot process another order as long as the current payment is not made. This simply means that if we

have only one person who does the job, a new order queued at "Wait Order" must really wait.

2. Another problem is the unclear semantics when a token pass from "Wait order" to "Receive order." In the absence of the representation of a "receive customer order event," we must accept implicitly that the activity "Wait order" expires automatically when a customer arrives.

3. This system is represented with an activity final node when exiting "Get payment" and "Ship order" instead of returning the token to wait order. This final node switches off any current activity and magically cancels all tokens in the network, so we must instantiate a new activity diagram and a new token for each customer. This is also an acceptable interpretation.

In Figure 4.14.2.2, at first sight, data flows and control flows are indiscernible in the diagram as the same graphical notation is used in both cases. When examined closely, an object flow is easily identified because it has an object (or a model element issued from *ObjectNode* metaclass) as node, at least at one side of the connection. For instance, the direct flow between two activities, for

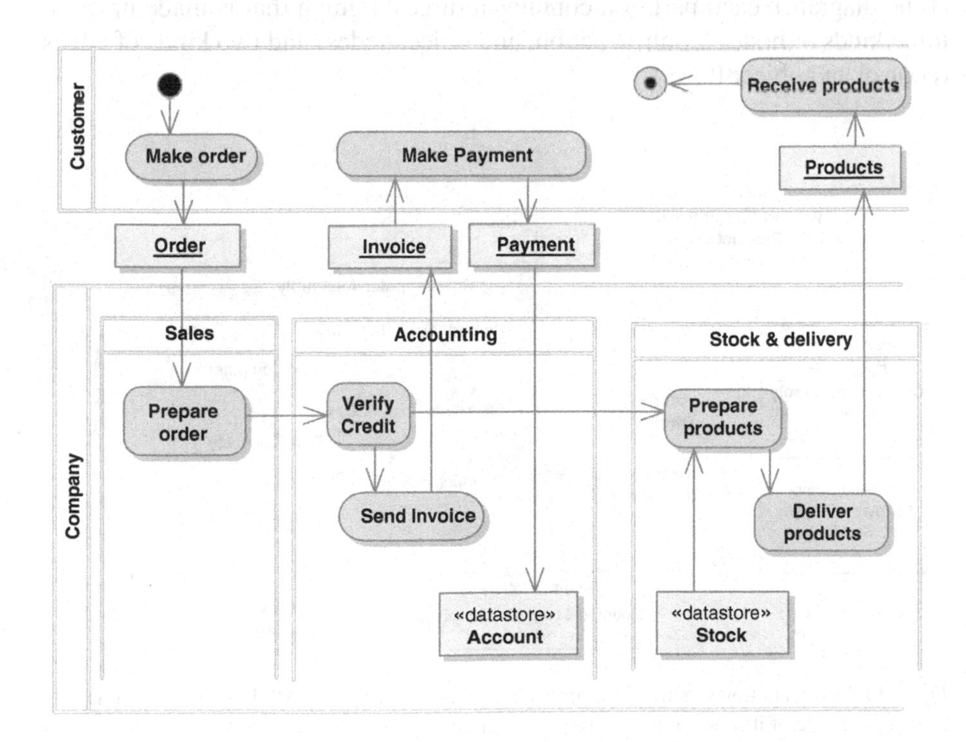

Fig. 4.14.2.2 Activity diagrams with partitions and identification of participating objects

instance "Prepare products" and "Deliver products" is a control flow. The flow connecting "Send invoice" to the object Invoice is an object flow.

4.14.3 Data and Control Flow: Execution of an Activity Diagram

In Figure 4.14.3.1, the *ActivityEdge* metaclass, issued from *RedefinabledElement*, is a common parent of *ControlFlow* and *ObjectFlow* metaclasses. These two flows have exactly the same graphical representation that is a simple arrow. If a *control flow* means simply a transfer of control from an activity/action to another activity/action, an *object flow* connects an activity/action (executable node) to an object or something that derives from an object node, for instance a "pin" (action pin), an activity parameter node, a central buffer node, or a datastore node.

The flow of data within an activity diagram is supported by object nodes asserting the availability of different kinds of data at various points of the execution network. The type of data is defined by the type of the object node (its classifier). To make reasoning about activity diagram, we can consider a general activity diagram as a complex workflow represented graphically with a contour containing many "partitions" (equivalent to "regions" in a state machine diagram), each partition contains a directed graph that is made up from three kinds of nodes (control, action, and object nodes) and two kinds of edges (control and object flows).

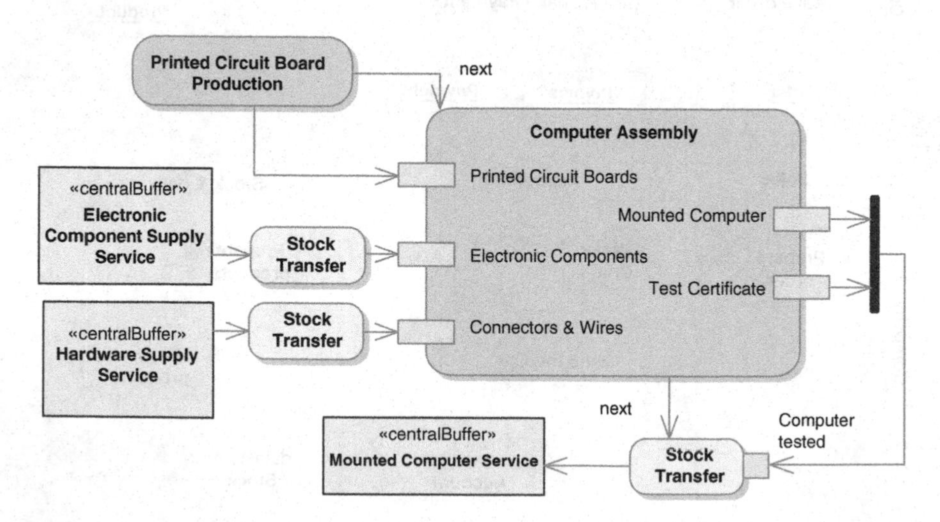

Fig. 4.14.3.1 Data flows, action pins and activity parameter nodes. All data flows must have at least, at one side of its connection, an object or an object node. The two "next" flows are control flows and they connect directly an action/activity node to another action/activity

To interpret the evolution of this graph, the UML version 2 has accepted for the first time the notion of "Petri token," kind of marker that locates where the system is at a given point of time. Tokens represent control, data/objects and could be imagined as "small colored dots" that move among activities/actions. Nodes store tokens and edges give the direction of the token flow. Specific control nodes (they do not have a behavior at their own but adopt a standard interpretation) support the specification of algorithm. *Fork* and *join* support concurrency. *Initial activity*, *activity final state*, and *control final state* decide when and where an activity/action must start or terminate. *Merge node* is like a join node but flows are not synchronized. *Decision node* allows the inscription of conditional flows. Object nodes which serve as the inputs and output of an action is called "pins" whereas object nodes used for activities are "activity parameter nodes."

In Figure 4.14.3.1, "Computer Assembly" and "Printed Circuit Board Production" are activities; three "Stock Transfer" are actions. Datastores, pins used with action, and activity parameter nodes used with activity, are all object nodes. All data flows in this example have at least, at one side of its connection, an object or an object node, for instance, when drawing an object flow from the synchronization fork/join node at the right to "Stock Transfer" action, we need a pin named "Computer tested" because, without this pin, this arrow could be interpreted as a control flow.

According to the interpretation of the movement of Petri tokens, for the two outputs at the right side, we must receive two tokens, the "Mounted Computer" and the "Test Certificate" fully filled before transferring this computer to its datastore. In programming at low level, generally, we need the presence of all inputs before being able to start the execution of a function. For a modeling concept as an activity, several data can arrive at the activities and the absence of an input bar means that this constraint is not activated. So, if a global rule is present to alleviate the representation, it is interesting to know when this global rule does not apply. Global rule like: "an action cannot execute when all input parameters are not present" and "output pins must be all filled for an action to terminate unless otherwise constrained" say that the join bar in Figure 4.14.3.1 is not absolutely necessary.

Both control and data tokens control the execution of an activity diagram. Theoretically, control flows and data flows are two independent networks but, if all activities are sequential and if there is no risk of misunderstanding, we can consider that data tokens carry and simultaneously control information, so it is not necessary to superpose a control network over a data network. In more complex systems with parallel or iterative processes, it may be necessary to draw two separate networks, on two diagrams or on a single diagram. The flow of a token is regulated by three constraints: the constraint at the source node, the constraint at the destination node, and another constraint imposed possibly on the edge itself. If these constraints are satisfied at the same time and if a token is present at the source node, the token will be propelled to the destination node, waiting for the next move. The notion of "weight" has been added in the CompleteActivities package and required that an edge flagged with N tokens must collect N tokens at the source node to be able to cross the edge. A weight is slightly different from a guard since a weight imposes a special rule on the number of tokens available at the source node and a guard is a more common condition attached to state transitions or control/object flows.

4.14.4 Object Nodes and Storage of Tokens

In Figure 4.14.4.1, the abstract *ActivityNode* metaclass is defined as a *NamedElement* and *ActivityNode* was a parent of the metaclass *Action*. In BasicActivities package (as shown in the Figure 4.14.4.2), the abstract metaclass ActivityNode has been redefined to make two metaclasses *ObjectNode* and *ControlNode* necessary to implement specialized object nodes like a *Pin* and other control nodes.

A central buffer node and a datastore node are repositories for persistent data. They are identified with their corresponding stereotypes. A datastore may represent a database as asserted in the standard:

> A data store keeps all tokens that enter it, copying them when they are chosen to move downstream. Incoming tokens containing a particular object replace any tokens in the object node containing that object.

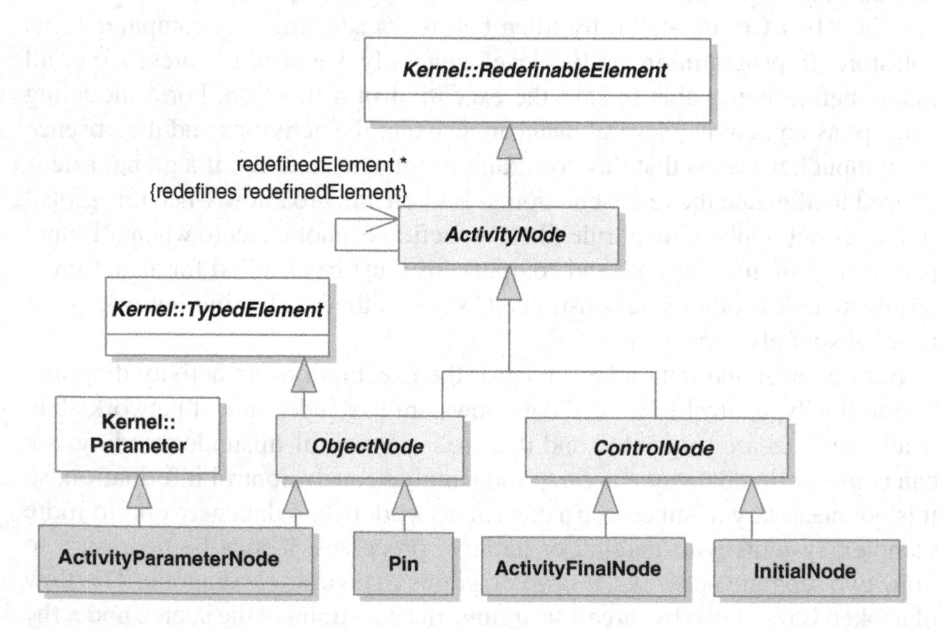

Fig. 4.14.4.1 ObjectNode and ControlNode metaclasses defined in BasicActivities package (Fig. 12.4 Superstructure Book)

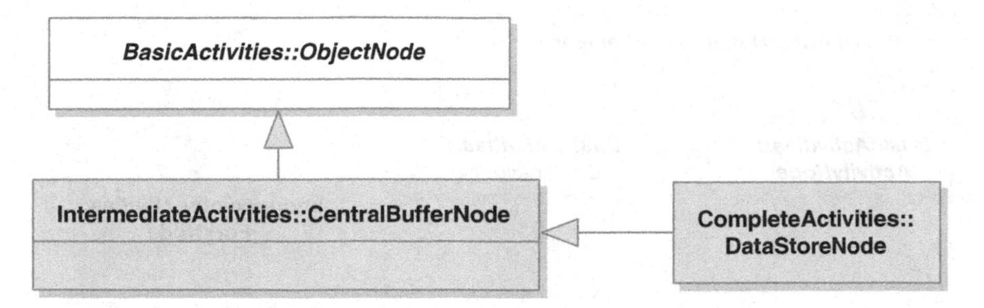

Fig. 4.14.4.2 CentralBufferNode defined in IntermediateActivities package and DataStoreNode defined in CompleteActivities package. They are ObjectNodes used to store tokens or persistent or temporary persistent data

The first part of this sentence refers to the persistency of all information entering a datastore. The second part refers to the data update process. Every object node must normally have an upper bound attribute that specifies the maximum number of tokens that it may contain.

> A central buffer node accepts tokens from upstream object nodes and passes them along to downstream object nodes. They act as a buffer for multiple in flows and out flows from other object nodes.

This definition of central buffer node prescribes its role as a storage space for in/out operations like a queue.

4.14.5 Control Nodes

Control nodes comprise initial, activity final node, flow final node, fork, join, decision, and merge. They rule the way tokens must circulate inside the control and data networks. Tokens cannot rest at control nodes (except for InitialNode). If data tokens must be collected at the end, a central buffer node or a datastore can be drawn for this purpose. Hereafter is the hierarchy of control node metaclasses with their corresponding packages (Fig. 4.14.5.1) as defined in the standard.

An *initial node* is the starting point for executing an activity diagram. The general rule is that tokens cannot rest at control nodes.

> Initial nodes are an exception to the rule that controls nodes cannot hold tokens if they are blocked from moving downstream, for example by guards. This is equivalent to interposing a CentralBufferNode between the initial nodes and its outgoing edges . . . Tokens in an initial node are offered to all outgoing edges . . . If an activity has more than one initial node, then invoking the activity starts multiple flows, one at each initial node.

A *flow final node*, represented with a circle with an X inside, terminates an individual flow and destroys all tokens that arrive at it. An *activity final node*, represented as a solid circle inside a hollow circle, stops all flows in an activity diagram. An activity diagram may have more than one activity final node. The

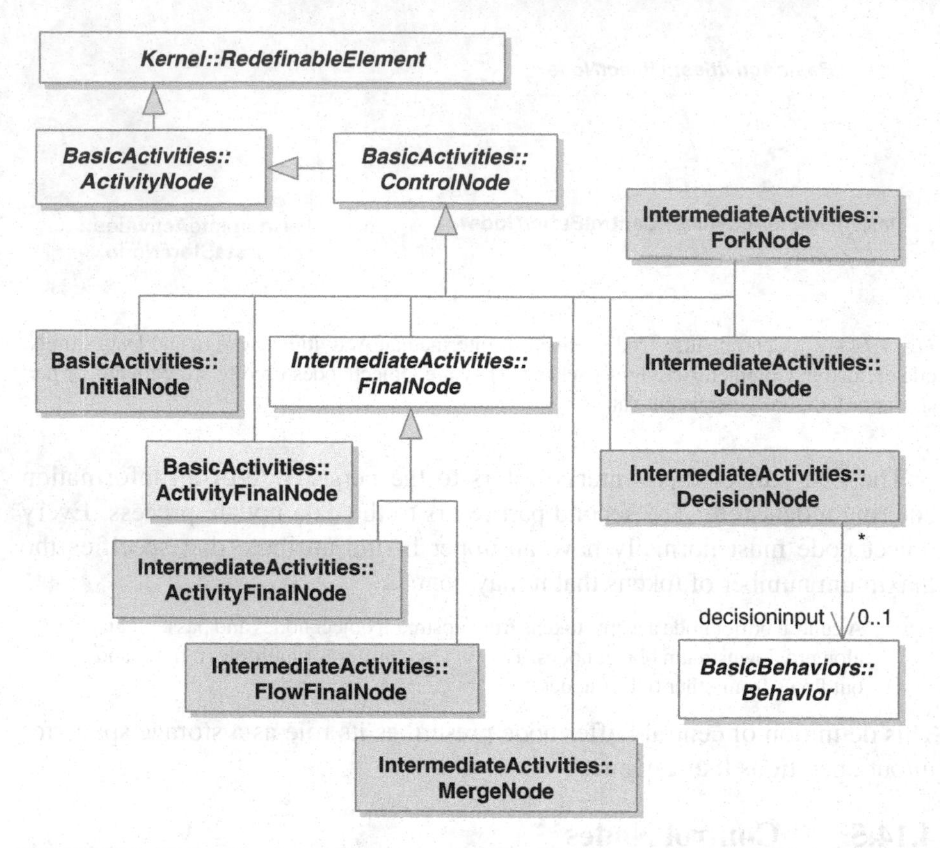

Fig. 4.14.5.1 Control nodes and their corresponding packages. ActivityFinalNode defined in BasicActivities is enriched in IntermediateActivities

first one stops all flows in the activity. If it is not desired to abort all flows, use a flow final node instead. As for the tokens, the standard "destroys all tokens in object nodes, except in the output activity parameter nodes." Tokens circulating in action pins or input activity parameter nodes are supposed to be destroyed as well. As for the tokens in datastores and central buffer nodes, it would be advisable to represent them outside the activity if tokens must remain and be used later.

A *fork node* has an incoming edge and many outgoing edges. A *join node* has several incoming edges and one outgoing edges. They are applied to both control and data flows, and the Standard allows the mixing of control and data flows on forks and joins. Moreover, fork and join nodes can be combined to have N incoming edges and M outgoing edges. A token coming to a fork edge is duplicated and passed to all outgoing edges with guards that accept tokens. For a join node, if all the tokens offered on the incoming edges are control tokens, then all incoming tokens are combined to offer one control token to the outgoing edge. So, there is a logical "AND" condition imposed on edges arriving at a join control node. For data, tokens,

> [I]f some of the tokens offered on the incoming edges are control tokens and other are data tokens, then only the data tokens are offered on the outgoing edge. Tokens are offered on the outgoing edge in the same order they were offered to the join.

The latter rule does not impose any "AND" condition for data tokens and just use the join node as a common collection path for them.

A *merge node* brings together multiple alternate flows. It is not used to synchronize concurrent flows but to redirect tokens to a common outgoing edge. A join needs a presence of all tokens available at incoming edges and a fusion of all tokens into one unique token, but not a merge node. A *decision node* chooses one among several outgoing flows. Tokens are not duplicated. Most commonly, guards on the outgoing edges are evaluated to determine which edge should be traversed. A decision node is normally used to make an IF or CASE conditions. A condition put on a decision node must be compatible with guards put on outgoing edges. The best attitude would be simply avoiding this kind of redundancy.

4.14.6 Exception Handling: Partition

To manage exceptions, the UML defines interruptible activity regions and exception handlers. In Figure 4.14.6.1, an *Activity* contains many ActivityGroups and an *ActivityGroup* is defined as an abstract metaclass that references many ActivityNodes. In BasicActivities package (Fig. 12.3, Superstructure Book), an *ActivityGroup* may reference many ActivityEdges. So, an activity group is an abstract metaclass that can group many activity nodes and activity edges inside one unique model element. In IntermediateActivities (Fig. 12.10, Superstructure Book), an *ActivityPartition* is derived from *ActivityGroup* and a *NamedElement* and this partition may reference many activity nodes and edges. So, there is a close similarity between an *ActivityGroup* and an *ActivityPartition*. Later, in CompleteActivities package, an *InterruptibleActivityRegion* is derived from *ActivityGroup*. In StructuredActivities package, an *ExceptionHandler* is derived from an *ExecutableNode* that is itself an *ActivityNode*. To summarize this complex hierarchy of metaclass definition, we represent hereafter a reduced and oversimplified version of this metaclass hierarchy in order to catch the true nature of interruptible regions, exception handlers, and partitions.

Exceptions in an activity diagram are processed through exception handlers. In current programming context, exception processing consists of two phases: throwing an exception at some point in the code inside an "interruptible activity region" or the "protected node," handling the exception by an "exception handler node." An example of representation is shown in the following table.

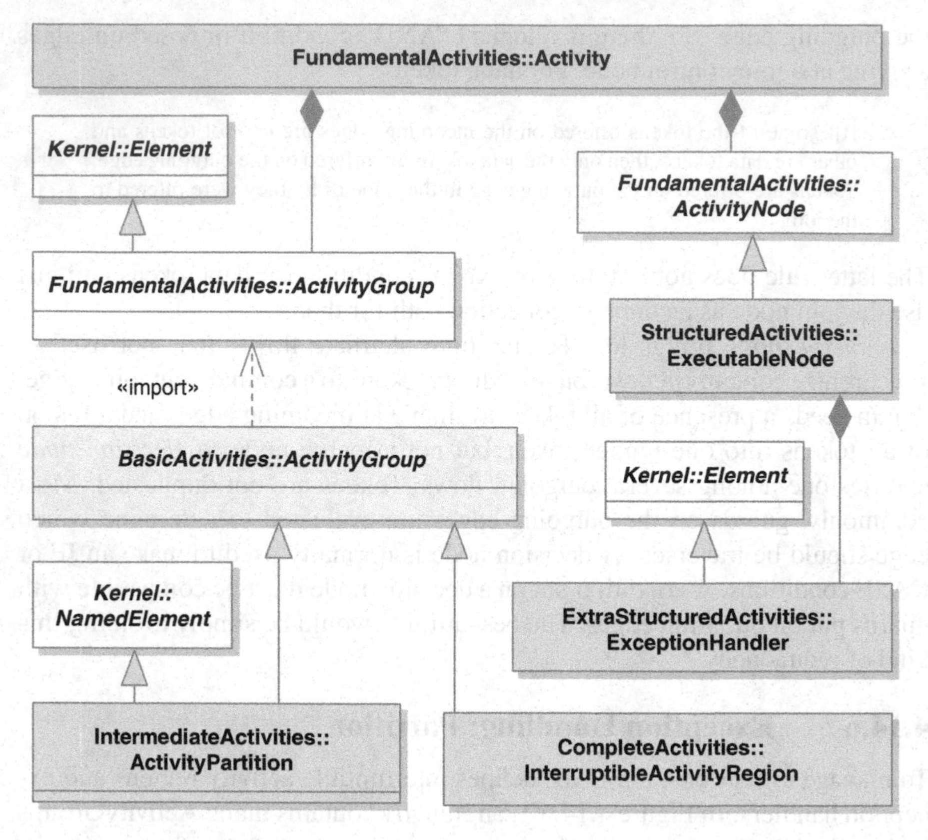

Fig. 4.14.6.1 InterruptibleActivityRegion, ExceptionHandler, and ActivityPartition metaclasses. An ActivityGroup is a generic grouping for activity nodes and edges. An InterruptibleActivityRegion derives from ActivityGroup and an ExceptionHandler derives from an ExecutableNode that is an ActivityNode

4.14.7 Graphical Notations

Table 4.14.7.1 Modeling concepts used in an Activity Diagram

Name and graphical notation	*Comments*
Activity, Action, (Event)	
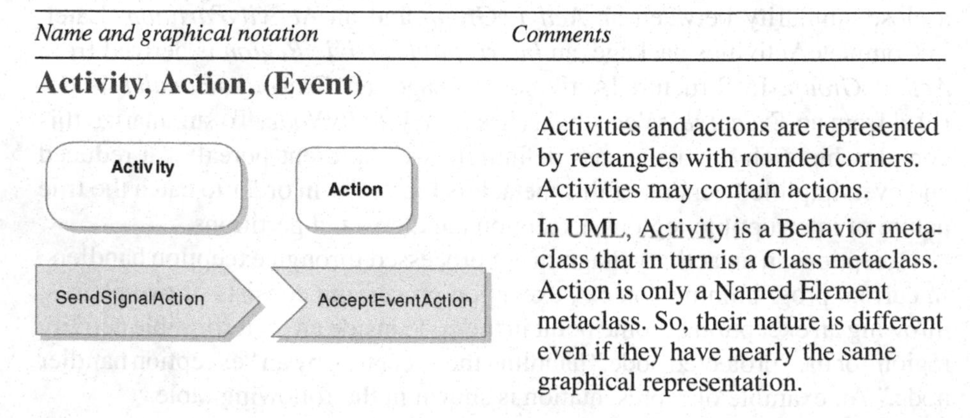	Activities and actions are represented by rectangles with rounded corners. Activities may contain actions. In UML, Activity is a Behavior metaclass that in turn is a Class metaclass. Action is only a Named Element metaclass. So, their nature is different even if they have nearly the same graphical representation.

Table 4.14.7.1 (Continued)

Name and graphical notation	*Comments*

An action is considered in UML as equivalent to an elementary function in programming language. An Action may contain other actions. Around 37 actions have been defined. Graphically, they differ only by their names. Special graphic representations are used to distinguish an action that accepts an event (AcceptEventAction) and an action that sends a signal at its output (SendSignalAction).

Another classification divides actions in three kinds: *Invocation actions* performing operation calls, sending signals and accepting events; *Read or write actions* for accessing objects values, modifying them if necessary; *Computation actions* transforming inputs values into output values. An action may be reentrant or not. If it is a reentrant, multiple copies of an action may be created at the same time.

The notion of *event* is more evident in the Interaction Suite (abstract *Event* metaclass and their derivatives) or in a State Machine (with *Trigger* metaclass), less in the Activity Diagram. *AcceptEventAction* is the way used to create an interface to accept events from the outside, and *SendSignalAction* to create events.

Object node, Pin

ObjectNode is a metaclass that derives from ActivityNode metaclass. Object Nodes are used for specifying inputs and outputs of activities, Pins are used to specify inputs and outputs of actions.

An activity object node or an action pin is represented by a small square located on the contour of an activity/action.

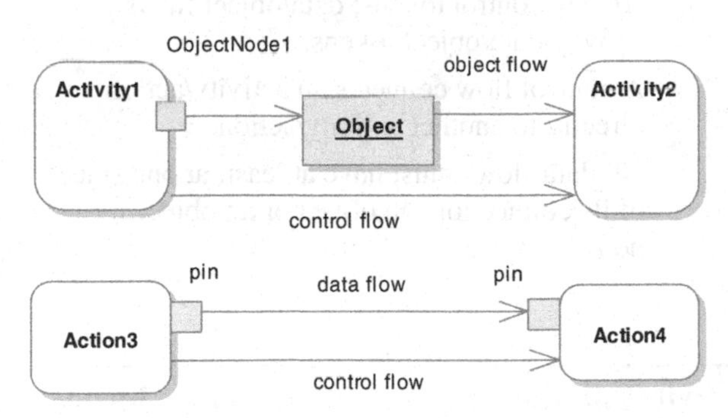

Object nodes are introduced to support the flow of objects, general inputs and outputs for activities/actions. In the first figure, Object is connected to Activity1 through an object node located on the contour of this activity. The connection is then identified as an object flow with object nodes at both sides. The flow connecting Object directly to Activity2 is also an object flow as its left side is an object.

In the second figure, pins are used instead of object nodes to support object flows between Action3 and Action4. A connection by data flow does not mean that the control will be transferred necessarily from Action3 to Action4; it means that Action4 needs data from Action3. If the context is not clear, a control flow is then necessary to specify both data and control flows.

(cont.)

Table 4.14.7.1 (Continued)

Name and graphical notation	*Comments*

An activity parameter node (ActivityParameterNode metaclass) is an object node for inputs and outputs to Activity. It is represented graphically as three object nodes side by side (same graphical notation as ExpansionNode).

Central Buffer Node (ObjectNode), Data Store Node (ObjectNode)

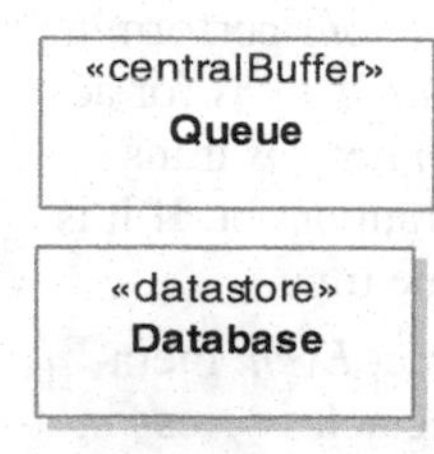

Central Buffer nodes carry the stereotype <<central buffer>>. They collect object tokens for multiple "in flows" and "out flows" from other object nodes. *"They add support for queuing and competition between flowing objects."* Buffering is useful when the inputs from multiple actions must be collected at the central place or when multiple actions extract objects from a common source.

A *datastore* is an element used to define permanently stored data. A token of data that enters into a datastore is stored permanently; updating token for that data that already exists. A token of data that comes out of a datastore is a copy of the original data. A datastore models a database in an activity diagram.

Flow, Control flow, Data/Object flow

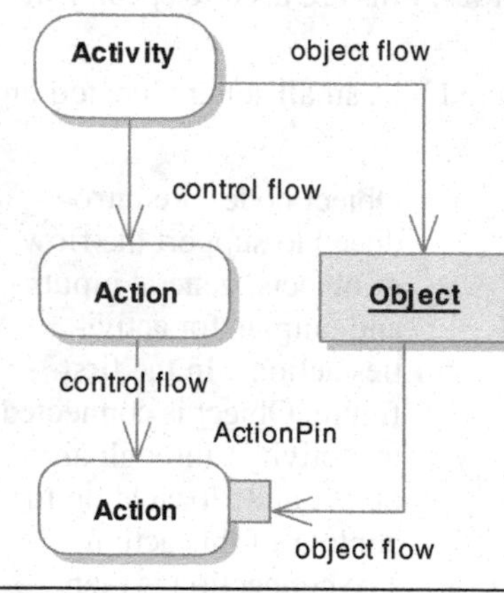

The arcs connecting nodes are called flows in an activity diagram. A control flow connects two actions/activities together. Control flows convey control tokens; data/object flows convey data/object tokens.

A control flow connects an activity/action directly to another activity/action.

All data flows must have at least, at one side of its connection, an object or an object node."

Control nodes: Initial, ActivityFinal, FlowFinal, Fork, Join, Merge, Decision

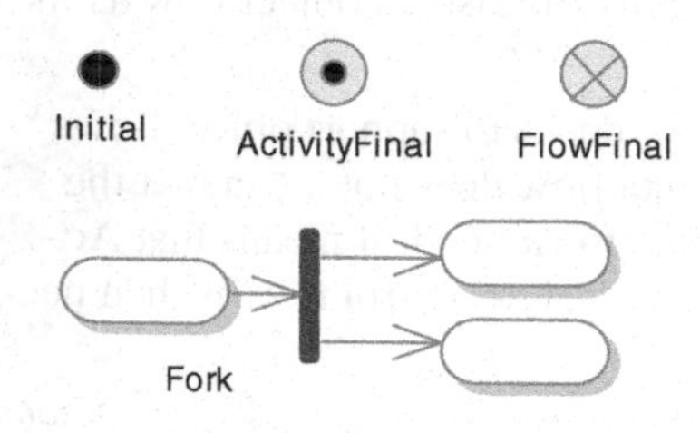

An initial node is the starting point for executing an activity diagram.

An activity final node, represented as a solid circle inside a hollow circle, stops all flows in an activity diagram.

A flow final node, represented with a circle with an X inside, terminates an individual flow and destroys all tokens that arrive at it.

Table 4.14.7.1 (Continued)

Name and graphical notation	*Comments* 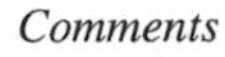
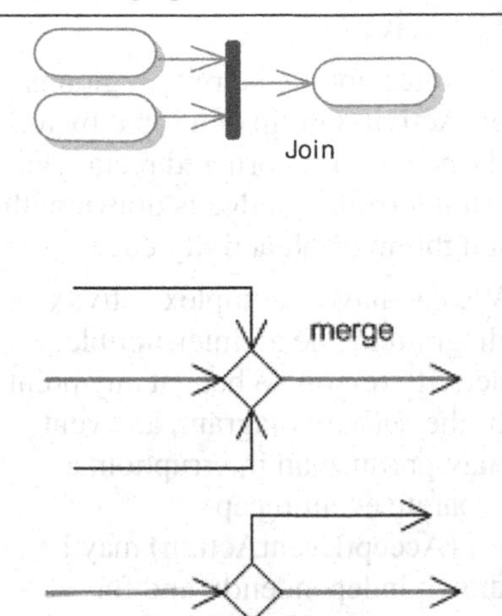	A fork node has an incoming edge and many outgoing edges. A join node has several incoming edges and one outgoing edges. They are applied to both control and data flows. UML allows the mixing of control and data flows on forks and joins. Moreover, fork and join nodes can be combined to have N incoming edges and M outgoing edges. A merge node brings together multiple alternate flows. It is not used to synchronize concurrent flows but to redirect tokens to a common outgoing edge. A join needs a presence of all tokens available at incoming edges and a fusion of all tokens into one unique token, but not a merge node.
A decision node chooses one among several outgoing flows.	

Expansion region, expansion node

An expansion region is shown as a dashed rounded box with one of the keywords *parallel, iterative*, or *stream* in the upper left corner.

An expansion region is a structured node that takes collections as inputs and produces results to output collections.

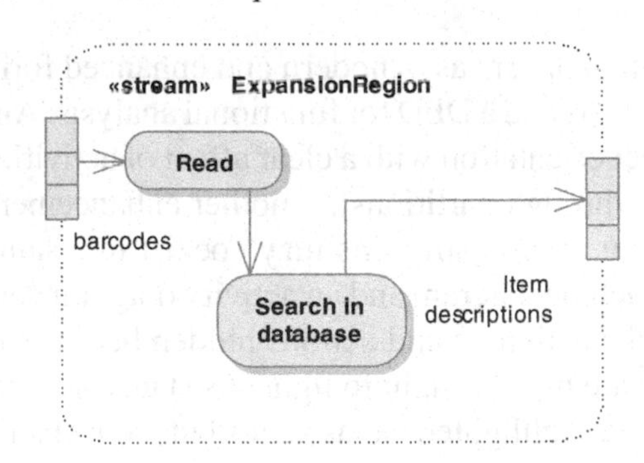

For *parallel*, the execution in the region progresses in parallel. For *iterative*, a subsequent iteration starts when a previous one is completed. For *stream*, we have only a single execution. Values placed in the collection of the input expansion node are extracted and processed. Output values are stored as collection in the output expansion node.

The expansion node is a particular case of an object node. A small rectangle with vertical bars suggests that data/objects form al list of elements (not limited to 3). If there is only one object/data, a pin is used instead.

(cont.)

Table 4.14.7.1 (Continued)

Name and graphical notation	*Comments*

InterruptibleActivityRegion, ExceptionHandler

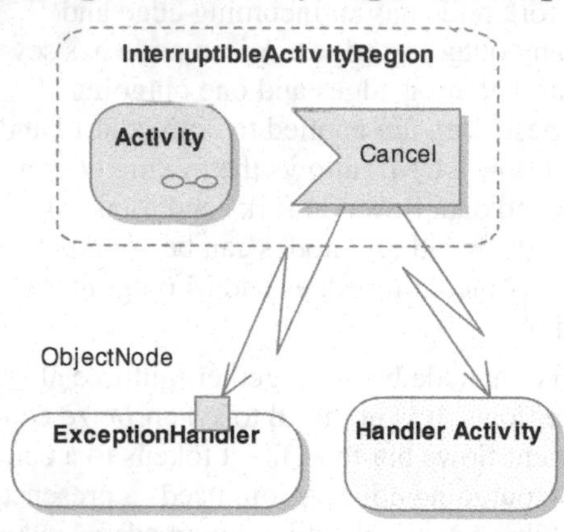

An interruptible activity region is an ActivityGroup delimited by a dashed, round-cornered rectangle. An interrupting edge is drawn with a lightning-bolt activity edge.

We can have a complex activity diagram inside an interruptible activity region. When, at any point of the activity diagram, an event may produce an interruption, a separate event receptor (AcceptEventAction) may be drawn independently and forwarded to an ExceptionHandler through an object node positioned in the contour of the handler.

ActivityPartition

An activity partition is a kind of activity group. Partitions can be used to represent activities of many synchronizing objects.

To summarize, an activity diagram appears as a modern and enhanced form of both a flowchart for workflow analysis and a DFD for functional analysis. Another important character is its object orientation with a clear affect of activities and actions on responsible objects (through partitions). Another enhancement is the explicit support for parallelism (many partitions may coexist in a same diagram). Fundamentally, a state machine diagram and an activity diagram deal with the same abstractions but all the activities and actions hidden behind the concept of states in a state machine are now brought to light. As states are intimately related to activities, if they are highlighted, so as states but, conversely, at this time, states are "behind the scene" and activities are more visible abstractions. Transitions in a state machine are replaced by control flows. An activity diagram must be compatible with a state machine diagram and the transformation could be smoothly done between these two important diagrams. The superposition of two networks, control flow network and data flow network, the adoption of the Petri markers (tokens), new adds (control nodes, object nodes, expansion regions, partitions, exception handlers) enrich the recent version 2 and transform the activity diagram into the topmost versatile dynamic diagram of the behavior suite.

4.15 Behavior: Interaction Suite

The Interaction Suite of the UML comprises four diagrams:

1. *Sequence diagram* (known in the past as "scenario" or "interaction diagram")
2. *Communication diagram* (known in the past as "collaboration diagram")
3. *Interaction overview diagram*
4. *Timing diagram*

To avoid confusion, older terminology like "scenario," "interaction diagram," and "collaboration" will not be used through our text to qualify diagrams; these terms are returned to the general vocabulary. All these diagrams describe interactions between instances of classifiers.

Interactions describe message exchanges between objects. In a sequence diagram, objects are represented with lifelines and the succession of messages on a time axis is highlighted. The communication diagram is an exact replica of the sequence diagram but the focus is put on the message exchange network and the objects that participate into the interaction. We can still retrieve the sequencing order of messages through a sequence numbering scheme."Interaction overview diagrams use activity diagrams notation where the nodes are either Interactions or InteractionUses."A timing diagram focuses attention on time of occurrence of events causing changes to an object or an interaction along a linear time axis. The following examples give an overview of these diagrams and their associated graphical constituents before exploring their metadefinitions.

4.15.1 Example of Sequence Diagram

The sequence of Figure 4.15.1.1 represents the interaction between various objects of a *Car* and the *Driver*. This diagram shows the main constituents of a sequence: *lifelines* and *messages*. Lifelines represent instances of classifiers (in most situations, objects). Lifelines have a vertical time axis and they are connected horizontally by messages, or calls (a call is a special form of message).

Each lifeline represents an object or an instance of classifier. A name may be added to the class Driver (for instance *John:Driver* but if the name is not pertinent to the current context and a generic *:Driver* is used instead (in our case,

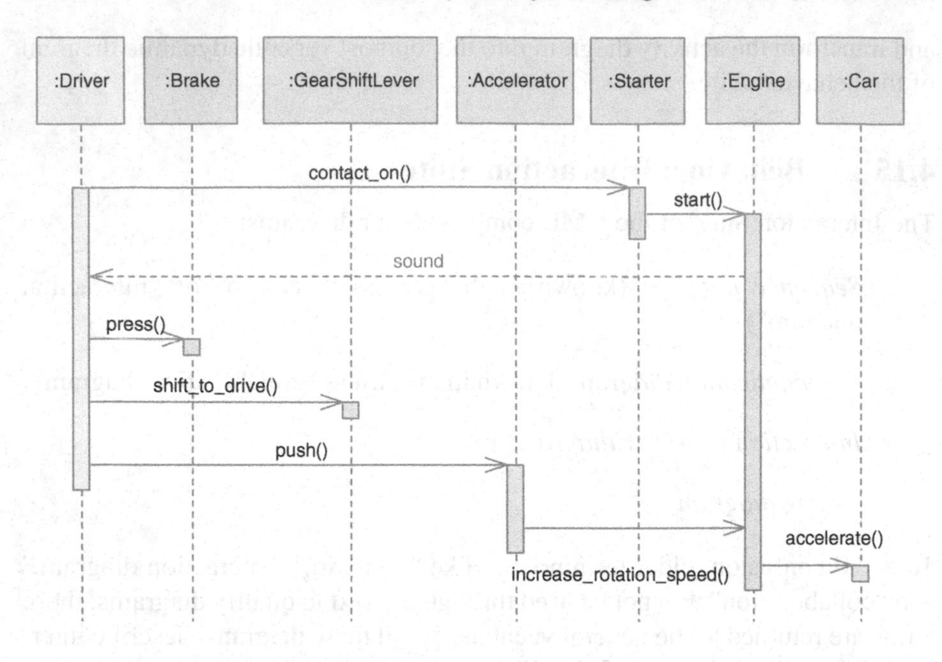

Fig. 4.15.1.1 Main constituents of a sequence diagram: lifelines and messages

all drivers will have the same reaction; only the name of the classifier is used to identify the lifeline). No measurement can be made on the lifeline and the time is not proportional to the axis length. Only the order of event occurrences is meaningful. To depict method activations, a thin and colored rectangle covers the corresponding lifeline.

For the driver, we have two phases, first the starting process then the acceleration phase and these two phases are represented by two different thin-colored rectangles over the lifeline. For the engine, it starts rotating after the starter. The car accelerates lately in the process so its lifeline starts only to the end of the time axis. We have the possibility to create lifeline "on the fly" without attaching it to any classifier as *Lifeline* derives from just a *NamedElement*. In the metalanguage, *Lifeline* could be compared to the notion of "Part" or "StructuralFeature" already studied in the Structure part. But, sooner or later, if the diagram is not used only for demo purposes, lifelines must be affected to their corresponding classifiers in any project.

Horizontal arrows represent messages exchanged between objects. We have the possibility to simply name them (without ending parentheses) or to relate the message to an operation of a classifier (with ending parentheses). The first choice is allowed because a *Message* is simply a *NamedElement*, so, if lifeline may be created on the fly, messages could be created on the fly as well. A name, if not created on the fly, is the name of a sent data, a returned data, or a signal. Data are *Parameters* (TypedElements package) defined within operations of classifiers and *Signal* is defined in Communications package as a *Classifier* (so the name of the *Signal* is either generic, using its class name, or an instance name).

To summarize, if different from an "operation," the name has no end parentheses. When the message is related to an operation defined in a classifier, then this name must come from one operation of this classifier and parentheses are automatically added at the end of the name. Most UML tools allow us to draw the sequence diagram and define at the same time classifiers behind their lifelines, so after the sequence has been established, we can collect all the defined classes and drag them into a class diagram without having to create them a second time. More importantly, this feature ensures that model elements in the structural and in the behavioral parts of a project are related together. In the previous example, only the "sound" (and vibrations) returned by the engine, that informs the driver of the success of the starting process, is a data or a signal. All other messages are operations of classifiers.

4.15.2 Example of Communication Diagram

The sequence diagram and the communication diagram are two alternatives of the same reality. They highlight objects that collaborate into an interaction and establish the sequence of events differently, along a time axis in a sequence diagram, with numbering scheme in a communication diagram (Fig. 4.15.2.1). The object *:Driver* has a *perceive_engine_sound()* that allow it to sense the sound signal produced by the engine. We can replace this operation by the signal "sound" as in the sequence diagram. The Car model captures both aspects but examples show two alternate representations.

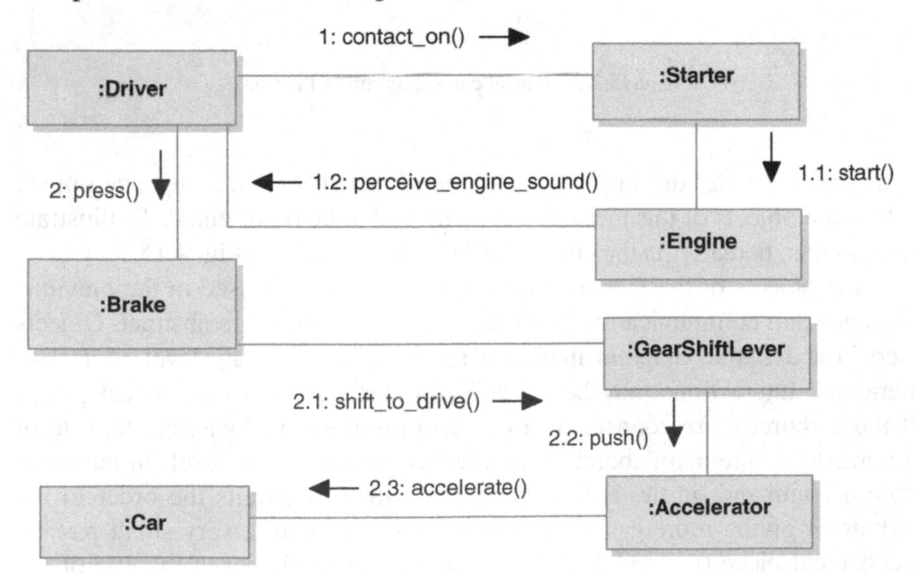

Fig. 4.15.2.1 Communication diagram, a replica of the previous sequence diagram

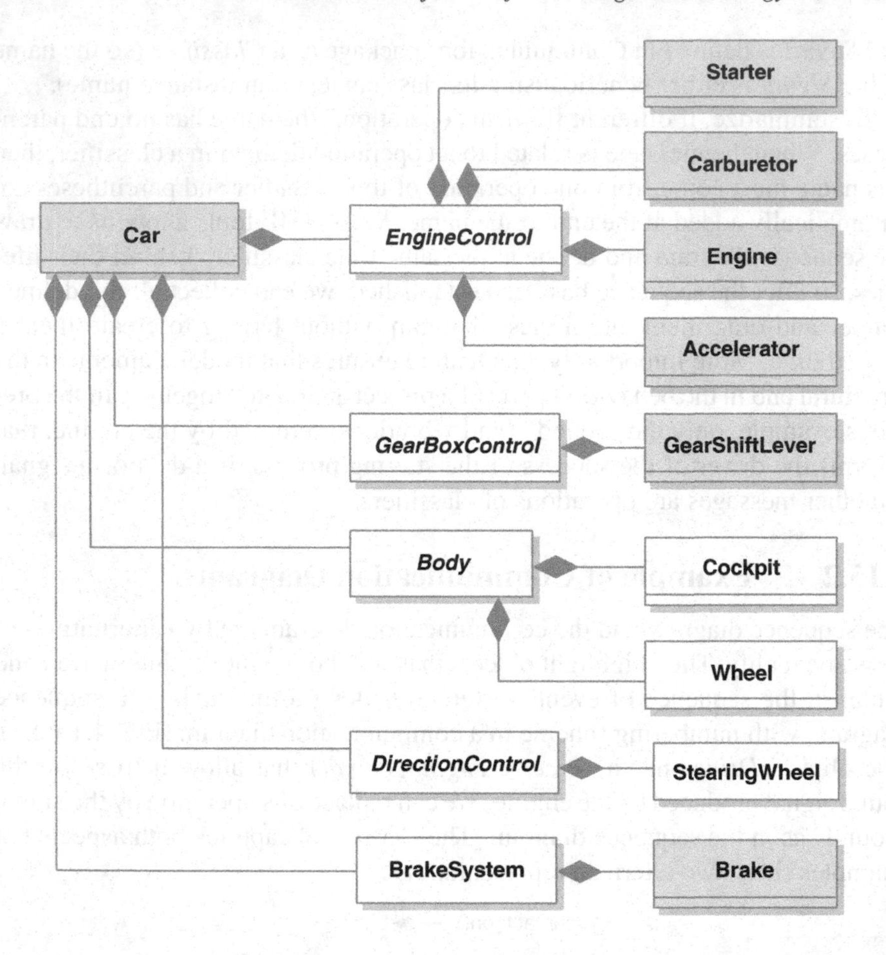

Fig. 4.15.2.2 Partial class diagram of the Car

In previous "Car driving" example, the object *:Car* is a composite object, only some objects of this hierarchy are selected to be represented. To illustrate this process, hereafter is the corresponding class diagram (Fig. 4.15.2.2).

Some objects of the *Car* hierarchy are selected to be used in the previous sequence and communication diagrams. Classes in italic are abstract. Objects used in a dynamic diagram may call for all objects at any level of a class hierarchy. Interactions may be "local" (for instance how mechanical pieces of the carburetor are connected to perform the dosage of gas) but high level "interactions" are a collaboration of various objects at any level. In our case, from a command on the accelerator pedal which transmits the order to the carburetor giving more gas to the engine, we start with a very small passive mechanical piece (the pedal object) to move the whole car at the end of the process.

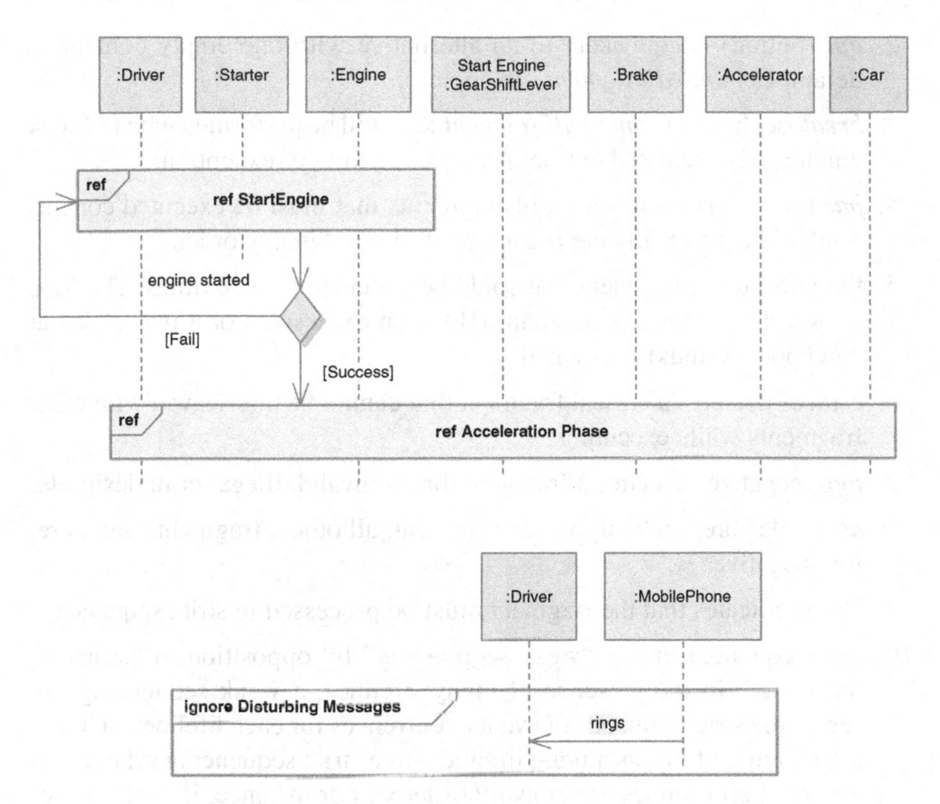

Fig. 4.15.3.1 Example of Interaction Overview Diagram. *ref* blocks are InteractionUses that are references to Interactions. Each ref corresponds to a separate sequence diagram (Interaction). The representation of lifelines behind ref blocks are optional as this information exists already elsewhere. The ignore block named "Disturbing Messages" is in its white box format (it can be created with a ref block too). This block shows messages that must be ignored explicitly in the current context

4.15.3 Example of Interaction Overview Diagram

An *InteractionUse* is a reference to an *Interaction* and is represented as a *CombinedFragment* with "ref" at the left corner side. A *CombinedFragment* is a specialization of *InteractionFragment* and the latter is an abstract notion of the most general interaction unit.

> An interaction fragment is a piece of an interaction. Each interaction fragment is conceptually like an interaction by itself.

The semantics of a *CombinedFragment* depends upon an *InteractionOperator* that may take the following values:

1. ***alt*** (alternatives) is one of the choices flagged by *alt* will be executed when its guard (condition) is evaluated to be true. Alternatives could be understood as classical IF or CASE in programming language.

2. ***opt*** (option) is equivalent to an alternative with one empty content as default *CombinedFragment* (no choice).
3. ***break*** declares a *CombinedFragment* that will be performed when a break condition is evaluated to true. A break is a kind of exception.
4. ***par*** (parallel) declares a set of fragments that must be executed concurrently. The set of fragments can be interleaved in any order.
5. ***loop*** declares a fragment that could be executed several times. The loop is executed as long as its guard (Boolean expression or a pair of lower and upper bounds) is satisfied.
6. ***critical*** declares a critical fragment that cannot be interleaved with other fragments while executing.
7. ***neg*** (negative) declares a fragment that is invalid, illegal or undesirable.
8. ***assert*** declares an "only valid" fragment, all others fragments are therefore negative.
9. ***strict*** indicates that the fragment must be processed in strict sequence.
10. ***seq*** (sequence) means "weak sequencing" by opposition to "strict sequencing". In the presence of many lifelines, a weak sequencing observes the strict sequence of event occurrences for each lifeline but admit interleaving of events among lifelines. In a strict sequencing scheme, all events of all lifelines are globally ordered. For instance, if we have two set of messages, a and c belonging to lifeline1, then b and d belong to lifeline2. If a strict sequencing scheme prescribes the ordering "a then b, then c then d," a weak sequencing admits, "a then c" and "b then d" in all possible orderings.
11. ***ignore*** operators specify how an interaction should deal with unexpected messages. Ignore fragments declare messages of no interest if they appear in the current context. The system must not take them into account.
12. ***consider*** fragment is an opposite case of the ignore fragment and declares that messages not included in a consider fragment should be ignored. The concept of consider/ignore finds an interesting usage for instance in debugging when we want to isolate specific messages to study them.

4.15.4 Example of Timing Diagram

Timing diagrams are used to show interactions when a primary purpose of the diagram is to reason about time. Timing diagram focus on conditions changing within and among lifelines along a linear time axis.

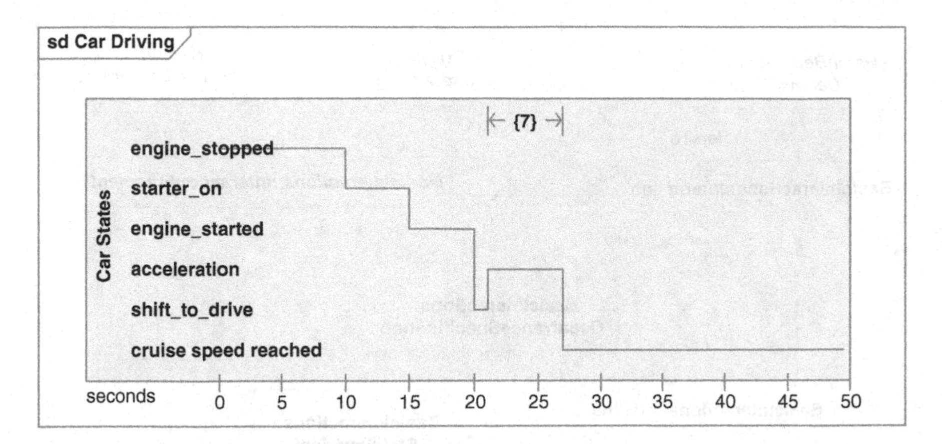

Fig. 4.15.4.1 Timing diagram for time specification. States in a timing diagram could be specified with duration. In our case, all the states have no duration constraint. But, if the driver is a sport car pilot, the maximum acceleration duration he must perform could be an important parameter

Timing diagrams (Fig. 4.15.4.1) are timings of objects or timings of a whole interaction involving a collaboration of several objects, focusing attention on time of occurrence and events. This diagram inherits from the timing diagram used by electronic engineers to study the timing of logic circuits. It is slightly modified to support the object orientation.

4.15.5 Metaclasses of the Interaction Suite

A sequence diagram is based on an *Interaction* that is a unit of Behavior (The keyword "sd" is often used in fragments to depict a "sequence diagram"). A sequence contains lifelines and messages exchanged between lifelines. To deal with complexity, an *Interaction* may contain many *InteractionFragments* (*NamedElements*) that are units or pieces of *Interaction*. If the sequence is developed outside the current diagram, it can be referenced by an *InteractionUse*. The following diagrams summarize main metaclasses encountered in the interaction suite.

In Figure 4.15.5.1, from *InteractionFragment* that is a *NamedElement*, *CombinedFragment* is derived. The *InteractionOperator* is a meta-attribute that can take values like alt, opt, break, par... previously cited in a list. Fragments flagged with "ref" are not *CombinedFragments* but *InteractionUses* as they refer to other *Interactions.* A message is a *NamedElement*; it has two *MessageEnds*, one sending the *Event* and the other receiving the *Event*. A *Gate* is a *MessageEnd*.

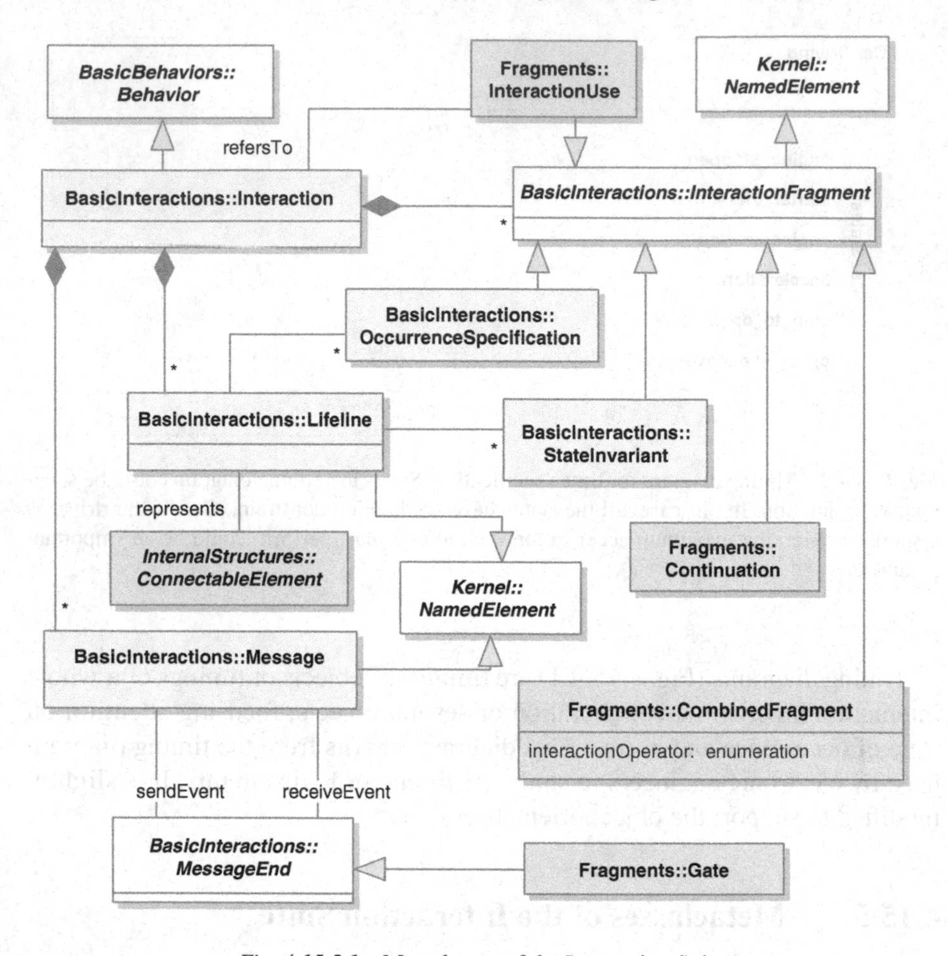

Fig. 4.15.5.1 Metaclasses of the Interaction Suite

4.15.6 Graphical Notations

Table 4.15.6.1 Modeling concepts of the Interaction Suite

Name and graphical notation	*Comments*

Frame

A frame is a solid-outlined rectangle with a pentagon at its left corner for storing the name of Interaction name. A Frame represents an Interaction. Frames are generally represented as a white box with any diagram of the Interaction Suite inside. Even if the term "sd" is coined initially from "sequence diagram", a Frame may be used with any diagram of the Interaction Suite as it acts a container for an Interaction. The term "sd" may be suppressed if there is no misinterpretation.

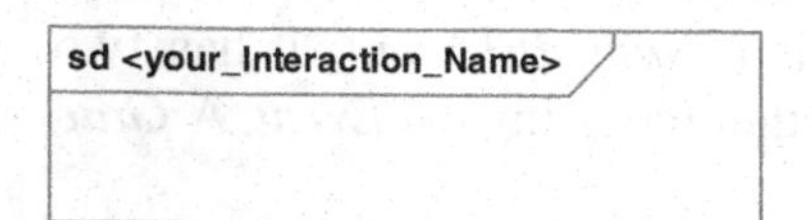

Table 4.15.6.1 (Continued)

Name and graphical notation	*Comments*

CombinedFragment

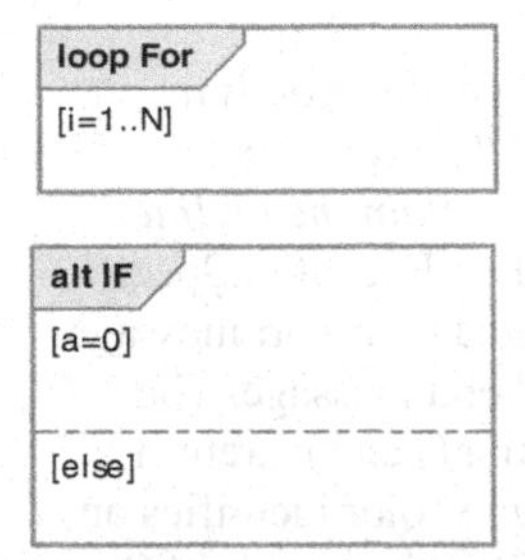

A CombinedFragment is a specialization of InteractionFragment and the latter is an abstract notion of the most general interaction unit. "*An interaction fragment is a piece of an interaction. Each interaction fragment is conceptually like an interaction by itself*" . The semantics of a CombinedFragment depends upon an InteractionOperator that may take the following values: alt, opt, par, loop, critical, neg, assert, strict, seq, ignore, consider.

InteractionUse or "ref" Fragment

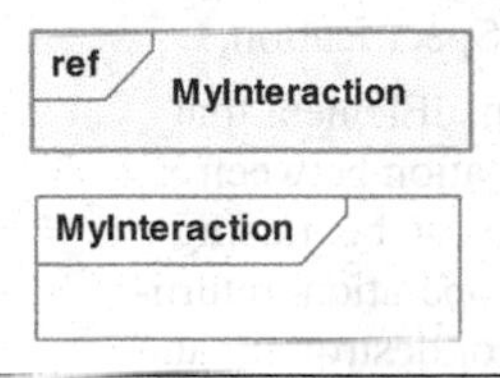

A "ref" makes use of the same graphical representation of a CombinedFragment but it is an InteractionUse that refers to an Interaction. Practically, an Interaction is already (or must be) drawn elsewhere and its reference is imported inside the current context, mostly as a black box.

Lifeline In SD and IOD

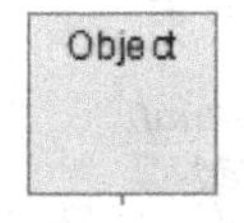

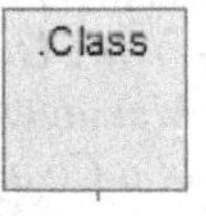

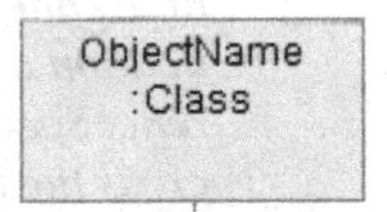

A lifeline is simply a NamedElement and represents the lifetime of a participant into an Interaction. It is shown with a rectangle forming its head with a vertical line (dashed or not).
The lifeline can be created on the fly and identified with a name. If it is related to a Classifier, it can be represented under its generic form (with only its classifier name) or with its instance name followed by its classifier name. The second alternative is mandatory when there is more than one instance for a same classifier.

Lifeline in Communication Diagram

name [selector] :ClassName

Lifeline in a communication diagram has just the "head" without the lifeline. It is identified by an element name followed by a selector (expression that identifies a specific part of a set) then a Class name.

(cont.)

Table 4.15.6.1 (Continued)

Name and graphical notation	*Comments*

Message, ExecutionSpecification in SD and IOD

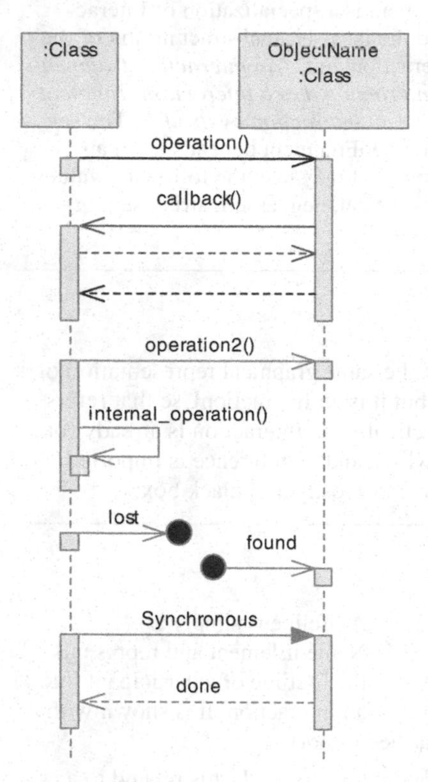

"An ExecutionSpecification is a specification of the execution of a unit of behavior within the lifeline." The duration of an ExecutionSpecification is delimited by a start message and an optional end message. The message operation() and its return arrow of the same color identifies an ExecutionSpecification. callback() and its return is another ExecutionSpecification overlapped inside the previous ExecutionSpecification.

A message is a NamedElement that defines a communication between lifelines. A message can be: raising a signal, invoking an operation, returning a data, creating or destroying an instance. "*A Message associates normally two OccurrenceSpecifications, one sending OccurrenceSpecification and one receiving OccurrenceSpecification. Messages cannot cross the boundary of a CombinedFragment*".

If a message needs a return value, then the return arrow is a dashed line.

A self message has two OccurrenceSpecifications on the same lifeline, e.g. *internal_operation()*.

A small black circle (undefined source or destination of the message) is used to specify messages that have an undefined OccurrenceSpecification. A *"lost"* message sinks into an undefined point and a *"found"* message is issued from an undefined point.

Asynchronous messages do not block, do not wait for immediate responses and are represented with open arrow heads. Synchronous messages block and wait for returned results. Synchronous messages are identified with filled arrow heads.

Message format in SD and IOD

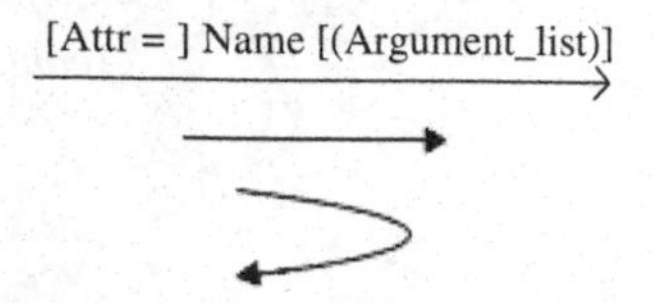

A message in sequence and interaction overview diagrams has at least a name. Others fields are optional.

Name only: Data, Object or Signal sent in the direction indicated by the arrow.

Table 4.15.6.1 (Continued)

Name and graphical notation	*Comments*
Return_Value	*Name()* with parentheses. The name corresponds to the activation of an operation of the destination lifeline. *Name (argument_list)* with parentheses and argument_list. Idem to the previous case but the operation needs arguments.
Attr = Name(). This syntax requests an operation of the destination side that returns a value that will be assigned to an attribute at the source side.	
Message format in CD For messages used in Communication diagrams, the syntax is identical to that used in SD or IOD, except that a particular numbering scheme could be used to order their time occurrences. Number : Sequenceformat	
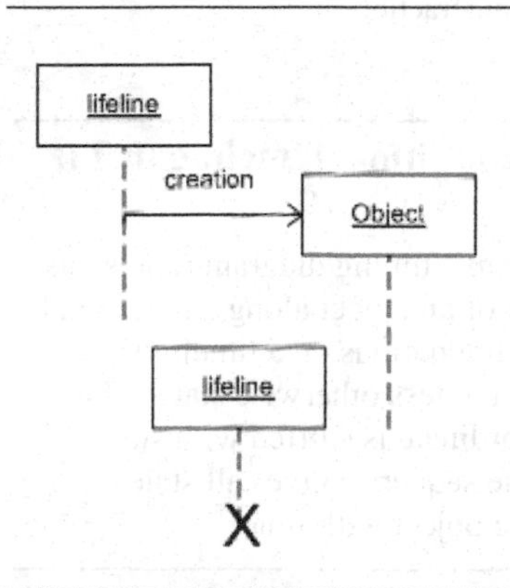	**Creating a lifeline, Destroying a lifeline (DestructionEvent) in SD and IOD** To create a lifeline, a creation message has the arrow pointing to the head of the lifeline or the object created. This message corresponds to a CreationEvent that derives from Event in Communications Package. To terminate a lifeline, just put an X on its lifeline. This corresponds to a DestructionEvent that derives from Event in Communications package.
Gate in SD and IOD 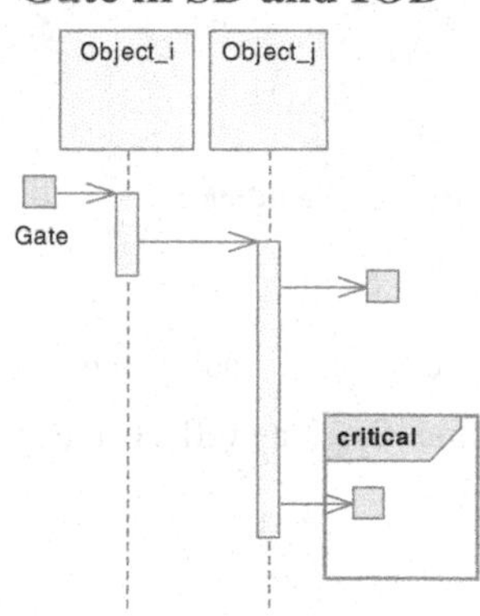	A Gate is a MessageEnd. A Gate is a connection point for relating messages outside fragments to messages inside fragments. Gates allow a diagram to be connected to other diagrams through messages. A Gate is a representative of an OccurrenceSpecification that is not in the same scope as the Gate. A Gate may have an explicit name or not. A same Gate may appear several times in the same or different diagrams.

(*cont.*)

Table 4.15.6.1 (Continued)

Name and graphical notation	*Comments*
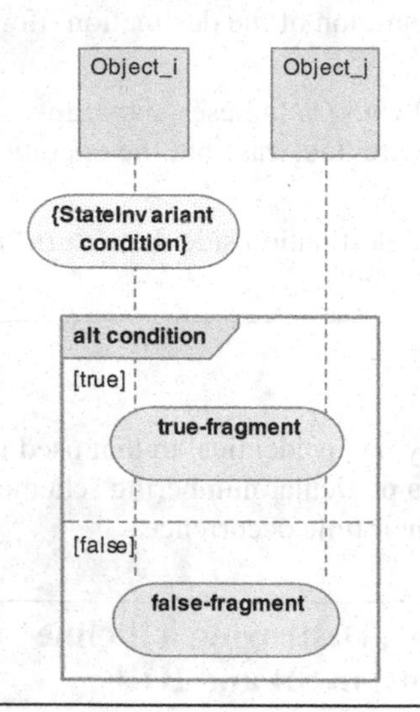	**StateInvariant, Continuation in SD and IOD** A StateInvariant placed in a lifeline is an InteractionFragment that adds a run time constraint to a participant of the Interaction. The constraint is evaluated immediately prior to the execution of the next OccurrenceSpecification. If the constraint is true, the trace following the StateInvariant specification is a valid trace. If false, the following trace is an invalid trace. A Continuation is an InteractionFragment. It is a way to represent different branches of an Alternative CombinedFragment. It is similar to a label in a flow of control. Combined often with "alt" fragment, the Continuation fragment allows jumping at the appropriate interaction.
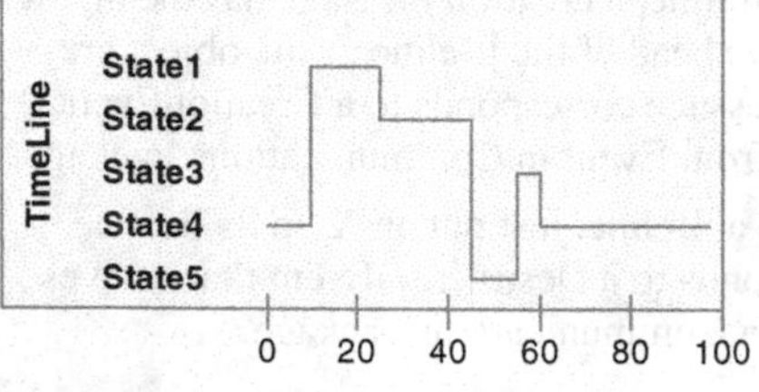	**State or condition Timeline in TD** A Timeline in a timing diagram represents the lifetime of an object along a horizontal time axis. Graduations on a timeline are proportional unless otherwise stated. The vertical coordinate is stuffed with states. The timeline sequence gives all states taken by the object with time.

General Value Lifeline in TD

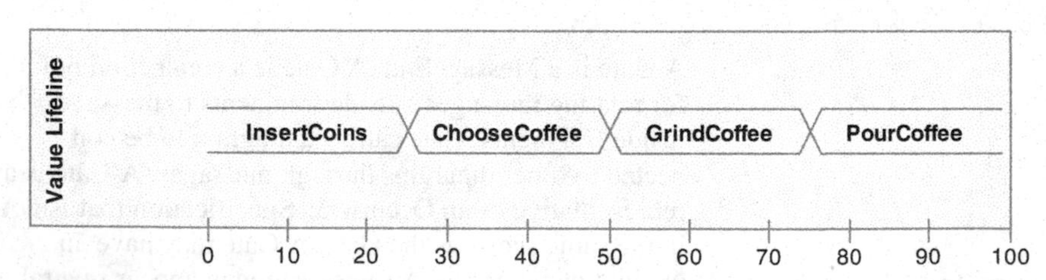

A Value Lifeline is an alternative representation of state timeline. Crossing reflects the event where the value changes.

Table 4.15.6.1 (Continued)

Name and graphical notation	*Comments*
Parallel lifelines, message, and message label in TD	

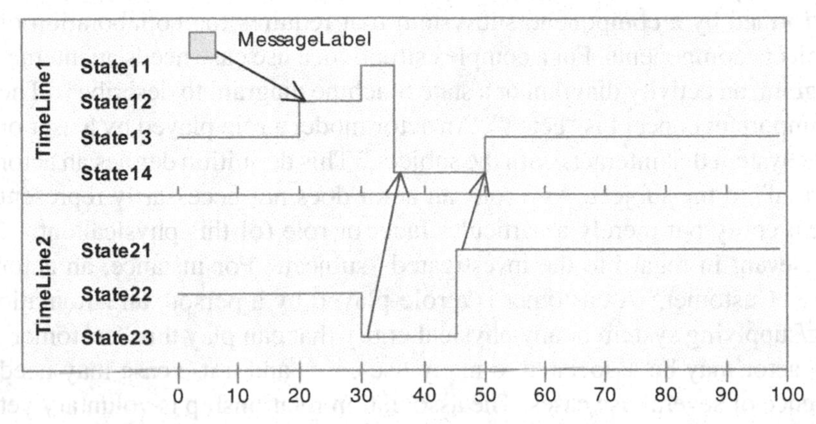

Many parallel lifelines may be represented with the same time axis to show dependences between their respective states. Arrows are messages exchanged between timelines. Labels are notation means used to avoid diagram crossing by arrows for lifelines that are far apart. Message labels convey signals in and out from other sheets.

SD: Sequence diagram; CD: Communication Diagram; IOD: Interaction Overview Diagram; TD: Timing Diagram. When the type of diagram is not indicated, the model element can be used in all four diagrams of the Interaction Suite.

Interactions will typically not tell the complete story but only interesting behaviors. As these diagrams take time to establish, it would be a very big contract to build a whole system with interactions although some project leaders request that all possible behaviors of a system should be documented through interactions, mainly with sequence diagrams. A sequence diagram represents a possible path through an execution graph, so documenting with sequences is equivalent to describing all possible paths of this graph in a graphical form. For complex system or low budget projects, it would simply be unfeasible.

4.16 Behavior: Use Case Diagram

Typically, use cases are used to capture the requirements of a system, that is, what a system is supposed to do. The key concepts associated with use cases are actors, use cases, and the system in the background ("*subject*" from UML terminology) to which the use cases apply. The UML defines the notion of "subject" as

> [A] physical system or any other element that may have behavior, such as a component, subsystem or class. Each use case specifies a unit of useful functionality that the subject provides to its user . . . The behavior of a use case can be described by a specification that is some kind of Behavior, such as interactions, activities,

and state machines, or by pre-conditions and post-conditions as well as by natural language text where appropriate.

This definition of use case shows that a use case may be a very simple chunk of functionality (low level function with pre and post conditions) or any complex task performed by a component, subsystem that requires the collaboration of many objects/components. For a complex situation, a use case needs an interaction diagram, an activity diagram or a state machine diagram to describe it. The second important concept is "actor." "An actor model a role played by a user or any other system that interacts with the subject." This definition defines an actor as "external" to the subject. As a role, an actor does not necessarily represent a physical entity but merely a particular facet or role (of this physical entity) that is relevant in regard to the investigated "subject." For instance, an actor may be a "Customer." A customer is a role played by a person, an automatic ordering/supplying system or any physical entity that can play this "customer" role. An actor may be associated to many use cases and a use case may need the presence of several use cases. The association relationship is voluntary yet not clearly defined in a UC diagram as this kind of diagram is intended for capturing system requirements at the starting point of a development process.

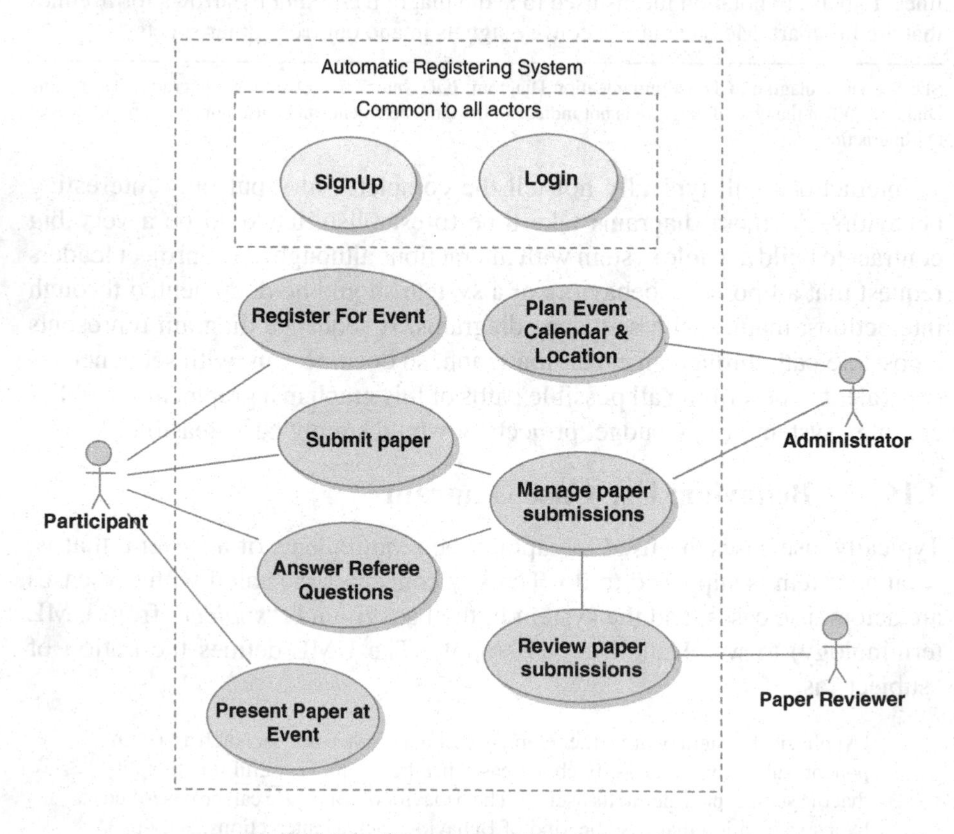

Fig. 4.16.1.1 Concepts used in a use case diagram

Very often, the association between an actor and a use case is not named as the kind of relationship cannot be fully asserted at this early stage. Hereafter is an example of UC diagram with main model elements.

4.16.1 Example

In Figure 4.16.1.1, use cases are represented as an ellipse containing its name. Actors are concerned with use cases by some kind of relationship. Use cases can be tied together by undefined dependencies. The "subject" is "Automatic Registering System." Boundaries can be used to delimit the frontier of a subject or for specific purposes (an economy of graphic representation in this example).

4.16.2 Metaclasses

All use case metaclasses are described in the UseCases package. A *UseCase* is derived from *BehavioredClassifier* that derives itself from *Classifier*. An *Actor* derives from *Classifier* and may own many *Usecases*. A *Classifier* can be a subject for an *UseCase*. A *UseCase* may contain two kinds of relationships, Include and *Extend* and many *ExtensionPoints*.

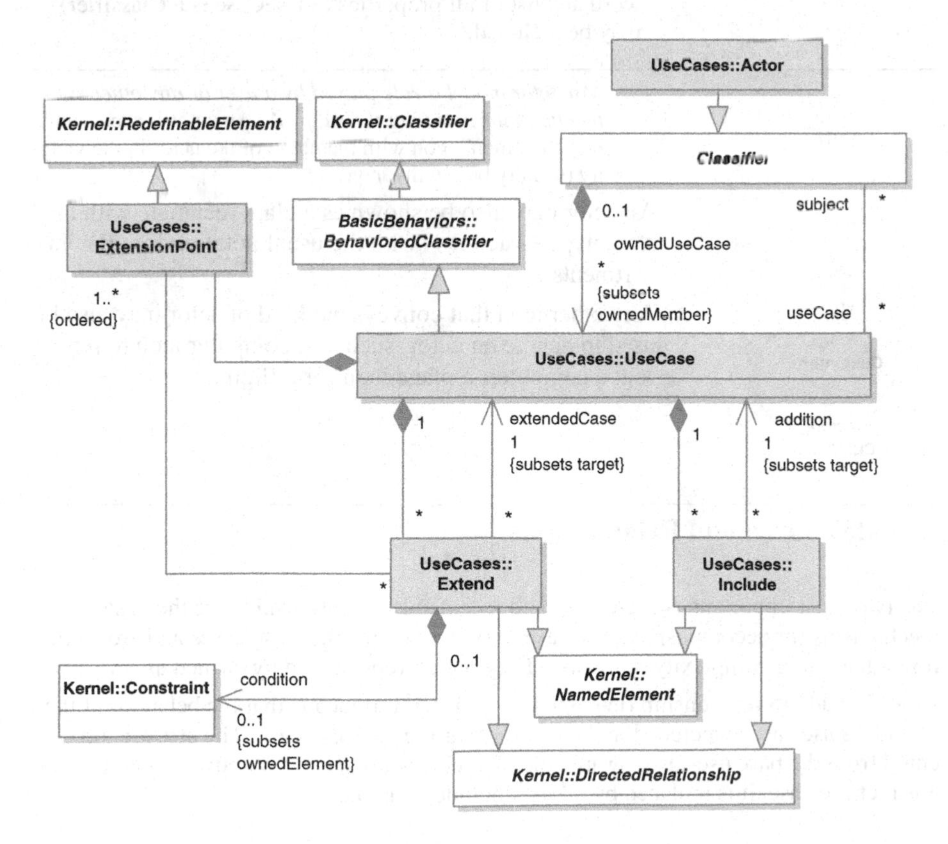

Fig. 4.16.2.1 Concepts used for modeling uses cases (Fig. 16.2, Superstructure Book)

> An extension point identifies a point in the behavior of a use case where that behavior can be extended by the behavior of some other (extending) use case, as specified by an extend relationship.

An ExtensionPoint can be understood as a reference point to other locations that are part of the current description.

4.16.3 Graphical Notations

Table 4.16.3.1 Concepts used in a use case diagram

Name and graphical notation	*Comments*
UseCase Use Case	Use cases are used to capture the requirements of a system, that is, what a system is supposed to do. It can be understood as a chunk of functionality associated to one or many actors. A use case is shown as an ellipse (or an oval). The name can be placed inside or outside. Optional stereotype keyword and list of all properties (a UseCase is a Classifier) may be included.
Actor Customer «actor» Customer	"*An actor model a role played by a user or any other system that interacts with the subject*". An actor is represented by "stick man" icon with the name of the actor in the vicinity (usually below the icon). An actor may also be shown as a class rectangle with the stereotype <<actor>> with the usual notation for all compartments. Any other icon that conveys the kind of actor may also be used to denote an actor, such as a computer icon to represent a computer, a plane icon for a flight, etc.

Include and Extend Relationships

The two most important <<include>> and <<extend>> relationships are the basic mechanisms for decomposing a use case into more elementary use cases and are therefore a feature of complexity reduction. They are all represented by dashed arrows.

An <<include>> relationship (unnamed in the diagram) means that the behavior of the including use case is included in the behavior of the base use case. The arrow is oriented from the base use case towards the elementary use case. "Withdraw money" is an abstract use case, it is replaced by all its "included" use cases.

Table 4.16.3.1 (Continued)

Name and graphical notation	*Comments*
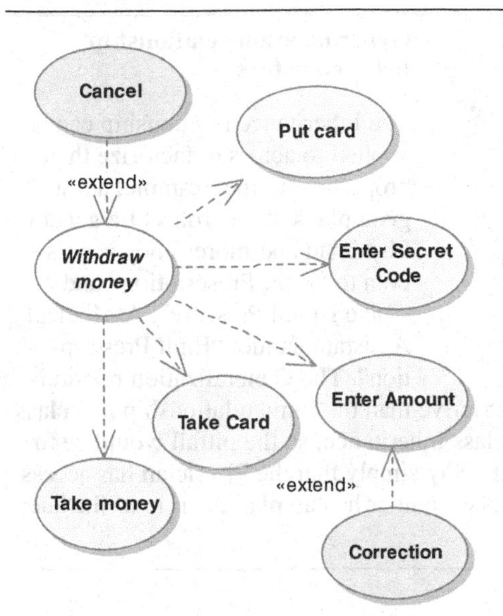	An <<extend>> relationship means that the extension use case extends the behavior of the base use case. If the base use case is meaningful and can be developed without this extension, the later adds additional functionality to the base use case. The arrow is oriented towards the base use case. For instance, if "Cancel" and "Correction" use cases are not there, the original functionality of "Withdraw money" would be operational but the user interfaces would lack user-friendliness.
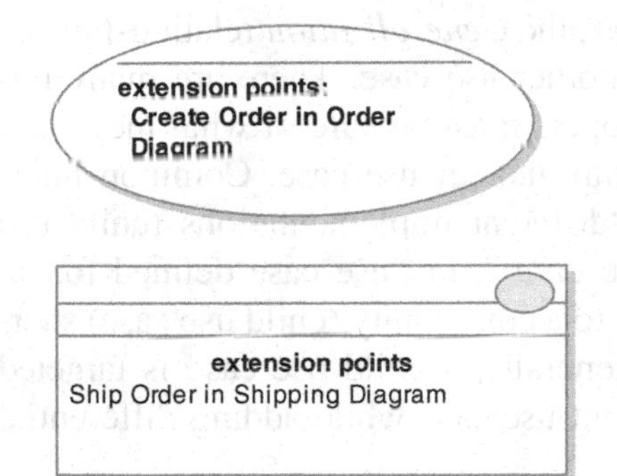	**Extension Points** "*An extension point identifies a point in the behavior of a use case where that behavior can be extended by the behavior of some other (extending) use case, as specified by an extend relationship*". An ExtensionPoint can be understood as a reference point to other locations that are parts of the current description. The restriction (or ambiguity) lies in the "extend" character. While waiting for a clearer context, we make use of extension points uniquely as reference points. Thus, an extension point references one or a collection of locations where other use case diagrams can be found to complete the current description.
Boundary 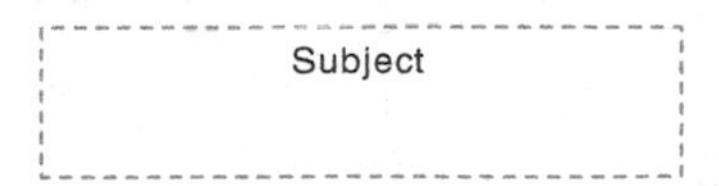	A boundary may be used to delimit the contour of a subject in a use case diagram. It does not have a corresponding metaclass in the standard but it is given as example in the Superstructure Book (Fig. 16.10).

(*cont.*)

Table 4.16.3.1 (Continued)

Name and graphical notation	*Comments*
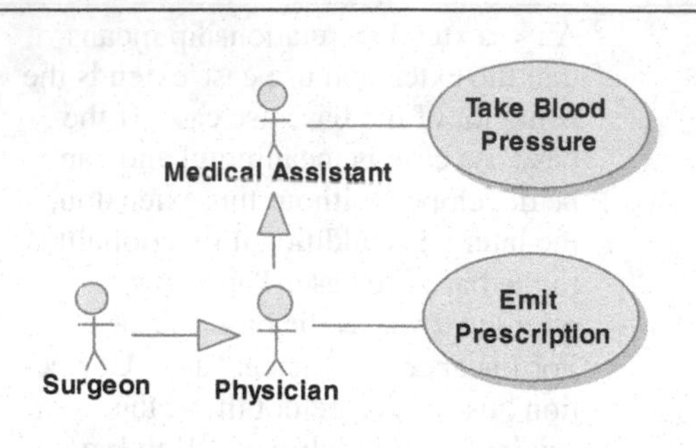	**Generalization relationship between actors** The Inheritance relationship can be applied to actors to factorize their properties. In the example, a Surgeon plays all the roles of a general Physician and more. They access both to "Emit Prescription" and "Take Blood Pressure". A Medical Assistant cannot "Emit Prescription". The Generalization relationship used in a use case diagram is more restrictive than the same relationship in a class diagram. They assert role inheritance, not class inheritance, so the pitfall would be to conclude "a Surgeon is a Medical Assistant". Say simply that the Physician has access to all functionalities defined for a Medical Assistant or he can play the role of the later if necessary.

As a *UseCase* derives from a *Classifier*, the *Generalization* relationship can be applied to derive a use case from another use case. There are many pitfalls associated to this process so developers must be sure of what they want to mean when inheriting a use case from another use case. Common interpretations seen in the literature were "different implementations (child use cases) of the same behavior (parent use case)", or "use case defined for an individual (parent use case) then applied to a community (child use case) sharing some common properties." More generally, a child use case is targeted to replace courses of actions of the parent use case while adding differential operations.

All dynamic diagrams (State machine, activity, and interaction diagrams) may be used to specify what a use case means or is supposed to do. As UC diagrams are mostly drawn at the requirement stage and they are by essence declarative, this possibility must be deployed with great care. A lot of books Armour and Miller (2001), Cockburn (2000), Kulak and Guiney (2000), Lefffingwell and Widrig (2000), and Robertson and Robertson (2000) have treated the complexity of use cases.

4.17 Auxiliary Constructs: Profiles

Profiles are lightweight or second class extension mechanism to the UML standard. They are based on additional *Stereotypes* and *Tagged values* that are applied to *Elements, Attributes, Methods, Links, Link Ends*, etc. A profile is a collection of such extensions (in the metamodel, *Profile* derives from *Package*, thus a profile is a package). A profile contains many stereotypes, imports elements from other packages.

> Stereotypes are specific metaclasses (in the metamodel, *Stereotype* derives from *Class*). Tagged values are standard meta-attributes that can be added to any model elements to pinpoint some properties of those elements.

Profiles must not contradict semantics defined in the UML metamodel, must be made interchangeable between tools and must offer a smooth integration of new modeling concepts with regards to the basic UML standard. From a practical point of view, a profile allows us to define new modeling elements and work with them in new projects as if they are initially part of the standard.

To summarize, profiles are lightweight extensions, so, a UML profile is a packaged set of stereotypes and tagged values that allows us to decorate a UML application with extra metainformation without changing the representation of the metamodel. These extensions are authorized by the UML and do not contradict the current standard in any way. Stereotypes can be regarded as subclasses of existing UML metaclasses whereas tagged values can be regarded as attributes of such new virtual metaclasses.

The Profiles package included in UML defines a set of artifacts to deal with specific concepts in particular application domains. The UML outlines several reasons for this need for extension:

1. The modeler desires to have a terminology that is adapted to a particular domain.

2. There are no constructs available in the standard.

3. The user wants to add special semantics.

4. The graphical notations are not suggestive enough for a special circumstance. Each stereotype may have a new image (the *Image* metaclass provides the necessary information to display an image in a diagram. Icons are handled through the *Image* metaclass).

5. There is a need to add constraints that restrict the way the metamodel is used.

4.17.1 Example of Profile

Figure 4.17.1.1 illustrates the principle of profiles.

Normally, to create a new profile, the user must create new stereotyped classes and use the ≪extend≫ relationship to connect these stereotyped classes to metaclasses of the UML. Constraints can be added to stereotypes, imposing restrictions on the corresponding new modeling elements. The constraint is defined in the profile but evaluated in the model.

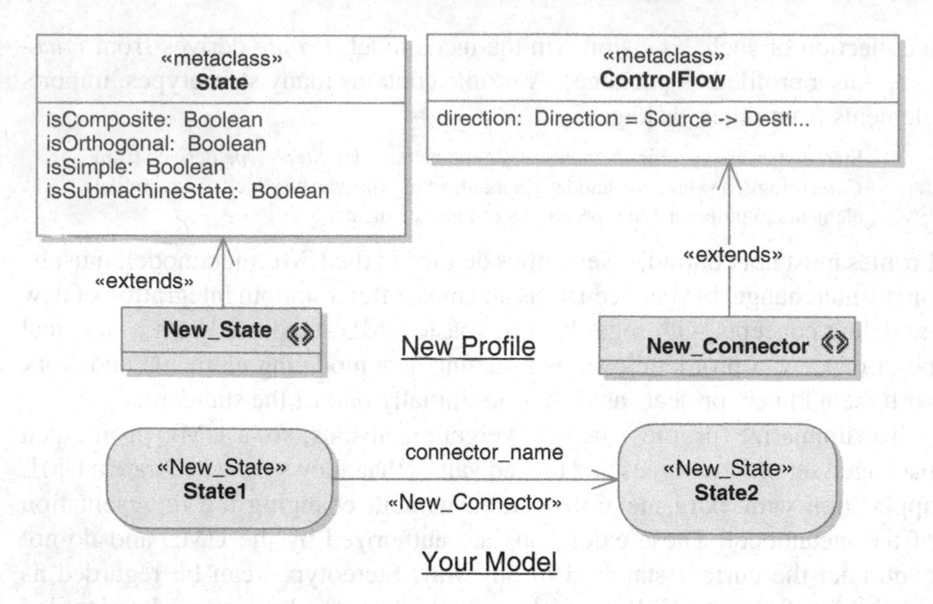

Fig. 4.17.1.1 Example of a UML profile. A new profile is created with two model elements, New_State and New_Connector. They are stereotypes linked to their corresponding metaclasses through the extension relationship. At the bottom, two new states 1 and 2 are instantiated from the profiles, as well as a new connector which connects them together

4.17.2 Graphical Notations

Table 4.17.2.1 Modeling concepts used in profiles

Name and graphical notation	*Comments*
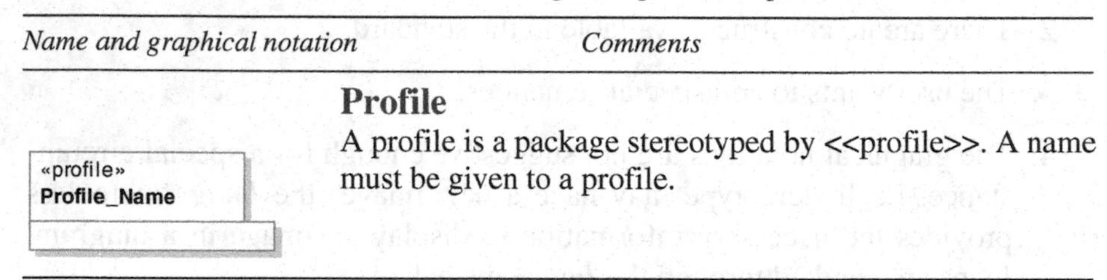	**Profile** A profile is a package stereotyped by <<profile>>. A name must be given to a profile.

Stereotype, Metaclass and Extension association

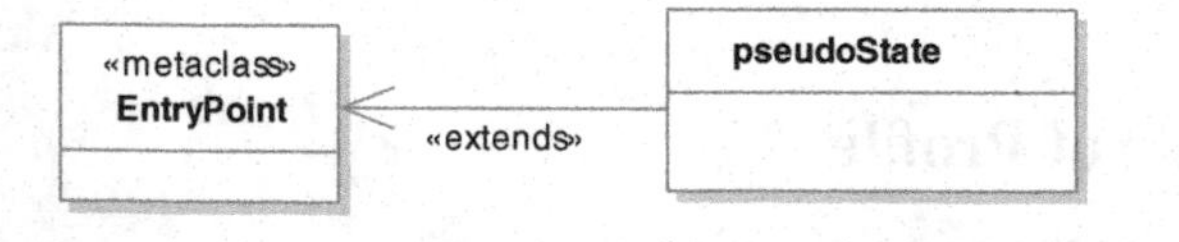

Stereotype derives from Class. Stereotypes are used to make classes in a specific domain. <<metaclass>> can be any modeling element metaclass, structural or behavioral, graphic node or link, defined in UML.

The extension association connects a stereotype to its metaclass. In the example, we define a stereotype <<pseudoState>> that extends the metaclass EntryPoint that is itself a pseudostate.

Table 4.17.2.1 (Continued)

Name and graphical notation	*Comments*
«profile» .Net «profile» Java «apply» «apply» web_application	**Profile Application** ProfileApplication metaclass derives from PackageImport so it is a relationship between packages. The notation is a dashed arrow pointing to the applied profile. In this example, new constructs. Net and Java, are used to build a web application. As profile is package, the application relationship may be draw between profiles.
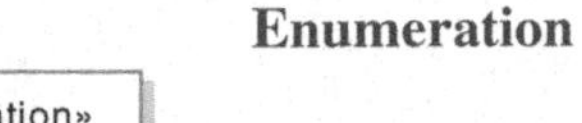 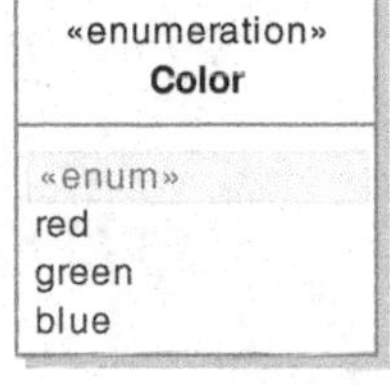	**Enumeration** Enumeration stereotype let us define new enumeration types in the same profile package.
Generalization	A stereotype can specialize or generalize another stereotype; a metaclass cannot participate in a generalization relationship with the stereotype.
Tagged values	Tagged values, or tag-value pairs, are a convenient way of adding additional information to stereotypes.

A profile is a very useful concept with regards to the facility it offers when communicating with people inside a specific domain. For instance, when talking of Web services, it would be more suggestive to stereotype main actions or activities as ≪WebServiceCall≫. When dealing with relational database developers, we can build a complete new profile with several definitions like ≪entity≫, ≪table≫, ≪index≫, etc. instead of using only classes of the bare UML standard to rebuild the older development environment with new object tools.

Stereotypes are more suggestive since, in a stereotype, we have two fields: the stereotype field that can be used now to typify the class and the class name will be now reserved for more contextual identification. In a bare class, there is only the name. Moreover, if needed, tagged values can be deployed to offer to stereotypes a set of tags-values to make them more informative and to add new criteria to their classification. So, the stereotypes offer roughly three levels of specification (one for the stereotype name, one for the class name, and one for the set of tagged values). It would not be surprising that profiles will be used intensively as metamodels for ontological purposes. This fact turns into a very unpredictable but profitable sideline of the UML. For object theorists, profile is only a very useful cosmetic makeup, a wrapping mechanism to help people retrieve objects of their domains.

Chapter 5

Fundamental Concepts of the Real World and Their Mapping into UML

5.1 Abstraction, Concept, Domain, Ontology, Model

5.1.1 Abstraction

Modeling is a complex task, an attempt to capture the intricacies of real-world situations, to describe the characteristics of real-world objects, their relationships, and the way objects communicate together to evolve. The modeling process is based on *abstractions* and *concepts*. An abstraction is a mental process of taking a thing, *material* (mobile phone) or *immaterial* (electromagnetic wave), *real* (person) or *abstract* (his emotional state), pruning all details that are not relevant for a *particular purpose*, naming it, giving it a short description to be able to manipulate and work with the *abstraction* (as a result) issued from this complexity reduction process. An abstraction is therefore both a process and a result of this process. The object or the idea that results from an abstraction cannot be a specific detailed thing ("Ferrari Car" cannot be an abstraction but "Car" can be an abstraction).

Another meaning of abstraction in engineering disciplines is *information hiding*. *Data abstraction* results in hiding data that are not required to be visible at the interface of an object or a component. This meaning of the term "abstraction" is captured as the process of making black boxes.

Another meaning of abstraction refers to the way models must be designed. At the early stage of the development process, it is necessary to abstract away implementation aspects and considering only on role that objects must play in a system. This process is called model abstraction. The MDA is based on this kind of abstraction.

The last meaning of abstraction refers to level abstraction that is central to managing complexity. Usual antonyms for "abstraction" are concretization, generalization, specialization, etc. according to its meaning variations. The following table summarizes all the meanings and their mappings in the UML.

D. M. Bui, Real Time Object Uniform Design Methodology with UML, 191–288.

Table 5.1.1.1 Mapping of Abstraction concept and different meanings in the domain of modeling

Name of the concept	*UML mapping*	*UML definition*
Abstraction	Relationship in Dependencies package	1. An abstraction is a *relationship that* relates two elements or sets of elements that represent the same concept at different levels of abstraction or from different viewpoints.

1 **UML Graphic Notation**

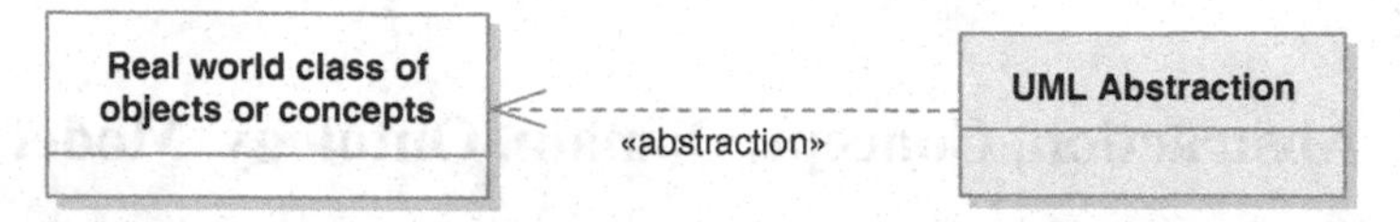

For each object or concept in the real world, we build a UML abstraction to model this object or this concept. The Abstraction metaclass is standardized in the Dependencies package of the Kernel (Fig 7.15, Superstructure Book). The dependency relationship is oriented from the model element towards the element being modeled.

Other meanings in modeling	Comments
2. *Mental process for conceptualization or the product that results* from this process	An abstraction is a mental process of taking a thing, material or immaterial, real or abstract, pruning all details that are not relevant for a particular domain, naming it, giving it a short description to be able to manipulate it in a model. The result of this process is also called an abstraction (or sometimes a concept). An "abstract" of a scientific paper, which states the key ideas without details, is an abstraction of this paper.
3. *Data abstraction* and information hiding	Concept of black box. Irrelevant data, not pertinent to users, are hidden at the interface of an object/component. The result of the data abstraction and information hiding is also called an abstraction, e.g. a class or a component is the result of data abstraction.
4. *Model abstraction* or *design abstraction*	At any stage of the development process, a model must ignore all details of subsequent models. For instance, implementation details are not considered at design phase. The result of a model abstraction is an abstract model taken at any level of the MDA concept.
5. *Level abstraction*	The idea of level abstraction is central to managing complexity. At any level, a set of abstractions is used to describe a system. Each abstraction taken at any level may be decomposed to a new set of abstractions for the next level.

The definition of the *Abstraction* relationship in the UML is mapped to the first common meaning in Table 5.1.1.1. Data abstraction or information hiding is currently one of the most important features of object paradigm. The model abstraction or design abstraction is recently substituted by the newly coined term MDA. The level abstraction governs all decomposition process and is central to managing complexity. We can add specific stereotyped dependencies for other meanings if necessary.

5.1.2 Concept

A *Concept* by its nature is abstract and a universal mental representation of things. An "abstract concept" is therefore a pleonasm. Knowledge is expressed in concepts and people use concepts in their daily lives. Many dictionaries give a general definition for concept as "An abstract or general idea inferred or derived from specific instances". This general definition make the term "concept" very close to the result of the mental process defined for abstraction (first meaning in Table 5.1.1.1). Any concept needs, in the first stage, an abstraction (as process) that formats it, and then the abstraction (as result) is later enriched with properties, connected to remaining concepts of its domain. A concept is therefore a result of a process named *conceptualization* (different from "conception" that is equivalent to design) and *conceptual* is the resulted state of something that can be considered as building from a network of linked concepts.

So, if an abstraction is a low-level process (or a result of this process), *a concept is a more finished and ready to use*. If the concept is very elementary, its meaning is very close to the term "abstraction." Any model element from the UML (*Class, Object, Event, Part, Component, Relationship, Association*, etc. are model concepts). All other meanings of the term "abstraction" (data abstraction and information hiding, model abstraction or MDA, level abstraction), except the first one (mental process for conceptualization) are therefore concepts. For instance, "data abstraction" is a concept.

Concept mapping was developed in early 1960 by Novak (1990) and was a technique for representing knowledge in graphs (network of concepts). The result is a *concept map*. All concept maps are not necessarily concepts but specific concept maps compose new concepts. Concepts are useful in all human activities and are fundamental tools to communicate complex ideas and knowledge, to help learning, to generate new ideas (e.g. *Object, Classification,* and *MDA*), to design complex communication structures (e.g. long texts, hypermedia contents), to diagnose understanding or misunderstanding, etc. Concept maps are very close to semantic networks developed by Quillian (1968) for formal knowledge representation.

The UML does not define explicitly a metaclass Concept as for Abstraction. In fact, all model elements of the UML are *modeling concepts* (or metaconcepts) and what we try to create with the UML are, for a large part, *application concepts*.

Concept maps, semantic networks, and recent ontology engineering can be supported by the UML with class diagrams. Conceptual modeling is the process of identifying, analyzing, describing, and graphically representing concepts and

Table 5.1.2.1 Mapping of a simple concept that necessitates just a categorization process with a Class

Name of the concept	*UML mapping*	*UML definition*
Simple concept	Class	None

Element Car

Example:
For each concept in the real world, a simple class (Element) or a composite class (Car) may be used for expressing this concept. The Concept metaclass is not defined formally in UML but all model elements in UML are actually basic concepts.
Element that is the metaclass of UML is a concept. *Car* is also a concept, but Car is a user-defined composite class made of thousands of elementary classes.

constraints of a domain with a tool like the UML that is based on a small set of basic meta concepts. Ontological modeling is the process of capturing, describing relevant entities or categories of a domain (in an ontology of that domain) using an ontology specification language that is based on a set of basic ontological categories or entities. The UML may be used as a support for ontological modeling but it is necessary in this case to prepare this set of basic ontological categories or entities. So using the UML as ontological tool necessitates some preliminary works. Moreover, the UML addresses only the knowledge representation of ontology and there is no support for inference or exploitation of this ontology.

Complex concepts need more elaborate representations. For structural concepts, *structural diagrams* (class diagram containing some connected classes, component diagram, composite structure diagram, and object diagram) are sufficient, but concepts explaining system dynamics may require both structural and behavioral diagrams. In fact, a complex concept may require a complete model to explain it. When we model a system like a Car, it is a composite and a complex concept, which involves a hierarchy of concepts underneath.

5.1.3 Domain

The term "domain" used in modeling means a field of study, an area of activity and knowledge characterized by a set of concepts and terminology understood by practitioners in this area. Physics, Mechanics, Electronics, Biology, etc. are examples of engineering domains. A domain may in turn be decomposed into subdomains. As the decomposition tree is rather large and the relative position is difficult to locate, we call any subdomain of any level simply a domain or a subdomain, and the prefix "sub" is contextual as a subdomain is simply a domain under the current domain taken as the reference level.

Table 5.1.2.2 Mapping of a "Complex Concept" concept and example

Name of the concept	*UML mapping*	*UML definition*
Complex concept	Class, Component, Object, and Composite Structure Diagrams. Behavioral diagrams may be required to explain dynamic concepts (complex concept may join "model")	None

Example of structural concept

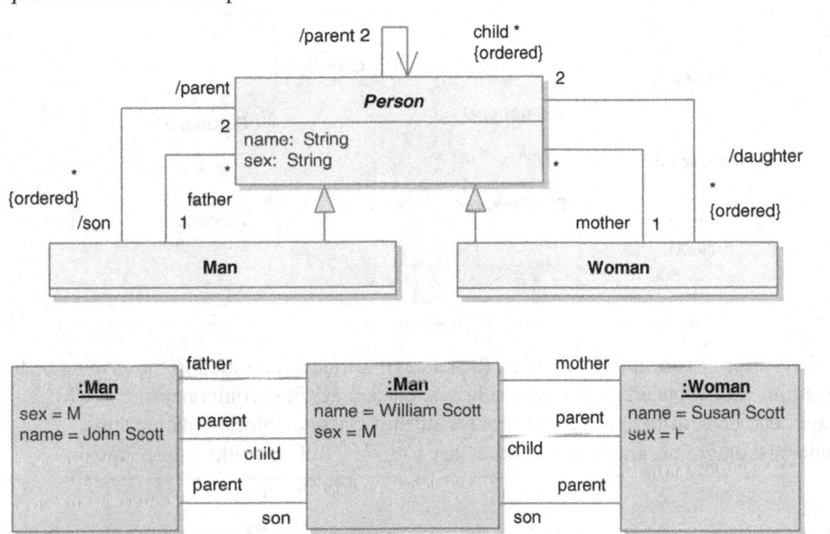

Class and object diagrams may be used to explain non elementary concepts. The concept of Family is built from three elementary concepts: Person, Man, and Woman, connected by a set of relationships. Person is an abstract class, both Man and Woman classes derive from the abstract class Person. Three objects, two instantiated from Man class and one instantiated from Woman class, together with the network of relationships, is a specific instance of the Family concept. In common language, the Class diagram represents the Family concept. The Object diagram represents a declarative knowledge.

In the domain of Computing, the ACM has classified various domains as: Literature, Hardware, Computer Organization, Software Engineering, Data, Theory of Computation, Mathematics of Computing, Information Technology, Methodologies, Computer Applications, and Computing Milieux. Subdomains of Methodologies are: Symbolic and Algebraic manipulation, AI, Computer Graphics, Image Processing and Computer Vision, Pattern Recognition, Simulation, Modeling and Visualization, Document and Text Processing, and Miscellaneous. So, the subject of this book is classified into the domain Simulation, Modeling, and Visualization, which is a subdomain of Methodologies.

Table 5.1.3.1 Mapping of a Domain concept in the UML

Name of the concept	*UML mapping*	*UML definition*
Domain	Package Diagram as a *Container* storing concepts (classes, components, artifacts, etc.). Tagged values may be added to give more precision about Package contents	None

Example:

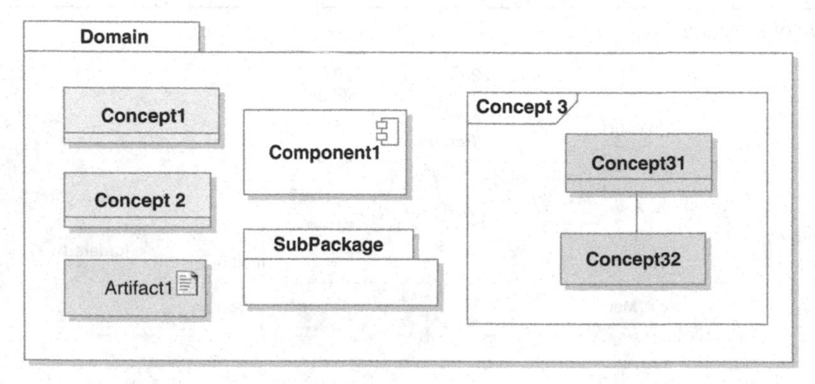

Domain is a set of concepts, knowledge (instances), artifacts identified for describing a class of applications. Concepts are mapped into classes, class diagrams, components, etc. UML packages can be considered as containers for storing classes, objects, relationships, components, diagrams, artifacts, etc. necessary to give a full description of a domain.

Domain Engineering is the activity of collecting, classifying, organizing, and storing concepts in a particular domain as assets that can be reused when building new systems. Domain Engineering targets a product and thus needs Domain Analysis, Design, and Implementation as all other products.

To describe an application, a *General Linguistic Domain* that affords scientific vocabularies, constructs will accompany every description. As that package is *omnipresent in any description*, it is not mentioned in the description or is implicitly included. When describing a multidisciplinary system, many packages are required. For instance, to describe a robot that is an electromechanical device controlled by a microprocessor-based software, we need at least three domains represented graphically with three UML packages (Mechanics, Electrical Engineering, and Computer Science) and a General Linguistic Domain.

5.1.4 Ontology

A domain provides only a vocabulary for identifying, communicating, classifying concepts and knowledge about a domain. Ontology is a modern term with various meanings. Ontology is originally a term in philosophy that means "theory of existence. Gruber (1993) defines ontology as a specification of a (domain) conceptualization. The web site www.owlseek.com provides an interesting definition of ontology:

> We can never know reality in its purest form; we can only interpret it through our senses and experiences. Therefore, everyone has their own perspective of reality. An ontology is a formal specification of a perspective. If two people agree to use the same ontology when communicating, then there should be no ambiguity in the communication. So, an ontology codifies the semantics used to represent and reason with a body of knowledge. Ontologies can be written in various forms or languages.

If we fusion all elements of the two previous definitions with what a domain is, an ontology does not include only a set of concepts plus a set of knowledge (domain) but it structures this domain, has a representation of meaning and constrain interpretation of concepts. Moreover, with the development of ontological tools in the market, ontology provides a basic structure and a framework around which a knowledge base can be built. Knowledge is always updatable to insure that ontology always reflect realities, facts, knowledge, beliefs, goals, predictions, etc. of the moment of a given domain.

Ontology Engineering is a very important technological issue. Ontology tools are software programs that can be deployed through many platforms. They have syntactical verification tool to verify new entries. Semantic errors can be elevated to the rank of knowledge so ontology must have some kind of retroaction and metrics to verify its validity when confronted with real applications. Ontology clarifies the structure of knowledge.

> When modeling a database of a university, we have students, professors, employees, females, males etc. but an employee may be student at part time. Further, ontological analysis show for instance that student or employee are not human categories like males or females but only roles.

So, ontology will clarify concepts and help reasoning. Second, ontology enables knowledge communication and sharing. The richer the ontology is in expressing meanings, the less the potential for ambiguities in creating requirements, in communicating knowledge. Today, ontology has grown beyond philosophy and has joined the information technology and AI.

If we cannot give an exact definition of ontology, we notice that ontologies are targeted to be very helpful in the near future. They are planned to be used as:

1. *Common vocabulary* for practitioners inside specific domains (usage identical to that of a domain previously mentioned)
2. *Terminology standardization* (close to the previous point with an official standardization)
3. *Conceptual schemas* for building relational or object databases exploiting existing schemas and patterns (database usage)

Table 5.1.4.1 Mapping of an Ontology concept in the UML

Name of the concept	*UML mapping*	*UML definition*
Ontology	Component Diagram as a *Container* and all of its contents. Component Diagram in the version 2 of UML is more abstract and is not implementation oriented, so it can be used to store heterogeneous structures. A Package Diagram may be used within the Component Diagram at any level.	None

Example:

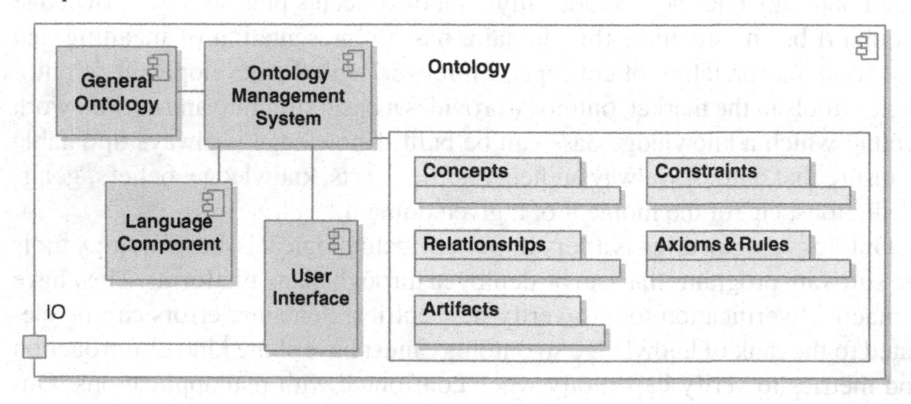

An ontology is first a domain but it is more than a collection of entities. Ontology is structured and incorporates semantics. It has properties, operations and is best represented by a Component Diagram. All kinds of diagrams or Packageable Elements can be inserted in an ontology.

4. For building and exploiting *knowledge bases*, answering competence questions
5. For supporting *model transformation* in the new MDA architecture, etc.

In fact, ontologies must assist us in every step in any engineering process needing a precise vocabulary, sharing existing knowledge, reusing available patterns, and artifacts. Tale 5.1.4.1 shows some possible mappings of ontology representation into the UML.

5.1.5 Model

In the software field, the perception of importance of model is increasing but traditional development, mainly code-centric, still predominates. Developers sketch their ideas with some diagram drafts but discard them quickly once the code has been implemented. Scientific disciplines use models to get a clear

Table 5.1.5.1 Mapping of a "model" concept into the UML

Name of the concept	*UML mapping*	*UML definition*
Model	A *model* in its simplest form may be a *concept.* In this case, all mappings for the concept can be applied to a model. In its complex form, a model can be a whole system. So, if a model can be represented by a UML model elements and diagrams, then: *All 13 diagrams of UML* may be used and combined at infinitum with artifacts to make any model	Omnipresent in the UML but there is no Model metaclass defined

Examples:

- A model of a *Person at the requirement analysis* is simply a role and represented as a model element with a pictorial "stick man" taken from the use case diagram
- A model of a *Person in a database at design phase* can be represented with a class with structural attributes
- A model of a *Person in a real-time system at design phase* can be represented with a class with structural and behavioral attributes and operations
- A model of a *Person considered as a social agent* at the design phase can be represented by a class, a component, or more
- The UML itself is a metamodel and need diagrams and artifacts to describe it, etc.

understanding of what they developed to predict the behavior of their systems, to communicate ideas among developers or stakeholders.

> For instance, an astrological model that makes statements on masses, positions, velocities of planets must be able to predict their positions x years after.

However, a model is not the reality but an abstract and close representation of this reality. If physical or mathematical models are based on formula or deterministic statements, most human models (medical, social models) give only some level of accuracy and their acceptation threshold fluctuates with space, time, and cultural standards. The validity of a model is its usefulness, its accuracy in predicting the future, and its reuse for similar problems. At the starting point, it must adhere to some standards of knowledge and rules to ease its communication and understanding. A model is therefore a container containing a rich set of artifacts, diagrams, etc. describing concepts, artifacts that describe and communicate the understanding of the "thing" been modeled.

A metamodel is a model to make models. It packs modeling elements, axioms and conventional rules that must be verified when establishing user models. The UML is a metamodel, defines model elements, and suggests graphical notations and a set of rules that must be verified for any valid UML models. As descriptions of the UML can be made using itself, the UML is a reflexive metamodel or self-descriptive metamodel. This capacity of the UML is outstanding as it can represent a model that need several metalevels to describe it (UML has been broken into four levels M0–M3 in Chapter 4).

The classical definition of a model must be adapted to the new MDA paradigm that separates application knowledge from specific implementation technology. The consequence is an enhancement of the possibility of reuse of proven high-level patterns, an improved maintainability due to better separation of concerns, and better consistency due to a better direct mapping between domains and models.

5.2 Structural, Functional, and Dynamic Views of Systems

Communicating architectures to stakeholders are a matter of representing them in an unambiguous, readable form with information appropriate to them. As stakeholders may be of different disciplines, the communication task is not easy. Architecture is the only means to an end. Information that stakeholders can infer at the end is more valuable than information about just the architectures. The challenge would be how to select the information to be presented and how to select the sequence to maximize the understanding. A complex system would be impenetrable if all elements are exposed at the same time. The strategy is therefore multifaceted and follows the general concept of top-down design. But, this layering process or this tree view process itself is not sufficient. The level is still too complex to be handled as a whole. At each level, we must introduce only one facet of the system known as its selective or partial view. "A view is there fore a reduced set of coherent concepts used at any level to describe a system." We can apprehend the view concept as a special lighting to bring out selective details. For instance, in medical context, X-ray machine tuned differently allows us to see different organs of a human body and ultrasonic probe complements some diagnoses. By combining all structural, functional, and dynamic views of any system, we force it to deliver all its secrets.

The UML has adopted solely two views to describe systems, Structure and Behavior, and melts functional and dynamic views inside a unique behavioral view. There is an interdependence of the functional and dynamic views and most diagrams are of combined nature as they mix these three aspects to some extend.

> The class diagram is first classified as a structural diagram but the functional view appears clearly when class operations are detailed. Activity diagrams mix both functional concepts (actions, activities) to dynamic and evolutional aspects (states, control flows). Use case diagram is clearly functional.

So, the description would be more precise if the behavioral aspect could be subdivided, at the application level, into two subdivisions, *functional* (responsibility, potential capacity of actions, and interventions) and *dynamic* (evolutional aspects).

After half a century of functional modeling, designers generally agree that function is the most important concept in determining basic characteristics of a product or a system.

> As a first reaction when approaching any system, we ignore the "How" and we want to know only if it "Can" or "Cannot" fit our needs (function availability).

Recently, the functional view is resuscitated by Business Processes that have emerged as an important aspect of enterprise computing landscape (B2B, Web Services,etc.). In the UML, the functional view is supported by main concepts as *Action*, *Activity*, and *Operation* metaclasses. There is no "Function" Metaclass and the term "function" is returned to the *General Linguistic Domain.* But, at the implementation level, this term reappears as fundamental programming concepts (C++, C#, or Java "functions"). Table 5.2.0.1 shows abstractions in three views and the impoverishment of the vocabulary passing from application domains towards the programming domain.

5.3 Concepts of the Functional View: Process and Business Process Modeling

The vocabulary of *process modeling* is rather rich: function, process, transformation, service, method, task, thread, action, activity, operation, datastore, dataflow, controlflow, etc. *Business process modeling* is a convergence of *process modeling* and *business modeling*. Process modeling is a general modeling domain and business modeling is an application domain that is business oriented. They have been melted to build a new concept "business process." Terms as vision, mission, influence, assessment, goal, objective, strategy, tactic, policy, conformity, resource, task, etc. are part of business application domain. At the UML side, we have only four metaclasses representing functional aspects: *Action, Activity*, class *Operation*, and use case. The mapping will depend upon the interpretation given to new merged concepts of the BP model. Figure 5.3.0.1 illustrates this process with an UML-compatible diagram.

In this figure, the BP model is built from three domains: Process, Business, and General Linguistic. Once the concepts of the BP model is clearly defined with a proposed semantic, the Mapping activity takes each BP model concept, applies the mapping rules, finds the correct UML metaclasses and proposes a UML metaclass (or a component or a diagram pattern) that maps each input concept in the BP model into a corresponding output "UML compatible concept". Most of the time, as this mapping process can be done manually and quickly, it is not interesting to automate it, but the mental work would be the same. We must be guided by a set of self compatible rules to insure that this mapping can work most of the time (if not anytime).

Table 5.2.0.1 Vocabulary depletion when evolving from application domains towards programming domain

	Application Domains		*UML Domain*	*Programming Domain*
Structural view	*–(Unlimited) description vocabulary for structural part of all systems —* System, problem, world, solution, software, data, database, data warehouse, data repository, knowledge base, object, class, agent, property, characteristic, variable, parameter, specification, rule, etc.		Class, association class, interface, object, datastore, component, package, node, artifact, part, composite structure, port, profile, stereotype, property, relationship, association, connector, dependency, namespace, multiplicity, role, expression, constraint, etc.	Class, object, attribute, variable, parameter, namespace, component, program, code, web service, etc.
Behavioral view	Functional View or Process view	Operation, function, service, procedure, action, task, process, business process, script, inference, transformation, etc.	Operation, use case, action, activity, event, condition, state, state machine, control flow, data flow, trigger, signal, message, time interval, time constraint, interaction, lifeline, etc.	Method, function
	Dynamic View	State, evolution, tendency, movement, evolution, collaboration, interaction, event, condition, interruption, time, etc.		Execution, execution path, thread, program state

5.3.1 Concepts used in Process Domain

Before mapping a modern application like BP model, let us start with the mapping of a more elementary Process Model.

This mapping will account for the mapping of all Process methodologies in the past to a unique UML-compatible Process Model. Some software companies still make use of Process methodologies (Stevens et al., 1974; Yourdon and Constantine, 1979; DeMarco, 1979; Gane and Sarson, 1978; Ward and Mellor, 1986) and it would be interesting to achieve such an exercise. Moreover, the exercise is interesting by several aspects:

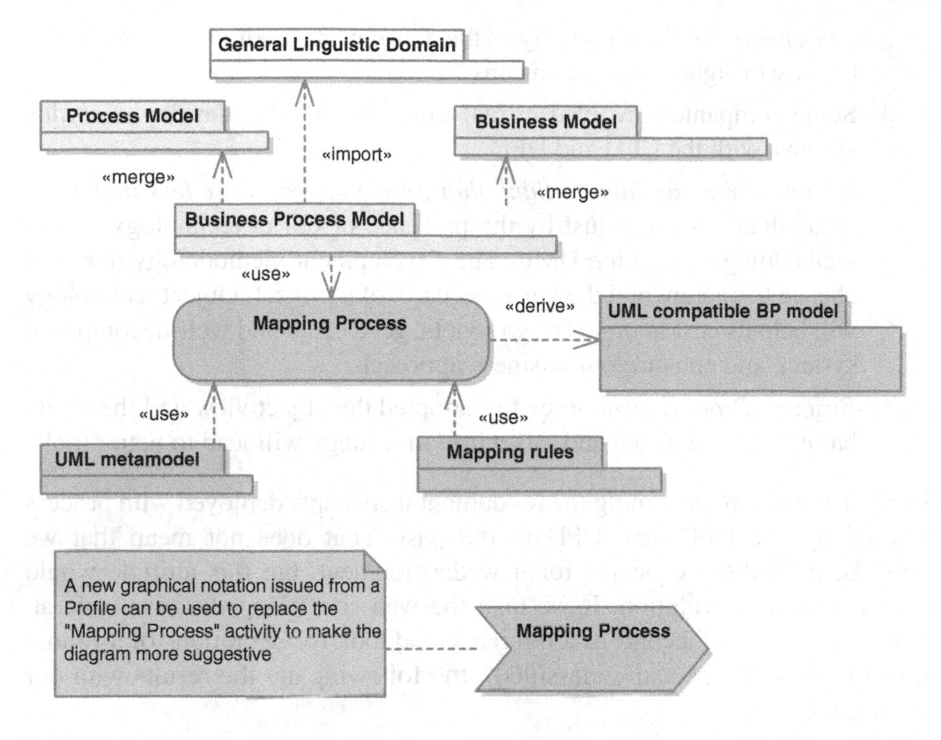

Fig. 5.3.0.1 Mapping of the Business Process Model into a UML compatible BP model. The Mapping Process BP is represented with a UML activity but could be replaced by a more suggestive graphical notation. All nodes are UML packages. All connections are stereotyped dependencies

1. Functional concepts are fundamental.
2. There is a *huge amount of projects* developed with older process methodologies, with DFDs, CFD (Control Flow Diagrams) that currently operate in industrial environment or in governmental organizations that *need*

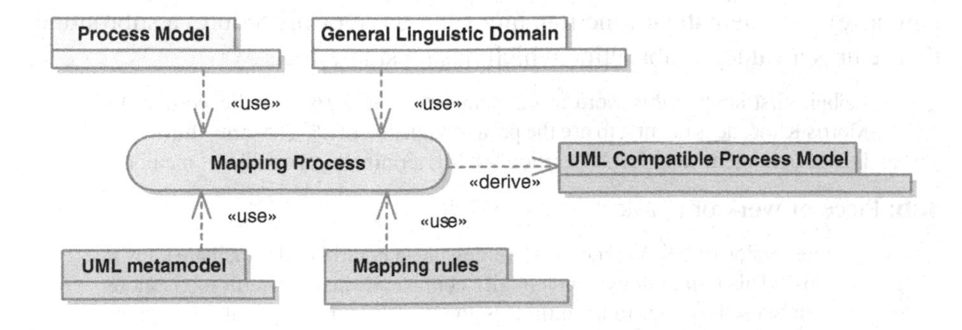

Fig. 5.3.1.1 Mapping of the Process Model directly into a UML compatible Process model

conversion, totally or partially, to the new object paradigm, only for maintenance or lightweight extensions.

3. Some companies involved in real-time systems still develop everyday systems with the CFD and DFD.
4. It would be a *bad idea to think that process aspects are less important* today than yesterday just by the presence of object technology, object methodologies, and the UML. The development methodology does not change the nature and the "process part" of a project. Object technology will help us best in organizing models, get a clean and well-decomposed system, and enhance our business approach.
5. *Business Process technology* has adopted the object view and the OMG has endorsed the idea and this win-win strategy will lead to better tools.

Hereafter, we start examining the fundamental concepts deployed with process technology, the DFD, and CFD in the past. That does not mean that we must keep all those concepts for new development, but this attitude would ensure a smooth evolution. If we take the web site Wikipedia (available at: http://www.wikipedia.com) as a reference and look for definitions of all functional terms (alphabetically classified), the following are the results with our comments:

Action: (Lack of computer-related definition, only in philosophy) certain kind of thing a person can do.

> Action is a metaclass in UML and means a "fundamental unit of behavior specification" or "fundamental unit of executable functionality."

Activity: Generally action

> Circular definition is very common with elementary concepts in General Linguistic Domain. In UML, an activity specifies the "coordination of executions of subordinate behaviors", using a control and data flow model. Activity is a metaclass defined in UML.

Function: (In computer science and depending on the context and programming language) a mathematical function, any subroutine or procedure, a subroutine that returns a value, a subroutine which has no side-effects

> Leibniz first applied this word to mathematics around 1675, and according to Morris Kline, he is the first to use the phrase "function of x". The term "function" is used to implement object "operation" (UML terminology) or object "method."

Job: Piece of work or a task

> In some version of UNIX, processes running under control of the shell are known as "jobs". This terminology is potentially confusing, since the term job is more commonly used to refer to an entire session, i.e. to everything that a computer is doing. The old gang knows that in FORTRAN environment, everything was a "job" in IBM machines (compiler job, linker job, execution job, etc.). This term is now obsolete and has been returned to the General Linguistic Domain.

Method: (Science) codified series of steps taken to complete a certain task or to reach a certain objective

In object technology, a method is the processing that an object performs. When a message is sent to an object, the method is implemented to take into account the message and perform the desired action.

Operation: In its simplest meaning in mathematics and logic, an operation combines two values to produce a third (e.g. addition). Operations can involve mathematical objects other than numbers

Operation is a metaclass in UML and correspond the object "method" in object technology. Common meaning of "operation" is a "planned activity". In Assembly language programming, operation is equivalent to "assembler instruction".

Process: Naturally occurring or designed sequence of changes of properties/attributes of a system/object

A process normally comprises a number of activity steps. If not assimilated to its elementary "activity" component, a medium size process has a goal, needs resources, data, can be assisted by computers, humans to progress through those activity steps. This term has been retained to open a new era for workflow management study and business process automation.

Procedure: Subroutine or method (computer science), a portion of code within a larger program

The PASCAL programming language made use of this term to designate a subroutine or a portion of code equivalent to a C function block. Some authors assimilate procedure to "algorithm". Wikipedia defines an algorithm as a procedure (a finite set of well-defined instructions) for accomplishing some task which, given an initial state, will terminate in a defined end-state. Notice the circular reference of "algorithm" to "procedure.".

Service: (Lack of appropriate definition for computer science) economic activity that does not result in ownership

We can say that objects provide "services" through operations (UML level) or methods (implementation level). In large and distributed systems, it is useful to view the system as a set of services offered to clients in a service-based architecture. "Service" is used mainly in Client/Server architecture (Web Services is specific services in Web environment). In any system, if an object provides an operation, owns this operation, and executes this operation only though external object requests, we are in the context of client/server architecture and objects operations may be called "object services."

Task: Part of a set of actions which accomplish a job, problem or assignment

In Operating System Domain, process and task are synonyms. Some authors make use simultaneously of three terms (process, task and thread) and distinguish between "a privileged task" and "a non privileged process" in an Operating System. Others consider only two concepts "processes and threads" as OS buzzwords and the term "task" is returned to the General Linguistic Domain.

Thread: Sequence of instructions which may execute in parallel with other threads

In Operating System Domain, a "thread" characterizes a lightweight process or simply a single path of execution through a program. If we refer to Microsoft terminology, a thread is the basic unit to which an operating system allocates processor time. Many threads can execute inside a process (e. g. multithreading).

In .Net Framework Developer's Guide, "an operating system that supports preemptive multitasking creates the effect of simultaneous execution of multiple threads from multiple processes. It does this by dividing the available processor time among the threads that need it, allocating a processor time slice to each thread one after another. The currently executing thread is suspended when its time slice elapses, and another thread resumes running. When the system switches from one thread to another, it saves the thread context of the preempted thread and reloads the saved thread context of the next thread in the thread queue. The length of the time slice depends on the operating system and the processor. Because each time slice is small, multiple threads appear to be executing at the same time, even if there is only one processor. This is actually the case on multiprocessor systems, where the executable threads are distributed among the available processors."

Transformation: (No general definition in computing domain). A data transformation converts data from a source data format into destination data

This term was used in the past when dealing with "data transformation" in Dataflow Diagram. It looses now its importance and is returned to common vocabulary.

Processes and all process-like entities needs *inputs*, *outputs*, *data flows*, *control flows* (connecting processes together), *data repositories* (acting as data sources and data sinks). To support this process view of system, all those entities should compose a minimum and coherent working set. We will clarify those notions later, for the moment, let us quote the minimum list to achieve a workable set of concepts for process modeling.

Datastore:

A datastore designates a storage element representing locations that store data, objects. The notion of datastore exists in UML under DataStoreNode defined in Activities Package. In UML, it is considered as "central buffer node for non transient information". So, datastore can be understood as a source or a sink of non transient information.

Terminator:

This notion was introduced in the past in some process methodologies to represent external entities with which the current system communicates. It can be a person, a group of persons, an organization, another system, a computer, etc.

In UML, Terminator will be mapped, according to this definition to Actor metaclass in a use case diagram, to objects "external to the current system". It would be interesting to notice that process methodologies admitted curiously in the past, the existence of "objects" outside their systems but not inside.

Data flow:

A data flow is a flow of data, information, objects between any combinations of datastore/process/terminator.

In Object Technology, data flow concept is replaced by object flow. In UML, there is no *Dataflow* metaclass but an *ObjectFlow* metaclass is defined in BasicActivities package and refined in CompleteActivities package.

Control Flow:

The notion of control flow has existed in some methodologies oriented to real time system development, e.g. SART (Structured Analysis Real Time). In UML, a metaclass *ControlFlow* has been defined in the BasicActivities package as an "edge that starts an activity node after the previous one is finished".

5.3.2 Examples of Process Modeling and their Conversions into UML Diagrams

Without adding application difficulties to the mapping problem, we choose a classic "order processing" to shows concepts used in Process modeling: processes, datastores, terminators (external entities), data flows, and control flows. We let in the background the correctness of the application logic itself and focus on the way the mapping is conducted to translate any process model into the UML model.

The Figure 5.3.2.1 shows a process diagram. Owing to the round circles used to represent processes, this kind of diagram was in the past, commonly called:

1. Bubble diagram
2. Bubble chart
3. DFD
4. Process model
5. Workflow diagram
6. Function model

Some methodologies (Gane and Sarson, 1978) make use of the same concepts but the round circle has been replaced by rectangles with rounded corners. So, graphical notations may vary from one author to another, but as long as the concepts are the same, all those diagrams are of the same type.

The terminator *Customer* is repeated twice but they are the same. This repetition avoids drawing complicated connections. Terminators are considered as external to the *Order Processing* system (so the contents of a Terminator cannot be changed). Dashed arrows represent control flows and regular arrows are data flows. Datastores or stores are named Repositories in this application and are

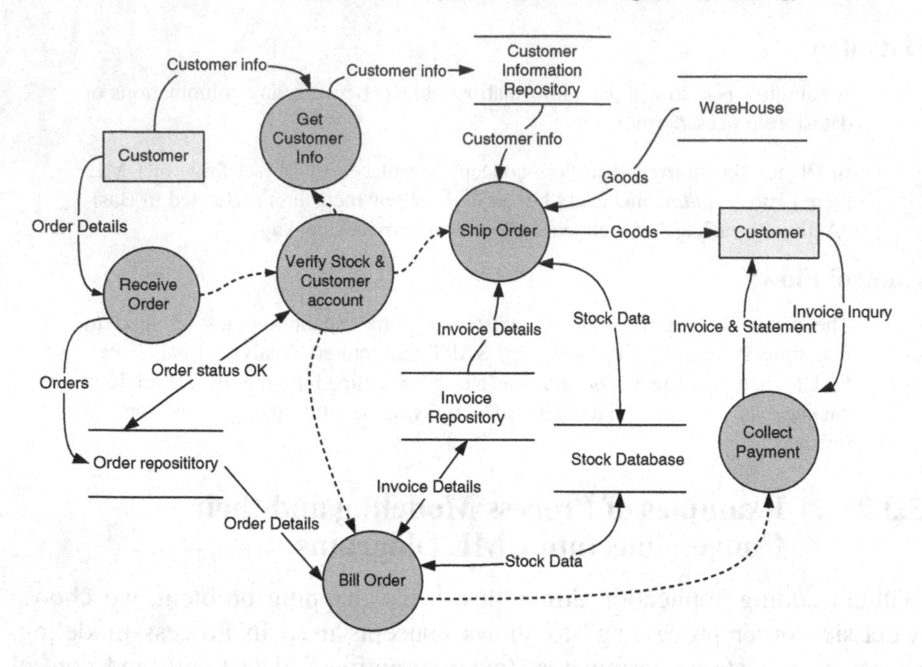

Fig. 5.3.2.1 Bubble process model of an Order Processing. Bubbles are processes, "Customer" is Terminator. Datastores are represented with two horizontal bars. Arrows are data flows and dashed arrows are control flows

represented by two horizontal bars. Datastores in this example depict an object repository (*Warehouse*) or a database (*Customer Information Repository, Order Repository, Invoice Repository*, and *Stock Database*). No precision is done at this step on the term *Repository* so a datastore may represent a database or a repository (metallic cabinets) for printed documents (orders and invoices).

The conversion from a DFD into an activity diagram is not straightforward and there exist some subtleties:

1. *Objects in an activity diagram are tied to object flows; they are not vertices but edges*. With the current version of the UML, the notion of "object" supported is object accompanying object flows. In this case, objects are attached to edges (connections) but not really to vertices (nodes), even if the diagram considers them as vertices. The activity diagram has also a notion of "partition" that identifies objects responsible of all activities located in this partition. Once more, partitions are not vertices but act as superposed layers on the activity diagram. So, to summarize, in the current activity diagram, we cannot draw an object freely as in an object diagram. For instance, we cannot illustrate that an object activates an activity. This difficulty can be patched for the moment with an object flow (edge) drawn directly between an object (vertex) and an activity

Table 5.3.2.1 Mapping rules of a simple Process Model intothe UML

Process Model	*UML model*	*Comment*
Process	**Activity, Action** Activity Action	A Process can be mapped into Action or Activity in the UML. Process model does not have an equivalent structural view of systems, so Operation of Class has no equivalent. The same problem occurs for use case. *Use Case* metaclass is reserved for Requirements Analysis, but Process model does not distinguish development phases.
Terminator	**Object** Object :Class	The Terminator can be mapped into an Object with eventually a defined classifier.
Datastore	**CentralBuffer-Node, DataS-toreNode** «centralBuffer» Queue «datastore» Database	UML proposes two options: CentralBufferNode and DataStoreNode. *Example*: A warehouse is better mapped into a CentralBufferNode and all databases into DataStoreNode.
Data flow	Activity1 ObjectNode Object ObjectNode Activity2	**Dataflow** Data flows are replaced in Object context by object flows. An object flow can be differentiated from a control flow as it has at least one side an object, an object node or an activity parameter node.
Data flow between Process and Terminator	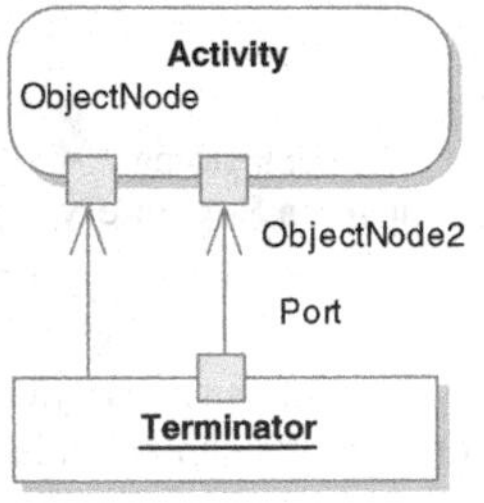	The UML allows drawing an object flow between a Terminator object and an activity (not control flow). But the UML interprets that the Terminator itself as the object of the exchange. So, to avoid any ambiguity, object node, activity parameter node, and Port can be used to create data flows between Process and Terminator.

(*cont.*)

Table 5.3.2.1 (Continued)

Process Model	*UML model*	*Comment*
Control flow	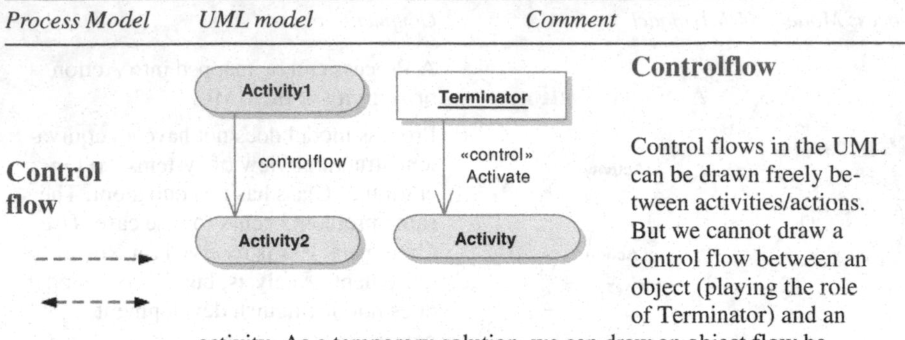	**Controlflow** Control flows in the UML can be drawn freely between activities/actions. But we cannot draw a control flow between an object (playing the role of Terminator) and an activity. As a temporary solution, we can draw an object flow between an object and an activity and interpret this connection as a control issued from Terminator. A stereotype can be added and if necessary a name to avoid misinterpretation.

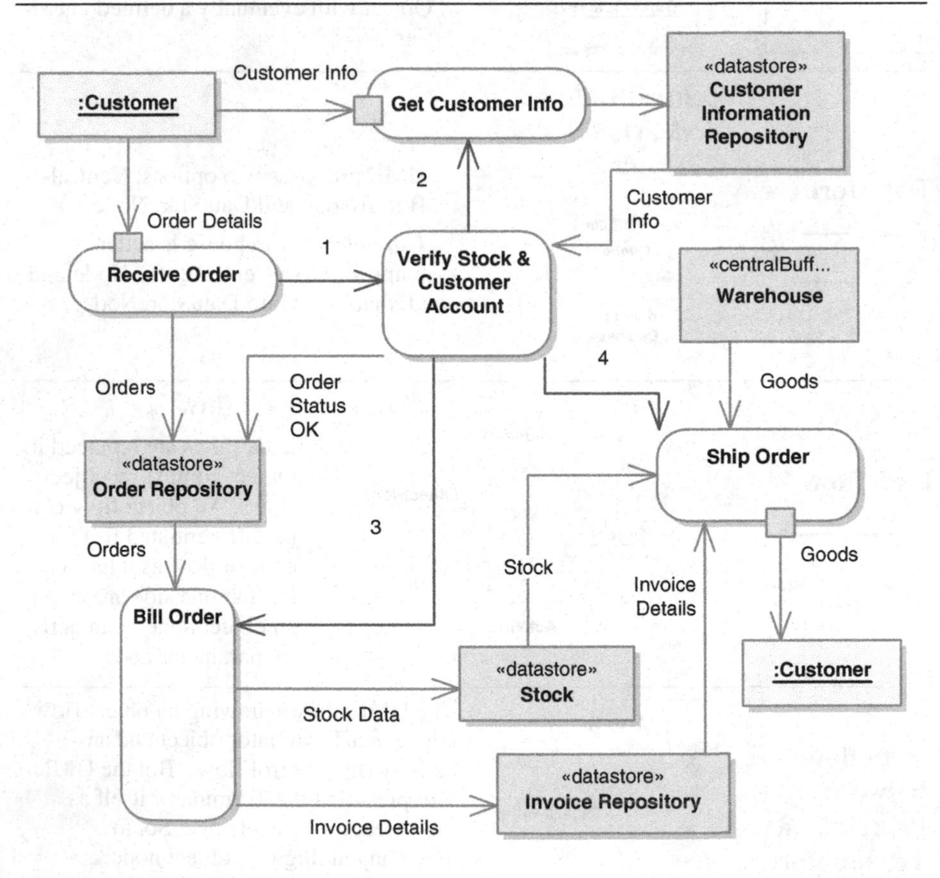

Fig. 5.3.2.2 Conversion of the DFD "Order Processing" into an activity diagram. Control flows are numbered to help interpreting good sequence order. Ports can be added to <<datastore>> and typified if needed

(vertex) but interpreted as a control issued from the object. A stereotype and a name can be added to avoid misinterpretation (see last line of Table 5.3.2.1).

2. *Objects flows cannot be bidirectional.* Normally, when the process "Bill Order" can write to Stock (for instance to reserve some stocks for further delivery), we supposed that the write permission entails automatically the read permission. If we adopt this rule locally (by a constraint imposed to some diagrams or to the whole project), we don't have to represent two object flow arrows, one in the read direction and one in the write direction. Object flows are directional in the current version of the UML and there is no bidirectional object flows.
3. The *CentralBufferNode* concept is used to represent the Warehouse as goods entering a warehouse are supposed to leave it quickly. All databases are represented by *DatastoreNode* concept that can be used as a data sink as well. There must be some difficulties in grayed zone when data or objects are stored with intermediate duration. In this case, the best would be using a stereotype describing exactly the type of datastore we want to create.

5.3.3 Building a Proprietary Methodology

It is usual to see that big companies have in the past designed their proper methodologies, their proper graphical notations, and interpretation rules. This situation has in the past given rise to several thousands of methodologies and it would be a useless task to create an inventory for all of them. The question for those companies would be "is it possible to align smoothly towards the UML without having to convert existing designs?" Two extension mechanisms are already mentioned. With the UML, the second class extension can be done via "Profiles" (set up a UML Profile with proprietary names). The first class extension can be realized at the Infrastructure level but the risk of encountering incompatibilities using UML tools must be weighted. We suggest a third strategy named the "compatible strategy" described in Figure 5.3.3.1. This strategy lets the UML semantics intact and suggests another way of interpreting fundamental concepts with available graphical notations and connection rules.

5.3.4 Business Process Domain

The functional view of systems remains at the foreground scene contrarily to some prognostics of object theorists of the first wave. Processes become more structured and find their strength mainly in business models. To deal with business processes, three levels of intervention are separated in the way processes are approached in an enterprise environment:

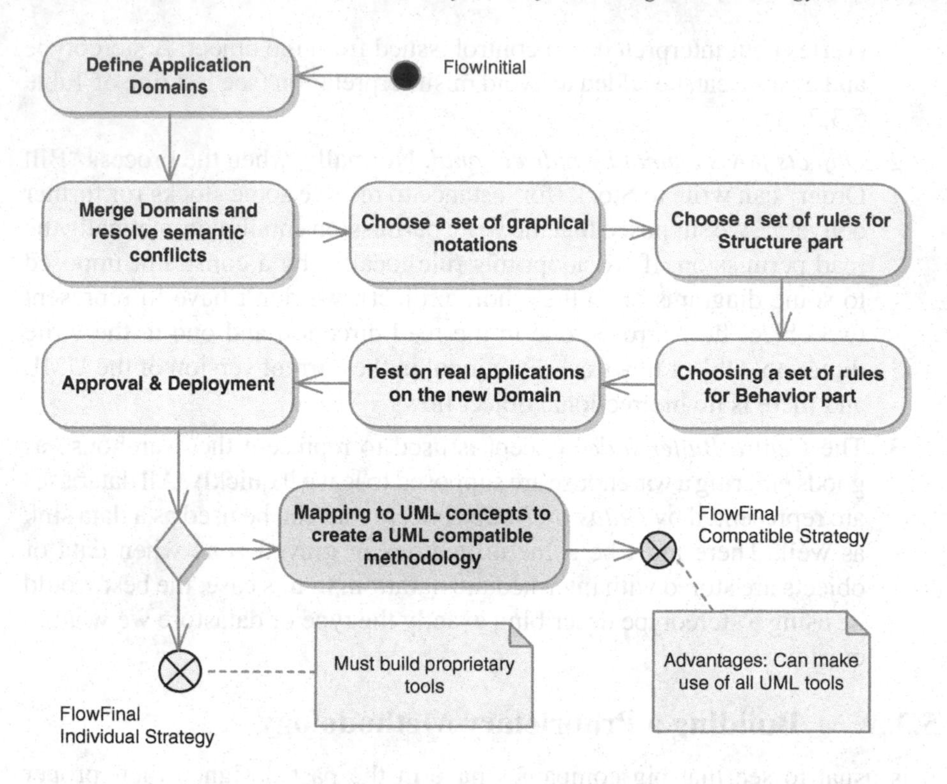

Fig. 5.3.3.1 Compatible strategy for adapting an existing methodology to the UML. While keeping the proprietary methodology, a mapping table can be created to convert concepts of the proprietary methodology into UML concepts (Profiles can be used in some steps)

1. The *operational level of business processes* that is the deployment of a designed process to manage daily operations with customers, suppliers, partners, etc. To deal with this level, a *Business Process Modeling Notation* (BPMN) has been standardized by the BPMI (available at: www.bpmi.org) and the document is currently available on the OMG site
2. The *conceptual level of business processes* known as *Business Process* or *Workflow Management*. This level specifies, describes, evaluates, uses metrics on workflows, optimizes, tries to automate parts of processes when possible, to adapt business processes to the moving environment (technology, competitor, supplier, customer, regulation, etc.)
3. The *metaconceptual level of business processes* identified by the OMG as the *Business Motivation Model* (BMM). This level takes into account higher-level concepts like organization missions, goals, objectives, strategies, tactics, policies, internal, and external influencers, organization assessments to influence the way business processes must be designed, taking into account rules at the workflow management level. The BMM

has been standardized and the document is currently available on the OMG site.

To explain the difference between these three levels, let us start with the first operational level of business processes.

A business process is a process in the business domain and a workflow is concerned with the information required to support each step of the business cycle. The BPMI has developed a standard BPMN and in June 2005, the BMPI and the OMG groups announce the merger of their BPM activities.

> The new Business Modeling & Integration Domain Task Force mission is to develop specifications of integrated models to support management of a an enterprise. These specification will promote inter and intra-enterprise integration and collaboration of people, systems, processes, and information across the enterprise, including business partners and customers.

The final specification of the BPMN has been adopted in February 1, 2006 by the OMG and can be consulted freely on the OMG site [available at: www.omg.org]. This specification is out of scope of this book, but we are concerned with the mapping of business applications into the UML. A limited but principal set is illustrated. The rationale behind the BPMN comes from the observation that business people are very comfortable with visualizing business process in a flow-chart format instead of the activity diagram proposed in the UML.

BPD (Business Process Diagram) is a diagram designed for use by the people who design and manage business processes. The BPMN provides a formal mapping for BPEL4WS (Business Process Execution Language for Web Service) that is a standard copyrighted by BEA Systems, IBM, Microsoft, SAP AG, and Siebel Systems.

Fundamentally, the BPD is much closed to the UML activity diagram, if not nearly the same. Graphical notations are enriched and oriented towards business process applications to make the diagram more appealing to business people that are not familiar with the stripped esthetics of a technical diagram like the UML activity diagram. Some concepts come from metaclasses borrowed from the UML Interaction Suite or State Machine (e.g. Event).

In BPD, three concepts have been selected to represent functional concepts: *process, sub-process, and task.* Process, subprocess are all UML *activities.* A task can be mapped into the UML action or sometimes into activity. The concept of *Process* is *intrinsically hierarchical* in the BPMN. Processes may be defined at any level from enterprise-wide processes to that performed by a simple person. A Business Process is a "set of activities that are performed within an organization or across organizations."

A task is reintroduced as an atomic activity that is included within a process. It is used when the work in the process is not broken down to a finer level of process model detail. This proposal of *task* for substituting to *elementary process* avoids the same sound for different things inside a same domain. A *sequence flow* and a *message flow* replace the control flow and data flow in an activity diagram.

To support all the Business Suite, SBVR (Semantics of Business Vocabulary and Business Rules) is explained in a 390-page book that defines the technical vocabulary and rules for documenting the semantics of business vocabulary, business facts, and business rules. The proposal of this book is to formulate technical English in such a way that *quantification* (each, some, at most one, at most n, etc.), *logical operations* (p and q, p or q, p if q, it is not the case that p, etc.), *modal operations* (it is obligatory that p, it is impossible that p, must, must not, always, etc.), and specific keywords (the, a, an, a given, that, is of, etc.) could be recognized and semantics extracted for computation.

In Figure 5.3.4.1, we have two *Pools* (Supplier and Buyer). *Pool* is equivalent to *ActivityPartition* in an activity diagram that represents Participants in the order process. Small round circles at the left are Start Event (equivalent to *InitialNode* metaclass in UML) and the same small circles with a more accentuated line thickness are End Event (equivalent to *FlowFinalNode* metaclass). All rectangles with rounded corners are processes or tasks (tasks are elementary

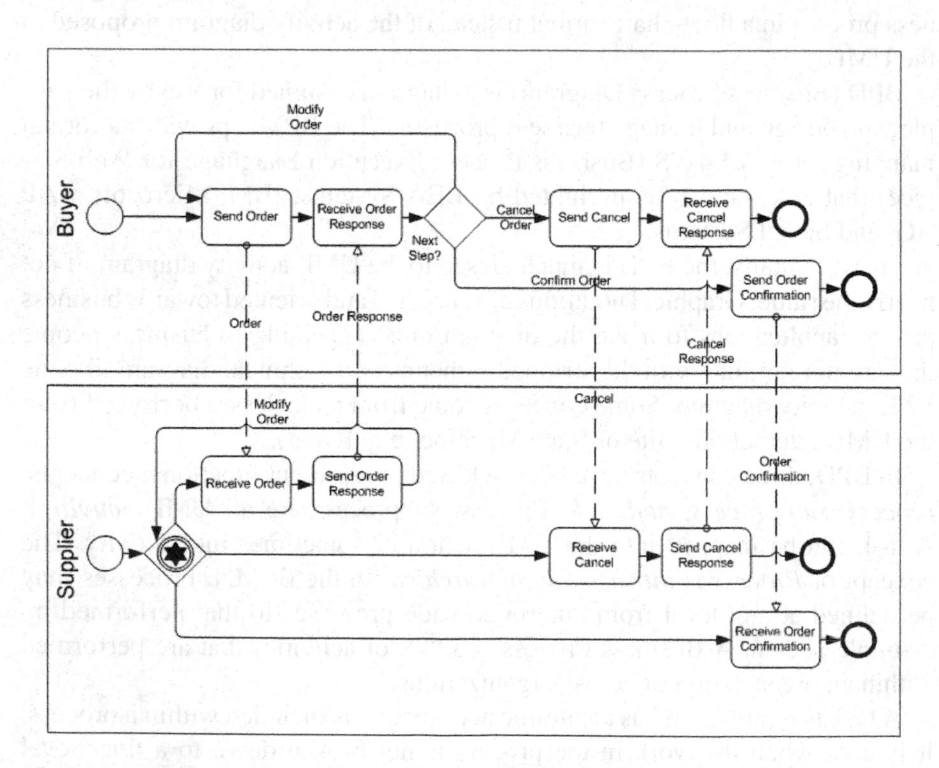

Fig. 5.3.4.1 Overview of Business Process Modeling Notation (BPMN) of an Order Processing in the site of www.bpmi.org

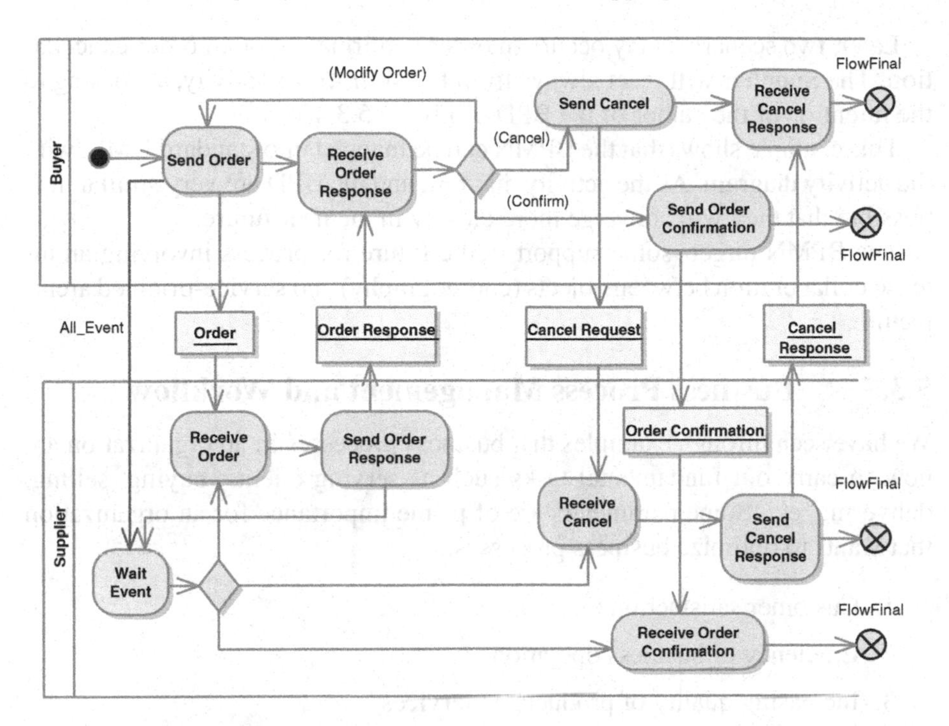

Fig. 5.3.4.2 Order Processing diagram converted from Business Process Modeling Notation (BPMN) to activity diagram notation. This diagram observes the original intent of its author

processes). They can be mapped into *Action* metaclass or eventually *Activity* metaclass.

DecisionNode metaclass in an activity diagram (represented with diamonds) are called *gateway controls* in the BPMN. The gateway in Supplier pool that contains a David star is a subcategory called "event based" (Exclusive OR type).

All solid arrows are *sequence flows* used to show the direction of process execution (equivalent to *control flow* in the UML). Dashed arrows crossing the pools are *messages* exchanged between Supplier and Buyer Pools. The notion of object flow in an activity diagram is replaced in the BPMN by *message flow* without requiring any object node placed somewhere on the trajectory.

We have converted the diagram in Figure 5.3.4.1 into its equivalent in Figure 5.3.4.2. All vertical arrows are object flows, except *All_Event* that is a control flow. When *Send Order* activity from the Buyer receives a first "token," it sends a token to *Wait Event* activity, waiting for an object *Order* sent by Buyer's *Send Order* activity and received by Supplier's *Receive Order* activity. The token moves afterwards in the Supplier's *Send Order Response* Activity that sends an *Order Response* back to the Buyer. Later, the token is refunded to *Wait Event*.

If the Buyer decides to modify the order, a new *Send Order* process will be initiated and the loop in the Buyer activity may be executed any number of times needed to obtain a full agreement between Buyer and Supplier.

Later, two scenarios may occur: an order confirmation or an order cancellation. The Supplier will react always from the *Wait Event* activity, according to the intention of the author of the BPD of Figure 5.3.4.1.

This example shows that the BPMN can be mapped into standard UML, with the activity diagram. As the activity diagram and the BPD are very similar, it is possible that they will converge more closely in the near future.

The BPMN targets some support in the future for process involving an intense collaboration between objects (choreography) and service-oriented architecture.

5.3.5 Business Process Management and Workflow

We have seen through examples that business processes in an organization are how to carry out fundamental tasks such as serving clients, buying, selling, delivering, etc. Some parameters are of prime importance for an organization that wants to optimize business processes.

1. Customer satisfaction
2. Efficiency of business operation
3. Increasing quality of products or services
4. Reducing production cost
5. Reducing operation cost, etc.

All those parameters need a constant *reengineering of business processes* to adapt themselves to the moving environment (technology, competitor, supplier, customer, regulation, etc.). *BPM* or *Workflow Management* is the branch that studies how well business processes are conducted in an organization. Sometimes, workflow management is considered more local or as addressing the same concept at a smaller scale than the BPM, but the frontier between these two concepts would be in the future a matter of convention.

Workflow Management (Georgakopoulos et al., 1995) is the activities of modern organizations. It involves coordination of daily activities, management of organizations resources, administration of internal information system; the whole process must comply with some business logic. Contemporary workflow management has been preoccupation in early 1970 with office automation (Ellis and Nutt, 1980), which were oriented towards the automation of human-centric processes. One of the famous implementation of the BPM was the Six Sigma concept initiated in the early 1980s at Motorola Company (Pande and Holpp, 2002). If we want to reach the workflow management ancestors, Fayol and Taylor in the first half of the 20th century were notorious pioneers in the domain of workflow management.

Another direction of workflow automation targets low level processes only. For instance, in the domain of Web Services, WSCI (Web Service Choreography Interface; available at: http://www.w3.org/TR/wsci/), WSCL (Web Services Conversation Language; available at: http://www.w3.org/TR/wscl10/), or BPEL4WS (available at: http://www.bpmi.org/) do not contains any human actors/agents and are rather focused only on technical coordination of interenterprise processes.

The market seems to seek a new standard for the BPM or workflow management. At the writing of this book, the last conference of this group (BPM Think Tank) was organized at Arlington, Virginia in May 2006 that gathered the business and technology experts in order to shape the BPM domain in the future. On the site of the OMG, people can still access *a Workflow Management Facility Specification* of 96 pages dated from April 2000.

5.3.6 Business Motivation Model

The notion of *motivation* is introduced to characterize this metalevel of business processes. If, at the conceptual level, we can say why a given business process is ill organized or takes so much time to execute, and what is the remedy to improve a given process, at this metalevel, we must be prepared to the question "if an enterprise prescribes a certain approach for its business activity, it ought to be able to say why." So the question is targeted to the "raison d'être" of a business process and not the fact of admitting its existence as a fatality and stumbling on it year after year. A motivation model is thus recently coined to be able to give a correct answer to this inquisition. In addition, modern organizations tend to take into account new criteria that overstep the bounds of its proper structure, its internal components, and resources. Modern organizations, besides being constantly concerned about reducing the cost of doing business and optimizing profit, are more sensitive to the following discourses:

1. Better understand business by supporting internal information systems
2. Better prepared to deal with global competition
3. Identify new business opportunities
4. Rapidly develop new services and products
5. Set a business apart from competitors

Examples of this metalevel of business processes can be found in the literature:

BMM, 2005

[http://www.omg.org/news/meetings/ThinkTank/presentations/T-10_Hall_2.pdf

www.bpmi.org

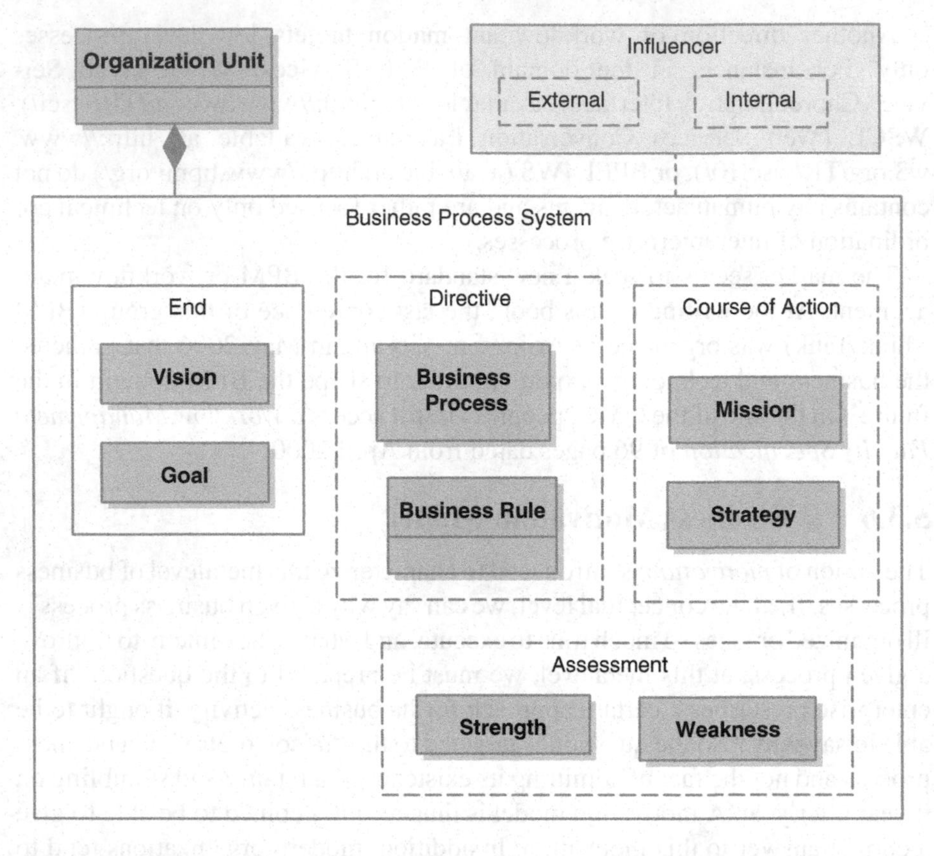

Fig. 5.3.6.1 Oversimplified version of the Business Organization as inspired from drafts on the OMG site to explain the mapping of business processes

www.businessrulesgroup.org]
Audris Kalnins and Valdis Vitolins
[http://melnais.mii.lu.lv/audris/Modeling_Business.pdf]

Once more, the BMM is outside the scope of this book, but its metamodel exemplifies a very interesting process model structured as a class diagram instead of a dynamic diagram like an activity diagram. So doing, from a fundamental viewpoint, it is not mandatory that processes must be mapped into class operations, actions, or activities. A whole process may constitute an entire class but this mapping must be justified by solid criteria.

Figures 5.3.6.1 and 5.3.6.2 summarize the BMM model. If we oversimplify the description of the BMM, a BMM starts often to define a *Business Organization* that is responsible for several *Business Processes*. Organizations are created with *missions*. To accomplish missions, organizations must have *visions* composed of *goals* quantified by *objectives*. Each mission is planned with a *strategy*

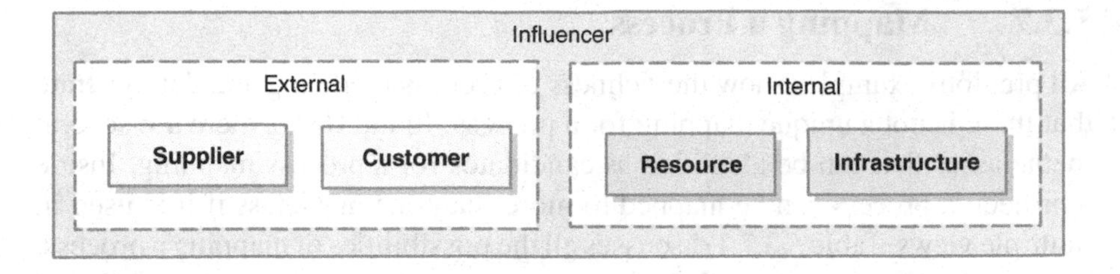

Fig. 5.3.6.2 Reduced version of the Influencer as taken from the OMG site

implemented by some *tactics*. Strategies and tactics must be compliant to *business rules* derived from *business policies*. Missions own some *risks* but have potential *rewards*.

The context of the organization is made of *internal and external influencers*. Influencers are objects inside or outside an organization that may affect the organization. They represent anything that has the power and at a less extent the possibility to influence and persuade a decision maker.

Customer assessments are beliefs (strength, weakness of the organization) that guide customer behavior. They may represent processes conducted to get these parameters. ISO 9126 defines assessment as a formal evaluation of a particular set of processes.

For a business organization, minimally, external influencers comprise two classes *Supplier* and *Customer*. Internal influencers are minimally *Infrastructure* and *Resource* classes. If internal classes must be developed, they can constitute abstract classes at the top of a derivation hierarchy to reach specific resources and particular piece of Infrastructure.

If an *Organization* owns a *Business Process Automation System*, Figure 5.3.6.1 gives some usual classes that could be defined. First a *Vision* class allows instances of a set of visions for the organization. *Goals* are business plans. A complete hierarchy of goals and subgoals can be built to realize missions that correspond to the visions of this organization.

The term *mission* can have many meanings (responsibility, even sometimes interpreted as vision) but the meaning adopted in the BMM is a series of planned operations to reach some goals. In this case, a mission can be modeled as a set of activities (or composite activity) within an activity diagram. To deploy a mission, we need a strategy instantiated from the *Strategy* class that gives recipes (knowledge) to build missions. Missions are ongoing operational activities of an organization.

Business processes realize courses of action (missions), provides processing steps, sequences, structure, interactions, and set up events. Business processes are guided by business rules. Business rules, organization units, and business processes are put into the BMM via placeholders but their exact definitions are in other business models.

5.3.7 Mapping a Process

All previous examples show the richness of process modeling and demonstrate that there is not a unique mapping for a process. In the UML, there are several metaclasses that can be identified as candidates for a process mapping. Inside a project, a process can be mapped to more than one metaclass if it is used in multiple views. Table 5.3.7.1 discusses all the possibilities of mapping a process, of any complexity, into the UML metaclasses. An object is only a "*role*" played by a real object, for instance, the Supplier or the Buyer is simply a role played by an object Person. That does not exclude the possibility that Person may buy manufactured objects that he sells.

To comment the possibility of mapping a process to a class diagram (e.g. business process), Figure 5.3.7.1 shows an example of a business process into a class diagram. A small process that can be executed individually by an object/component, map into an operation hosted by a class or a component. For collaborative tasks that need the contribution, sometimes equal of all participants, it would be hard to find a hosting object. For instance, if an enterprise starts a product line that needs the full cooperation of its staff, we can easily determine who must take the decision. But, is it really natural to affect this process to the Director object, the Board of Directors objects or the enterprise itself? The production process comprises a complete hierarchy of activities and the same problem of affecting operations to objects may be found as well at intermediate levels.

Affecting the business process to a Director means that this person is the only person, irreplaceable that could execute or instantiate the product line unless the operation is designed as a service in which case the Director works under the authority of another person (either the situation is very specific for this enterprise, either the solution exposes some incoherence). Moreover, if the Director is the only person that can start the production line, he holds all business processes of his company. This design is very centralized and may put a hard burden on the Director without any flexibility.

If we affect the business process as an operation of the Enterprise object and decide that one person belonging to the Board of Directors may instantiate the production process by invoking an operation of the Enterprise object, this choice seems to be an acceptable solution too. But, operation is a piece of behavior so states of objects depend upon current states of all their operations. For an enterprise that has hundreds of production lines, facing complexity, we cannot really combine all states and behaviors of each production line to build up composite state or behavior of the Enterprise object itself. So, if the solution could be acceptable for an enterprise with only one production line, it does not hold for an enterprise with several production lines. If a solution fits in very few

Table 5.3.7.1 Mapping of process into the UML metaclasses

UML metaclass	*Diagram*	*Comment on the mapping*
UseCase	Use case diagram	This metaclass is used mostly in requirement analysis, it must not be considered as being part of a final design or a run-time component, so it can represent all process or collaboration or service while specifying any system.
Operation of a class	Class diagram	An operation is a behavioral feature of a classifier and as such can represent any process or service of an object. The condition is the natural and easy identification of the object that is responsible for the execution of this operation or this service. *Example*: send_order() is an operation of the Buyer and *receive_order()* is an operation of the Supplier. This category of mapping includes both private and public operations of a class. A *private operation* is accessible by the object only. It can be called by a private or a public operation of this object. A *public operation* is an operation that can be called by any object including itself.
Operation of an interface	Class diagram	If the operation cannot be affected to any class and needs a collaboration of many objects to achieve it, possibly, it could be interpreted as a service accomplished by an interface from the Client viewpoint. *Example*: *take_call()* is an operation that must be affected to a generic <<interface>> Vendor if the company is a long-distance call Provider. Each time we call to 1–800 numbers, a different person takes the call. Thus, *take_call()* is an operation of an interface and it could be realized by any Employee of this company. Normally, interfaces are designed with public operation only by its semantics.
Collaboration and CollaborationUse	Composite Structure diagram	*Collaboration* is functionality, a macroscopic task that is executed by calling for collaboration of more elementary instances. A *CollaborationUse* (collaboration occurrence) is a collaboration applied to a specific context. A collaboration shows the roles and connections required to accomplish a specific task. In the context of a composite structure diagram, a process can be mapped to represent an activity executed by the collaboration of many objects. *Collaboration* or *CollaborationUse* belong to the structural view and shows only participating objects (parts) in the "process" but do not give any dynamic or evolutional information. *Example*: See Section 4.9.4

(*cont.*)

Table 5.3.7.1 (Continued)

UML metaclass	*Diagram*	*Comment on the mapping*
Action	Activity diagram or State Machine diagram	As a functional unit of executable functionality, an action can map an elementary process. At the same time, it corresponds to a class operation in a class diagram. Typically, an action is executed by an object (the action can be initiated by the object itself or an external object), a well-identified composite object, or even a component. In the case of collaboration between several objects, it is recommended to map into an activity.
Activity	Activity diagram or State Machine diagram	All processes not identified as actions could be represented in the dynamic view as activities. Activity can be found in the definition of state (Section 4.13), but activity or process is not state (a state is a set of clichés of a system taken while executing the process).
Interaction	Sequence and Communication diagram	An interaction is a unit of behavior so it can be part of a process or eventually a whole process.
Fragment	Sequence diagram	An interaction fragment is like an interaction itself so it can be part of a process or a whole process.
Complete sequence diagram		May represent part of a process, a process or collaborative process between many objects (lifelines).
Complete activity diagram		May represent part of a process, a process or collaborative process between many objects.
Complete use case diagram		May represent part of a process, a process, or collaborative process between many objects in the requirements specification phase.
Complete class diagram		*Example*: business process As a class contains slots to store class operations, theoretically, any operation could be transformed into a class. This possibility could be misused as we can be tempted to create artificial classes just as a container for operations. However, there exist several situations that this mapping is justified.
Component or complete system (set of diagrams and artifacts)		*Example*: Inscription process in a university. Project management, military operation, etc. In the business domain or in organizations, the initial view of all systems is mostly functional. Terms like "process," "management of . . . " show that some people reason uniquely in terms of processes, tasks, and functions. So systems are built to achieve processes and objects and components are only resources. This fact justifies why high-level processes could be mapped into a component or a whole system composed of all UML diagrams and artifacts to explain the process model. A whole system can be implemented to perform a single process and the role of the process is highlighted instead of the supported pool of objects.

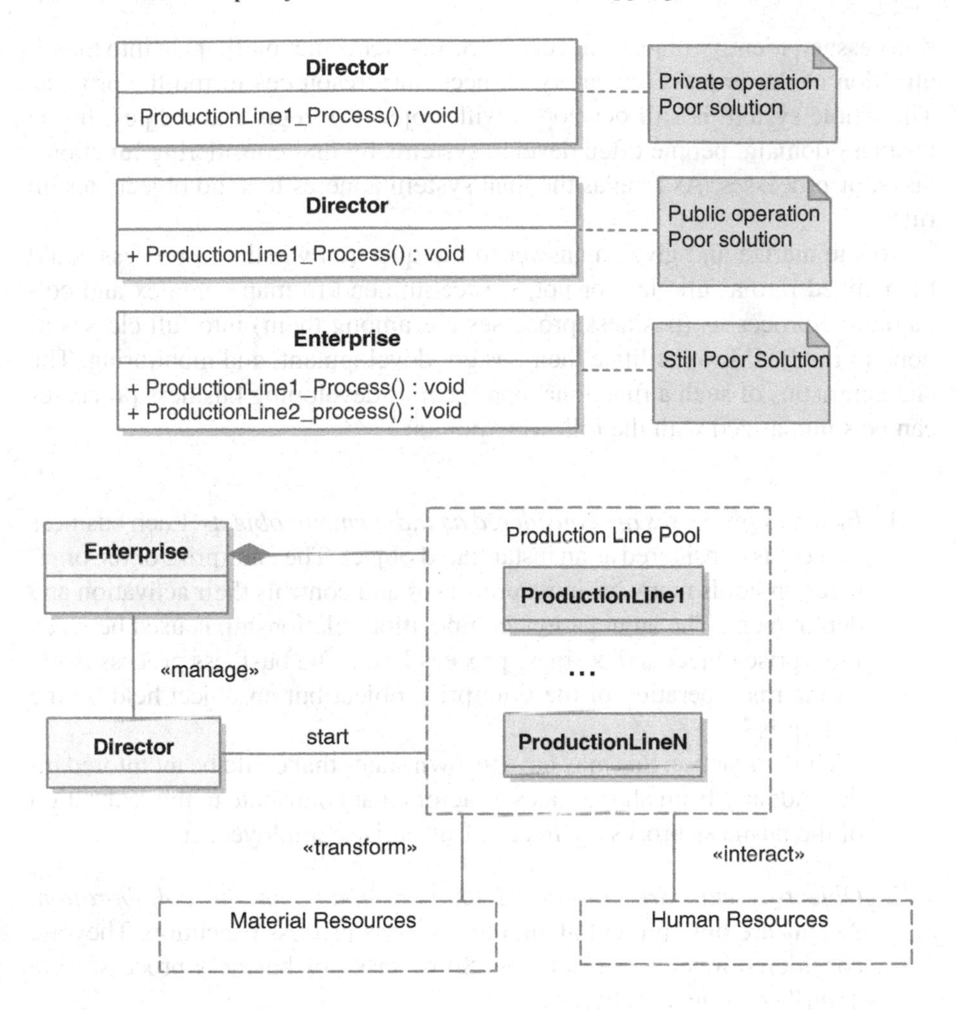

Fig. 5.3.7.1 Mapping Business Processes in an Enterprise (see text)

cases, it must be systematically suspected as a poor solution though not really incorrect.

The last solution retained is a *pool of business processes*, deriving from one or many classes. This design gives to a business processes many flexibilities and the view of systems being developed is *functional* at the top level instead of object. But this choice does not mean that we backtrack to the old functional methodology. Instead of considering how to activate structures and how to coordinate all the existing objects/agents/components to execute business processes, we start viewing a business process as a whole and independent entity. A business process can have attributes, holds various state variables to account for its evolution. We lay out all the steps to perform the process, decompose those steps if necessary, identify objects, agents, or components that participate into the realization of the process, inventory all necessary resources to run the process. The whole system is still developed with object concepts and recipes. In the business domain, people often develop systems by first considering functions, tasks, or processes. As long as the final system adheres to solid object, design rules.

To summarize and give an answer to the question whether a process could be mapped into a full class or not, we recommend to map complex and collaborative processes (business processes are among them) into full classes as done in the BMM to facilitate their design, development, and monitoring. The characteristics of such a functional approach of developing business processes can be summarized with the following points:

1. *Business processes are considered as independent objects*. Each business process is considered as an instantiated object. The enterprise or the organization holds many business processes and controls their activation and deployment. The aggregation–composition relationship is used between Enterprise object and business process Pool. The business process is not an internal operation of the Enterprise object but an object held by the Enterprise.

   The production line may have its own states that could be monitored independently from all the states of actors that contribute to the realization of the business process (Director, Enterprise, Employee, etc.)

2. *Objects/Agents are considered as participants in the collaboration*. Agents are omnipresent along the business process execution. They are considered as human resources. Some tasks of business processes are described as agent activities.

   The Director agent starts the production line. He may control and monitor the execution of the production process but the process cannot be part of his definition. He has an important "role" to play but nothing really belongs to him. Employees who participate in the production line are collaborative agents. They offer their services against salary.

3. All other objects are products, *material resources*, or *services*. We do not distinguish products that the enterprise manufactures from resources it consumes to produce products. In fact, all material objects are combined, and transformed, and the final products are only last states or results of a long transformation process so the distinction between final products and resources used to produce them is only a matter of personal preference.

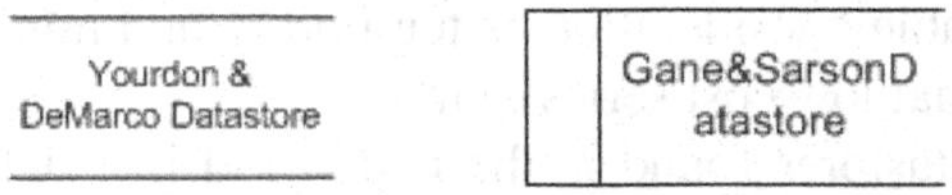

Yourdon & DeMarco Datastore

Gane&SarsonD atastore

Fig. 5.3.8.1 Datastores used in the past with DFD. There is no datastore defined in the BPMN and BMM

5.3.8 Mapping a Datastore

The datastore models a collection of data. In a DFD (Fig. 5.3.8.1), the graphical notation of a datastore is two parallel lines (Yourdon-DeMarco) or some variants (Gane-Sarson). For a software engineer, it is tempting to look at datastores as short- or long-term memories, files, databases, or objects like optical disks or mechanical cabinets to store paper documents.

The UML has two concepts, <<DatastoreNode>> and <<CentralBuffer Node>> to map datastores. Central buffer nodes collect object tokens for multiple "in flows" and "out flows" from other object nodes. "They add support for queuing and competition between flowing objects." Buffering is useful when the inputs from multiple actions must be collected at the central place or when multiple actions extract objects from a common source.

A datastore is an element used to define permanently stored data. A token of data that enters into a datastore is stored permanently, updating tokens for that data that already exists. A token of data that comes out of a datastore is a copy of the original data.

A datastore models a database in an activity diagram. A central buffer node materializes an object queue (for instance a system receives objects from a process but those objects are quickly consumed by another process and there is always a predetermined or limited number of objects inside the store). This number depends upon the size of the central buffer node. We can easily imagine the central buffer node as a "stock of merchandises" or a "printer buffer memory" that receives, in one end, characters to be printed from the processor and evacuates the same characters to the printer at the other end. Those two models of datastores are differentiated solely on the persistence of data inside the store. In a datastore node, the persistence is much longer (data can still be deleted in a database). In a central buffer node, we have just a "temporary" persistence. Objects in an enterprise warehouse are managed in such a way that input flow compensates output flow and there is roughly a constant number of products ready to deliver for a couple of days. This number is calculated by financial and business parameters like the delay imposed by the supplier, the intrinsic cost of the product (cash flow aspect), the speed of the input and output flows, or some imponderable consideration as temporary and non recurrent business opportunities (special low cost series), etc.

The range of datastores found in the real world is a rich set starting from a dumb and unstructured memory towards intelligent storage systems. Table 5.3.8.1 gives some possible mapping to the UML metaclasses.

Table 5.3.8.1 Mapping of a datastore into the UML concepts

UML metaclass	*Diagram*	*Comment on the mapping*
(*Non human*) *Actor*	Use case diagram	The metaclass Actor in the UML can be stereotyped with a *nonhuman* form to represent later a datastore. «actor,datastore» **Stock** — **Store**
Class	Class diagram	Class can be used to represent all objects starting from a dumb and unstructured memory towards intelligent storage systems. For instance: Dumb and unstructured memory Hardware cabinet, shirt pocket, desk drawer, etc. Files, directories, DVD units Any kind of tank, pool, or container Databases, databases with RDBMS or ODMNS (relational or object Database Management Systems) Storage space for enterprise stock SAN (Storage area network) Any intelligent storage system with internal operating system.
Component	Component diagram and Composite Structure Diagram	A component differs from a class in that it has an internal structure that could be eventually explained (white box view), but generally kept hidden (black box view). A class can be used instead of a component if the internal structure is not known or can never be reached or detailed. A component is therefore generally used for datastores that has a minimal structure, intelligence, or internal operating system.
Node	Deployment diagram	A node is a computational resource upon which artifacts may be deployed for execution. Node can be used in PSM (Platform Specific Model) or in implementation model to represent for instance a SAN or a server dedicated to a storage or a database.
Device	Deployment diagram	Device is a physical Node and can be used for physical datastore with implementation details.

5.3.9 Mapping Control Flows and Data Flows

In the UML, the data flow concept has been converted into the more general object flow concept, so data are considered as objects that embrace both data and objects. Control flow still remains in the UML with the same name but does not have the same meaning if compared to control flow in classical SART (Structure Analysis Real Time) methodologies.

Control flow in the UML (Fig. 5.3.9.1) is used to indicate the direction of the flow of control, to implement fork, join conditions, etc. Its nature has evolved towards the notion of transition (transfer from an activity to another activity) in a Petri Net (Peterson, 1981) but does not only mean a signal that activates or wakes up a sleeping or inactive process. Let us reproduce an arrangement of the DFD diagram in an acquisition experience before discussing subtleties in detail.

In this acquisition arrangement, we have two proccsses "Acquire and Convert" and "Read and Display" that are not decomposed. Full weight arrows are data flows. They are pathways through which data are transferred between processes, external entities, and datastores. Control flows are represented as dashed arrows oriented from the source of control towards the process to be controlled. Control flows are differentiated from data by the fact that they convey signals or data used to implement a protocol, to control events in the previous data flow networks. In the past, sometimes, analysts perceived the needs to separate the data flow network from the control flow network and drew two different diagrams (SART methodologies).

It is worth noticing that some communication protocols implement dynamics of protocols by reserving a specific set of data used as signals controlling the

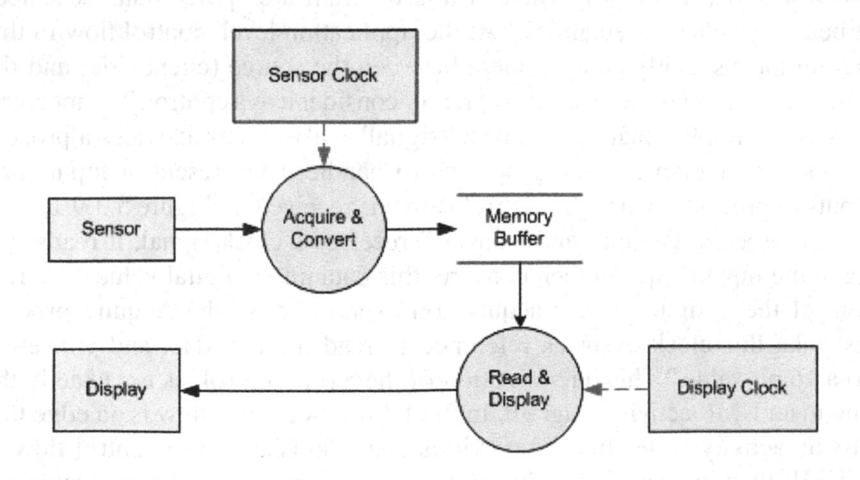

Fig. 5.3.9.1 Control flows and data flows in a DFD diagram

flow of data exchanged between a transmitter and a receptor. So, by examining the nature of data that are transmitted through communication channels, we can capture control information and differentiate them among a mass of data. For instance, in the well-known RS-232 communication lines, EOT/ACK (End of text, Acknowledge) and XON-XOFF protocols make use of some specific ASCII characters in the control block (from 00_h to $1F_h$) to regulate the flow of ASCII data through TD (Transmit Data) and RD (Receive Data) channels.

But, if we look at a large range of applications, sometimes, the distinction between data and control flows in the classic DFD is not so clear and sometimes, the process that receives data must analyze them to determine if there is some *control semantics that accompany data.*

For instance, if a clothing company receives frequently regular orders of ten thousands shirts and considers this kind of orders as regular data, the same company must reorganize its structure, seeks temporary workers if it receives an exceptional order of 200 thousands shirts. This exceptional order has simultaneously data and control meanings. First, this order is regular data if we consider that customers need what this company always produce, in occurrence shirts. Second, as the number exceeds it normal manufacturing capacity, the company must set up very quickly a temporary manufacturing structure in order to absorb this exceptional order. So, this order of "200 thousands shirts" bears some control aspects as it wakes up several uncommon processes of this company.

At the opposite end, we can find real applications that consider control signals as regular data.

For instance, all calls to 911 numbers are considered from the client side with "control semantics" but are treated as regular data by any employee working at 911 organizations.

To summarize, classic DFD diagrams allow us to display two kinds of concepts, data flow and control flow. Data in this diagram are "pure" data as it does not bear any "control semantics." At the application level, control flow in this diagram means tacitly an agreement between the source (client side) and the destination (server or service side) to really consider it as a control. The meaning of "control" implies that the received "signal" wakes up or activates a process or some uncommon internal processes to handle data present at inputs and outputs of processes when the control signal is received. In Figure 5.3.9.1, each time the process "Acquire and Convert" receives a clock signal, it reads raw data at the input (acquire) then converts this data into a digital value. The real intent of the designer in this acquisition experience is "the Acquire process must take this clock event as reference to read the raw data and convert it into a 16 bit value." This interpretation of the term "control" is not exactly the same in an UML activity diagram. In the UML, "A control flow is an edge that starts an activity node after the previous one is finished." The control flow in the UML then indicates what the next process to be activated is and supports the construction of elaborate control structures. If we come back to the DFD

Table 5.3.9.1 Meaning of UML control flow, object flow, interrupt edge

UML metaclass	*Diagram*	*Meaning and indication for the mapping*
Object flow	Activity diagram	*An object flow is an activity edge that can have objects or data passing along it.* *Mapping*: This flow replaces all dataflows in a DFD (Data Flow Diagram) and most control flows in a DFD (see text for discussion).
Control flow	Activity diagram	*A control flow is an edge that starts an activity node after the previous one is finished.* *Mapping*: This flow is a kind of "sequence flow" and indicates how activities are "fired" with time. This edge is used to implement conditions like fork, join, etc. to handle parallelism in an activity diagram.
Interrupting edge	Activity diagram	This edge is used to connect two activities together (one activity is interrupted by another activity). An InterruptibleActivityRegion may be defined to group all interruptible activities.

and try hypothetically to interpret all control flows with the UML meaning, it appears that the *Clock Sensor* must first start then the control is transferred to *Acquire and Convert* by a control flow. *Acquire and Convert* is a sink for tokens. So, *Acquire and Convert* can do the job if it receives tokens from *Sensor Clock* that is supposed to be a generator of tokens. This interpretation with *Petri net logic* (towards that the UML was oriented with its version 2) fails in most DFD we try to convert. Control flow in the DFD diagram can be mapped as control or object flow and the structure of the DFD must be modified in an important way to support token movement. In this sense, the term "Sequence Flow" used in BPMN to replace "Control Flow" could sound more appropriate as the term "control flow" presents some dangerous misleading interpretation when confronting it with the common meaning interpreted by a generation of electronic engineers and process modelers.

The conversion of the DFD of Figure 5.3.9.1 to an activity diagram is more appealing as in activity diagram, partitions can be used to separate objects and clearly show the parallelism of two processes (acquisition executed by the Acquisition Card and display process executed by the Display Processor). These two objects act independently and communicate through a *Double Access Memory.* The Memory Buffer is now characterized by its type <<CentralBufferNode>> that pinpoints its buffering nature.

The most important point is the conversion of all control flows in the past to object flows (arrows connecting two objects *:Sensor Clock* and *:Display Clock* to their corresponding processes). Object nodes "raw data," "16 bits," "analog,"

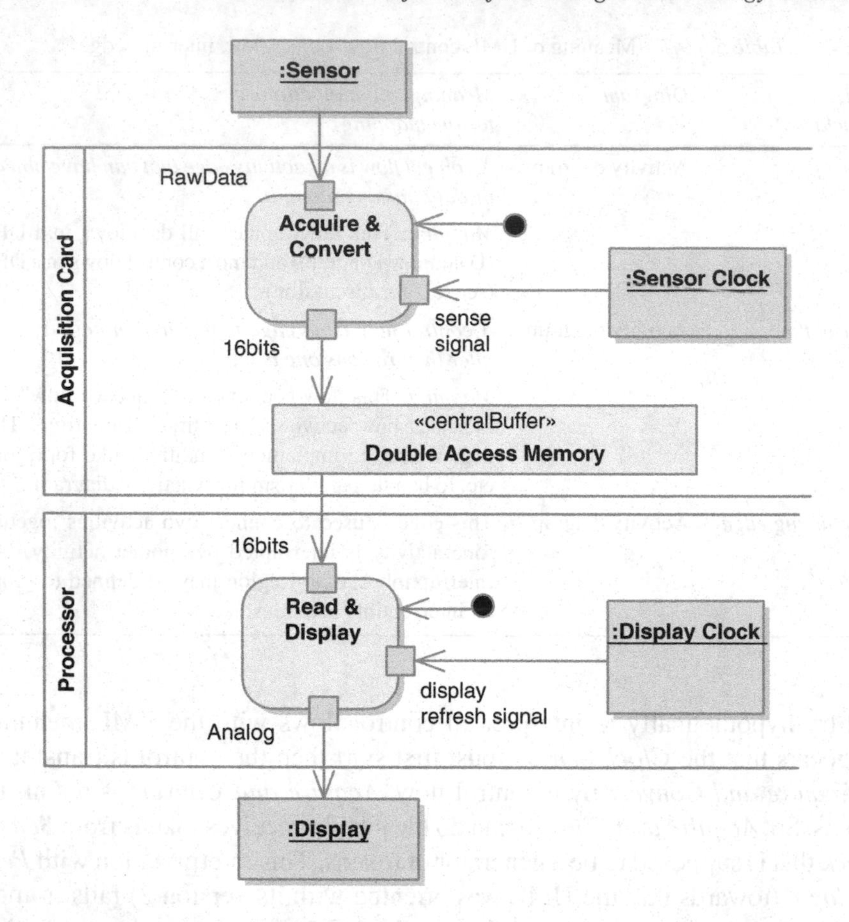

Fig. 5.3.9.2 Conversion of the DFD of Figure 5.3.8.1 into activity diagram. Signals "sense signal" and "display refresh signal" sent by clocks are used as time references to acquire and display. Signals are considered in this example as regular objects. If we represent them as control flows, when tokens traverses from Clocks to Processes, it appears as Sensor Clock and Display Clock are deactivated, that may cause interpretation problems. Actually, Clocks are source of tokens and the two processes must have token sinks. The conversion with control flows (with Petri interpretation) is more complicated

"sense signal," and "display refresh signal" are added. This conversion seems very uncommon (may eventually surprises process analysts) but it highlights one of the many differences between a classical process view and the object paradigm.

5.4 Fundamental Concepts of the Dynamic View

As dynamic view is part of system behavior, it would be natural to continue with fundamental concepts of dynamic view as they are closely related to processes, actions, and activities. The dynamic view of a system reveals temporal, behavioral, and evolutional aspects of systems. Traditionally, this view is complementary to the functional view (process view) in the study of systems, mainly real-time systems. If database development gives more importance to the data view (structural view), i.e. conceptual schemas of data, if business process engineering is focused on processes (process view), real-time systems need a complete investigation of the dynamic view to be able to build systems. This view determines:

1. States of each object in the system (set of snapshots showing how an object behaves with time or under stimuli, events, etc.)
2. Composite states of many objects taken as a whole
3. List of all events, conditions that shape the behavior, reaction of objects
4. List of actions available to each object
5. List of activities available to objects or pools of objects
6. How objects collaborate together to accomplish tasks
7. Exception processing (how systems react to error or abnormal but predicted situations)
8. Possibility of concurrence

As results, we have a set of diagrams, documents, and artifacts that allow us to explain and communicate our visions of the system to be developed. The UML has identified three complementary axes in the dynamic view:

1. Activities (activity diagram)
2. Interactions (diagrams of the activity suite: sequence, interaction overview, communication, timing)
3. States (state machine diagram)

The choice of diagrams, thus axes, to represent the dynamic view of systems depends upon the nature of the system, its best format for communication purposes, tools available at the development time, policy of the company that takes over the development, and finally some dose of personal taste of developers involved in the project.

5.4.1 States and Pseudostates

A *state* is an interesting step, phase, and evolutional stage of a system. The term "state" belongs to the UML domain and has a defined metaclass. A state of an object is defined by activities it is executing and a state does not mean necessarily "static".

A *person is walking*, a *plane is flying* are states of Person and Plane objects.

A process in an operating system may evolve through many known states as: *ready* (runable and waiting to be scheduled), *running* (currently active and using CPU time), *blocked* (waiting for an event or an IO signal to occur).

When a Vendor in a shop is waiting for customers, he is in an idle state.

Table 5.4.1.1 Description of the UML concepts related to the real world concept of "state"

UML metaclass	*Diagram*	*Comment on the mapping*
State	State diagram	*A state models a situation during which some invariant conditions hold.* This definition of the Superstructure Book can be translated into fixed or bounded values for state variables. Metaclass attributes of State metaclass are: *isSimple* : Boolean. If isSimple is true, the state is non decomposable (elementary state) isComposite : Boolean. If isComposite is True, a state is a composite state. In this case it contains at least one region. Many states, transitions, and pseudostates may fill up regions. isComposite means that the state is not elementary and could be decomposed into more elementary states if needed. *isOrthogonal* : Boolean. An orthogonal state contains two or more regions. "Orthogonal" implements AND-states logic, so the main state combines all active states, one in each region. is SubmachineState : Boolean. If true, the state is a complete submachine state so must be decomposed if details are needed. A submachine state may have regions too, so decomposition may be realized in both horizontal (regions) and vertical (substates of state until elementary states are reached).
Activity	State diagram	In a state diagram, states are defined by main activities written under the Do Action keyword.
Action	Activity diagram	Actions in state machine diagrams appear as *On Entry* action, *On Exit* action, actions executed when transitions occur. We may consider them as transient states. If the context of the problem requires that *transient states* must be studied thoroughly as people suspect abnormal behavior of the system, actions may be eligible to become regular state. *Example*: A driver may decelerate a car that passes from *cruising speed* state to *deceleration* state by just *releasing the gas pedal* (very short action) and *pushing the brake pedal* (action). For a specific study "reactions of drivers over 70," actions may become critical and justify in this case full state consideration.

(*cont.*)

Table 5.4.1.1 (Continued)

UML metaclass	*Diagram*	*Comment on the mapping*
Operation	Class diagram	When a system executes an operation (operations are defined in class), we talked of execution states. *Example*: An order management process may have a series of operations as *take_order(), verify_stock(), prepare_delivery(), prepare_invoice()*. As operations are behavioral concepts, when a system is executing a given operation, it is in a "state" of executing this operation. This fact shows a thorough coupling between the functional view and the dynamic view. When a system is examined at higher level of decomposition, it is best apprehended as *behavioral*, as done in the UML.
(full diagram)	State machine, activity, sequence, communication, timing diagrams	Notice there is no metaclass "diagram" so the term "full diagram" is put into parentheses. A *state machine diagram* can be a submachine state before decomposition process, so it is a candidate for a composite state or submachine state. An activity diagram may be used to explain state evolution. Activities have a close relationship with states. A state can be defined by an executing activity and a complex activity can go through many states when executing. A sequence or a communication diagram is a unit of interactions so they contain state evolution of a system. A timing diagram show timings of systems so it also describes system states.
Fragment	Sequence diagram	A fragment is a unit of interaction. So, fragments describe system states.

States may be categorized by *state variables* that are mathematical elements used by a designer to model a system or an object. State variables can be defined officially as object attributes. In this case, attributes are not structural but behavioral (functional or dynamic). A state occupies a time slot. Inside this time slot, we suppose that all values of state variables are fixed or bounded by an interval.

Let us take a simple LED (Light Emitting Diode) used as signal lamp in a wide range of applications. It has structural variables as *dimensions, color, and electrical intensity*. Dynamic variables are reduced to *state* that can take two binary values *On* or *Off*.

A mechanical switch or a push button plays the role of input devices for numerous systems. It has a dynamic variable that can take two binary values *On* or *Off*.

A shift lever of a car has normally a dynamic variable with the following values: *Park, Reverse, Neutral, Drive* and *D*3 or *D*2, etc. Therefore, we have a state variable with an Enum (enumeration) value.

Water may be considered as having three physical states: *Ice, Liquid and Vapor*. If its temperature T is taken as state variable, water is *Ice* when $T<0^\circ$ C, *Liquid* when 0° C $< T < 100^\circ$ C and *Vapor* when $T>100^\circ$ C. Values of water temperature are bounded and made discrete buy imposing limits.

To handle analog values, the same conversion to discrete values to ease the description of states is very frequent. For instance when the speed of a car passes from 0 to 65 mph, the state of the car is *accelerating*, when reaching this ceiling value; the car has reached its *cruising speed*. When passing from 65 mph to 0, the car is *decelerating*.

Dynamic attributes are not so evident to find out. Object actions and object activities must be identified before starting a dynamic study as states are directly tied to activities.

Before knowing that a state attribute of a Shift Lever object may take these values: *Park, Reverse, Neutral, Drive, D3* and *D*2, we must now that the Car object can park, can drive, can move backwards, can be in a Neutral state, etc. So, a preliminary functional study must be conducted to be able to define state values.

Pseudo states in the UML have already been studied in Section 4.13. The UML has defined *initial, final, deep history, swallow history, join, fork, junction, choice, entry point, exit point, terminate* pseudostates, adding diagrammatic artifacts used to guide the interpretation of the diagram and implement its logical structure. For instance, *initial* pseudostate in a diagram shows where to start reading the diagram and the next state after *initial* is a real state of the system.

5.4.2 From Elementary State Towards Global System State

The *State* metamodel of the UML (Fig. 4.13.0.2) defines a state with four parameters *(/isComposite, /isOrthogonal, /isSimple, /isSubmachineState)*. Hence, full-featured states may have regions, can be decomposed indefinitely towards more and more simple submachine states. At the bottom of the decomposition hierarchy, we reach a simple state. When analyzing system states, we have many scenarios:

1. When an object executes an operation, states of the operation are also object states relative to this operation
2. If an object performs simultaneously many operations, we still consider individual states relative to each operation unless operations are dependent. In the latter case, combined states may be required to describe object states
3. A collaborative task is executed by many objects. In this case, each object contributes to this task with their proper operation set. Normally, the task

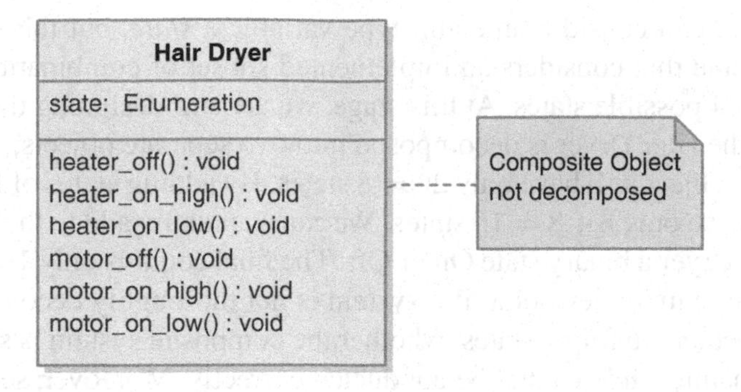

Fig. 5.4.2.1 States of the Hair Dryer not decomposed. As many as 64 states are possible at the beginning, based on six binary operations (activated or not)

has its hosting object (for instance "Production Line") and each object has its proper states (case 1 or case 2)

Figure 5.4.2.1 represents states of a *Hair Dryer* composed of two elementary objects, a *Motor* and a *Heater*. If the object *Hair Dryer* is kept as composite object and not decomposed into elementary objects, it will have operations like *heat_off(), heat_on_low(), heat_on_high(), motor_off(), motor_on_low(), motor_on_high()*. Each operation offers two states *True* and *False*. With six binary variables, we reach 2^6 or 64 possible states. Some combinations are impossible and must be rejected after analysis but we focus here on the initial count.

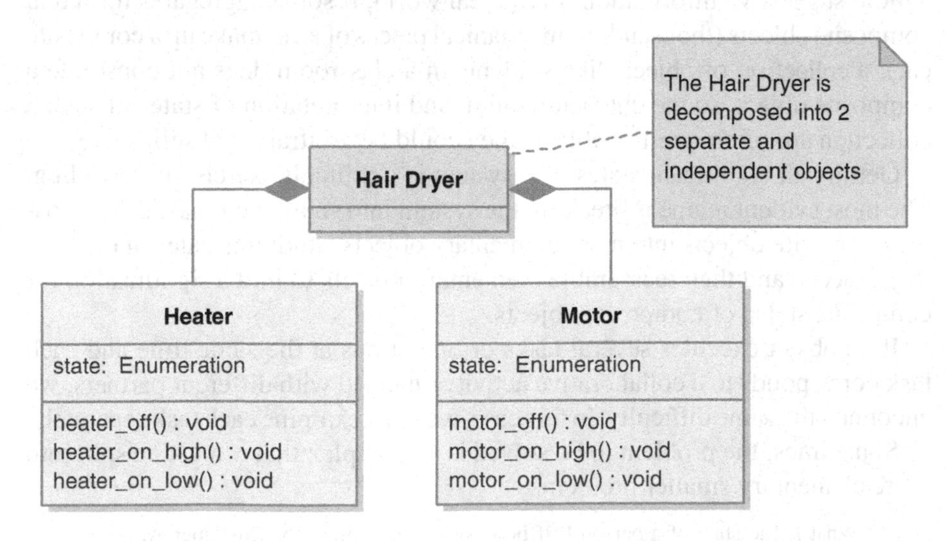

Fig. 5.4.2.2 States of the Hair Dryer decomposed into two separate objects. The number of states is 16 (8 + 8)

(Note: We can consider an enum type variable a *state*, but this variable is a final variable that considers an implemented subset of combinations chosen among the 64 possible states. At this stage, we are still designing the system.)

Now, if the Hair Dryer is decomposed into two separate objects, Heater and Motor, each object will have only $2^3 = 8$ states. By adding states of Heater and Motor, we have only $8 + 8 = 16$ states. We can eventually add to the composite object Hair Dryer a binary state On or Off. The final count is only $16 + 2 = 18$. So, from a dynamic viewpoint, if a system is not thoroughly decomposed and studied, whether we forget states, whether the composite system is so complex that the dynamic study cannot be conducted correctly. Moreover, semantics of composite states is more difficult to grasp. For instance, in the case of the Hair Dryer, we can only say that the Hair Dryer composite object is On or Off. Other enumeration states like "full power," "full heater power," "low fan," etc. need communication details.

> For an enterprise that has several running production lines, it would be easier to quote states of elementary production lines than asserting the overall state of the Enterprise object through all of its production lines. But, saying that this enterprise is at 70% of its production capacity gives sometimes helpful information for investors, stakeholders. But this information are not technical, it is synthesized by often an unknown or undefined algorithm.
>
> In a classroom, how to quote the state of a classroom if 20% of the students are sleeping, 10% are listening to music with their ipods, 30% is listening to the teacher and the rest are doing their homework with their laptops?

"Composite states" of a composite object resulting from the combination of more elementary states of all of its components may need some rework to offer a more suggestive information. In the real world, if some aggregates form real composite objects (thousands of mechanical pieces of a car make up a composite car), a collection of objects like students in a classroom does not constitute a composite object, so the characterization and interpretation of states of such a collection are often questionable as they could be arbitrary and subjective.

Getting all significant states of a system is a difficult exercise in modeling. The most evident recipe is breaking the system into smaller chunks, decomposing composite objects into more elementary objects; studying states of elementary objects, and then reassemble elementary objects to find a signification for composite states of composite objects.

If an object executes several tasks or operations at the same time and each task corresponds to a collaborative activity engaged with different partners, we encounter the same difficulty. In this case, we must examine each task separately.

Sometimes, the problem to be solved is so complex that it must be split into more elementary smaller problems.

> What is the state of a person P if he is verifying a stock for Customer A, taking an order for B, and preparing an invoice for C? In fact, P can do several

operations *take_order(), verify_stock(), prepare_delivery(), prepare_invoice()*. If we are studying the business process starting from the reception of the order to the collection of payment, it is recommended to consider the states of this person P through interactions with a generic customer that could be A, B or C. States retained through this long process are: *idle, taking_order, verifying_stock, preparing_invoice, preparing_delivery, getting_payment.*

Now, if the person P is able to work apparently on several tasks at the same time means that we can explore some parallelism with the resource used to implement this business process. A new scheduling problem of "evaluating how many customers could be handled by one person during one day" will be superposed to the previous problem. The second problem is typically a workflow management problem. If we have n Customers, new states could be added as: *idle, busy_with(i).* The busy_with(i) is a global state that encompasses *taking_order(i), verifying_stock(i), preparing_invoice(i), preparing_delivery(i), getting_payment(i).* A priority scheme can be added to take into account real conditions. For instance, if P is currently busy preparing a delivery for Customer i and if P receives an phone call from Customer j, he can postpone the task for i as the activity of taking a new order is more important than preparing a delivery. Final output of the workflow study could be for instance estimating the average number of customers the person P can serve during one day.

It is worth noticing that the conventional state used for state machine diagram is not elementary as it defines several components inside.

Figure 5.4.2.3 gives the structure of a standard state. It is composed of an activity that defines the state (*character/handle character* that represents the *Do Action*). While entering the state *Typing Password*, an action has been performed *On Entry* (*set echo invisible*). *On Exit*, another action *set echo normal* is activated. An action can be defined *On Event* (*help*). An equivalent activity diagram shows the equivalent of the standard state in the UML.

If *handle character* is candidate for a state, *set echo on* and *set echo normal* are also candidate for *transient states*. Transient states are temporary states that a system passes through during an insignificant time (compared to the time

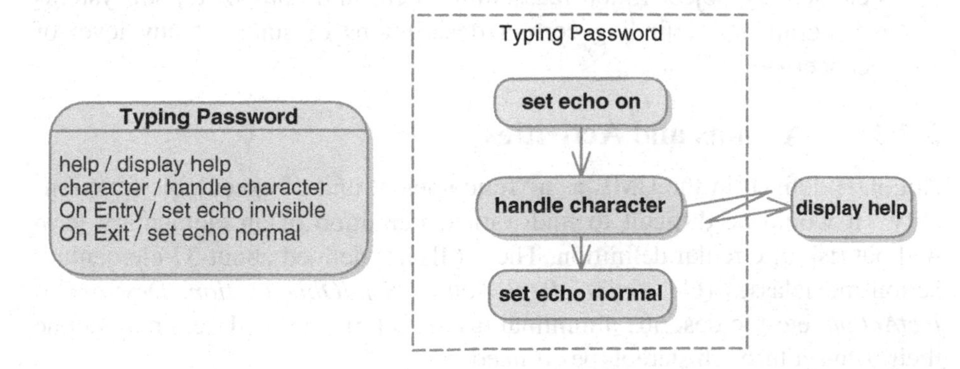

Fig. 5.4.2.3 Conventional state used with the UML is not elementary state (Fig. 15.32 Superstructure Book)

scale of the current system) before reaching an equilibrium state. Sometimes, transient states must be studied if some behavioral problems are suspected during transients. In this case, Figure 5.4.2.3 must be decomposed into three different states. This is the reason why the standard state of the UML is not really elementary.

To summarize, it is worth noticing some important points when working with states:

1. States are a very elastic concept that depends upon the level of decomposition of a system. States can be tied to object operations at high level, assimilated to activities in general cases, and possibly to elementary actions when transient states are investigated.
2. As states are states of objects through their operations and activities, the separation into functional and dynamic view are for description purposes only. Functional and dynamic study must be conducted simultaneously. The UML talked of behavioral study.
3. When decomposable, states are composite or submachine states. Composite states are states of several objects identified in regions ("horizontal" decomposition) connected through AND-states or OR-states logic. Submachine states can be decomposed "vertically" into more elementary submachine states or elementary states.
4. If object or agent states are easily understood concepts, states of a collaborative group of objects or a collection of objects (without any collaboration) are harder to describe. Interpretation is contextual and subjective if not done correctly. Generally, a precise description ends with a description of states of constituting elements.
5. To get a thorough description of a system, it is recommended to always decompose the system to its utmost elements, study states of elementary objects, then reassemble them into composite, subsystems or components, finding correct descriptions of states at any level of concern.

5.4.3 Actions and Activities

Action is defined in the UML as a "fundamental unit of executable functionality." It would be difficult to find a clear definition of an elementary term without risk of circular definition. The UML has defined about 37 elementary action metaclasses (*CallAction, ReplyAction, SendObjectAction, DestroyObjectAction*, etc.) to describe a minimal usable set of actions. Users may define their own set through stereotypes if needed.

In object/agent system, we consider an action as a nondecomposable act of doing something performed by an object/agent that can modify the state of itself or the state of its environment (other objects/agents). An action results in a *world move* and constitutes a unit of analysis while working on human interactions. Actions are contextual and they cannot be understood outside this context.

> A man can perform the following action "make a weapon". But we don't know if he makes a weapon for hunting, for killing somebody or for selling it to get money.

Contrary to actions that are contextual, an activity is composed of a series of actions (repetitive or not), and has a more advanced structure, particularly if it has motives and goals. If actions may be unconscious, activities that are coordinated actions are goal-directed. When an activity performs, it needs *subjects* or *actors* (objects/agents or doers responsible of the activity), *objects* (what to change), operations and actions (procedural steps, processes), and all *artifacts* and *resources* needed to execute the activity. All ingredients constitute the context of this activity but after execution the activity reshapes its context. Context is not only external but is internal as well.

> If an enterprise "proposes a new product", "promotes it", perceives the reaction of customers, changes its strategy, etc. Then, activity may change the goal of this enterprise, so may change its internal context.

There is a close relationship between an activity and the state of the object that executes this activity. Less frequently, there is also a close relationship between the definition of a state and actions. A state can be defined by rich combinations of actions and activities:

1. State defined by a *continuous action*
   To maintain a speed of 65 mph, the driver must act constantly on the gas pedal. We can eventually say that the state is defined by a fixed parameter that is the cruising speed of the car, but this constant speed is controlled by the continuous action of the driver.
2. State defined by *repetitive actions* during a state
   If a fax calls at an occupied number, it will enter in a *retry state*. In this special state, it will try five times to make a redial before deciding that the number cannot be reached.
3. State defined by intermittent actions
   Software shows intermittent bugs. As long as programmers cannot find where bugs come from, the software will be in the debug state and cannot be sold.
4. State defined by several activities
   A company enters a liquidation state and needs to clear away stock in Canada and in the USA. If the warehouse in Canada is cleared, the company is still in its liquidation state globally.

5. State defined by a *date*
   Spring starts in March 23 and marks the blooming period for several spring flowers, but all species will not give flowers at this date.

As there is a close relationship between action, activities, and states, please refer Table 5.4.1.1 built for states in order to map real world actions or activities.

5.4.4 Events

The UML defines the *Event* metaclass as "An event is the specification of some occurrence that may potentially trigger effects by an object." An event is a fact occurring in time, can capture attention of a system and cause its state change. It is always the result of a direct or indirect action or activity of an object (or system or agent). This object may be internal or external to the current system and sometimes, we cannot really identify the object that creates the event.

Turning the car key contact is viewed by the object Car as an event, but the action has been performed by the Driver object.
A click on a mouse is received as an event by the software but the action is executed by the User object.
An intruder acts as an event for an alarm system.
When the phone rings, we receive the ringing as an event but the object Phone is responsible for producing the ringing tone.
The arrival of the 737 or an Airbus or Boeing flight is an event for people waiting for their relatives.
Other events are for instance the opening of the door, the arrival of new expected software on the market, the change of the tax ratio, the disappearing of dinosaurs on the planet, etc.

An event may occur at *regular* time (clock chimes), may be synchronous to another event. "After each rain, insects appear. The appearance of the insect is synchronous to the event 'rain'." An event that appears regularly with time is sometimes called synchronous. In fact, time is perceived as something regular, and any event that occurs on a regular basis is perceived as synchronous with time, thus simply synchronous.

An event is said asynchronous when we cannot relate it to anything, to any other event or any timing reference. The occurrence of an asynchronous event may be possibly predictable but we cannot say exactly when this event will occur.

Synchronous and asynchronous may bear another meaning in modern communication.

In a meeting, when all parties involved in the communication are present at the same time, we have a synchronous communication. Examples include a telephone conversation, a company board meeting, and a chat room. Asynchronous communication does not require that all parties involved need to be present and available at the same time. Examples of this include e-mail service (the receiver does not have to be logged on when the sender emits the message), discussion threads, which allow exchanges over some period of time.

Random events are events with unpredictable outcomes. Random events may come from undefined sources, but some random events come from well-known sources (mathematical random number generators) but, for those who ignore how sources create or generate events, they appear as random. Random generators are a very interesting research concern as they are useful in simulating complex systems, such as the spread of diseases, the evolution of financial markets, or the flow of Internet traffic, besides constituting the brain of Las Vegas slot machines.

An event is modeled as a fact without any time duration. From a theoretical viewpoint, this hypothesis is false as nothing is instantaneous (even the light needs 1 ns to travel 1 ft), but in modeling, sometimes, systems must be studied with a rough model, then, when dynamic studies are necessary, we reiterate through models several times to refine them. This attitude allows us to ignore, at first approximation, all transient states and consider only most important ones to have an approximate view of the project before starting more meticulous dynamic studies to reach final algorithms or a procedural way of solving problems.

In the UML, the term "event" means the type and the word "occurrence" means the instance. When necessary, "event type" and "event occurrence" will be used. It could be surprising but the UML has no graphical notations to represent events, but, as events are created by actions, in some tools, AcceptEventAction and SendEventAction may be used as sink or source of events (see Table 5.4.4.1). In fact, events are omnipresent in dynamic diagram of the Behavior suite, is part of the state transitions, lifeline interactions, and messages in timing diagram. Events are the main driving force of system change and world movement. For instance, in an interaction diagram, when a lifeline sends a message to another lifeline, the receiver interprets the message implicitly as an event. In a state diagram, most transitions occur as a result of an event. The event appears in the "trigger" part of the transition specification.

5.4.5 Condition

The UML defines "A constraint is a condition or restriction expressed in natural language text or in machine readable language for the purpose of declaring some of the semantics of the element". In the UML, a condition is therefore some constraint put on a model element to shape its behavior. A condition must be evaluated to be true in order for the constraint to be satisfied. Preconditions and postconditions are used intensively in operation or protocol specifications. Condition nodes or test nodes are nodes used specifically to determine if a

Table 5.4.4.1 Support for concept of event in the UML

Diagram	*Comment on the mapping of event concept*
State machine diagram	In a state machine diagram, events appear in the "trigger" part of the transition specification. Trigger [guard] / effect
Activity Diagram	In an activity diagram, each action or activity, dependent on their nature, may potentially generate an event towards another action or activity. The following examples quote some possibilities: 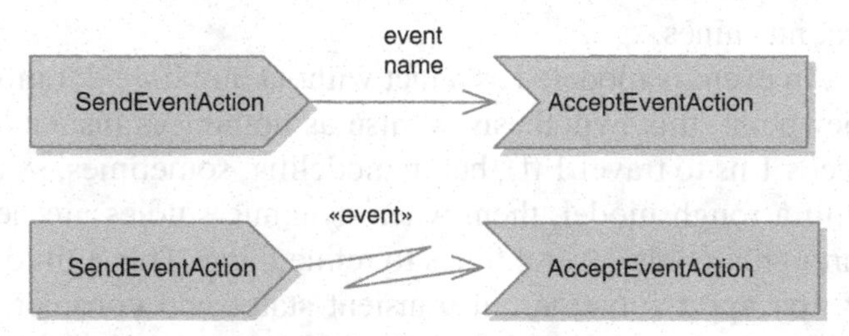 In the figure, SendEventAction acts as event source, and AcceptEventAction as event receptor. Control or object flows can be used as supports for events. 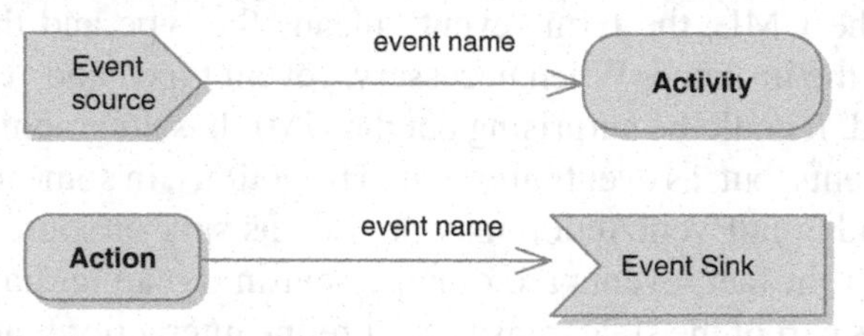 Regular activities and actions can replace specific graphical notations. Control or object flows can be used as supports for events. 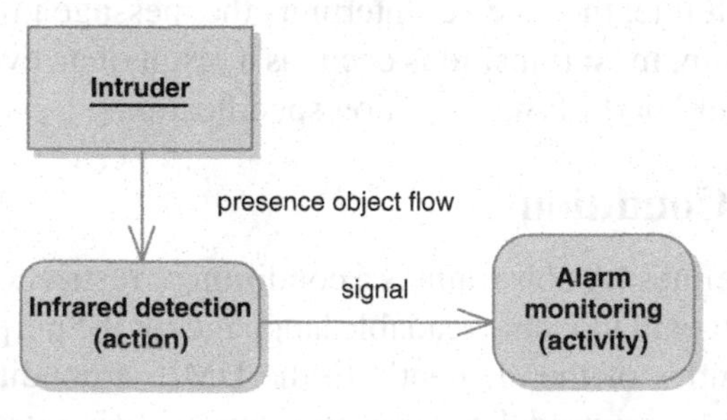 Objects can send events to actions that can send other events to activities. Object or control flows may be used as supports for objects, signal, events, or messages.

Table 5.4.4.1 (Continued)

Diagram	*Comment on the mapping of event concept*
Sequence diagram	When a lifeline sends a message to another lifeline, generally, the receiver interprets the message implicitly as an event. So, events are implicitly encoded in interactions. Please notice that, comparing with the example in the activity diagram, actions and activities are converted into objects (infrared detection to Infrared Sensor and Alarm monitoring into Alarm Detector). 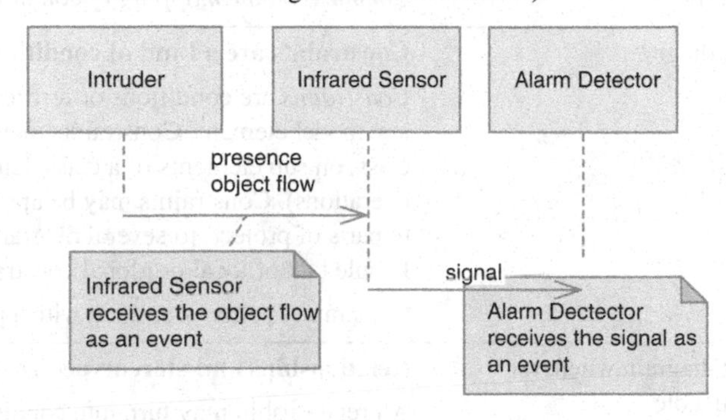
Communication diagram	Idem for communication diagram that is a replica of the sequence diagram. 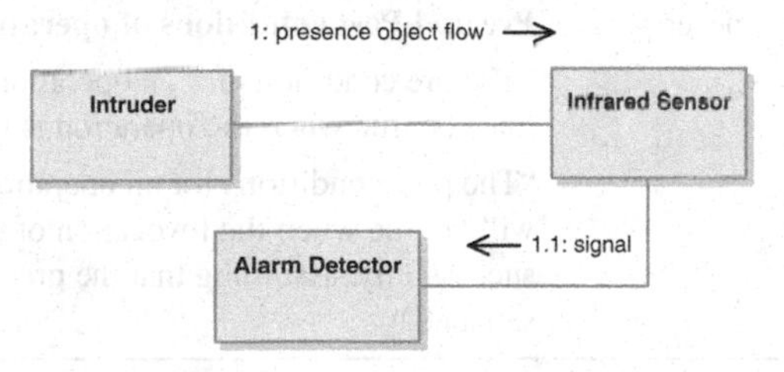

condition is evaluated to be true. Conditions appear as guards that determine if transitions must occur or not in state machine diagram.

Condition in spoken language has a rich set of meanings. It can be assimilated to a state."The Car is in a bad condition." Or, when defining states, the UML declares "A state models a situation during which some invariant conditions hold." Initial or final conditions are effectively states as conditions do not evolve. The slight difference is that conditions focus the description on some details of states but are not intended to completely describe the states.

Some imprecise uses of "conditions" say for instance "When entering a state, *On Entry* actions must be executed or when exiting a state, *On Exit* actions must be executed." In fact, some "mentioned conditions" are implicit to the definition of states and this imprecise prose seems to add supplementary conditions that must normally be pruned off.

If the Car must enter its deceleration state, the Driver must release the gas pedal and pushes on the brake pedal. So, the action on entry of the state is not a condition but an regular action of the state if from the beginning, states are designed with both On Entry and On Exit inside. However, if On Entry and On Exit actions are defined outside the state, their activation may be ruled by conditions, e.g. "Want to decelerate?" Yes, No, etc.

Table 5.4.5.1 This table explains how the UML deals with conditions at various levels

Diagram	*Comment on the mapping of* conditions
Any diagram	**Constraints are a kind of condition** *Constraints* are conditions or restrictions imposed on nearly any model element. Constraints may appear for instance on class, on sub elements of a class definition (attributes or operations). Constraints may be applied to a whole project, to parts of project, to several diagrams, to one diagram, etc. People talk of local or global constraints. Constraint is often identified with a pair of curly brackets {}
Any diagram whenever applicable	**Relationships and stereotyped relationships** All relationship may turn into condition ruling the association between two modeling elements as a relationship can tie a class to another class.
Any diagram whenever applicable	**Pre and Post conditions of operations** "The pre conditions for an operation define conditions that must be true when the operation is invoked." "The post conditions for an operation define conditions that will be true when the invocation of the operation completes successfully, assuming that the preconditions were satisfied."
State Machine diagram	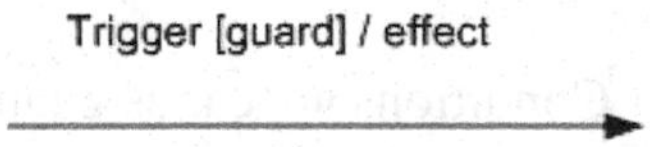 Trigger [guard] / effect **Guard part of transition specification** A guard imposes a condition that, if true, authorizes the transition to take effect.
Activity diagram Interaction Overview Diagram	**Some control nodes of the activity diagram** A *decision node* imposes a condition to the system, selecting the next activity to be executed. A *merge node* imposes that all flows must converge to a unique output. A *fork* imposes that a set of tokens is distributed simultaneously at all outputs (Petri Net logic). A *Join* imposes that all tokens must arrive before the next activity could be activated (Petri Net logic). 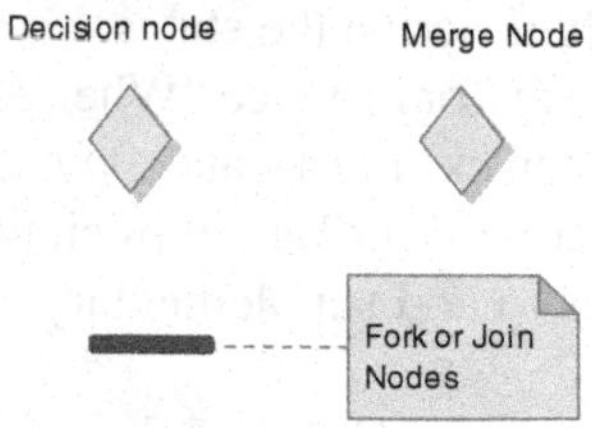 Decision node Merge Node Fork or Join Nodes

Table 5.4.5.1 (Continued)

Diagram	*Comment on the mapping of* conditions
Sequence an Interaction Overview diagrams loop For [i=1..N] alt IF [a=0] [else]	**Fragments with predefined conditions** Fragments are predefined conditions put on units of interactions for definition of standard fragment types: alt (alternatives), opt (option), break, par (parallel), loop, critical, neg (negative), assert, strict, seq (sequence), ignore, consider. *Example*: Fragment for implementing a FOR loop condition and an alt fragment for an IF condition.
Use case diagram	**Include and Extend Relationships** The two most important <<include>> and <<extend>> relationships are the basic mechanisms for decomposing a use case into more elementary use cases and are therefore a feature of complexity reduction. They are metaconditions imposed on the way processes are connected together.

When a sequence of actions (for instance, A1 then A2 then A3 must follows in this order), there is an implicit condition on the sequence order. Otherwise, we must add everywhere in the activity diagram constraints that A1 precedes A2 that in turn precedes A3. If we have an automatic code generator, it will translate the actions in this order. So, to avoid a very complex specification, it is recommendable to keep conditions specifications only in situations enumerated in Table 5.4.5.1.

Moreover, it would be useful to distinguish conditions and constraints imposed by users to their models at the application level from conditions ruling the interpretation of the UML diagrams themselves or *metaconditions* (conditions ruling the interpretation of the UML itself). On Entry and On Exit defined in states are metaconditions and some of them are inherent to the way diagrams are laid out and need to be combined to user conditions.

5.4.6 Messages

In object/agent architecture, the dynamics or behavior of the whole system is defined in terms of interactions between objects. Interactions can be described entirely with message exchange mechanism. When an object S (sender) sends a message to object R (receiver), S executes a *send message action* and R also executes a *receive message action*, so the two objects collaborate through this messaging process.

The real purpose of a message depends upon the semantics of applications and is theoretically limited only by our imagination. S can inform R, requests a service from R, wakes R up, fire a chain reaction in which R plays only an intermediate role, asks R to compute a mathematical formula and returns to S the final result, kills the object R with a delete object request, etc.

If there exists more than one typed message between S and R, R must interpret the message from S to analyze the sender request and performs the corresponding action or activity. If we model a message theoretically, we must have at least the following information:

1. *Sender identity*. This information may be used to inform R of the identity of the sender S.
2. *Receiver identity*. This information allows to a message dispatching system to direct the message to the qualified receiver.
3. *Address of the receiver.* If the receiver identity is redundant information, the address completes the identification of the receiver and make his identity unique in the system.
4. *Envelope.* The envelope is the *message wrapper* that allows the message to adapt its format to the format of the media used to convey and deliver the message (for instance, TCP/IP packets in the Internet).
5. *Content.* In programming context, *Content* = *Service* + *Parameters.* A service is called a method or an operation at programming level. This information of highest importance is generally embedded in the envelope, sometimes, encoded. The receiver is normally the only object that can read and interpret the content, unless the message is not encoded and broadcasted. In the context of object/agent architecture, this content is the service that R must execute and all the parameters help R to perform correctly its task. So, the content can be split into *Service* + *Parameters*.
6. *Timestamp*. This information is not intended to the receiver. It is often used to manage the envelope in the messaging media, to determine the lifetime of the message, and to avoid congestion of the messaging media in case of huge amount of nondelivered messages.
7. *Media*. The media used to deliver the message may have various speeds and real-time parameters.
8. *Returned message*. The return message could be everything (nothing, data, object, pointer or reference, etc.) depending on what the sender asks the receiver to do. In the general case, the return data is a new message with the same parameters.

So the format of the message M contains the following data:

M = (Sender, Receiver, Address, Envelope, Content, Timestamp, Media, Returned data)
(formula 5.4.6.1)

If we take the US or Canada Post Office, an instantiated message is composed of:

M = ("Bob", "Mary", "1000, Ocean Drive, CA 92100", protecting envelope, letter inside the envelope, "1 July 2020", ground service, none)
Constraints: The names of the Sender, Receiver, the address, the timestamp are put on the protecting envelope and the letter is inside the envelope.

Anyone who has programmed in C++, Java, or C# knows that if we want to invoke a function *compute()* declared in an object *Obj* (message receiver), we write for instance:

result = Obj . compute (a + b)

If we translate this implementation into the general messaging format, we have a reduced version of message M with numerous information removed from the initial format.

Returned message = Receiver . Service (Parameters) (formula 5.4.6.2)

Notice that Content = Service + Parameters
Here is the explanation.

1. The *Sender* is implicit. The call is made from the current program code of the current object, so, the compiler knows implicitly who is the sender so this information is absent from the format
2. *Obj* stands for the *Receiver*
3. *Address* is not necessary as the name of the object *Obj* identifies unequivocally the receiver object in the current small system. If the system is for instance the World Wide Web, an IP address like 132.203.26.21 would be necessary to complete the receiver specification
4. The *Envelope* is always opened in current object programming languages and there is no special protection for the receiver except that *compute()* can be made *public, private*, or *protected.* In the present case, *compute()* is public to allow the sender code to call the function
5. The *letter inside the envelope* contains the name of the Service *compute()* and the two parameters a and b necessary to satisfy the preconditions of the *compute()* function
6. *Timestamp* is not necessary in current programming environment unless a special debugging mode is required and a timestamp is asked systematically when entering the call stack

7. The *media* is automatically the Windows, Linux, etc. Operating System, so the media is also implicit and absent from the message specification
8. The *returned message* is now moved to the start of the formula and represents the data returned by the function compute()

Actually, the message model (formula 5.4.6.2) at programming level (implementation or platform specific level) used in this example lacks generality and cannot serve as generic messaging format.

> To show the problem of this format, let us consider a citizen that wants to be served in a Ministry. He must emit a request like
>
> Bob . Receive (my_application_form)
>
> This model of requiring a service cannot stand for a Ministry. It supposes that the citizen, when going to the Ministry, can go through all the barriers, reach directly the person who is in charge of the application form, knows that his name is Bob and his role in this Ministry is to receive forms. The Receiver cannot select the appropriate operation to execute

One of the fundamental principles of modeling is to respect the full correspondence between the reality and its model. Thus, to model real world, it is recommended to start with the generic formula (5.4.6.1) and adapt it to specific usages for mapping real messages into the UML.

5.4.7 Relations between Action, Activity, Event, State, Condition, and Message

Through the mapping of fundamental dynamic concepts, we discovered some interesting results. Fundamental concepts are much closed to each other and they are all related. The difference of terminology used comes from the observer viewpoint, where we are on the *cause–effect chain*.

1. *Action and event depends on the observer viewpoint.* When an object S (source) executes an action, the receiver object R perceives it as an event. When a Car A hits a Car B, Car A has done an action and Car B receives it as an event. As action is generally of short time, the result takes the form of an event by its rapid and sudden character.

   In the given modeling context, we can find that some actions do not lead systematically to an event.

   When a person sneezes, apparently, if we are observing this person, we cannot perceive the "event" character of this act, but, if we are sleeping or if we are in another room, this fact may be perceived as an event. So, an action can give rise to an event or not according to the context.
2. *Activities are basic ingredients for building states.* An Activity takes place across a time interval defined by its start and end points. Activities are more structured than actions and we can identify their goals.

Table 5.4.6.1 Support of Message concepts in the UML

Diagram	*Comment on the mapping of* messages
Object or Communication diagrams	**Messages on Link** *Links* in an object diagram are pathways for exchanging messages between objects. *Messages* are based on some operations defined in classes. 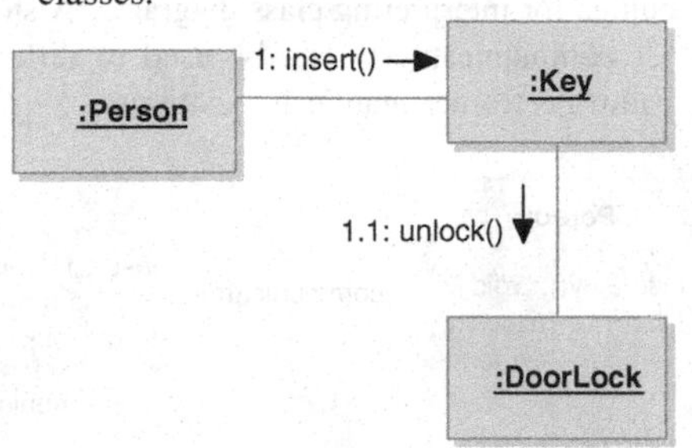 In this example, the object Person insert objects Key in the object DoorLock to unlock it. In object world, *:Person* sends a message *insert()* to *:Key* that in turn sends another message unlock() to *:DoorLock.*
Class diagram	**Operations in classes** Some operations in classes are designed for supporting communication (all communications between objects are realized through messages). *Example*: Classes in previous example. When object *:Person* sends a message to *:Key*, it activates the *insert()* operation of object *:Key*. 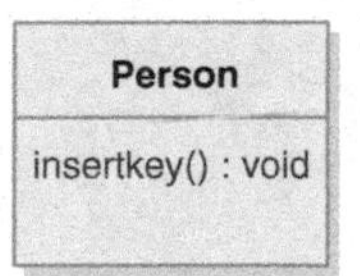 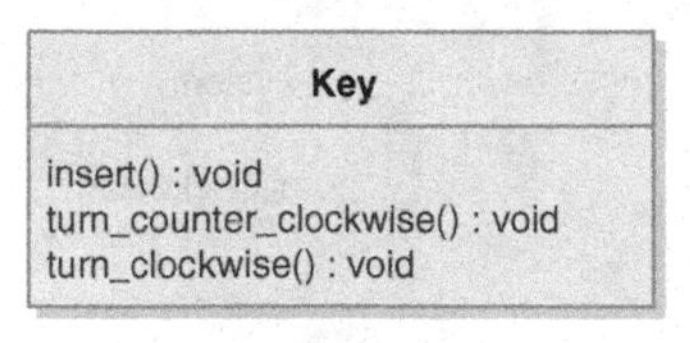
Class diagram	**Class associations** Messages are exchanged between objects and objects derived from classes. To show that classes Key and DoorLock manufacture objects *:Key* and *:DoorLock* that can communicate together through messages, some designers draw an association between class Key and class DoorLock to mean that their objects may communicate together.

(cont.)

Table 5.4.6.1 (Continued)

Diagram	*Comment on the mapping of* messages
	From a theoretical viewpoint, the UML accepts that objects are instantiated from classes and links are instantiated from associations, so this way of doing things is acceptable. However, as associations mean relationships between classes, they could create some difficulties for interpreting class diagrams. A stereotype <<com>> or <<communicate>> can be used to differentiate communication pathways from semantic associations. 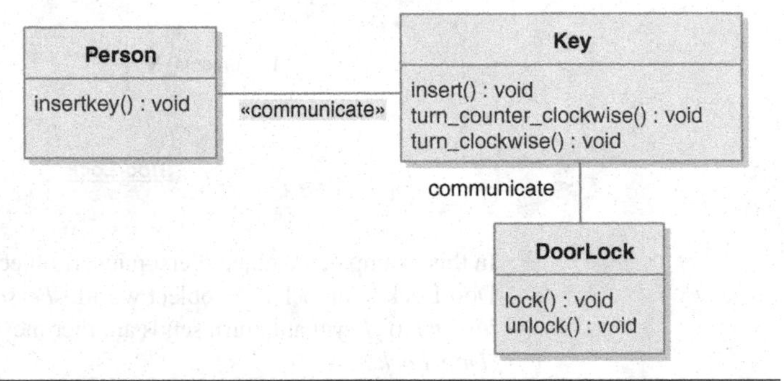
Sequence diagram	**Message between lifelines** In a sequence diagram, messages are represented by solid arrows sourcing from the sender lifeline and ending on the receiver lifeline. 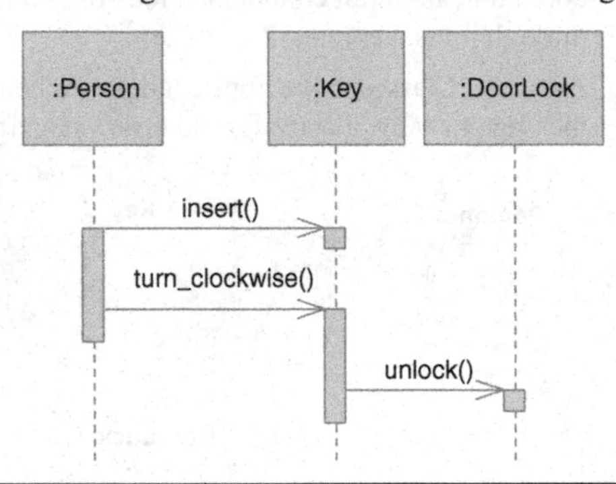
State machine diagram	**Actions in states, on entry, on exit, in state, on transition** All actions in a state machine diagram (on entry, on exit, in state, on transition) are potentially messages exchanged between objects. 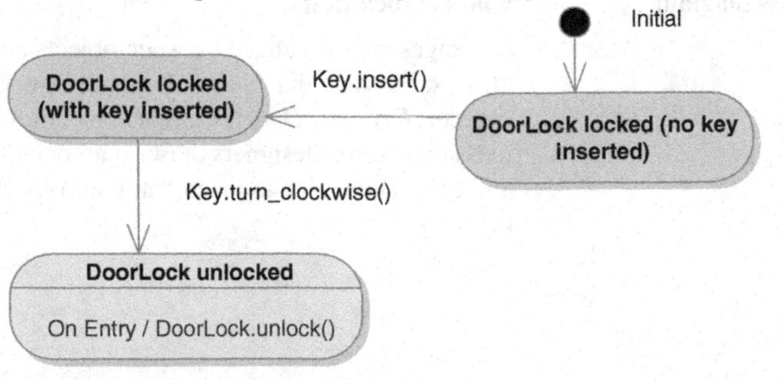

Diagram	*Comment on the mapping of* messages
	In this example, only state of the DoorLock is represented. Events received by the DoorLock can be named with passive voice like "Key inserted" or "key turned clockwise." We can make reference directly to actions responsible for generating those events *Key.insert()* or *Key.turn_clockwise()*.
Activity diagram	**Actions or Activities may send messages** Objects are behind "partitions" in an activity diagram. Control flows and object flows are responsible for conveying messages between objects. *Example*: Person sends two successive messages to Key, one for inserting and the other for turning the key. DoorLock receives two messages from Key and performs two actions, receive_key() and unlock(). The representation of the concept of message is oversimplified in the UML as the complete chain is not represented. For instance, the action *turn_key_clockwise()* executed by the Person is absent from the model. Only, *turn_clockwise()* is represented as operation in the Key object. We will come back to this aspect in exposing the Uniform methodology).

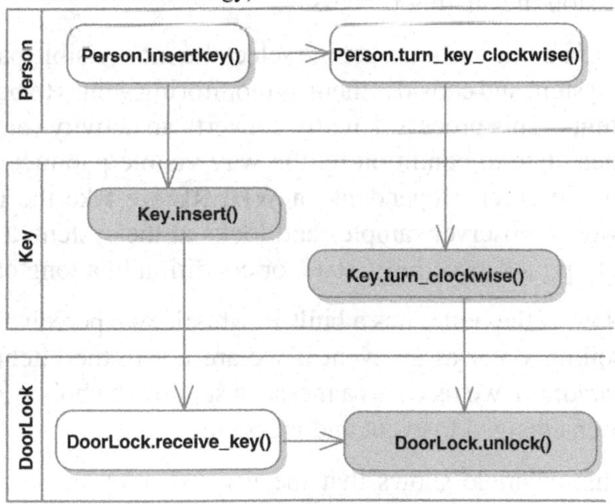

The time interval between its start and end points may be used to build a corresponding state. If many observable changes are evidenced between these two points, we can decompose activity in subactivities and consequently corresponding substates. An activity may contain several actions ("in state", "on entry" and "on exit" actions of UML states). If the model needs more precision, transient and short states may be isolated from the main activity for a more accurate investigation.

3. *Actions and activities*. If actions and activities are well defined in the UML, in real world, the term "action" is a strong political term, is

overused so as to mean "reaction," "position face a situation," etc. So be careful for the mapping.

A phrase like "The action of the government in agricultural markets is translated into various activities in order to improve the productive efficiency of the whole agribusiness chain" for instance reverses apparently all what was said about action and activity.

4. *Actions can be considered sometimes as transient states.* Some real time and safety-critical systems are very sensitive to short actions and transient states. In these cases, actions could be candidates for building short but important states.

5. *Activity, state, condition, event are related in a cause–effect chain.* For making tea, we need boiling water. The water receives its energy from the electric heaters. The Heater object performs an activity "Transmit heat to water." The state of water changes with time and boils after 5 min. The temperature that starts from 20°C then reaches 100°C after 5 min is the most visible parameter. We look at the condition (or state) of the water and decide when the water is sufficiently hot for making tea and decide to stop the heating process.

   So, in some systems, people select the most visible parameters to monitor a system and consider them as monitoring values to be tested against some limits. This process actually converts an activity (or action) to state, and then state to condition by the way visible parameters are processed by the observer. Dependent on WHERE we take the information and the way an observer samples and looks at the system, a same reality can be interpreted as activity, state, or condition in a long cause–effect chain.

   Now, if the boiler has a built-in whistle, we perceive the condition of the boiling water as an event if we are not in the kitchen. From the object paradigm, we received a message sent by the boiler. A condition has thus been changed to event and message.

   This example shows that the way we interpret a phenomenon or fact depends upon the nature of the dynamic interaction chain, the way information is captured and analyzed, where we are on the cause–effect chain, etc. In human organizations, things are more complex as cultural and instructional parameters are added to the batch and influence the observer conclusion.

6. *Actions and activities are executed by objects and messages are data/objects/signals* exchanged between them to coordinate their activities. If in an activity diagram, the object flow is present in the current standard, there could be some difficulties in the way message concept is implemented in sequence and communication diagrams.

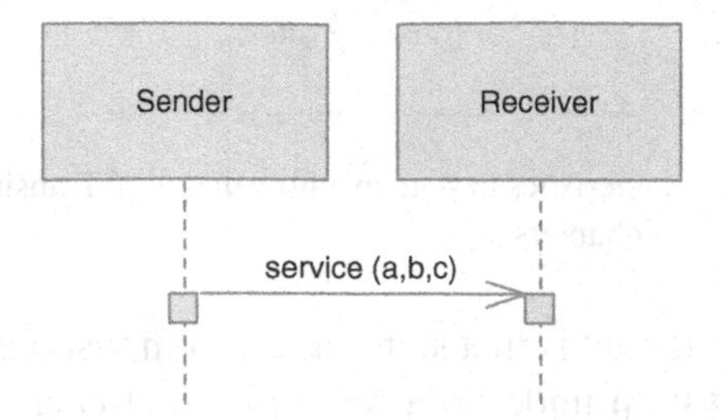

For the moment, we cannot, in the sequence or communication diagram, define a message object between two lifelines, showing that two processes (one at the sender side, another at the receiver side) collaborate with this object exchange to evolve. The activity diagram has already implemented an object flow between two activities/actions and is more appropriate for handling this case.

From a theoretical viewpoint, it would be more flexible and natural to propose that the content of the message suggest (or not) to the receiver a given operation for answering his message but it is up to the receiver to choose the most appropriate operation to serve the sender's request. The way object programming languages implements message concept is really not universal and cause distortion with reality when modeling at high level.

5.4.8 Transitions: Relationship with Control Flow

This question very often comes back with students. A transition in a system is often named *system change, system movement, system evolution,* or any term that accounts for an observable dynamic change.

The UML has an equivalent metaclass bearing the same name that can serve as a direct mapping. *Transitions* are used in the state machine diagrams to show evolution of a system from state to state. In Figure 5.4.8.1, a transition is composed of three concepts:

1. *Event or combination of events* that is the cause of the state change specified by the transition
2. *Guard* that is the internal or an imposed condition that must be evaluated to true to authorize the state change
3. *Actions* that accompanied the transition (*actions on transition*). Those actions are specific to the transition and cannot be affected to any states

Transitions are drawn as arrows connecting boxes representing states. Arrows also indicate the direction of the transition. A state may be represented mathematically by a set of state attributes. A UML transition must therefore change at least one of the state values in order to have a state change.

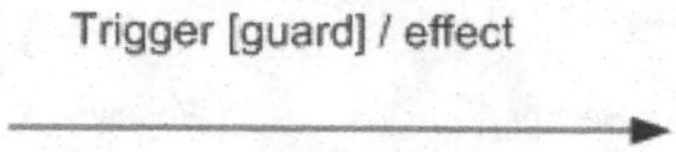

Trigger [guard] / effect

Fig. 5.4.8.1 Transition characteristics in state machine diagram. Transitions are made of known concepts exposed in previous chapters

One more interesting fact is that states are decomposable. So, state attributes that characterize a system undergo a redefinition at each level of refinement. So, we may have many state changes in lower levels of decomposition and the state at higher level will stay unchanged.

> If a Car is characterized only by two global states, "at rest" or "moving", the "moving" state may be decomposed into "accelerating", "cruising" and "decelerating". State attribute at high level is a Boolean "Moving" attribute. At next level of decomposition, we must add another Enumeration "Speed" attribute to make sub state description. When the car goes through three states "accelerating", "cruising" or "decelerating, it is in the same high level state "moving", so at high level, it appears as the system does not change its state.

So, some transitions that occur at low level are hidden from the high-level view of a system, in submachine or composite states. Without entering into philosophical considerations, in modeling, sometimes, "systems seem stable at high level when scrutinizing in more details, variations could be important."

As said earlier, standard states defined in the UML may contain activities and actions (on entry, on state, on exit). Transitions between states are complex model elements (*trigger*, *guard*, and *effect*). At the difference of transition, a control flow in an activity diagram is a more elementary concept. It is used to make up control rules (fork, join, merge, etc.), to indicate the sequencing order of processes. So generally, transitions are not equivalent to control flows.

Table 5.4.8.1 Mapping of system change, system evolution, system transition, or system movement into a UML transition and UML control flow

Diagram	*Comment on the mapping of* transition *or* system change *to UML*
State machine diagram	**State transition** System change, system evolution, system transition, or system movement in real world are mapped into UML transitions in a state machine diagram. *Trigger* explains the cause of change. *Guard* explains the condition of the system that allows this change. *Effect* describes action that results directly from this change.
Activity diagram	**Control flow** Control flows in activity diagrams are used to indicate the movement of tokens (activity marker) between activity nodes. Activity nodes can be actions, activities, and control nodes (fork, join, etc.). So, control flows in an activity diagram can show system change or evolution as well.

5.5 Fundamental Concepts of the Structural View

Structural view describes systems as a collection of objects, their characteristics and relationships. The structural view was qualified in the past as *static view,* by opposition with the dynamic view that describes essentially the evolutional aspect of a system.

In the domain of real-time systems, the vast majority of software developments are PSM. Notions like *reuse*, *logical model*, and *PIM* are not widespread; we often found in a structural description objects specific to applications, disseminated among objects like computer, operating system, software components, deployment nodes, and even hardware mounting devices, were all mixed together in an inextricable manner.

In the domain of database development, things were better organized. Conceptual data models made with E-R tools tend to behave like PIM (however, not all PIM are conceptual data models). Conceptual data modeling is the first stage in the process of database design. Traditionally, it identifies *entities* (persons, objects, data, etc), their attributes (information that describe these entities), and relationships that link entities together into a network that gives guidelines for developers to interpret the model. If we compare the conceptual data models with the object structural view, the only difference seems to emerge only from the replacement of the terms *entities* by *objects*.

However, there exists a deep difference between the E R model (Chen, 1976) and the object paradigm. Maybe, the semantic models (Smith and Smith, 1977; Hammer and McLeod, 1981), which provided a modular and hierarchical view of data, can be considered as precursors of object data model. The primary components of semantic models were the introduction of the *ISA* ("is a") *and aggregation relationships. ISA* relationship (e.g. a square or a triangle *is a* polygon) is equivalent to *object inheritance* and the *aggregation relationship* has its equivalent *aggregation/composition* in object model. The composition relationship is the basic architectural concept that supports complexity reduction mechanism and allows a system to manufacture composite objects or components from smaller objects and thus makes all systems manageable at any level.

However, semantic models are not object models. Essentially, semantic models encapsulate only structural aspects of objects, whereas *object paradigm encapsulates both structural and behavioral aspects of objects.* Moreover, E-R models lead to relational models that, in turn, are implemented with *tables of values*. Tables are value based or value oriented (Khoshafian and Copeland, 1986; Ullman, 1987). Objects are identified by their *surrogates* (system-generated globally unique identifiers) rather than their values, objects have OID (Object Identity) and do not need to create value based strings to serve as primary keys for indexing.

To explain the value based problem for identity, in E-R modeling, the class Person is for instance identified with a specific string (its NAS or National Assurance Number). Codd (Codd, 1970) called this attribute "user-defined identifier key". There are several problems with identifier keys as the concepts of data value and identity are mixed together. If a value is served to identify an entity, people cannot change this value easily as this change may affect all dependent applications based on this identification. For instance, if a database encodes the department name "ABC" in the product identification "ABC12345678", if we change the department name "ABC" for "DEF", all products that have an encoded identity must be updated to reflect the new name change. So, values interfere with identities and add substantial data manipulation operations.

In an object system, it is not necessary to have a special attribute for distinguishing one object from another. A surrogate is automatically affected by the system, doubled by a more easily reminded identity proposed by the user for local and short term use. If surrogates and their synonyms are unique at object creation, they may serve at runtime as object identity. If this identification method based on surrogates avoid the previous problem of value-based identity, there could be some difficulties in a distributed environment (Leach et al., 1982), for instance in the Internet. Many solutions are available to handle distributed databases on the Internet and we will come back later in the "object identity" chapter.

Object technology places the *encapsulation* of attributes and operations and *class inheritance* among its natural and basic characteristics. When digging inside inheritance, once more, inheritance in object technology is deeply different from inheritance in semantic models as it encompasses behavioral inheritance (presence of object operations) instead of inheritance of attributes only as observed in semantic models. For detailed information and a survey paper about semantic models, please consult Hull and King (1987).

To build the structure of an object system and later provide a description of its structural view, we need to consider the following points:

1. *The adhesion to the reuse concepts* (see Section 2.8) will structure differently the application.
2. *The identification of application domains* being developed and the adhesion to ontologies defined on these domains give us a rigorous description and a coherent vocabulary. If ontologies do not exist, a small glossary with precise definitions would be an advisable minimum.
3. *Developers must be convinced of the importance of models and MDA guidelines* that partition the development process into activities of building PIMs and PSMs. The development is therefore a multistep or multidimensional process. When decomposing a PIM, complexity reduction and decomposition still respect PIM principles and must not end up with platform-specific components. A description of a structural view is said to be model-based in this case.
4. *It is advisable to model even the PSM in the form of the UML model* instead of the running code, despite the fact that the PSM may most of

the time contain the same information as its implementation counterpart. So doing, automatic code generators can take UML models and translate them into code with interface definition files, configuration files, makefiles, and other file types. If generators are not available at the developing time, we must handcraft a code, however, we still have a good model for the future, so the development is said to be *future proof*. The major trend in software engineering adds one or several layers of middleware, so possibly there could be more than one PSM layer.

Middleware is the "glue" between software components or between software and the network (Internet or Local Area Network), and roughly the "slash" in a Client/Server architecture. In today's corporate computing environment, many applications have to share data. Putting middleware in the middle means each application needs only one interface to the middleware instead of separate interfaces for each connection. Middleware can keep captured data and hold them until they are processed by all applications or databases. This capacity of "buffering" adds another dimension to middleware. Middleware add several services needed nowadays in most distributed environments (checking data for integrity, printing out, converting data, reformatting, etc.). From a functional viewpoint, middleware allows multiple computers to do multiple things across a network and allow one computer to do many things or one complicated thing across a network (link a database system to a web server across the network, allow users to access the database via a web browser, link multiple databases in multiple servers, online games, etc.) Middleware can be a single application, or it can be an entire server. Middleware are often invisible or transparent to some categories of developers. Well-known middleware packages in the Internet Domain include the Distributed Computing Environment (DCE) and the Common Object Request Broker Architecture (CORBA). There is another marginal trend that considers middleware as an adapter device (for instance connecting a new printer to an older computer. In this case, middleware are not reserved only for web or distributed applications.

5. *Nested classifiers* are an extremely powerful concept for structuring object system. In the UML, almost every model building block (classes, objects, components, behaviors such as activities and state machines, and more) is a classifier. So, we can nest a set of classes inside the component that manages them, or embed a behavior (such as a state machine) inside the class or component that implements it. This capability also lets us build up complex behaviors from simpler ones. For example, we can build a model of an Enterprise object, zoom in to embedded site views, to departmental views within the site, and then to applications within a department.

6. *Use of inheritance relationship when appropriate*. The major advantage of object-oriented programming is the ability of classes to inherit the properties and methods of their parents. Inheritance enables to create objects that already have built-in properties and functionality.

5.5.1 Class, Type, Object, and Set of Objects

5.5.1.1 Object. In an article dated from 1987 Wegner quoted:

> An object has a set of operations and a state that remembers the effect of operations" Objects may be contrasted to functions which have no memory. Function values are completely determined by their arguments, being precisely the same for each invocation. In contrast, the value returned by an operation on an object may depend on its state as well as its arguments. An object may learn from experience, its reaction to an operation being determined by its invocation history.

This definition reinforces concepts used in pioneer languages like SIMULA and SMALTALK. Object is implicitly an entity, has operations but cannot be assimilated to them (operations and functions are synonyms in this definition). Objects have states through the execution of their operations, may learn from experience (this fact draws objects towards common characteristics of "agents") and actions deployed by objects are not of combinatorial type but have a "historical" component. So, the reaction of objects depends upon current inputs and experience learnt from the past. Objects therefore own some kind of memory and can use them to elaborate new behavior.

Passing in review all the definition of the term "object" in the literature will be a very annoying task, if not controversial. In fact, the definition of a concept is a complex task that is dependent of many factors:

1. The intended *audience*. Cultural and education level may influence the definition
2. The *domain* of interest
3. The *view* adopted inside this domain.

   For instance, while modeling, an activity in a dynamic view may correspond to an operation in the structural view. So, terms used will change accordingly
4. *Where the concept is located* in the ontological architecture built on the current domain. Surprisingly, an elementary concept is more difficult to define than a complex one. Very often, the definition of elementary concepts leads to circular use of elementary terms.
5. The *purpose* of the definition. More insidiously, if a definition must fit into an existing system, sometimes, the definition must be adapted to the context where this definition must be inserted, and in this respect, the result can depend upon the purpose of the exercise. The problem is currently referred as the contextual interpretation of terms.

6. The *validity range* of the term to be defined
   The definition of the term "object" is very difficult as an object can be very elementary object as a passive "piece of chalk" used by the instructor and at the opposite end, it can be the whole earth seen from the moon, passing through social object like agents, or abstract objects like our mind or human memory system.

The term "object" already defined in the Section 2.4.1 is closed to the definition given by Wegner. It includes agents, humans, organizations, etc.

5.5.1.2 Classes, Attributes, and Operations. The notion of class is ubiquitous in science and is central to the representation of structured knowledge and database development. Wegner (1987) defined class as:

> A class is a template (cookie cutter) from which objects may be created by "create" or "new" operations. Objects of the same class have common operations and therefore uniform behavior. Classes have one or more "interfaces" that specify the operations accessible to clients through that interface. A class body specifies code for implementing operation in the class interface.

Even this definition is closer to the implementation model; the main idea is that a class is a kind of mold for creating objects that have common characteristics and behavior (objects do not have all identical states at run-time). Class is not a "set of objects" but just a *mold of creation*.

> We can have two switches *SW1:Switch* and *SW2:Switch* issued from the same class *Switch* in a same system and SW1 is On and SW2 is Off.

But, the notion of class and classification in real world goes beyond the concept of mold for object creation and code sharing in object-oriented programming languages. Classes are often considered as results of the classification process. To simplify object description, we can see the *classification* as the process of mapping objects into categories, using some *classification rules*. Rules are based on *classification attributes* values.

There are several kinds of attributes in a class that need comments:

1. *Classification attributes* or *global attributes*. When we make a class Person, a class Sensor, a class Alarm_Detector, etc. to solve a real-time problem of home security system, we make our own "user classes" or "application classes" in the UML and suppose that all classes of our application are already there and magically "classified" by our imagination or design skill. In fact, it appears as the whole world has all its real classes already made, each corresponds to a real object or a conceptualized object. If somebody asks us the question "How are classes made and on what basis?" we could be very embarrassed. At this time, we look for some criteria that may help us to justify the way we make our own classes at the

application level. The most obvious solution is "taking an enumeration variable and affecting values as Person, Sensor, and Alarm_Detector to it." Even if concepts are not well explained (because we do not have to explain evidence), classes are there, differentiated by value of an enumeration type. If nobody asks us embarrassed questions, the mentioned attribute is "invisible." It is a *classification attribute* or a *global attribute*. The term "global" comes from the fact that we suppose that the whole humanity agrees with us on our classification.

In applications like pattern recognition, text recognition, biology, oceanography, biometric parameter matching, ontological development, etc. where we must take objects, analyze all their characteristics, classify them into categories, classification attributes are not "invisible" but must be defined and justified by the classification process.

So, classification attributes or global attributes are used to classify objects of the world or more restrictively in a given domain of interest. Classification attributes are "invisible" for people who develop from scratch systems with the UML.

2. *Local attributes or application specific attributes*. Local attributes are defined inside an application. If a class Person is defined in a database application with attributes like *firstname*, *lastname*, *age*, etc., those attributes are local to the application. Local attributes are partitioned into structural, functional, or dynamic attributes and can be defined freely for modeling the application.

   Firstname, lastname of a class Person is a structural attributes. Functional attribute may affect for instance the availability of operations defined in this class and condition the way they must be activated or executed. For instance, some mobility functions cannot be activated for a handicapped person (handicap attribute). Dynamic attributes are defined in the context of the application to store states of the system (e.g. a home alarm system may be found, when powered, in the following dynamic states: idle, intruder_detected, alarm_on, central_calling, phoneline_occupied, etc.). Some attributes are hybrid, for instance, the age of a person is naturally structural, but, at some high values, can inhibit some of his functions, for instance drive_car().

The UML defines itself meta-attributes (for instance *isComposite* for a state). Meta-attributes dictate how model elements are represented or interpreted in a diagram and is not concerned directly with the semantics of concepts we want to model.

Objects works through their *operations*. Operations are called *methods* or *functions* in object language programming. Operations are modeling concepts that represent processes or tasks in the real world. Hereafter are some characteristics of operations:

1. *Visibility.* The visibility of the operation or its access rights show who can invoke operations.

   (a) The UML defines four default visibilities:

   - Private ("-"): Only the current object can invoke the operation
   - Protected ("#"): Only object of the current class or inherited classes can invoke the operation
   - Public ("+"): Any object can invoke the operation
   - Package ("~"): Only classes within the same package (namespace, container) can invoke the operation

   (b) As visibility is defined as an enumeration, we can go beyond the default number and define another visibility. If more complex visibility is required, we can always put a "public visibility" frontal service desk, make verification and redirect or reject the request and, in doing so, achieve any form of visibility.

2. *Operations may be of any complexity.* The complexity of operations derives from the complexity of the wide range of objects.

   (a) Starting from a most simple private operation that changes only the value of an internal attribute (e.g. a clock that changes at each second its internal time), we can activate a complex operation of an agent who must realize a mission. During this mission, he interacts with other agents by activating their operations and so on.

3. *Operations parameters.* To execute any operation, an object needs data obtained from local attributes, data coming from external sources, or in complex situation, it must request data from another objects, different from the invoking object

   (a) If we ask an agent to compute the conversion of 100 CAD into US currency, we must give, with our request, the value 100, and invoke the operation *convert()* of this agent. This agent must eventually request data from another agent to get the most up-to-date conversion ratio at the moment where our request arrives.

4. *Operation returned results.* The result returned by an operation can be of any complexity.

   (a) The model of programming languages let us return only a simple value (integer, real, string, etc.). In a complex case, we can return a pointer to an object or a pointer to any arbitrary complex structure.

5. The call to an object can be synchronous or asynchronous. Generally, when invoking a service from another object, the source or caller object

will lock itself and wait for the returned result. This kind of behavior is qualified as "synchronous" in the UML.

(a) In an "asynchronous" call, after asking the service of other objects, the caller will continue with its own tasks without waiting for the returned value. As the caller may make many calls and the returned results may come back to the caller in any order, the returned results are therefore considered as new messages and must be identified correctly with a full message content and envelope. Effectively, in stateless protocols like HTML, the second scenario prevails.

(b) The previous example describes the software implementation model. In real world, we can have more complex situation to model. There can be no bound on the amount of time that can elapse between process steps, no bound on the time it can take for a message to be delivered. Semi-synchronous model of communication is a synchronous model that set up the upper and the lower bounds for time. This model serves often as a basis for real-time reasoning.

The number of slots defined in a class is usually 3. The first slot is reserved for storing the *class name*. The second slot groups *local attributes* defined by designers and the third slot gathers all *operations*. The metamodel does not fix any number of slots, so user-defined slots are expectable feature in UML tools for organizing various categories of attributes and operations.

5.5.1.3 Type. Type has been extensively studied in Computer Science Domain (Cardelli and Wegner, 1985) (Nierstrasz, 1993). Type is a complex issue and the meaning may vary considerably with domains. Common meaning defines type as a "category of being". In this sense, typing is classification.

Human is a type of things and Animal is another type of thing. When basic types are defined, subtypes can be added to enrich the tree structure commonly used in classification.

The type verification with this common meaning needs a more precise definition with multiple criteria that encompasses *both structural and behavioral aspects* of objects. If everything is well defined, normally we can make type verification.

In computer language domain, when programming, we manipulate datatypes like integer, real, string, double, etc. They are primitive datatypes and their values ranges are well defined. Linear data structures (arrays, lists, etc.), graph data structures (B-tree), heterogeneous structures (structure, union) can be built from primitive datatypes.

Starting from the simplest definition, a *type* A is a set of all values that has that type A. *Int* is a set of integers. *Float* is a set of IEEE floating point values coded with 32 bits. To say that a particular variable is of type A, we write

$$a : A$$

By extension to record types, we write

$$\{a_1, a_2, \ldots, a_n\} : \{ A_1, A_2, \ldots, A_n\}$$

Functions are also variables but functions need arguments, so the type of a function f is its extended signature (see Section 4.6.4) that comprises the type of all arguments P plus the type of the return value V.

$$f : P_1 \times P_2 \times \ldots \times P_p \rightarrow V$$

Languages in which all expressions are type consistent are called *strongly typed* and a compiler can guarantee that programs accepted by it will execute without type errors since verifications at compile time can discover inconsistencies. This verification at compile time is called *static typing*. Static typing may lead to loss of flexibility by constraining objects to be associated to a particular type. Dynamic typing allows datatypes to be changed at run-time.

Now, what is the type of an object?

An object type is that which can be seen from the outside, so we must take into account only the attributes and operations that are declared *public*. If A is used as generic letter for attribute type and O generic letter for operations, the type of an object comprises n attributes and m operations (each operation type is itself another record type of its parameters and return value).

$$\{A_1, A_2, \ldots, A_n, O_1, O_2, \ldots, O_m\}$$

The static type verification so defined and based on the type of all public local attributes and the extended signature of all public operations is a low-level process. If two objects passed the type verification, we can ascertain that they belong to the same class if in our local system. If there is no collision (declared objects that bear the same appearance but their interfaces are not built with the same classes). Moreover, this type verification does not verify anything about the behavioral aspect of objects. We can have two operations that take no parameters, which return *void* (nothing) value, but behave differently and finally do not come from the same class. In a global or distributed system, the collision of two objects that have the same appearance is not negligible.

To summarize, an object type is a record type of its public attributes and operations and the object type verification cannot say in all cases that they are necessarily issued from the same molding process. This verification is nevertheless the first step for any identity checking.

An object-oriented system is not necessarily class based. We can work with objects in a system that does not contain any definition of class concept. Systems can be built just by composition of objects. A car is made of a composition of tens of thousands of mechanical pieces. We can make a new object by designing it from scratch or by taking an old object and then adding by composition some new attributes and operations.

To summarize, the way the notion of types was defined in programming languages is not universal. With systems having the class concept, classes play the role of molds (of creation). *Object type* can be seen from class interface (the term "interface" used here is more general and does not mean only UML interface) with its overall signature or record type. A class contains private attributes and private operations, so *a class is not always a type*. They are fundamentally different concepts as a mold is not the appearance of the object it molds. When discussing about inheritance, inheritance or subclassing is not subtyping.

5.5.1.4 Set of Objects. A set is a collection of objects, usually with some common properties. For instance, we can divide a class of students into the set G of all girls and the set B of all boys. We can also divide the same class of students into three sets A-F, G-N, and O-Z, consisting of students whose first letters of names are, respectively, in those intervals. We can combine sets together and find all girls with names beginning from O to Z.

A set of objects is therefore a collection that shares some attribute values in common. A set is not a class, because a set is a collection of instantiated objects and a class is a mold. The mold is unique but the cardinality of the set is different from 1. However, there is an interdependence of class and set of objects in the way classes are defined in an application as illustrated in the following example.

> If we make a class Car as shows in Figure 5.5.1.1, we can instantiate a set (or group) of cars manufactured in USA, other sets from Japan, Korea, Germany, etc. The class Car, as defined, can be used for registering all cars that are in circulation in California State.
>
> Now, let us consider the Car class created by a manufacturer of US cars. He is not interested to make a class with local attributes *origin_country* and *constructor* as all values would be initiated by default to "USA" and "GM" (if the constructor is e. g. GM). Normally, he must name his class "US_Car_from_GM" but he decides to call his class Car to simplify things.
>
> Now, let us consider a vendor of "made in US only" used cars. He must declare a class Car with *constructor* attribute as he must differentiate between GM, Ford, Chrysler, etc. If he changes his mind and decides to sell Japanese cars together with US cars, he must reengineer his database and insert *origin_country* as local attribute in the definition of his Car class.
>
> The possible ambiguity is the fact that the dealer does not call his class "made_in_US_Car", and the manufacturer does not call his class "US_Car_from_GM" (This example is hypothetical only since GM may decide to create all attributes, for instance to record a GM car made in Japan). In fact, there is implicitly a transfer of local attributes to the block of global or classification attributes.
>
> "US_Car_from_GM" is a subset of "made_in_US_Car" that is itself a subset of "State_Registration_Car" that is itself a subset of "Generic_Car". Subsets based on values of attribute are often isolated to create independent classes. This fact is

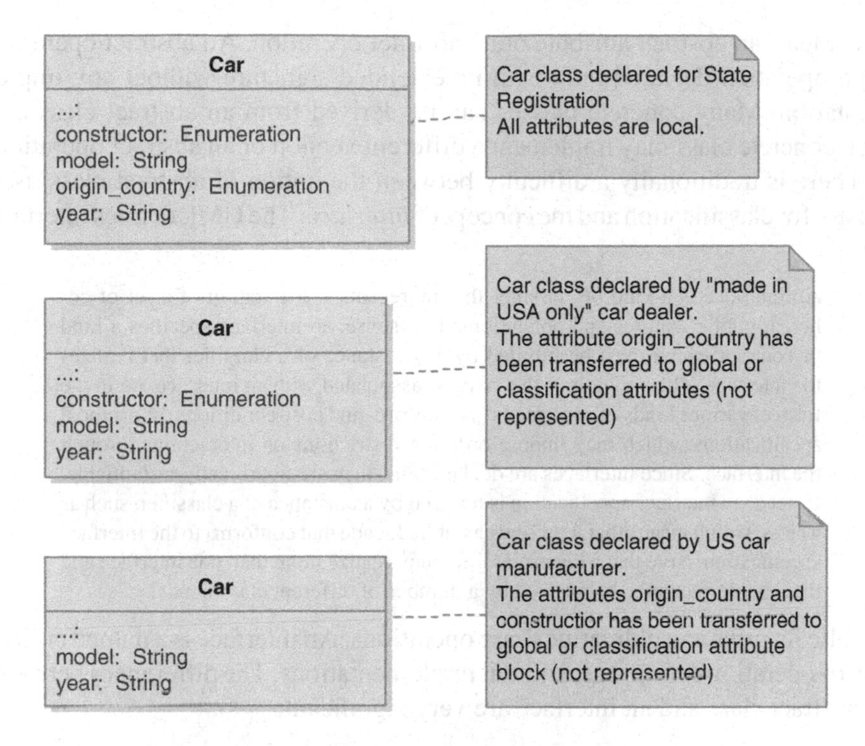

Fig. 5.5.1.1 Car class declared in three different contexts

often observed in real systems. Dependent on the number of enumeration values, this fact could create a very rich set of classes disorderly named if we look at the problem globally. But each local database is still consistent if no data merging, no data fusion, no data cross referencing are necessary.

If we consider global attributes and local attributes as a whole, the number of attributes has not changed but according to the context or the domain, there is a *bilateral transfer of attributes between the global and the local blocks*. This transfer has an impact in the classification mechanism as we can make theoretically any class based on any combination of values of attributes. This fact explains the subtle relationship between sets of objects and classes.

5.5.1.5 Abstract Class: Difference with Interface. Abstract class is defined as a class that cannot be instantiated to create objects. Why do we need them?

Food? What is an instance of food? So, food is an abstract thing that we can eat. But we can derive salad, orange, apple from food. So, abstract class Food is useful as an abstract element in the hierarchy of food for structuring a global description.

Abstract classes are very useful in defining domains and ontologies, as they are semantic nodes or semantic reference points. A class becomes abstract if it has at least an abstract attribute or an abstract operation. An abstract operation is an operation declared only with its extended signature without any implementation. Many concrete classes can be derived from an abstract class and each concrete class may implement a different version of an abstract operation.

There is traditionally a difficulty between the notion of abstract class used mostly for classification and the concept of *interface*. The UML defines interface as:

> An interface is a kind of classifier that represents a declaration of a set of coherent public features and obligations. In a sense, an interface specifies a kind of contract which must be fulfilled by any instance of a classifier that realizes the interface. The obligations that may be associated with an interface are in the form of various kinds of constraints (such as pre- and post-conditions) or protocol specifications, which may impose ordering restrictions on interactions through the interface. Since interfaces are declarations, they are not directly instantiable. Instead, an interface specification is realized by an instance of a classifier, such as a class, which means that it presents a public facade that conforms to the interface specification. Note that a given classifier may realize more than one interface and that an interface may be realized by a number of different classifiers.

"Public features and obligations" are operations. An interface is a named collection of operation definitions, without implementations. The differences between an abstract class and an interface are very significant:

1. An interface cannot implement any operation, whereas an abstract class can implement some common operations for all derived classes
2. An interface is not part of the class hierarchy
3. Classes derived from an abstract class share commonalities whereas interfaces can be built from completely unrelated classes.

5.5.1.6 OID, GUID, and UUID. OID has been investigated in the past (Khoshafian and Copeland, 1986). Identity is a property that distinguishes an object from all others. Object identification is a general problem in real life (complicated names, collisions, etc.) and is not specific to computer sciences. If in a local system, we can adopt a strategy to name unequivocally things, modern communication networks need a well-designed identification in order to have a global and unique ID (GUID) and at the same time a shortest ID, avoiding communication delay.

> When we name a directory of our file system, we are facing the problem of object (directory) identification. Every path taken from a computer must be unique, for instance *C:\data\book\chapter1*. If we have a second computer connected to the first one and the file directory contains the same name *C:\data\book\chapter1*, we are facing a collision problem of names. Happily, standard operating system has resolved this difficulty by adding, for instance, the name of the second computer to the directory name, e.g. *\\forgetmenot\ DiskC\data\book\chapter1* (*forgetmenot* is the name of the second computer and *DiskC* is mapped to *C*

on the local computer). So doing, we don't have to rename all directories of the second computer to be able to see those two file systems at the same time.

Internet is a global communication medium. Any server in the net is located with an IP address composed of four fields starting from 000.000.000.000 and ending at 255.255.255.255 allowing more than 4 billions servers to be identified in the Internet (some addresses are however reserved). As this number is hard to remember, an equivalent textual address is used to map a server address to a more easy retained character string. For instance, www.omg.org is translated into 192.67.184.5 (just emit with the DOS shell a "ping www.omg.org"). www.omg.org is a reserved and protected string (as a name of a company), but the server underlying this string address may change every time. If OMG wants to shut down their server for maintenance, technicians can map temporarily www.omg.org to another IP address.

OID has been investigated independently in object programming languages and in database management, either relational or object. In the past, object practitioners claimed that they do not need to create an artificial key to identify any record since the object system maintains itself an OID, so their system was automatically indexed. Before discussing this "advantage" let us come back to the original problem.

Identity is normally invariable and is independent of whatever can change with the object (value, properties, location, etc.). The trend towards giving a global identity to every object come from the fact that all networks in our modern world tend to be fully interconnected and there is a real need to identify objects globally in a distributed environment. We thus need a kind of universal ID called GUID (pronounced "gu-ID") by Microsoft. The GUID is also known as Universally Unique Identifier or UUID. GUID is ID that is guaranteed to be unique in space and time. An object with a GUID can be replicated in many servers for optimizing the access time but remains invariably the same object (the object coherence problem between copies must be solved in this case but it is another problem).

Object systems and object databases, by their nature, have a low-level identity generator designed for handling the object system. They can handle true clones (in an object system, we can have two "John Smith," same firstname, same lastname, same age, etc. but recognized as two different persons without any artificial primary key). But, this identity is not always accessible and the way it is encoded shows that it is a "machine" identity, not really human readable. So, from an application viewpoint, sometimes, a given identity, easily remembered (phone number, NAS, etc.) is still very useful in everyday life systems. So, the presence of the two identity systems, one low and one high level, is probably an optimized solution.

Some relational databases make use of surrogate keys that have the same property of a low-level ID in object context. The two worlds are connected through this initiative. Surrogate keys provide a number of advantages. The most significant one is that this key needs never to change. We can now easily change the NAS attribute of a Person in such a context. The surrogate key in relational world has its equivalent OID in object world. They are system-generated keys. At higher level, the "business key" or "user-defined key" in both worlds bears some semantic for human beings.

5.5.2 Database Objects, Real-Time Objects, and Their Relationships

Real-time objects are typically full-featured objects with:

1. *Identity*
2. *Structural attributes* that describe them
3. *Behavioral attributes*, so objects have states
4. *Operations* that make objects dynamic workers

The construction of a real-time object system is typically based on two important relationships:

1. *Ownership* or *aggregation/composition*. Objects are hierarchically structured in an object system, Objects and composite objects collaborate together to accomplish complex tasks.
2. *Inheritance*. Objects are classified and new classes can be defined from existing classes, in other words, it is possible to incorporate structural and behavioral features of existing objects into newly created objects.

Unless a database is used, for instance, to record the evolution of a scientific experience (in this case, behavioral attributes are relevant), records in relational databases describes only structural attributes. Each record is a list of values. To make a correspondence with real entities, a given identity is fabricated to mark each record unequivocally. So, relational database is concerned only with points 1 and 2 of the preceding list of four points.

> If our system describes for instance all the workers of a very large hospital that include patients, medical and administrative personnel, we can study how fast this hospital provides services or some targeted operations of this hospital by defining and simulating systems with patients, medical doctors, nurses, and so on. At the same time, this hospital must keep records of thousands of patients, data of its personnel. We have therefore a real time system and simultaneously a database system that may share common structural attribute values.
>
> Traditionally, for the database, the data model can be established with an E-R (entity-relationship) diagram and implemented with a RDBMS. User can access to the database by using SQL (Structured Query Language) in two steps, the first step is to define the database structure based on E-R schemas, and the nest step is to query this database to obtain information. Between the two, the database must be populated with data.

> The real time simulation can be developed with an object programming language (C++, C#, Java, object Basic, etc.). Objects will be created in the simulation program with classes. Some classes bear striking resemblances with entities in E-R diagram if we consider structural attributes.
>
> Now, how to make those two systems developed in parallel, to communicate together, i.e. allowing object programming to access relational datastores (for instance, if an object in this context has a long list of structural attributes, it would be natural to desire to store those attributes in the database, and use an object with the same identity in the real time system populated with behavioral attributes, having some kind of connection with the object structurally described in the database). There are known modern methods as JDBC (Java Database Connectivity) or ADO (ActiveX Database Objects) that make the glue between two independent systems. From the programming platform, SQL requests can be sent to RDBMS and answers are wrapped then returned to the requester. The method is slow but direct access implies some form of mapping between an object system and a system based on tables of values (O/R or Object Relational mapping). Some tools in the market are targeted to solve this issue but generally programmers prefer to hand code database access through JDBC or ADO.

One of the crucial questions is the opportunity to use objects as the front end, at the design phase, for developing relational databases, replacing the E-R diagram by the UML class diagram. The mapping is nearly straightforward as each entity can be replaced by a class and all attributes or fields in the E-R diagram are mapped into class attributes. Relationships in E-R diagrams are converted into associations in a class diagram that fully support all concepts (roles, multiplicities, navigability). Attributes of relation are converted to class associations. Inheritance or aggregation/composition is translated into tables as well. If the object model are well structured, the resulting structure issued from tables is blurred and information are got by an expensive table joining process to find things that need merely some navigation hops in the corresponding object implementation model. Nevertheless, the mapping is possible, the implementation is rather awkward but it works. So, in waiting for a better proposal widely accepted, we can use the object model as the front end, even if the implementation is purely relational. For more information on the mapping, please consult (http://www.service-architecture.com/).

The last effort made by the database community to adhere to object paradigm was the famous Object RDBMS (ORDBMS) that extend relational database systems with object functionalities to support a broader class of applications. This concept bridges the relational and object-oriented paradigms. SQL 1999 was an enhanced version of SQL intended to support this extension. ORDBMS was created to handle new types of data such as audio, video, and image files. One advantage of ORDBMS is its compatibility with core RDBMS, so organizations can continue using their existing R-systems while trying new O-systems in parallel.

In summary, it would be unwise to base everything on the small market share of the ODBMS (about 1% in 2005 of the whole database market) and hastily conclude that the object database has marked its end. It is still, at the start of the new century, an immature technology but the object paradigm is omnipresent and has let its marks everywhere in the database industry during the last two decades. Today relational databases are able to store complex data structures (BLOB: Binary Large Object; CLOB: Character Large Object) and the ORDBMS is present as a historical witness of the object footprint.

5.5.3 Dependencies, Relationships, Associations, and Links

Terms like "dependency, relationship and association" are defined in the UML with their metaclasses. They are compared to more common meanings.

Relationship

UML: Relationship is an abstract concept that specifies some kind of relationship between elements (Relationship metaclass from Kernel package)

AskOxford: 1. The way in which two or more people or things are connected, or the state of being connected. 2. The way in which two or more people or groups regard and behave towards each other.

Dependency

A dependency signifies a supplier/client relationship between model elements where the modification of the supplier may impact the client model elements. A dependency implies that the semantics of the client is not complete without the supplier. The presence of dependency relationship in a model does not have any runtime semantics implications; it is all given in terms of the model elements that participate in the relationship, not in terms of their instances.

Cambridge Dictionary: a state of needing something or someone, especially in order to continue existing or operating.

Association

An association specifies a semantic relationship that can occur between typed instances. AskOxford: "a mental connection between ideas".

Link

UML: A link is defined in UML as an instance of an association.

Connector

UML: Specifies a link that enables communication between two or more instances. This link may be an instance of an association...

As graphical notation, we read A connector is drawn using the notation for association.

In the current language, *relationship* and *link* are often taken as synonyms. *Connector* sounds hardware and is mainly used for electrical connections or graphical link between graphical objects. *Association* is used to relate more

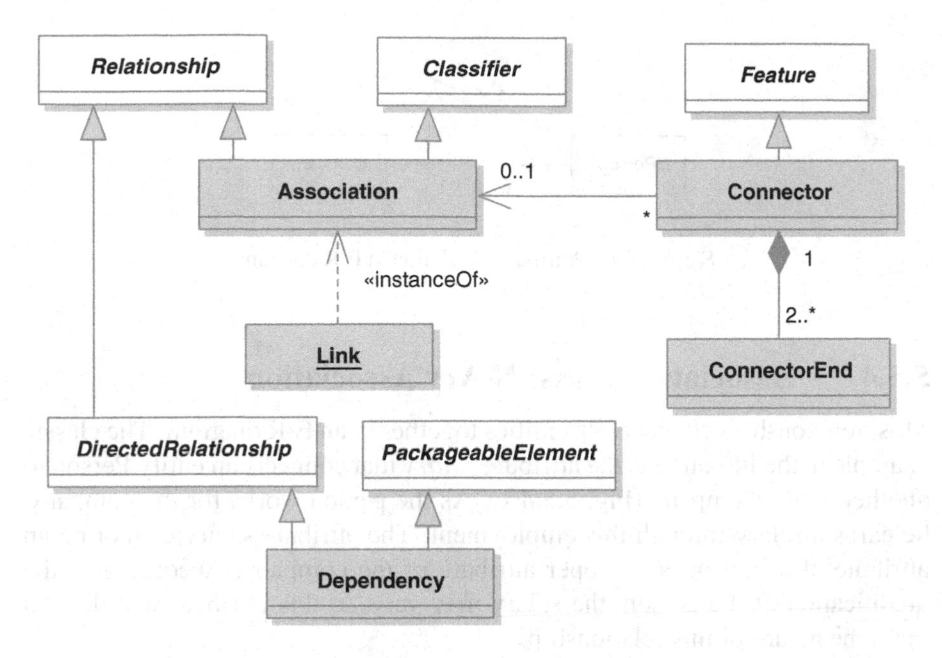

Fig. 5.5.3.1 Relationship between Relationship, Association, Link, Connector, and Dependency

abstract concepts. *Dependency* depicts an asymmetrical relationship between humans or things.

In the UML, the classification is more precise so the mapping will depend upon the meaning we can deduce from real contexts. The Figure 5.5.3.1 summarizes the relationship between those concepts in the UML. In this figure, *relationship* can be used as a general term as there is an abstract (noninstantiable) metaclass attached to it. So, relationship is the most elementary brick in this hierarchy. Connector is just a feature but it can be related to something more important as an association.

Association is a richer concept as it derives from both Relationship and Classifier. This view complies with the common meaning of this term in real world except that association is used at the class level in the UML. A link is an instance of an association and links connects objects (instances of classes) together. As, there is no link metaclass, association may be used both to link classes or objects, even objects to classes.

Primarily, *connectors* represent interactions among objects and provide communication paths between two or more instances (parts, ports, etc., see Section 4.9.5). It is used mainly in a composite structure diagram.

Dependency is a directed relationship and can be used to enslave a model element to another element. The exact definition of this dependency must be specified by a constraint or a significant name of this dependency.

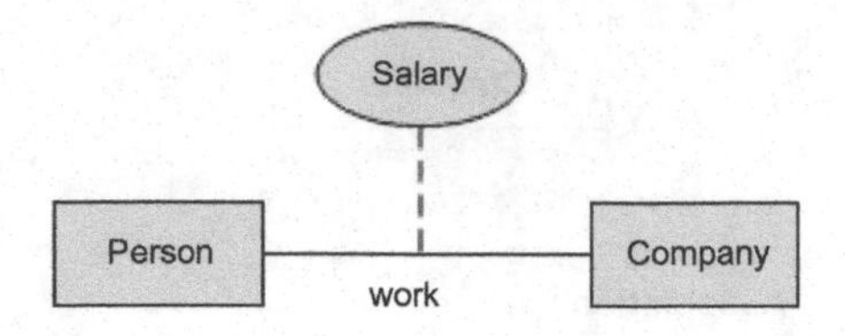

Fig. 5.5.4.1 Attribute of relation in E-R diagram

5.5.4 Association Class: N-Ary Association

Most relationships connect two entities together in an E-R diagram. The classic example in the literature is the attribute *salary* that connects an entity Person to another entity Company (Fig. 5.5.4.1). As the person works for the company, he earns a salary through this employment. The attribute salary cannot be an attribute of a Person or a proper attribute of the company. According to the qualification of the person, the salary may vary, so this attribute will depend upon the nature of this relationship.

If we map now this E-R diagram into a class diagram, orphan attributes are not tolerated in object world (each attribute must have a host object), so a class association replaces the attribute of the relation in this context. This class is instantiable, so it must be qualified for an independent existence. Actually, when we think of how human relationships are interweaved, very often, we run business by contracts (Andrade and Fiadeiro, 1999), the Person and the Company agreed on a Job contract. The job has a proper existence as independent class as the company may publicize the job to find a candidate. Later, when a person accepts the job, the job contract consolidates the relationship between the person

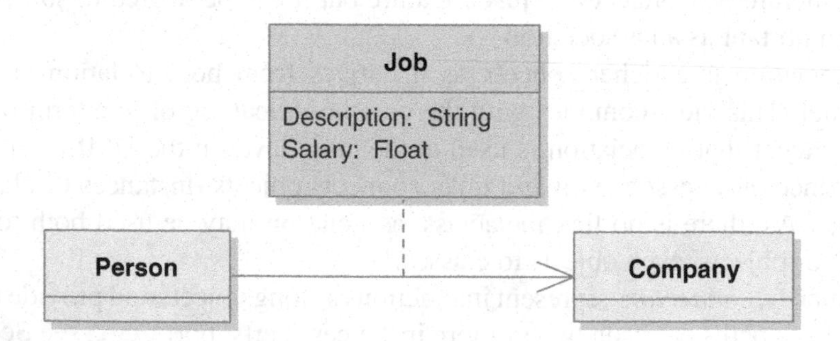

Fig. 5.5.4.2 Association class. When a person takes a job in a company, the two sides are tied by a contract that is represented as a class association. We can instantiate the Person alone, the company alone, and the Job alone. If the person decides to take the job, a contract must be signed with the company

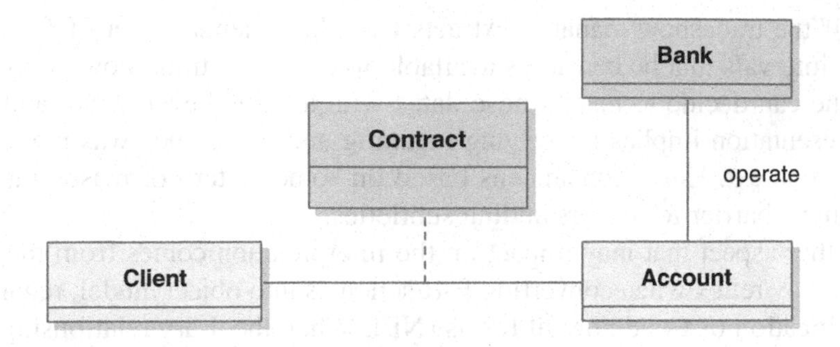

Fig. 5.5.4.3 Association Class. Association class represents a contract in business domain

and his company. The contract has the same importance and a proper existence as the classes Person and Company themselves (Figs 5.5.4.2 and 5.5.4.3).

An n-ary association is an association between three or more classes (a single class may appear more than once). The multiplicity for n-ary associations may be specified but is less obvious to interpret than a binary association. In the absence of a special constraint, it is considered as a potential number of instances in the association when other (n–1) values are fixed.

If we follow the logic of business, when x persons are implied in a contract or association with nearly the same responsibilities, an n-ary association can be conceived to represent this collaboration. If we suppress a member of this association, we can threaten its existence, weaken or destroy the n-ary relationship. This fact could be used to test the "solid" foundation for an n-ary association.

N-ary associations may be misleading and sometimes overused. The presence of multiple n-ary associations in a project may be a hint showing that this project is not correctly analyzed or not analyzed at all, as the analyst announces the problem in its most primary form. The following example demonstrates through an example.In the context of very large trade shows that must be held at several locations, the 4-ary association announces at the beginning that a participant may attend several trade shows; each of them can be held at more than one location at various dates. We therefore have the most complex association of type 4-ary with multiplicities N (or *) at each role end.

After the first examination of the 4-ary relationship, most analysts will correct and consider that FromTo_Date cannot be considered as class (or entity in E-R analysis), since each tradeshow has its proper set of *from* and *to* dates (notion of *weak entities* in E-R modeling). So, they are better considered as attributes of TradeShow class than as a standalone class.

Hereafter, we adopt this argument as well, but point out that transformations that results from hunting systematically "weak relationships" may change the semantics and the context in which this relationship has been designed. For instance, FromTo_Date may have a full existence as independent objects (so as class) if the tradeshow manager extracts from his calendar a set of *from* and *to* date intervals that he considers available to accept all tradeshow proposals. Later, he can decide to match those dates with accepted events. So, a choice of representation implies underlying meaning and this choice was made in a specific context. Hasty conclusions based on some pattern of reasoning may constitute a barrier for understanding subtleties.

Another aspect that may impact on the interpretation comes from the possible Pavlov reflex when converting E-R schemas into object model, regarding the verification of Codd normal forms (NF). When the 4-ary relationship was presented, it was a high-level model and not an implementation one so there is no need to verify all Codd *NFs* when modeling at a high level (even if at the implementation level, it must be transformed for evident reason of efficiency).

As the context is not specified, we may erroneously feel, in presence of n-ary relationships and weak entities, that the problem has not received a thorough investigation. This feeling is confirmed in the case of nonuniqueness of interpretation in the presence of n-ary association with high multiplicity values. The first correction reduces a 4-ary association in a ternary association between Participant, TradeShow, and Location. FromTo_Date class is suppressed and replaced by two separate attributes of TradeShow. Actually, the context has changed: we are not in the model of the tradeshow manager but we are managing all subscriptions (another model and another concern). Within this ternary association, we can track what participant is subscribed to what tradeshow and each location this participant may visit.

In E-R modeling and relational implementation model, if we make a table to display information stored by this ternary relationship (Table 5.5.4.1), some lines show violations of Codd NF.

In relational databases, as table is the only data structure used to store information, NFs previously help us detect some implementation problems. Pure object database (not *object relational*) store objects as they appears at high modeling level and use intensively a network of pointers to implement associations. Information retrieval in an object store is fundamentally *navigational*; there is no table to be joined. So, object databases are not really concerned with Codd NFs. In fact, it violates even the first 1NF.

> 1NF specifies that all attribute values must be atomic (indivisible). By the virtue of encapsulation, object concepts accepts at the starting point any complex structure (audio, video, image files etc.) so it violates the 1NF as the structure of a component is decomposable and not atomic.

Other NFs are exposed in (Codd, 1970, 1990). The ternary relationship in Figure 5.5.4.4 violates NFs defined by Codd and will be transformed by E-R analysts. The last version of the same Figure 5.5.4.4 eliminates the direct connection of Participant class to Location Class. Participants subscribe to multiple trade

Table 5.5.4.1 Table containing all possible entries of a ternary association between Participant, TradeShow, and Location

Line number	*Participant*	*TradeShow*	*Location*	*Comment*
1	Adams	Computer Peripherals	North Tower	Computer Peripherals and Location form an independent multivalued association (4NF violation)
2	Adams	Computer Peripherals	South Tower	
3	Bob	Computer Peripherals	North Tower	
4	Bob	Computer Peripherals	South Tower	
	...	...	...	
101	Mary	Software	South Tower	idem
102	Mary	Software	Central Tower	
103	Noemi	Software	South Tower	idem
104	Noemi	Software	Central Tower	
105	Bob	Software	South Tower	idem
106	Bob	Software	Central Tower	
	...	...	...	

shows (1st binary relation) and trade shows are held in multiple locations (2nd binary relation). We can eventually make a search through multiple tables and find out where a participant is loitering at a given time.

Object models are not really concerned with E-R schema transformation techniques, or, in other words, with relational Codd normalization forms. One of the strengths of the object implementation model is the ability to define references that concretize relationships between objects at high level. At low level, references are pointers or memory addresses. When storing an object with references to other objects, OID is used. Once reloaded in memory, the OIDs are reconverted into pointers. Navigating with OID in the storage structure or with pointers in memory is the natural way used by objects to retrieve information on their relationships.

The normalization in object model is by nature iterative, in which objects are restructured until their models reflect user's interpretation (Tari et al., 1997). Object normalization matches interpretation in the UoD (Universe of Discourse) to that of objects in the schema. An interpretation is a set of dependency constraints that provides a given "understanding" of the semantics of the underlying application.

So, when working with object models, we must systematically analyze high-order association (ternary to n-ary) to make sure that we really need them, or there exists a simpler way of modeling things and interpreting them, Secondly, the user's interpretation must be straightforward and fast.

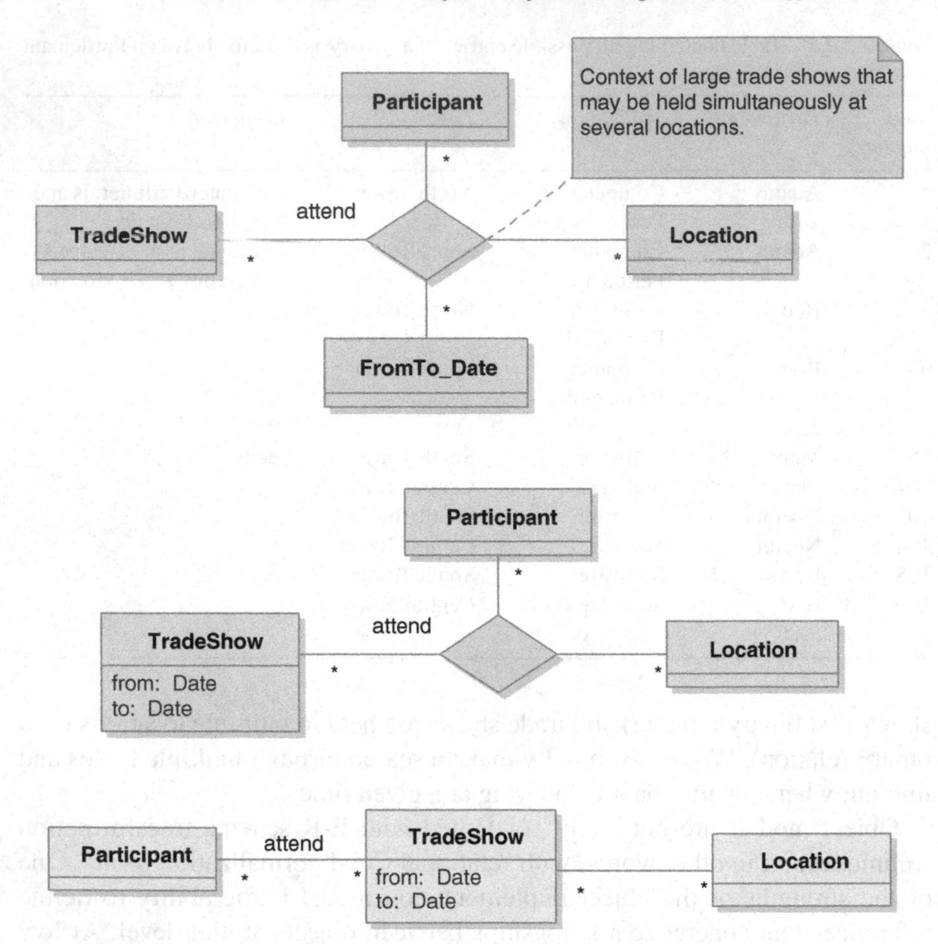

Fig. 5.5.4.4 The first example shows a misuse of 4-ary relationship, corrected in the first step to a 3-ary relationship, then corrected in a second step to transform the ternary association into two binary associations (see text for discussion)

5.5.5 Aggregation/Composition in the Object/ Component Hierarchy

Aggregation and inheritance, angular constructions and among the most important features of object paradigm, are often problematic. This fact can be confirmed by any instructors teaching object modeling. Moreover, aggregation has been differentiated from composition in the UML.

> An association may represent a composite aggregation (i.e. whole/part relationship). Only binary associations can be aggregations. Composite aggregation is a strong form of aggregation that requires a part instance be included in at most one composite at a time. If a composite is deleted, all of its parts are normally deleted with its. Note that a part can (where allowed) be removed from a composite before

Table 5.5.5.1 Aggregation is considered as the loosely coupled form and composition of the highly coupled form between parts and whole

Aggregation		Composition
Cataloguing purpose. Parts are grouped by convenience or for simplifying a description. Description of the whole needs description of parts as well	Treating collections of concepts as higher concepts	*Monolithic Component definition.* Suppressing details of parts to emphasize details of the whole
No emergent properties and functionalities	With *emergent properties* and functionalities at various degrees	Individual properties and functionalities are invisible
No lifetime constraint	With loosely coupled lifetime constraint	Lifetime constraint
No propagation of operations		Propagation of operations (whole to parts or parts to whole or mixed scheme)

the composite is deleted, and thus not be deleted as part of the composite. Compositions define transitive asymmetric relationships (their links forms a directed, acyclic graph). Composition is represented by the *isComposite* attribute on the part end of the association being set to true.

We must choose one alternative in the UML (between aggregation or composition) when representing this relationship. As the modeling world is richer and cannot be partitioned into two blocks, voluntarily, we call this relationship all over this book as aggregation/composition (or whole/part) and let the designer specify the correct constraint that accompanies each relation. In doing so, we class this relation into one unique category and let the constraint decide on the nature and the contextual interpretation of this relationship. More details can be found in Smith and Smith (1977), and Henderson-Sellers and Barbier (1999) that redirects to a rich literature.

Aggregation/Composition has some common characteristics: *binary association, asymmetrical* (if p is part of a whole w, then w cannot be part of p), *transitive* (if p is part of w, and q is part of p, then q is part of w).

Transitivity may cause some funny semantic problems in the loosely coupled side of this relation. A professor is part of a university. His brain is part of his body so his brain is part of his university.

Figure 5.5.5.1 represents this concept as a *continuous segment* with aggregation as the lower end of the concept and composition in its higher. At the lower end, this relationship can be served for cataloguing and at the higher end as a basic mechanism to build monolithic components.

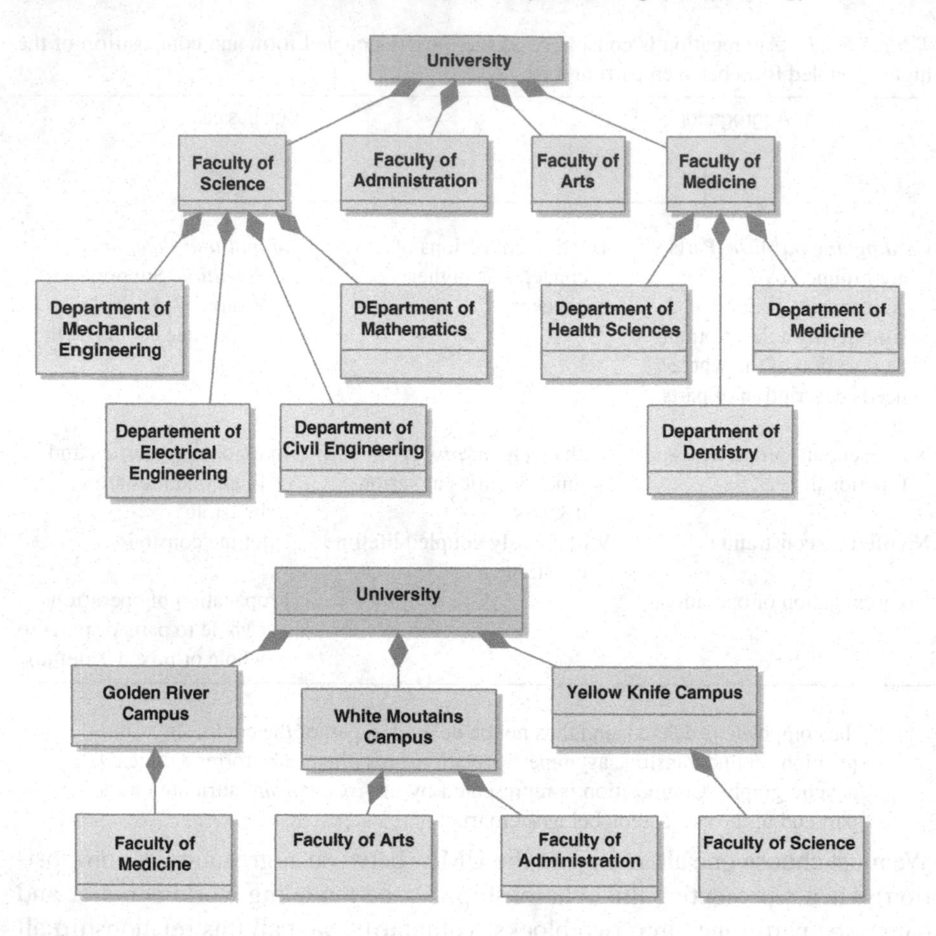

Fig. 5.5.5.1 Same objects but two different aggregation criteria that are "Administrative Structure" and "Geographical Localization"

Components are for composition and components are made of parts. Components have contractually well-specified interfaces and are made of other components, objects. A business component may contain humans or organizations. Aggregation/Composition is therefore a relationship that makes whole ensembles from parts of very different types. The result can be typed itself. Composition is seen as a strongest form that involves consideration on the lifetime of part objects in the UML.

> Considering a main window that spawns several underlying windows. If, in an application, we want, with a single click on the "X" of the main windows, delete all spawned windows, the main window is considered as owning all other spawned windows. In this case, the delete operation, activated in the main windows, has started a tree structured delete operation starting from the leaves of this tree structure. We have therefore an operation that is instantiated in the "whole" object and its execution begins from the lowest parts in the tree structure.

At aggregation side, an organization may suppress or reorganize an administrative Unit without any impact on other Units. In this case, aggregation is used as a grouping concept. Parts are grouped by convenience or for description. But, in the same organization, we can have a sub component that suppression may impact on the existence of this organization. In this case, composition is more appropriate. Inside a same application domain, a same relationship has different "degree of composition" or different "degree of aggregation". Only specific constraints can make things clearer.

Some system is able to work with reduced functionalities even if some components are declared as flawed. Even a flawed component may continue to offer some services. So, it would be very difficult to decide, by a binary decision, whether a system is still usable or not by examining only its lifetime criteria.

A human may lose a limb (an important part of a whole) and is still considered as a human, unless the relationship is not really a composition.

Giving a set of objects, we can have more than one aggregates built on this set of objects, according to the criteria used to establish the aggregates. Figure 5.5.5.1 shows that the tree structure changes completely with the criteria.

From an administrative viewpoint, a university is composed of faculties that are composed in turn of departments. If the criterion is the geographical localization, we have another aggregate since the number of campuses is not always equal to the number of faculties. Many faculties can coexist inside a same campus. So aggregations must be accompanied by their criteria. These criteria must be explicated to avoid non unique interpretation.

When decomposing a system with an aggregation tree or forest of trees, the rule of *level semantic coherence* must be preserved so as, we can stop at any level in the decomposition tree and still find coherence in the description at this level.

Saying, for instance, that a computer is composed of memory, processor P4000, a graphic card, an operating system, a mouse and a printer, etc. is probably the worse description since we find several sub domains at the same level of decomposition. The correct description that respects the level semantic coherence is depicted by the Figure 5.5.5.2.

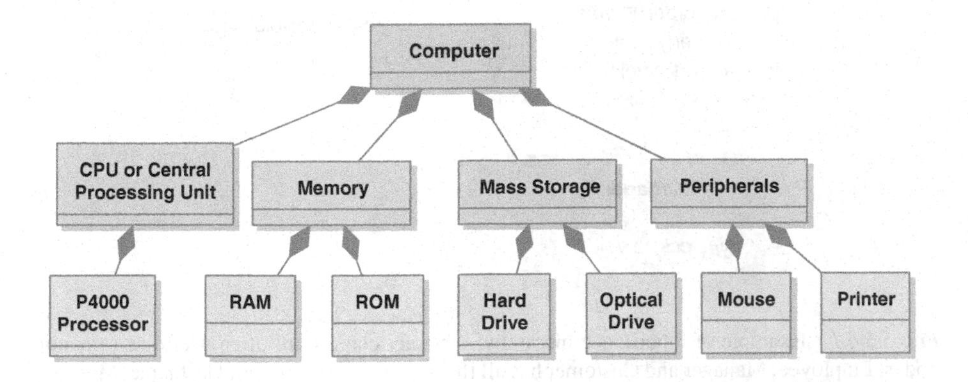

Fig. 5.5.5.2 Decomposition that respects the level semantic coherence

5.5.6 Inheritance

Inheritance is one of the topmost important features of object technology that impacts the reuse mechanism along the whole development process, starting from the inheritance in a UC diagram between actors towards code sharing at the implementation phase. Moreover, it simplifies considerably the way we catch the structural view of all systems. In semantic networks, its name was "is-a" relationship.

Inheritance (Snyder, 1986) works at the class level and factorizes properties and operations in parent classes. *Child* (*sub, derived, specialized*) classes must define only differential properties and supplemental operations that will be added to characteristics of parent classes. Qualifications as "parent" or "child" refer to the naming of two adjacent classes at a given point in the long chain of inheritance hierarchy. An *ancestor class* is a class that is far away from the current definition, starting from the parent, then the parent of the parent, and so on.

Operation in an inheritance hierarchy can be *surcharged*, i.e. the content of the operation in the child class can be changed but the same operation name is preserved through the whole hierarchy. For the signature or extended signature, local rules may apply.

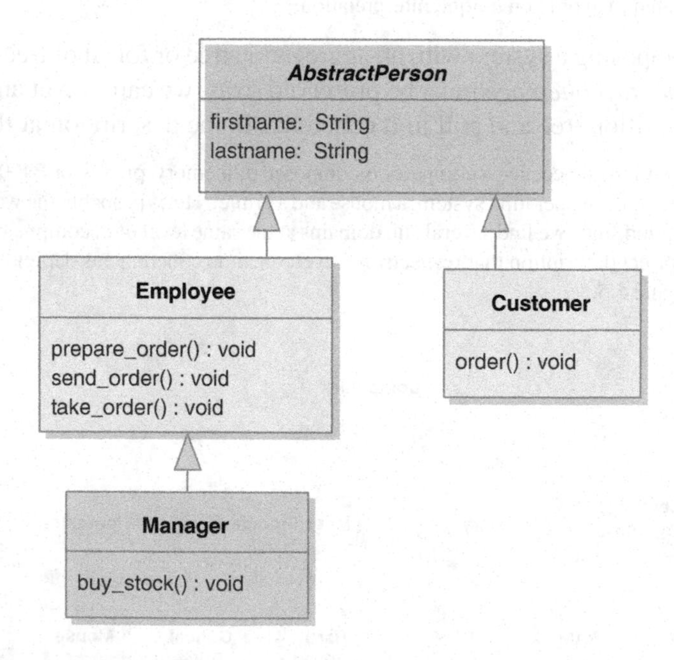

Fig. 5.5.6.1 Example of inheritance hierarchy. Abstract classes are often useful as semantic nodes. Employee, Manager and Customer has all the attributes firstname and lastname. Manager can execute buy_stock() and all operations defined for Employee

Most implementation language does not permit a derived class to "exclude" an inherited operation from its parent. If the operation of the child is redefined to signal an error or to enter an exception treatment if invoked, we realize the equivalent of operation exclusion. This artifice will subtract an operation from the parent.

> Excluding operations is an attractive feature. It corresponds to some situations in real world as well. Children cannot always inherit full properties of their parents. But, a hierarchy with some properties added, others subtracted is very difficult to work with as we must keep an eye on every step of the inheritance tree and keep track of all transformations.

When operations are redefined in the child, an interesting implementation feature would allow us to make use of any operation version of any ancestor by explicitly naming this ancestor. This possibility of surcharging an operation is interpreted mostly as a quality as it allows two or more versions of a same program to coexist in a same system (one for supporting older applications and a newer for new applications), compatibility issues are stringent and must be handled in some way. If operations are surcharged many times, it would be very difficult to recognize a class from its faraway ancestor. Structural clarity is decreasing because so much less can be inferred about the properties of descendants. In extreme cases, it would be questionable about keeping a hierarchy of inheritance if classes share only common names for their operations. In general, if a system accepts operation surcharge, there is no guarantee that the objects of the subclass behave like the objects of the superclass.

> Inheritance is *subclassing*. Subtyping is theoretically different from subclassing. Subtyping is a relation of inclusion between types. If
> $B \leq A$
> B is a subtype of A. We can supply a value of B whenever a value of type A is required. For instance
> Square $\leq$ Rectangle $\leq$ Polygon
> In computer science, subtyping is related to the notion of *substitutability*, meaning that computer programs written to operate with a given *supertype* can also operate with its *subtype*. Any given programming language may define its own notion of subtyping. In most class-based object-oriented languages, subclasses give rise to subtypes. If B is a subclass of A, then an instance of class B may be used in any context where an instance of type A is expected. In this case, subtyping results from subclassing.

5.5.7 Multiple Inheritance

Inheritance cuts down the description of child classes to the strict minimum and allows defining new classes from existing ones. When a system allows a unique parent class, it supports *simple inheritance*. If more than one parent class can be used to define the child class, the system supports *multiple inheritance*

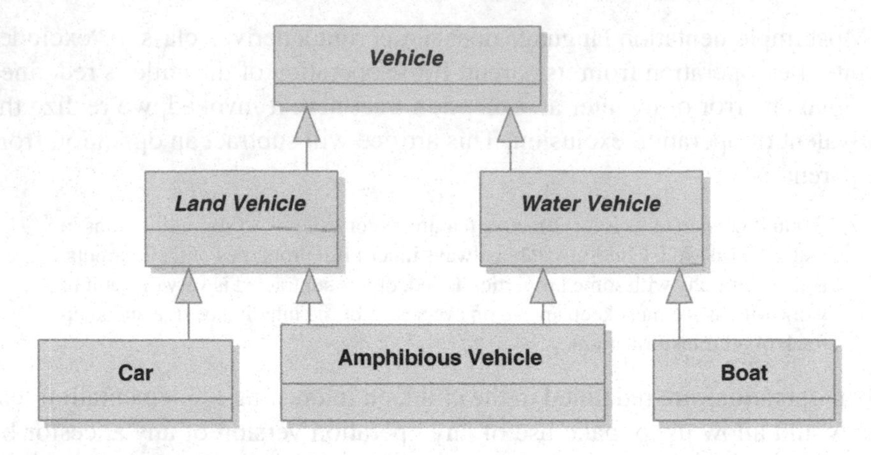

Fig. 5.5.7.1 In object technology, multiple inheritance use union of attributes and operations (not intersection), so the child class is essentially the union of the behavior of the two parent classes

(Cardelli, 1988; Scharli et al., 2003) that has its specific batch of conceptual and implementation problems.

For instance, the class *Amphibious Vehicle* has simultaneously attributes of *Land Vehicle* and *Water Vehicle* (e.g. it has four wheels and helix propeller, it has two maximum speeds, one speed as a car, and another speed as a boat), has both operations of the two parent classes (e.g. it can move both on land and on water). Figure 5.5.7.1 gives its UML representation.

Multiple inheritance is a difficult concept to apply and people can easily confuse it with other relationships, specifically composition. Designers of modern languages as Java and C# have decided that the complexities of multiple inheritance far outweighed its utility. C# supports only multiple inheritance of interfaces but not at class level. The reasons for omitting multiple inheritance from the Java language mostly stem from the fact that Java's creators wanted a language that most developers could grasp without extensive training and multiple inheritance just was not worth the headache.

Multiple inheritance (Fig. 5.5.7.2) additionally poses a classification difficulty when a class derives from two classes coming from different ancestor trees in a system having a forest structure. The easiest way to solve this problem is to accept that a class can belong to multiple categories. Another direction of solution is a relaxation of the famous *is-a* relationship. If D derives from B and C, then D is *like* B and D is *like* C where the form and the degree of similarity (*like* relationship) will depend upon local constraints.

As inheritance is a complex issue, it is often defined in a specified context. In an MDA, it will depend upon the model. So, at high level, inheritance and multiple inheritance can be used to structure and to provide an ontological

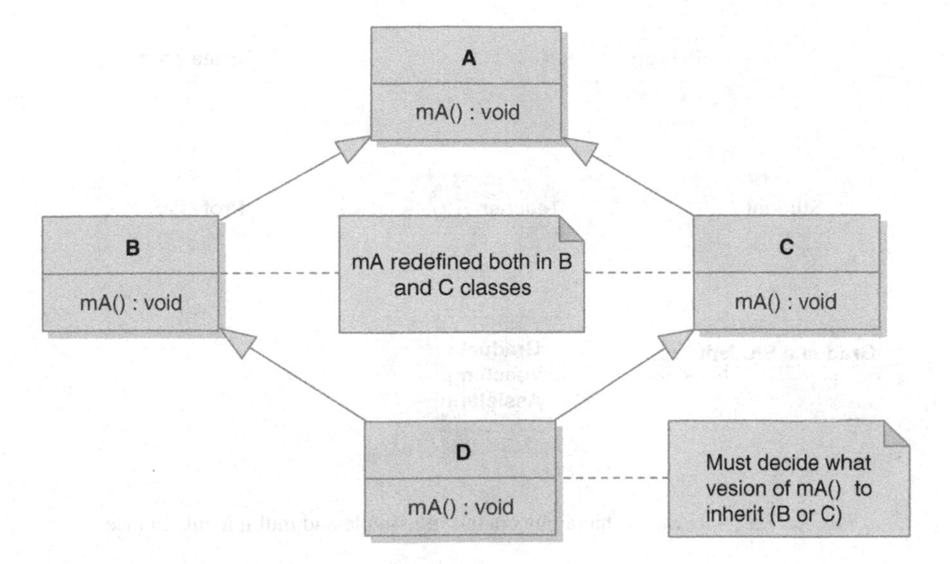

Fig. 5.5.7.2 Classical problem known as the "diamond" problem (Snyder, 1986) in multiple inheritance. Class A has ma() [A] operation implemented in A ("[]" indicates in what class, the implementation has been realized). Class B and C inherit from A and redefine both ma(). So ma() has two new versions under the same name ma() [B] and ma() [C]. If class D inherits from B and C, what version of ma(), from B or from C, the system must choose while deriving a multiple inheritance

description of a domain. In lower level, inheritance supports the reuse of existing classes for the definition of new classes. The mapping between levels or between models needs the engineering of intermediate models to bridge the two worlds. At any level of the MDA, we must observe behavior patterns of reasoning at this level and avoid distorting the reality perceived at this level. So, if a model needs a multiple inheritance, avoiding to model as such arguing that the implementation model does not support multiple inheritance is likely to distort high level models with low level model considerations.

To afford some guidelines for intermediate models, it would be interesting to notice that inheritance can be realized without classes (*classless inheritance*). Classless inheritance makes use of the mechanism of delegation. This mechanism is used to describe the situation wherein one object defers a task to another object, known as the delegate.

Single inheritance is known as a form of "class-class inheritance". Class instantiation is sometimes considered as "class-instance inheritance" since instances of a class inherits a common interface and an initial state from the class. Delegation is considered in the literature as "instance-instance inheritance" in which an object requests another object to execute the task for it (so it looks like as if there is a transfer of behavior and state attributes to another object).

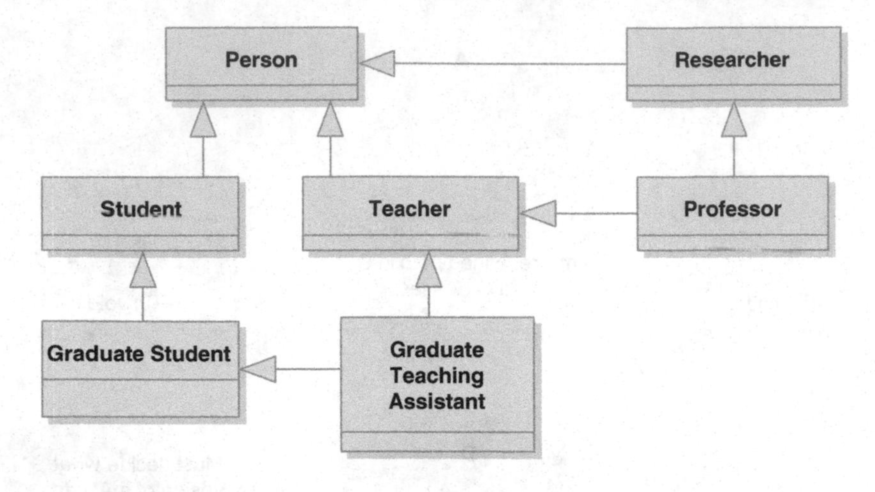

Fig. 5.5.7.3 Inheritance hierarchy containing simple and multiple inheritance

When using inheritance or multiple inheritance, the main thing to remember is the *principle of substitutability*, which essentially says that an object of a derived class should always be able to be used in place of (i.e., substituted for) an object of any ancestor class, whenever an object of that ancestor class was called for. To illustrate this idea, let us consider the example of Figure 5.5.7.3.

In this example, the *Graduate Teaching Assistant*, by the principle of substitutability, may replace anytime, *Graduate Student*, *Student* or *Teacher*. A Professor may replace a *Researcher* or a *Teacher*.

The substitutability rule is necessary but not sufficient to detect early incorrect structural constructs, as shown in Figure 5.5.7.4. A plasma display may be substituted to any of the three classes *Luxury Item*, *Electrically Powered Item*, and *Breakable Item*, but deriving immediately any final item from those three classes may create an awkward and bulky structure.

Another example of Figure 5.5.7.5 shows the difficulty of deciding between a composition and a multiple inheritance structure. The choice of a solution is often contextual.

> A company wants to have a person for a multidisciplinary project (for instance robotics) that needs the skill in the following fields: Electrical Engineering, Mechanical Engineering and Computer Science. By deriving a *superman Multidisciplinary Engineer* class and solving some minor conflicting parameters, we choose a multiple inheritance approach. By hiring three different engineers and building a team, we choose the composition approach. Doubtlessly, the first solution is more economical for this company as it has only one salary instead of three, no team work and potential communication or human problems to solve, but it would be difficult to find such a perfect candidate.
>
> If we transpose the same problem in the context of an expert system, if the computing resource is not a limitation, the choice between the two solutions will

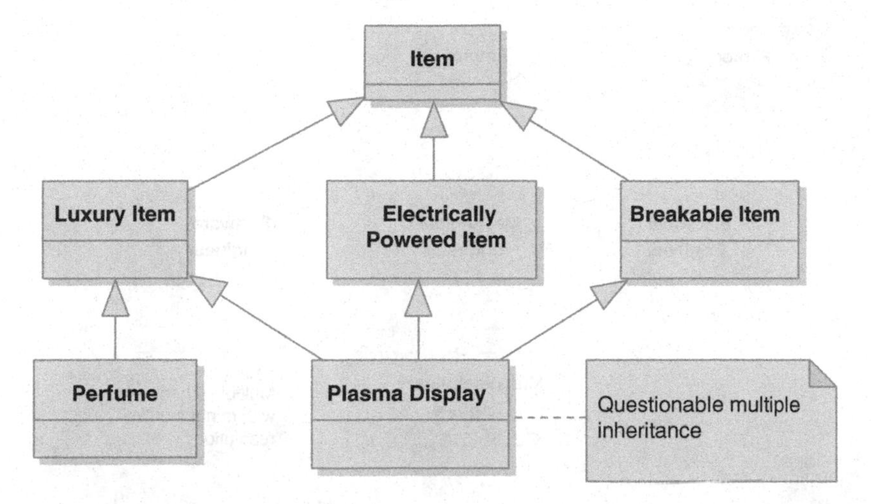

Fig. 5.5.7.4 Example of questionable multiple inheritance hierarchy. It would be advisable to create an intermediate class Luxury and Electrically Powered and Breakable Item class either by multiple inheritance or directly from Item if not all attributes and operations of Luxury, Electrically Powered or Breakable classes are needed

be less evident. We can examine also a hybrid solution that considers a conceptual model (PIM) different from the implementation model (PSM).

5.5.8 Objects and Roles

A role is a part played by an object in a relationship. A given object may participate in a system fulfilling any arbitrary number of roles. In the absence of specific constraints, there is no upper bound for the number of roles an object can play. Roles can be *dynamic*, so objects may change their roles in run-time (during their lifetime) and this fact could be modeled appropriately if we dissociate the role and the object that fulfill this role. If all objects keep the same roles during their lifetime, the system assigns *static roles*. The role concept has been used early in data modeling (Bachman and Daya 1977). The notion of role proposed in Object Role Modeling (ORM) (Halpin, available at: www.orm.net) is a different concept and methodology. (Steimann, 2000) has recently reviewed the notion of role. The notion of role used in this text simply means a part played by an object in a relationship.

Student is a role played by a human at a given period of his existence. *Electrical Engineer* is both a title (value) and a role. A person with this title may never exercise the corresponding role and may be doing something completely outside its role, for instance teaching music. A teacher who takes complementary courses plays both the role of a teacher and a student at the same time.

A company may be customer and supplier at the same time.

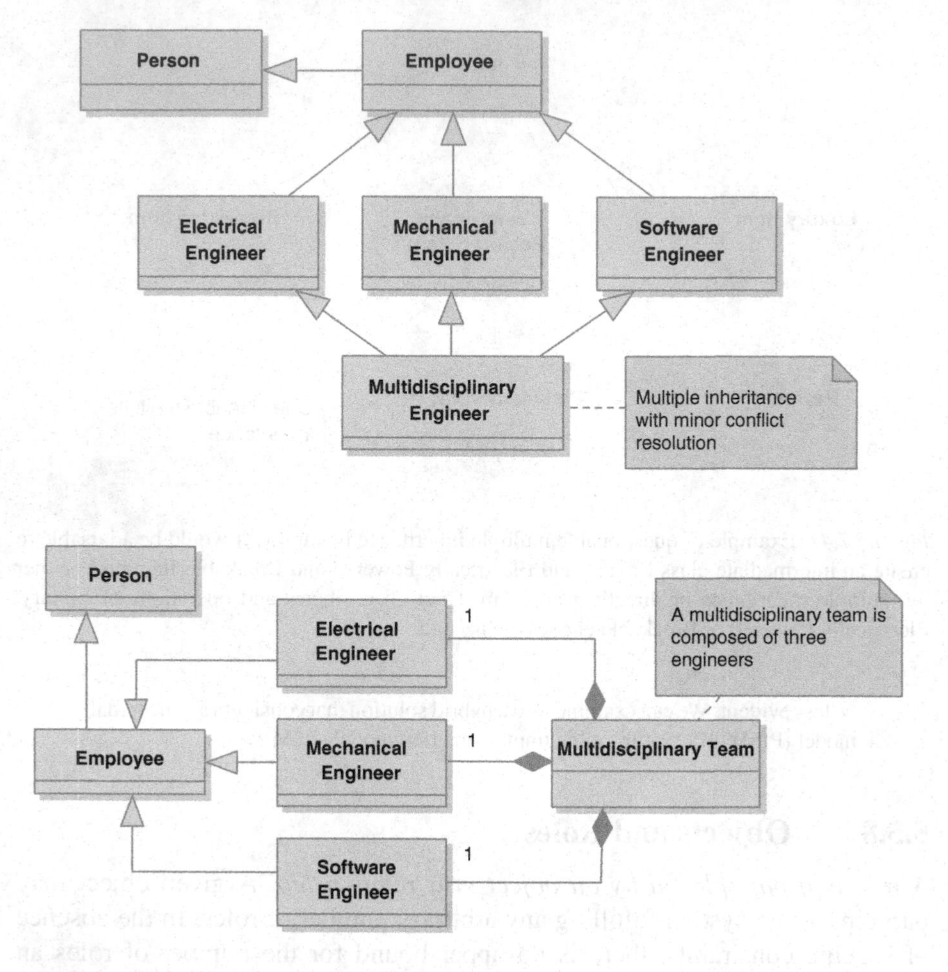

Fig. 5.5.7.5 Composition or multiple inheritance? The first solution is a multiple inheritance with some minor conflict resolution and the second is a team of three engineers

> A job in a company is also a role. Once instantiated, the state of this job can be suspended until there is an employee that accepts the job. If this person is laid out, the job will be suspended again until another candidate accepts the job.

The role concept describes a component of an organization, specifies skills, functions, tasks, and all features (a kind of contract) needed to perform any activity defined within the role. Even if, behind a role, there is an object, a role has its own set of state variables. Normally, an object that executes a task defined in a role keeps an identical copy of state values of the role it fulfills (more generally, it can access or keep an *image* that may reflect the reality or not). If it does not have enough resource to keep a copy, at least, when reentering the role, it can reload this copy in order to be able to continue to play its role.

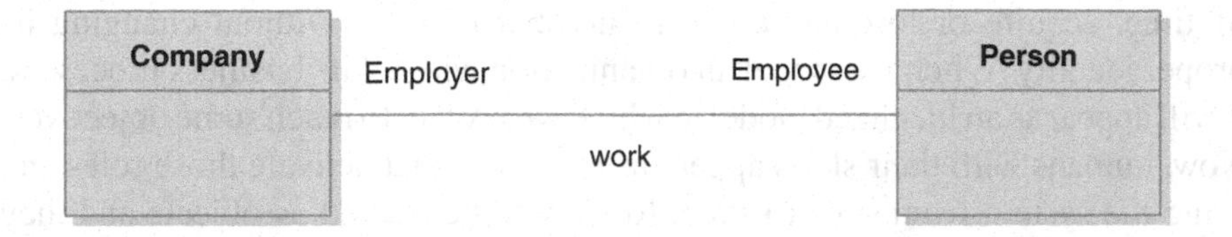

Fig. 5.5.8.1 Representation of role in the UML as name of an association end

Static roles are applicable to a small category of problems. Dynamic role attribution is the general model. Roles can be instantiated, have states, behaviors, and identities. An object and its roles are then related by a special relation *play* or *played by*. At the outside, the object and its roles are aggregated (Kristensen and Osterbye, 1996) inside a unique and indivisible *subject*. If an object drops a role, this does not necessarily mean in all cases that the role must disappear.

As discussed previously for a job, a role can be suspended. It is not necessary to suppress a role even if there is no object that takes over temporarily this role.

As for the identity of the role, some authors admit that only objects may have identities, not roles. Once more, this restriction works for some classes of problems. A role can have a temporary identity. A student in a university may have a registration number for this university. The proper identity of the person is different from that of a role. We can eventually make a search in a database with the identity of the person or the identity of his role.

If we dissociate roles from individuals that fulfill roles, the model would be more complex but very precise. Any real or abstract entity may, during its

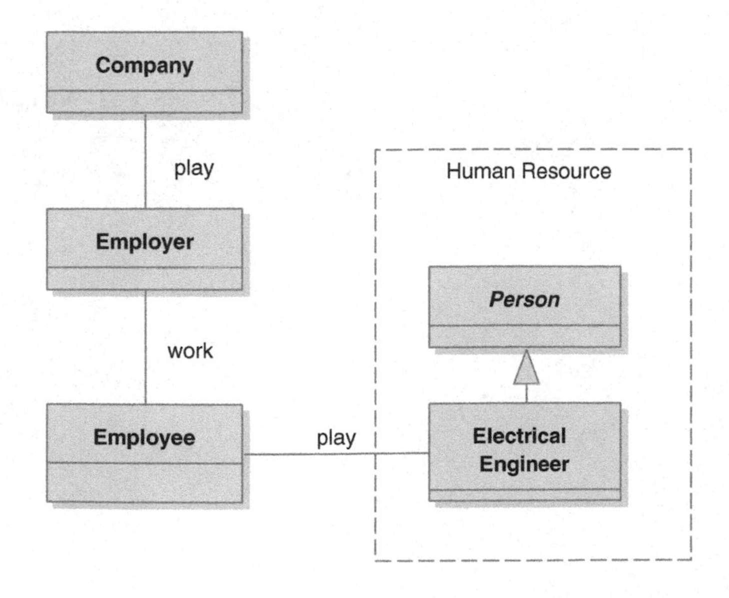

Fig. 5.5.8.2 Roles are modeled as separate classes

lifetime, acquire or lose any arbitrary number of roles without changing its proper identity. When we model an organization with all its business processes, it will appear as an idealized model in which we project to reach some objectives. Now, humans with their skills appear as resources that activate those roles and bring the system from state to state. Roles will be treated as objects and they have their own classes. Every system will now appear as been duplicated into two hierarchies. When we say for instance "somebody has not fulfilled its role," we now have a way to encode how we expect the system to behave and what is the result observed with objects we assign to some roles.

The UML proposes a way of representing roles as names of the association ends, so there is no duplicate hierarchy. It appears as the role delimits a subset of objects that participate into the collaboration and there is no "aggregate." This notation is heritage from the E-R diagram and the OMT (Object Modeling Technique) methodology.

Chapter 6

The Uniform Concept

Complex systems consist of a large number of objects/agents/components that interact in an intricate fashion. The dynamical behavior of such systems can be understood only with a correct model thoroughly decomposed into manageable chunks. Nowadays, with the power of computer simulation and software, more and more complex systems are attempted in many fields, stretching across natural, economic, social sciences, etc. Furthermore, complexity reduction techniques needed for analyzing complex systems are fundamentally the same across disciplines. There is currently great interest in bringing together researchers with different backgrounds for a truly multidisciplinary collaboration. The question is "could we really conceive a unifying modeling formalism that can be used across multiple disciplines?" Our research attempt targets to make use of the object paradigm as the starting point. As the UML implements a graphical notation to be used with the object paradigm, it is natural to include the UML inside the original package.

At the starting point, we have tried to answer some challenging questions as, for instance, "is there a solution to model for instance a whole robot made of electrical, mechanical and software components inside a same and smooth modeling tool like UML to reach an in-depth interpretation of the whole system?" or "is there a uniform way to describe a biometric installation in an airport, taking biometric information on humans than compare to data in a database and detect potential threats?" Doubtlessly, we cannot draw a mechanical plan, nor an electronic circuit with the UML, but we can explain for instance how a software decision is translated into a path change in the tool tip of a welding torch held by a robot and how a deviation in the trajectory of the same tool tip is translated into an information that alerts this machine to trigger a path correction algorithm. In this closed loop control experiment, there is a series of interactions between objects that must be described, not only in terms of vocabularies proper to each participating discipline, but with new revisited concepts of objects and messages. This approach is currently in use in computer simulation as all objects can be reproduced in a computer screen and the whole simulation world works only with messages between objects.

D. M. Bui, Real Time Object Uniform Design Methodology with UML, 289–379.

This proposal does not mean that engineers or practitioners belonging to other disciplines must change their way of thinking and describing things in their proper disciplines, but suggest that if a multidisciplinary team desires to understand and later simulate a complex system involving objects from various domains, there is possibly a way to access this unified description, via the object concept, augmented by some fundamental proposals of the uniform concept exposed in this chapter. It is worth noticing that the idea is a research proposal, and all interesting contribution of researchers in the future to enhance this concept would be highly appreciated.

6.1 Elements of the Uniform Concept

6.1.1 Elements of the Real World: Objects and Their Interactions

In a real world, the notions of class and role are inexistent. There are only objects and their interactions. To reach multidisciplinary objects, we need to clarify some attributes:

1. *Material or immaterial*
   (a) A dog or a house is a material object that we can touch and feel. Knowledge or soul is immaterial object. We cannot relate immaterial objects directly to matter (they can have a host object instead).
2. *Visible or invisible* (e.g. air, light, electric, or magnetic field)
   (a) A car is a visible object. The air or the oxygen we breathe is an invisible object. An invisible object can be immaterial, e.g. an electric or magnetic field. The light is considered sometimes as material (photons), sometimes as immaterial (waves), sometimes visible (from red to violet), sometimes invisible (ultraviolet or infrared).
3. *Inactive or active*
   (a) A pen, a computer mouse, and a roof tile are inactive objects. An inactive object does not solicit any service from other objects. They cannot instantiate any change to the surrounding world. An active object is an object that may initiate change to itself or to the surrounding world. A clock is an active object since it changes its own internal state (the time displayed). A powered robot instantiates changes to its world. When not powered, it becomes an inactive object. An inactive object can become active by induction if it receives energy (induced energy concept explained in Section 6.1.3).

4. *Human or not human*

   (a) This distinctive feature is used as an example in a UC diagram to differentiate two classes of actors (human and not human). In some agent systems, developers may desire to make the distinction between human agents and organizations.

5. *Inert or alive*

   (a) An object is inert if its state does not change with time. An object may be immobile, inactive but alive. Its internal state may undergo a number of internal changes without affecting the external environment. A bottle of wine let open in the kitchen at room temperature is transformed into vinegar so the liquid is an alive object as its state changes chemically with time.

6. *Complex or simple*

   (a) From the modeling viewpoint, a simple object is trivial so we do not have to make a dynamic study of this object as its dynamics is also trivial. This fact explains why we bypass the dynamic study of some objects (they are simply too simple). An object is qualified as complex if its functionality and general characteristics cannot be easily understood without a minimum effort of decomposition and a brief study of its dynamics. Complex objects are often composite objects but it is not always the case. A monolithic object (not composite) can be a very complex object.

7. *Object with missions*

   (a) The mission is simply an operation that is activated when an object is instantiated inside a system. A mission differs from a normal operation since it is a kind of survival algorithm activated at the object creation (this fact is not mandatory as an object may receive many missions that can be activated at run-time). For instance, when we put a battery and reset a clock, the mission of the clock is to maintain and display the correct time until its battery goes out. In most system, the program activated after the initialization and reset operations is a mission. Agents in organizations are often activated with missions. They do not need any regular activation signals or messages to accomplish what they are programmed to do.

   (b) Real-time systems are programmed mainly as event-triggered systems. For instance, when a battery is put into a pocket PC, the latter undergoes a series of initializations to finally reach a state where it enters a loop and waits for commands issued by users. This starting program is a mission for this pocket PC. When a phone call is received by the pocket PC, the mission activates a servicing program that allows users to answer the call. Then, after the call is ended, the control will be returned to the mission.

8. *Autonomous objects*

   (a) Stated simply, an autonomous object is not controlled by other objects or submitted to other forces. It has an independent existence. This does not mean that an autonomous object does not interact with other objects during its lifetime. It can be client or server for other objects but it can decide on its own of the moment or the opportunity to interact with other objects.

   (b) Sometimes, autonomous is understood as independent or self sufficient. But, an independent object can be neither autonomous nor self sufficient.

9. *Others*

   (a) The list is not closed and we can always add more attributes to account for the richness of concepts accompanying objects in a large application.

Figure 6.1.1.1 shows, for instance, an alarm system that illustrates invisible objects that must be modeled in the system to explain some retroactions that a system is supposed to receive.

> The particularity of this model is the explicit representation of the media that connects the alarm system to the neighborhood and explain why this system is only useful when there are some neighbors that make an effort to call the police when the owner is absent, instead of turning up their televisions to drown out the sound. This does not make sense to install an alarm system if the house to be protected is located inside an unpopulated region without any neighbor in the vicinity.
>
> This system use the sound as the invisible transport medium that is considered as an object that receives a message from the siren. If the owner is at home, he will be waked up by the alarm and normally he must stop the siren after a while. This explains why the alarm expects a reaction of the owner before triggering another action (for instance calling itself the police).

Each object of the real world, even invisible, must have a representation if it participates effectively in the feedback loop.

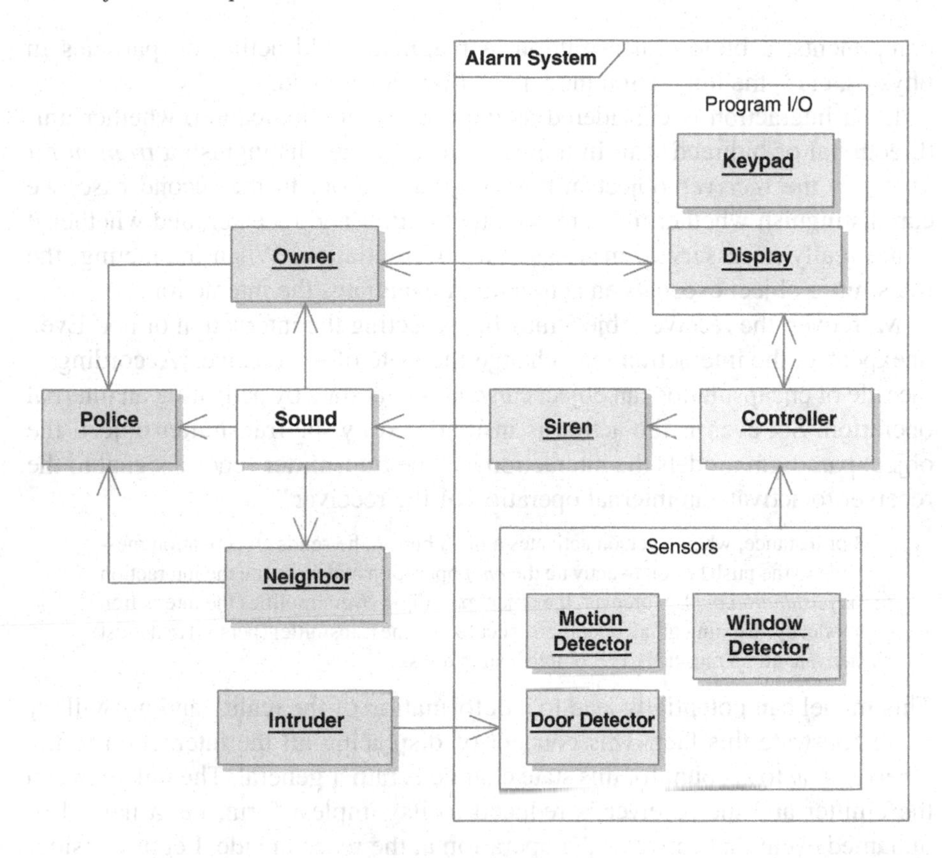

Fig. 6.1.1.1 Object diagram of a Home Alarm System showing the modeling of visible and invisible, human, and nonhuman objects. This representation shows that the Sound produced by the Siren has drawn the attention of the Owner and one of the possibilities to suppress the Sound is the rearmament of the alarm system by the Owner itself

6.1.2 Extension of the Message Interaction Model

Object interactions in software are, for instance, sending a signal to activate a process, communicating data. The interaction may imply the exchange of data or control signals, be responsible for the evolution of object states, and may generate new interactions.

> In computer sciences, the well known client/server for instance is a specific model of asymmetric interaction. A client issues a request to the server that processes the request then sends the result back to the client. When both can be client and server, we tend towards a symmetric *peer to peer* model of interaction.

If we extend this interaction model to other disciplines now (for instance, interaction between two mechanical pieces, a chemical reaction between chemical components, a biological evolution, a magnetic field acting on particles in physics, etc.), the interaction model must be reexamined.

Each interaction is considered as the act of communication, whether unidirectional or bidirectional. In the first case, we can distinguish a *transmitter* object to the *receiver* object in this communication. In the second case, we can distinguish whether they are both transmitter and receiver, and whether it is not really necessary to make such a differentiation. When interacting, the transmitter object executes an action that instantiates the interaction.

Moreover, the receiver object may be expecting the interaction or not. Even unexpected, the interaction may change the state of the receiver. According to the rule of encapsulation, an object can change its state by activating an internal operation. So, even if the action is instantiated by the transmitter object, the object paradigm models the interaction as "the transmitter sends a signal to the receiver to activate an internal operation of the receiver."

> For instance, when a person activates a push button, he sends an activation message to the push button to activate the *on()* operation and we write the interaction as *pushbutton.on()*. Moreover, the object paradigm oversimplifies the interaction model by stripping off all operation executed by the transmitter (person) that must activate all its muscles to accomplish such a task.

This model can potentially lead to a deformation of the reality and we will try to demonstrate this fact. This concept of displacing all the interaction to the receiver side to account for this state change is rather general. The link between the emitter and the receiver is reduced to its simplest form, i.e. a named or unnamed event that activates the operation at the receiver side. Let us consider a simple example with a human interacting with a mechanical and electrically controlled device to illustrate this modeling process.

> A driver tries to get a ticket in a car parking system after pressing a button at the entering bar. When the ticket is out of the slot, a signal is sent to the gate control that opens the gate bar. In this arrangement, the driver is an object that instantiates a series of action. In object world, it will be described as:

(a) The object *Driver* presses the object *Button*

(b) The *Slot* object delivers an object *Ticket*

(c) The object *Driver* takes the object *Ticket*

(d) The object *Slot* detects that the object *Ticket* has leaved the slot

(e) The object *Slot* sent a signal to an object *System* that opens the object *GateBar*.

If we represent conventionally this arrangement with an UML interaction diagram, from a user viewpoint (not as a designer viewpoint as discussed a little further), we get the following diagram (Fig. 6.1.2.1).

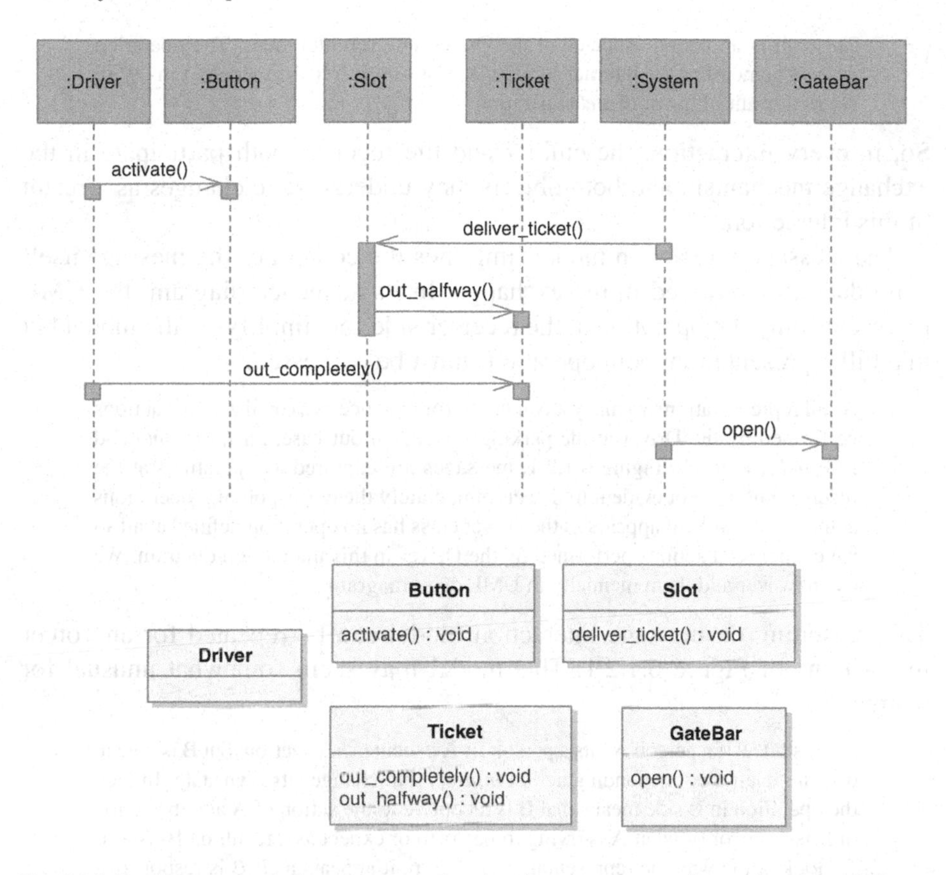

Fig. 6.1.2.1 Sequence and class diagrams showing the arrangement of a parking ticket delivery system from the user viewpoint

The user viewpoint is often incomplete as it models only elements visible at the user interface. This incompleteness appears in this example as discontinuities of the interaction chains. The first time when the system detects that the *Button* has changed its state to on, the second time when the System senses that the *Driver* has taken the *Ticket* out off the *Slot*. The system "behind the scene" is hidden from this representation.

When a Driver pushes on the mechanical button, it appears as the Button has its own operation *activate()*. The driver just sends a signal and the Button activates itself. In this interaction, the Driver has executed an action *push()*, but, unfortunately, this action cannot be represented in a sequence diagram in most UML tools of the market.

> If we simulate this interaction with a program, in the Driver program, we emit a call to the Button object with the instruction:
>
> B.activate () B is the name of a particular instantiated button

This call is an action executed at the Driver side (emitter side). The operation *activate()* modifies the internal status of the Button B (receiver side) from *Off* to *On* as a result of the *activate()* operation.

So, in every interaction, the emitter and the receiver both participate in the exchange mechanism and both objects may undergo state changes as a result of this interaction.

The message interaction model simplifies the content of the message itself if no data are conveyed in the exchange. In its sequence diagram, the UML represents only the operation at the receiver side for simplifying the model but in a full representation, both operations must be expressed.

A full representation has many advantages, for instance, we can list all the actions performed by the Driver at the parking system, in our case, *push_button()* and *take_ticket_out()*. In Figure 6.1.2.1, messages are captured as operations at the receiver but it is not evident to detect immediately the corresponding operations at the emitter side. It appears as the Driver class has no operation defined at all so we cannot read actions performed by the Driver in this interaction diagram. We can however add them manually in UML class diagram.

This reasoning on message interaction model can be repeated for any other interactions of Figure 6.1.2.1. This model may seem somewhat unusual for many of us.

For instance, if a person A hits a person B, A executes *hit()* action, but B is able to execute itself a corresponding action *injure()* that changes its own state. In fact, the operation in B side means that B is receptive to the action of A and the action in B is a mirror of what A is trying to perform or expect as a result on B. But, if we look at the way the representation is drawn, it appears as if B is responsible for its own misfortune.

This example shows clearly a deformation of the reality if we extend this low programming model to express high level problems.

6.1.3 Induced energy

Messages in a software program convey activation signals and data through communication channels to wake up processes. Messages in noncomputer systems can convey energy. Energy is then another form of data and is mostly "induced."

When the Driver in Figure 6.1.2.1 touches the button, he communicates a tiny chunk of energy to the button, so the later is able to move and make an electric contact. The button is naturally a passive object and normally does not have any mean to close the contact itself. But, the tiny induced energy it received from the Driver index transforms temporarily the mechanical button into an active object so it is able to induce change to its microworld.

Similar reasoning abounds in mechanical world. Roof tiles do not have energy but the wind can induce energy to tiles so they can hurt people. A ball in a gun

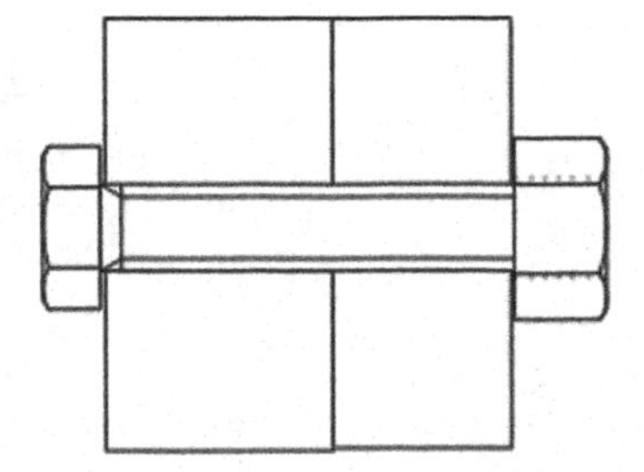

Fig. 6.1.3.1 At the starting point, we have four independent pieces Mechanical_Piece_1, Mechanical_Piece_2, Bolt and Nut. At the end, we have a monolithic piece shown in the figure

receives energy from a chemical reaction and can kill people. We can easily model the domino effect with the same concept of transfer of a chunk of energy from object to object. So, based on the principle of induced energy, an object, initially inactive, may be temporarily transformed into an active object and it can induce state changes to the surrounding world. This principle is the basic hypothesis to extend our reasoning to other disciplines with the message concept of object interaction.

> The model of induced energy can be more sophisticated than the example of the Driver pushing on a button at the parking system. If we want to tighten two mechanical pieces together (Figs 6.1.3.1 and 6.1.3.2), we must bring, with one hand a bolt, into facing holes made in the two pieces, and with other hand, communicate a circular movement to the nut. In this case, the bolt receives a series of induced translation movement and the nut receives an induced circular couple.
>
> The two mechanical pieces send bilateral messages to indicate the pressure during the tightening process. When the right couple is reached within the nut, this information is retransmitted to the right hand of the operator and the person will stop rotating the nut when a correct couple is sensed by his right hand.

6.1.4 Use of a Monitor Object

Electrical simulations deal with circuits, active and passive components, circuits, control elements, power devices, actuators, etc. Tools for simulating electrical circuits, both analog and digital, are rather crowded in the market. Without emphasizing any particular tool in this very competitive domain, simulation software has paved the way for helping generations of practitioners to design electrical circuits. Today's circuit designers use automated tools to model and simulate electronic circuit designs prior to fabricating them. However, for handling *multidisciplinary problem*, there are no or few facilities to model nonelectrical systems or interfaces between electronic and nonelectronic systems. In the absence of support for various types of physical devices (mechanical,

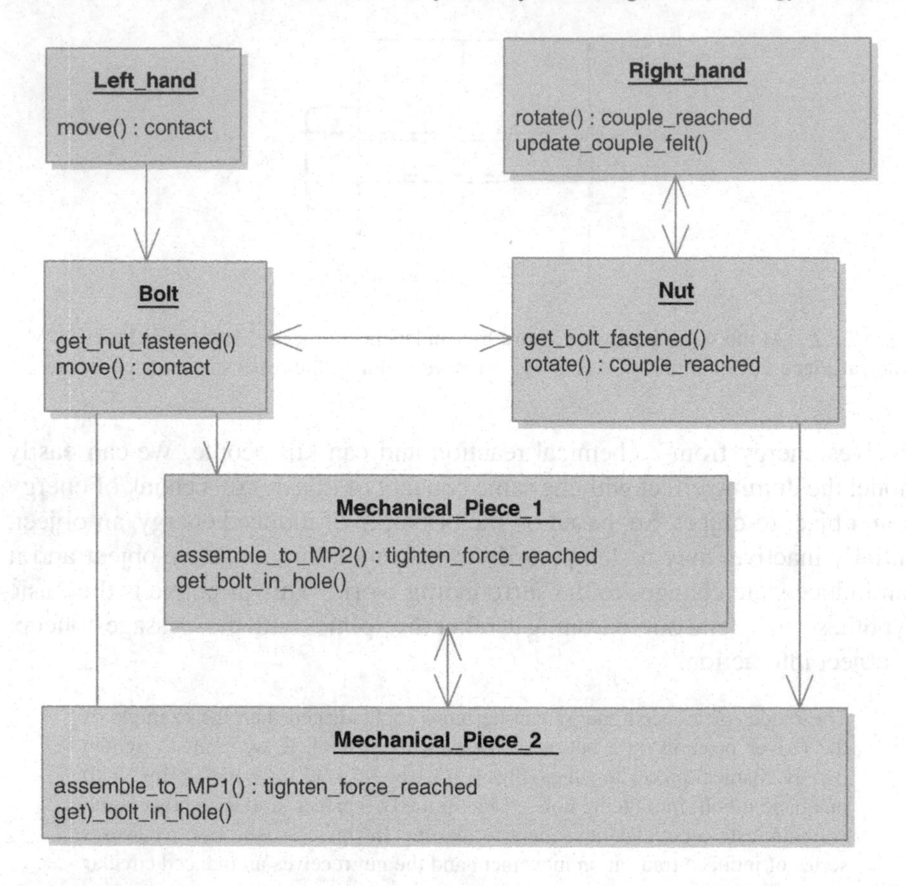

Fig. 6.1.3.2 Translation of mechanical assembly into objects and messages. The Bolt can move because it receives induced energy from the left hand. The Nut can rotate because it receives small rotations from the right hand. Couple sensed by the right hand is induced by couple sensed on the Nut

biological, chemical, etc.) and their coupling, we can make use of the bare object concept and the UML to design and model such interactions.

Figure 6.1.4.1 shows an example of electrical circuit with a switch, an electric fan coupled mechanically to a turbine, a variable resistor, a power supply (voltage source protected by a fuse), and an operator that turns the fan on by closing the switch. When changing the variable resistor from its highest value to its lowest value, we can see the variation of the current and in occurrence, the volume of the air flow. Figure 6.1.4.2 shows the object model of the electric circuit of Figure 6.1.4.1. The interaction model is very different from the corresponding electric circuit as it shows how objects interact together instead of how the current flows through electric components.

As stated earlier, the UML model of interaction is a simplified model as it does not represent operations in the source objects but only at the receiver

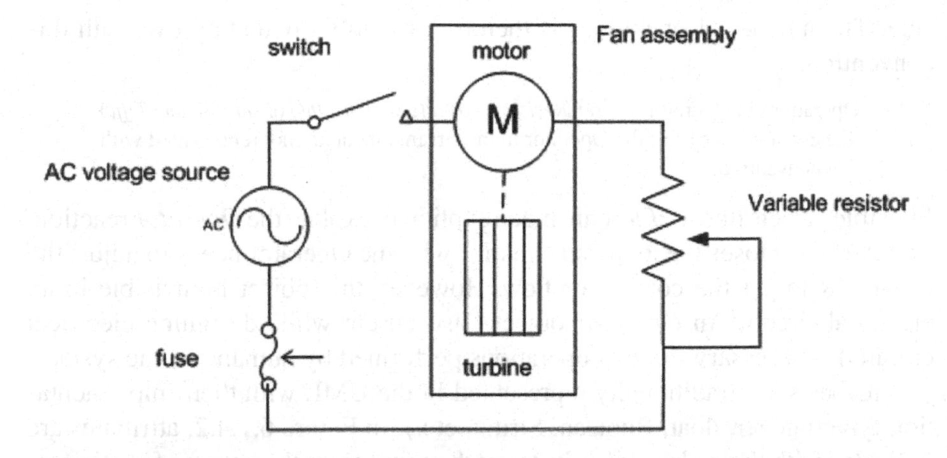

Fig. 6.1.4.1 Variable speed electric fan with its power supply. In real application, a variable resistor is not a good solution. A simple TRIAC is better but we make use of a variable resistor for illustrating the modeling principle

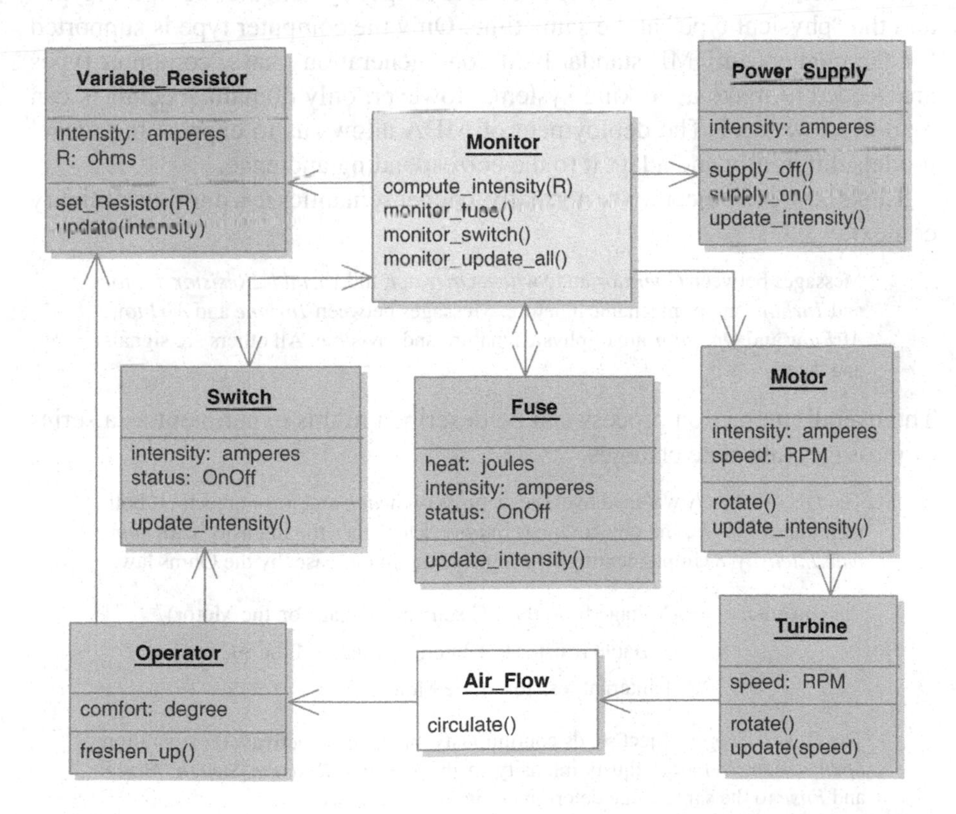

Fig. 6.1.4.2 Object diagram representing a control model. The invisible AirFlow object closes the loop and explains the action of the Operator on the variable resistor. The object Operator was added to evidence the source of interactions (see text)

side. The number of operations is therefore roughly divided by two with this convention.

> Operations in Operator like *set_Resistor()* or *set_Switch_On()* or *set_Switch_Off()* that explain the role of the Operator in this arrangement are not represented with this convention.

Invisible objects like *AirFlow* arc made explicit to explain the *Operator* reaction. The AirFlow closes the loop and explains why the Operator needs to adjust the resistor R to get the correct air flow. However, this object is invisible in an electrical circuit. An *Operator* object (also absent while designing electrical circuits) is necessary to show operations performed by humans on the system.

Attributes are traditionally represented in the UML with their implementation type (integer, float, Boolean, String, etc.). In Figure 6.1.4.2, attributes are indicated with their *physical units* to catch at first sight the nature of participating objects, instead of their computer-oriented types (integer, float, string, etc.). So, *Power Supply* and *Fuse* objects have *intensity* with "user defined" *amperes* type. In fact, it would be desirable to be able to specify both the "computer type" and the "physical type" at the same time. Only the computer type is supported for the moment in UML standard. At code generation phase, computer types are needed to make a working system. However, only domain specialists can read this low level. The deployment of MDA allows us to express high-level models differently and adapt it to the corresponding audience.

Links between objects have normally a richer semantics in a multidisciplinary context.

> Messages between *Operator* and *Switch*, *Operator* and *Variable_Resistor*, *Motor* and *Turbine*, are of mechanical nature. Messages between *Turbine* and *AirFlow*, *AirFlow* and *Operator* are of physical nature and invisible. All others are signals and data.

The overall interaction process can be described in this experiment as a series of messages and state changes.

> The *Power_Supply* watched over the statuses of *Switch* and *Fuse* objects. If both their status are *On*, the object *Power_Supply* determines the intensity with *compute_intensity(R)* implementing a formula given, in our case, by the Ohms law:
>
> intensity = (Voltage from the AC source – Voltage on the Motor)/ (variable resistor R + internal resistor of the AC source + internal resistor of the Motor)
>
> The *Power_Supply* object sends continuously messages to activate the operation *update(intensity)* that adjusts intensity in the *Variable Resistor, Switch, Motor* and *Fuse* to the same value determined by *Power_Supply*.
>
> *Power_Supply* monitors at the same time the status of *Switch* and *Fuse*. If the intensity reaches an intolerable threshold, the fuse is overheated and it will activate the *blow_up()* operation that chains with the *supply_off()* operation to suppress

the electric current when it detects that the fuse is blown up. Through the operation *update(intensity)*, the current in all devices will be suppressed. All those phenomena are instantaneous.

Figure 6.1.4.2 describes a model of interaction, giving to the *Power_Supply* some "active" properties (sensing *Fuse* and *Switch*, updating intensities in other passive devices). This attitude, based on "real" and "physical" elements present in the system, is acceptable as a power supply is a source of energy in most system and as such, can be considered as an event instantiator. However, the attribution of "active properties" to an element as *Power_Supply* may be perceived as arbitrary.

An alternative style of modeling (Fig. 6.1.4.3) considers an invisible object *Monitor* that is responsible for executing all actions that result from the application of physical, chemical, electrical, biological, etc. universal laws. These actions must be "nonimputable" to any visible objects of the system. The use of a Monitor object for concurrency control has been described intensively in software literature (Buhr and Fortier, 1995). The notion of monitor used in simulation and modeling of multidisciplinary object systems extends the functionality of this important concept towards to kind of invisible object whose main role is to apply physical laws at real time to all the components of the system, to ascertain that their states will react correctly, in compliance with rules or formulas that can be implemented as operations of the Monitor.

In Figure 6.1.4.3, the operation *monitor_update_all()* is a particular operation called previously a mission that is activated by instantiating the *Monitor*. At the beginning, *the Monitor* senses an opened switch, so the initial current will be reset to 0. The operation *monitor_update_all()* executes three elementary operations *monitor_switch(), monitor_fuse()* and *compute_intensity(R)* that in turn can be decomposed to find appropriate algorithms.

6.1.5 All Objects are Made Independent Even with Tightly Coupled Interaction Scheme

Components are independent chunks of hardware or software. All objects, considered at high-level models, must be made independent too, as if they own their proper "active component" internally. In the implementation model, cost optimization and other constraints often downsize the system to only one active component (controller, processor) that controls everything, either on a time-sliced basis and/or a priority scheme.

The suppression of the interdependence between objects in higher models allows us to *well perceive fully the role each object plays in a system.* Moreover, it simplifies the logical model as there is no interference scheduling aspects. In doing so, we separate the logical model from its implementation model.

In the previous example 6.4.1.3 with a *Monitor* object, we have 7 inventoried objects. Each object has inputs, outputs. Internal mission of passive components

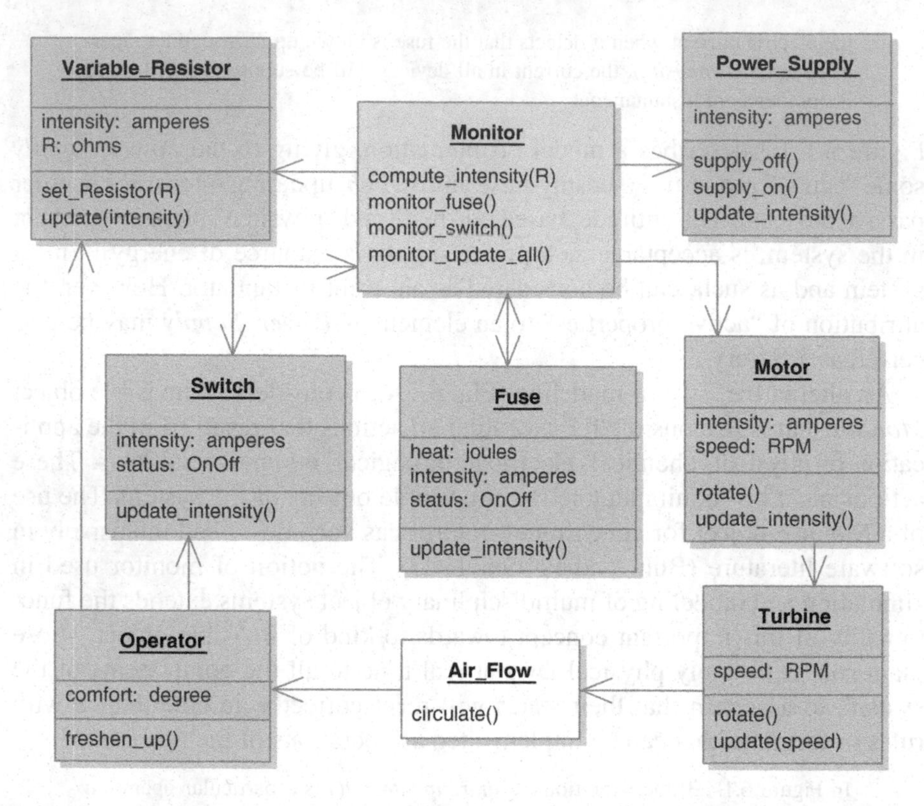

Fig. 6.1.4.3 Use of a Monitor object. When the Operator adjusts the variable resistor R (visible action), the intensity in the motor changes accordingly (invisible action accomplished by the Monitor). The whole chain of reactions that follows the setting of a new R value will be executed by the Monitor as well (computing the new value of the intensity, updating all the intensity values)

> consist of monitoring their inputs and reacting appropriately to events according to their nature and the way they are designed to behave. So, if the *Monitor* updates the intensity in the *Motor*, the later will activate its *rotate()* operation to respond appropriately. By mechanical coupling, the *Motor* activates in turn the *rotate()* operation of the Turbine, and so on.
>
> The *Monitor* is considered as having its proper processor that activates its mission *monitor_update_all()* when instantiating the *Monitor*. This processor works at the lightning speed to apply physical laws. This vision corresponds to the reality. If we set the switch off, automatically, all currents in the electrical circuit will disappear instantaneously. But, at simulation, as we have only one processor to manage the simulation tasks of 7 objects plus regular tasks of the operating system, the simulation process will share time with other computing tasks.

This independence of objects at the logical phase of design plays an important role in complexity reduction at *process level* (the decomposition of a system in many chunks is a complexity reduction at *spatial level*). This independence can be applied to passive components as well.

> If an object *Controller* in an alarm system has three sensors to monitor (window detector, motion detector and door detector), in a monoprocessor implementation, it must visit successively these three sensors (e.g. in the enumerated order). So, there could be a delay if a given system has dozens of sensors with various speeds to monitor. But, in logical design, it would be important to consider, in a platform independent model, as if each sensor owns a processor and all changes are detected immediately with zero latency. So doing, the operation of each sensor is made independent from the presence of other sensors. The latency problem in real time system will be taken into account later at implementation phase when scheduling the task of the processor.

Tightly or loosely coupled are terms used initially in multiprocessing system to characterize the communication flows and the degree of dependence between two or more computers (or processing nodes) in their achievement of collaborative tasks. Tightly coupled communications occur, for instance, between a processor and its memory, between two processors inside a biprocessor system, between a processor and its DMAC (Direct Memory Access Controller).

In a tightly coupled communication, computer nodes need to be built close to each other in order to support a high bandwidth of data exchange. At the opposite end, loosely coupled computers need only to communicate erratically, so a network topology is often deployed in this case. Internet communications are of loosely coupled type and intermediate computing nodes can be inserted between two communication points whereas tightly coupled systems often need direct connections without any data replication to optimize the speed of data exchange.

> However, the two schemes may be mixed intimately in a system. A network of computers may contain tightly coupled components (computers) connected through a network that supports loosely coupled communications.

For modeling, objects are always considered as independent, despite the fact that they are involved in tightly or loosely coupled schemes. The flow of interactions is denser and faster in a tightly coupled than in a loosely coupled scenario.

6.1.6 Cause–Effect Chain

Cause–effect chain analysis is important in computer debugging (Zeller, 2002) and is generally a main concept used in all branches of sciences, particularly in environmental studies. A system is made of a large number of objects. A coarse grain analysis highlights only main objects so the conclusion may be erroneous as the decomposition process is not completed and all objects are not brought into light. The search of a failure in a system needs a thorough decomposition since the simplest piece of a system may be at the origin of a failure. To diminish the number of test runs, the search may be started by a coarse grain analysis and a thorough decomposition is made in the neighborhood of the suspected origin of the failure. Causal knowledge is a preoccupation of cognitive science as well (Waldmann and Hagmayer, 2005). Sometimes, correlation or conjunctions of events replace the notion of causality if the number of parameters is abnormally high or cannot be identified practically.

The cause–effect chain is a concept that highly impacts the modeling process as the latter is directly dependent of the decomposition process or divide and conquer mechanism. The main conclusions that could be drawn when modeling with cause–effect chain in mind are:

1. *Rush conclusions are systematically suspected in a coarse grain system.* When a system is not decomposed and all elements are identified, the real cause is unknown and erroneous conclusions are highly improbable. When designing a solution to solve a problem, the design phase should be brought to its end.

   Let us model a conversation between two persons A and B via the public telephone network that we believe is highly reliable. In a very hot discussion, if the conversation is canceled abruptly, both may conclude at this moment that the opposite person is impolite and has hung up. But, if we model the phone line with two wireless telephones, we may discover that a dog has broken the line in another room while playing with the child.

2. *The cause of failure may exceed the frontier of a system.* Often, when a failure occurs with a product in the market, the natural tendency is to incriminate internal components of the product itself, less by the way it interacts with other products. This short view may come from the bounded frontier of the investigation.

   If the grass is wet, it has rained. But if we look at the state of grass of the neighborhood, we can discover that somebody in our home has inadvertently put the sprinkler on.

3. When getting back to the cause of an event or an action, the nature of the initial cause may have nothing to do with the current problem observed. Bronchitis may cause cough that in turn causes insomnia. The mechanical domino is the only process that may have the initial cause similar to the last action but more generally, the nature of effects are very different from step to step in the cause–effect chain. Described in terms of domains, a causal model is made of a set of nodes, a set of links between nodes, and a conditional probability distribution for each node. The nodes can thus cross the frontier of several domains before reaching the current failure or problem observed.

6.2 Requirement Analysis Model

6.2.1 "What to Do?" Phase

A rather rich literature covers this phase of development (Zave and Jackson, 1997; Sommerville and Sawyer, 1997), our intention is studying how to deploy this phase with UML tools and packing it into the most condensed but comprehensive form using graphical tools instead of huge amount of texts. This phase is an important part of the system design process involving the client, the whole team including business analysts, system engineers, software developers, and managerial staff. The primary purposes of this phase are:

1. Organizing and analyzing requirements into a form that can be verified and accepted by the Client
2. Producing comprehensive documents and diagrams that *can be used by the engineering staff as input to design*
3. Elaborating *criteria for validating the end product*. Generally, these criteria are put in contract between the Client and the Developer
4. Estimating *project Cost, Resource needed, and Time* for each activity involved in the process (CRT parameters).

To achieve these goals, business analysts and system engineers have to listen to user needs, and then create a complete and unambiguous specification document. Actually, this phase is performed very differently from one organization to another according to the organization culture and habitude.

From the Client side, many scenarios are possible; we mention only two extreme limits. Current practices will be located somewhere between these two drastic positions.

1. *The Client prepares a whole book specifying all the requirements.*

   This attitude is often seen in governmental institutions and large companies. When large and costly projects are concerned, the Client generally takes all the effort necessary to document itself on the feasibility, available technologies, and at the limit, he has chosen himself the technology and the Developer side acts as simple Execution side.

   From a technical viewpoint, the effort at the Developer side is oversimplified as the requirement analysis phase is already or partially made and the only thing is to compete with other developers for the project cost, timing, and QoS.

2. *The Client emits only vague ideas of the final product and explain his project informally* (e.g. orally).

This attitude is often observed with small and short-term projects, with companies that want products but have no internal human resource to establish a requirements document. Sometimes, the utmost effort of the Client side leads to some pages with a list of brief descriptions disorderly organized.

The second case does not mean that the Client will accept everything as design and difficulties could occur for the Developer side at the horizon, so the role of the cautious Developer is to build a complete specification to join the first case. Of course, the Developer may propose many proven solutions currently offered by his company to lower the development time or to deploy reuse mechanism to cut down his proper investment and optimize his profit. He can eventually charge the Client for building the specification and this practice is rather common and recommended as the client of the second type are likely to drop out the project after knowing its real cost than clients of the first type that have already a good idea of the project, its cost and has already committed to go ahead with the project.

6.2.2 Parameters of the Requirements Engineering Step

The success of a product or a system is the degree to which it meets the purpose for which it was initially intended. This "purpose" goes beyond the simplistic definition of goals or objectives in natural language prose. The result of this phase is materialized by a business contract signed between organizations before entering the design phase. This contract proves their commitment to go ahead with a project. Everything must be defined in depth in the "what to do" field and occasionally in the "how to do" field if specific or technological constraints are embedded in the project.

There are no theoretical laws or theorems that govern the management of this phase but the conclusions drawn from this phase impacts largely on material, human resources, and even existence of organizations. In the literature, requirement analysis is seen as inquiry-based (Potts et al., 1994). The term *requirements engineering* is often used and is mainly goal oriented (Dardenne et al., 1993; Lamsweerde, 2001). It embraces *requirement analysis*, *requirement specification*, and *production of textual and technical specification documents, artifacts*. So, for large projects, the requirements engineering can be nearly considered as an independent subproject. In the MDA context, it is considered as a *requirement model*. As this phase needs a complex interaction between human actors, recommendations of the requirements engineering step are:

1. *Talk the language of the captive audience*. As said before, this phase involves the client, the whole team including business analysts, system engineers, software developers, and managerial staff. As people come from different horizons, not everything is suitable for viewing.

(a) For instance, the UC diagram may appear as a very complex formalism and cannot be easily understood by everyone, so, even if a project is analyzed using this tool, conclusions must be reformatted into a more easily digested form.

(b) The business process diagram that bears some resemblance with older DFD with inputs and outputs is more suggestive and less problematic than the UC diagram.

(c) Managerial staff and client are more sensitive to project cost, resource needed, and timing. Those points must be particularly formatted with a suggestive and convincing presentation.

2. *Conduct the necessary inquiry phase.* The Client is not a technical person. He is often unable to formulate all relevant requirements precisely, and has limited knowledge about the technical environment in which the system will be developed or even used. This is the reason why he relegates the task of building the system or solving the problem to the Developer staff. Even if the Client gives the impression to specify all, experiences show that a lot of inquiries will be necessary to avoid misunderstandings.

   (a) Content *analysis* (Berelson, 1952; Kassarjian, 1977) is a research technique for the objective, systematic, and quantitative description of the manifest content of communication. Content analysis is used in political science, social psychology, survey techniques, communication, advertising or propaganda campaign, etc. Textual information is imprecise, may convey tendentious clauses, and is systematically suspected by content analysis researchers. Techniques in this domain are applicable to formalizing and quantifying the universe of discourse engaged between Client and Developer sides.

   (b) Very often, the Client ignores the regulation and it is up to the Developer to inform him and updates the specification accordingly.

3. *Fight against the natural tendency of some developers to shortcut to requirement analysis phase.* The technical staff at the Developer side is highly technical, is often biased, and has a clear tendency to consider that they lose their time in endless discussions.

   (a) Experience in our university shows that students that often jump directly to the programming phase are more receptive to requirements analysis based on technical diagrams like UC diagram and BPD instead of textual clauses issued from content analysis technique. The two aspects are all necessary in a requirements engineering process: we can start with diagrams and end up with text.

4. *Reach a complete specification.* At the end of the requirements engineering, specifications must be validated by comparing the whole specification against Client objectives. From a practical perspective, the result at the Developer side is a set of activities decomposed into their utmost details, where Developer can put resources, estimate time, estimate components to be built, components to be acquired, the overall cost, his markup. So, the perception of completeness varies with the experience of the Developer in the domain. Experienced developers can quantify very quickly whereas newcomers in the domain must work hard to reach the same result for the first time (they will acquire experience later).

5. *Build unambiguous specifications that can be used as inputs to design phase. Several techniques or several views may be required.* Unambiguous specifications need often technical tools. As said before, the Client could get lost with the technical level of UC diagram and BPD. Ideally, a comprehensive set must contain both technical diagrams for engineering purpose and accessible textual clauses that can be served as communication support with nontechnical stakeholders. Many views or submodels of a same system may be needed to get a thorough understanding.

   (a) TROPOS (Bresciani et al., 2004), a tool devoted to agent-based system design, builds the early requirement models based on activities centered around:
   (b) *Actor modeling*: Identifying and analyzing both of the environment and the system's actors and agents
   (c) *Dependency modeling*: Identifying actors which depend on one another for goals to be achieved, plans to be performed, and resources required
   (d) *Goal modeling*: Analyzing actor goals, from the point of view of actors, by using means-end analysis, contribution analysis, and AND/OR decomposition
   (e) *Plan modeling*: (technique complementary to goal modeling)
   (f) *Capability* modeling: Identifying individual capabilities of actors.

6.2.3 Requirements Engineering

Figure 6.2.3.1 the summarizes most important steps of the requirements engineering phase with an UML activity diagram.

Step 1: Have a very good idea of the Client needs (Elicitation phase). We start with two possibilities corresponding to the two extreme cases, one with a Client that has established a requirement document and another without any prepared document for the Developer side. As we must, in all

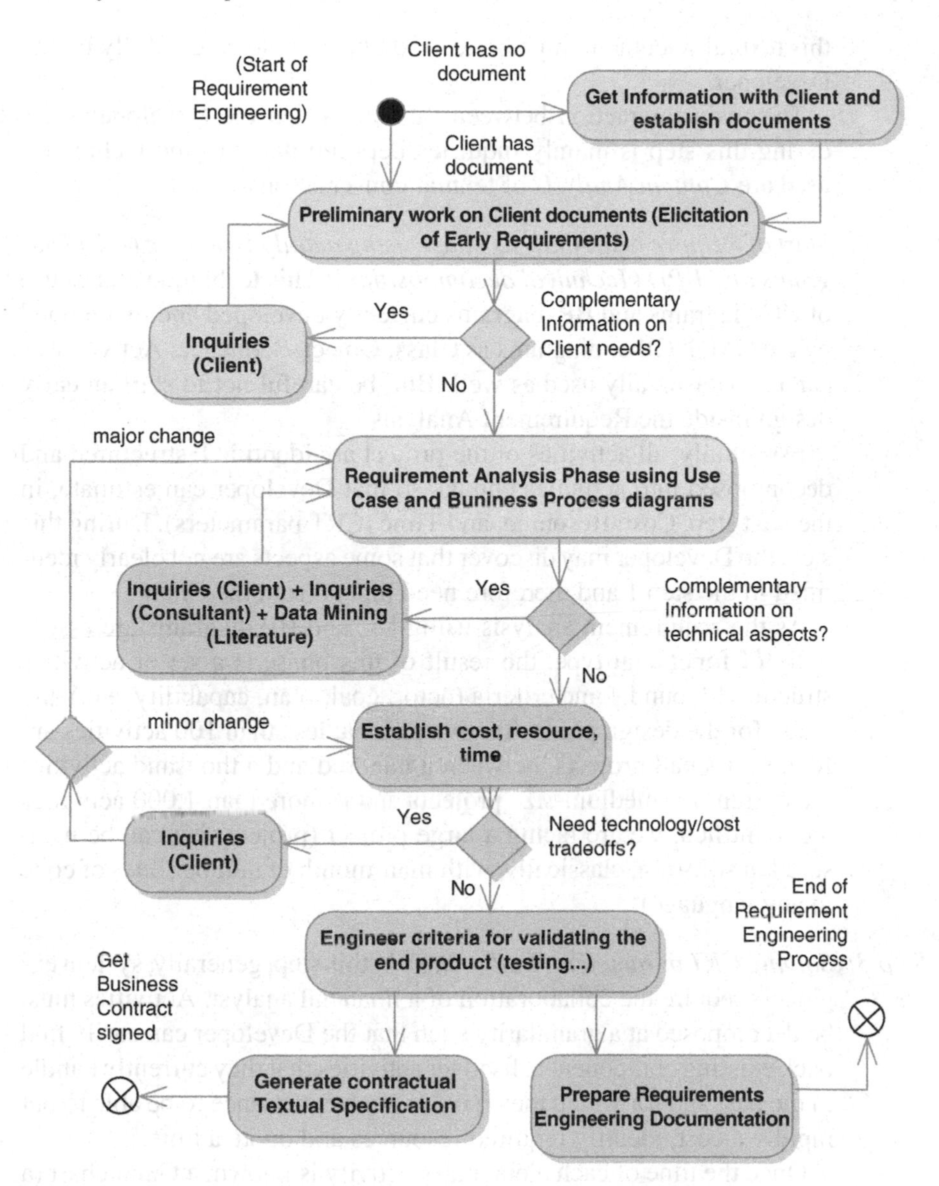

Fig. 6.2.3.1 UML activities diagram showing activities of Requirements Engineering phase. We make use of use case (UC) and Business Process (BP) diagrams for the technical part of this phase

cases, reach the same result and a thorough understanding of "what the Client needs," preliminary textual documents can be used to explicit the Client intentions at the beginning of the process.

Even in the case where the Client has established himself a Specifications document, this work can never be established with all the precision of a document targeted to be used at the Developer side. So, this textual document must be revised and completed carefully by the Developer.

The main interaction between the Client and the Developer sides during this step is mainly Inquiries between humans and techniques used are *Content Analysis* of textual conversations.

Step 2: *Start of Requirement Analysis phase using mainly Use Case (UC) Diagrams and BPD (Technical decomposition).* This technique makes use of UC diagrams and BP diagrams currently developed and maintained by the OMG. Other diagrams as Class, Object, Sequence, Activity etc. can be occasionally used as well. But, be careful not to start an early design inside the Requirement Analysis.

Essentially, all activities of the project are identified, structured and decomposed into activities chunks so that Developer can estimate, in the next step, Cost, Resource, and Time (CRT parameters). During this step, the Developer may discover that some aspects are not clearly identified in the step 1 and therefore need inputs from the Client.

As the requirement analysis using UC and BP diagrams are essentially of functional type, the result of this phase is a set of activities structured around some criteria (actor, goal, plan, capability, etc.) and ready for the design phase. To give an idea, less than 100 activities are found for small projects, between a hundred and a thousand activities are current for medium-size projects and if more than 1,000 activities are identified, we are facing a large project (project size can be measured in software, classically, with man-month of number lines of code in any language).

Step 3: *Solving CRT parameters (Budgeting).* In this step, generally, system engineers require the collaboration of a financial analyst. Activities must be decomposed at a granularity such that the Developer can easily find back existing components, discover activities that they currently handle in the past and for which they have enough experience to be able to put rapidly a cost, identify required resources and quote a time.

Once the time of each elementary activity is known, a Gantt chart (a way of visualizing a series of tasks and their dependencies) describing the project schedule can be built for estimating the overall timing of the project. Sometimes, it is necessary to simulate a *work breakdown structure.* The project schedule is just the combination of all the breakdown structures with allowance made for interdependencies between tasks. Specific project management software could be of great help, instead of using only spreadsheets.

It would be advisable at this step to contact the Client and give him some insights about the approximate overall project cost and time. If early adjustments are needed, they will save efforts to generate detailed specifications.

Step 4: Prepare criteria for validating the end product (Validation of conformance). The validation testing is introduced into a project to assure that the final product will carry out the functions described in the requirement specification. Normally, it protects the two parties, the Client by forcing the Developer to engineer the right product or solution, and the Developer by forcing the Client to accept the same product if it passes all validation tests.

Testing scenarios have to be well defined at this step so that conformance is possible to demonstrate and validate. Terms as "appropriate response" or "perform satisfactorily," if not accompanied by more precise and testable scenarios must be banned. One of the directions that can save a lot of time later would be trying to write the User's Manual early at this phase as a starting point for validation. But, User's Manual is not enough for validating a product since this manual mostly gives standard scenarios with "normal usage" of the product, while validation may test for limit cases, error handling scenarios, functional metrics and all mirrors of specification clauses. A validation document, more accurate than the User's Manual, is something that a Developer can give to the Client and the latter can give it to his technical service to support customers having trouble using the product. Some products in the market include this technical document at the end of the User's Manual in order to help customers diagnose themselves their problems.

Step 5: Generate contractual textual documents and prepare instructions for the design phase. Documents required for the contract varied greatly according to the engineering domain. Generally, we find: overview description of the project, architecture of the system to be built, product specifications, product validation, financial, and legal documents.

Documents for the design team include: overview description of the project, architecture of the system to be built, product specifications, product validation all relevant diagrams developed at step 2.

6.3 Requirements Analysis with Technical Diagrams

This phase can be started once the elicitation process of textual and early requirements is completed and the developer knows what the client wants. Normally, we have a lightweight document containing an enumeration of early requirements containing *goals*, *unstructured description of elements* entering the composition of the system to be built. Such an "unordered structure" is perfectly normal as Client and Developer want to go forward with an idea (or stakeholders'

Table 6.2.3.1 Subphases of the Requirements Engineering process. We focus only on the first two subphases. The requirement analysis (subphase 2) makes use of Use Case for representing graphical representation of the system to be built. The subphase is unavoidable as we must make a good text content analysis to understand the Client needs.

Subphases	*Techniques used*	*Targeted concepts*
1. Elicitation of early requirements (unstructured and unordered data)	Text Content Analysis	Goals. Context description. Early product description. Features of the system to be built.
2. Requirements Analysis	All UML diagrams, mainly *use case* and *BPDs*	Activities, actors, their relationships. Actions, activities and their inputs, outputs.
3. Budgeting	Spreadsheet or specialized software	Determine CRT (Cost, Resource, Time) parameters from activities
4. Definition of validation of conformance criteria (Tests specifically designed for acceptance process)	Requirements Validation Tools or manual list of test scenarios	Limit cases, error handling scenarios, and functional metrics.
5. Document Preparation	CASE tools for generating technical and textual documents from diagrams	Legal, contractual and technical documents (Differentiate between "soft goals" and "rigid" constraints)

intentions) but the domain is not still correctly mastered. They have put some main ideas into preliminary documents but the true analysis work has not been started. So, the project needs to be investigated more systematically in order to identify components and activities, organize, and structure them.

The UC diagram (Jacobson et al., 1999) in UML standard has being recognized as formalism well adapted to this investigation phase. The BPD already studied in Section 5.3 can be used to describe inputs, outputs, events around important processes that need to be specified. Both these diagrams are of functional flavor. *Why an alignment on functional view with object technology?* The answer is: This is a phase where technical considerations have to be balanced against social, organizational, and financial ones. We need to get the CRT parameters, so, going further than a rough identification of activities will propel us directly into the design phase, or at least in the first strata of the design phase. As requirements engineering is an incursion into the functional view of a system and not a complete study, it was recognized as a phase where the most and costliest errors are introduced into a system. One of the directions to negate this effect would be a flexibility of the two parties involved to accept some adjustments to the project even during the design phase. To illustrate the Requirements Engineering process, let us consider early requirements of a Remote Monitoring System of Health Conditions (RMSHC).

> *Early requirements of a Remote Monitoring System of Health Conditions:* A Client desires to manufacture a system that limits a patient's health risk, reduces the time and frequency of check-ups in a hospital, and may occasionally serve as pharmaceutical field tests. This technique can bring peace of mind to people giving them the insurance that they are constantly monitored by qualified people. The system is a kit containing a wristwatch Blood Pressure Monitor (BPM), a pill box, and a mobile phone that automatically transmits patient data to an internet center where doctors and nurses can monitor the patient at will. The BPM takes at determined frequency systolic and diastolic blood pressure and pulse readings. The patient can activate manually the BPM with only one push button command to inflate, measure and display. The request can be sent by the doctor or the nurse through the telephone. The pill box is of electronic type. It must keep track of the time a pill is removed from the box. If the patient forgets to take a pill, a reminder is sent through the mobile phone or the Nurse can contact the Patient orally.

The next step will be the analysis of those early requirements.

6.3.1 Requirements Analysis with Use Case Diagrams

Some readers may be obfuscated and perplexed about some ideas advanced with the way we make use of UC diagrams; but in the requirements engineering phase, as the system is not built, as we need a *succinct tool* and formalism dealing with fuzziness, it would be more important to be able to answer quickly "what is the project cost?" "how long it takes to develop?" or "what is the existing resource (or to be built) that is not considered as being part of the project but that must be deployed to support it?" than satisfying mathematical considerations.

UC diagrams are used as techniques for formalizing roughly functional requirements and factorizing actor roles (Sandhu et al., 1996). Their syntax is already explained in Section 4.16. This diagram is based on three main concepts: *Actor, Use case*, and *Relationships* connecting use cases to use cases or use cases to actors. The four most important relationships in use case concepts are *include, extend* and *communicate*, and *inheritance*. *Inheritance* relationship is used to factorize actor roles. Table 6.3.1.1 summarizes our limited interpretation of concepts in a UC diagram.

Using use cases as requirements engineering tool is a technique has some limitations (that can be seen as "advantages" if it can discourage premature design) as it addresses only the functional view. Relationships and actors is an attempt to organize and structure this functional view, however, their semantics is rather limited. In order to make this diagram more expressive, some interpretations must be extended and enriched, so some additional extensions are necessary:

1. *Use cases* can represent *goals, plans, rules, or requirements* that all need activities to be realized. Goals and business rules are integral parts

of organization behavior and guide their day-to-day business decisions. Some authors consider them as nonfunctional requirements. In fact, the relationship between goals, rules and activities is rather complex as people cannot really verify how products, manufacturing processes, solutions to conform to business goals and rules if they are not mapped or inserted in

Table 6.3.1.1 Convention used in this book: basic concepts in use case diagram and extensions of concept interpretation

Concepts and graphical notations	*Comments and examples*
Stickman	**Stick form Actor** A stick form actor is used to represent a *role* accomplished by a human, a group of humans, a human organization, or a mixed system involving humans.
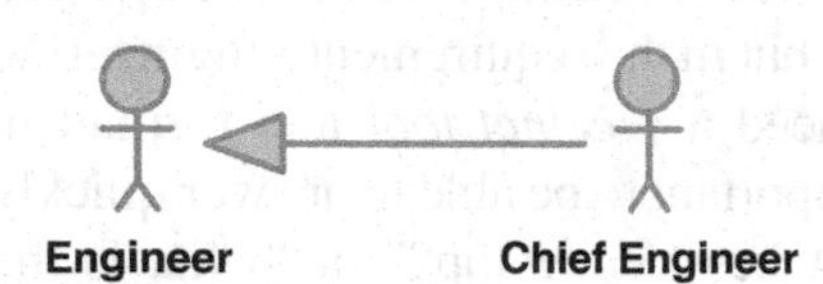	**Inheritance relationship between Actors** In the example, a Chief Engineer can accomplish all the tasks assigned to an Engineer and more. This inheritance relationship is viewed exclusively from the role consideration. A physical person may assume an unlimited number of roles or a role can be played simultaneously by many humans.
Nonhuman or object/system Actor 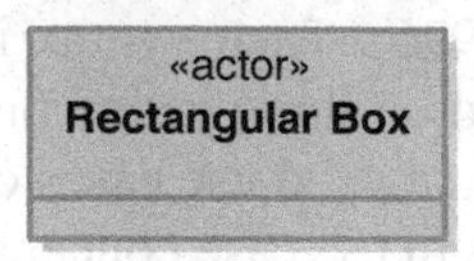	A rectangular box actor represents an actor role, the entity behind the actor can be human or nonhuman (this case embraces the human actor role). Generally, it depicts a system, a subsystem, a component, an object, and an undefined system (black box) to be developed.
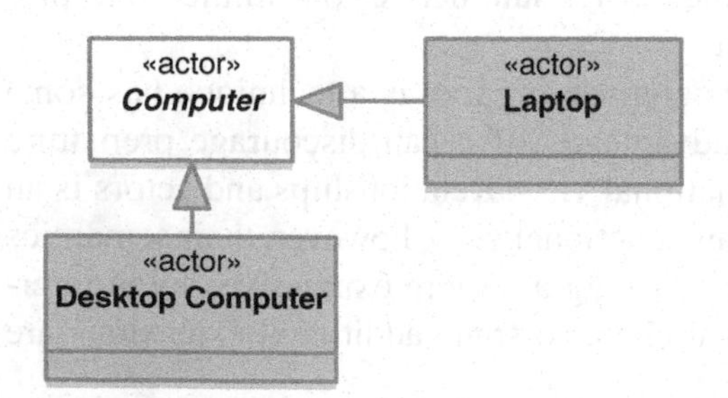	**Nonhuman actors involved in inheritance relationship** In this inheritance relationship, a Laptop and a Desktop Computer have all the properties and operations of a generic abstract Computer.

(*Cont.*)

Concepts and graphical notations	*Comments and examples*

Use case representing an action, an activity, a task, an operation, a business process, a collaboration, a scenario, etc.

An *action* represented by a use case is an important action that can take few CRT parameters, or at the limit zero, but, if not represented, may obscure the interpretation of a use case diagram. (Example: *Log On* use case)

An *activity*, decomposable or not, is the most frequent and usual use for a use case. It can be executed by any general actor, a collaboration of mixed actors. An activity can be a scenario describing a sequence of actions/activities executed in a certain order.

Tasks and *operations* are also activities in other domains. In business modeling, *business processes* can be represented as use cases at requirements analysis.

Example: Scenario represented by an abstract use case and four use cases representing activities in a parking system.

Driver
Entering Parking
Stop at the gate
Push the button
Take ticket
Enter when gate opened

Entering parking is an abstract use case that is "replaced" by four other use cases. This representation avoids counting *Entering Parking* twice.

***Extension:* Use case representing a goal, a plan, a requirement, etc.**

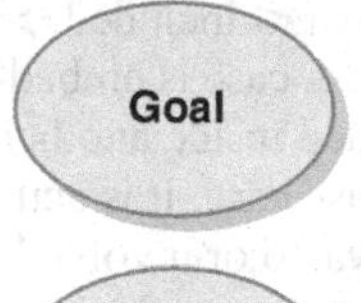

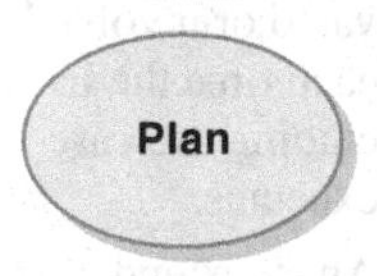

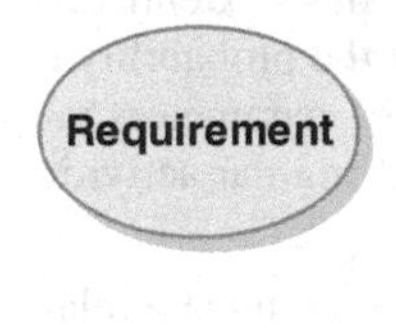

...

Example: Goals

Goals influence the way products or solutions are designed in a system and impact heavily on all activities of an organization.

Rules will be mapped somewhere into manufacturing, business processes or verification processes.

Plans are a collection of activities to reach some goals.

Requirements will be mapped into specifications that must be realized in the design phase.

So, all these concepts are of functional flavor even if the relationships are hard to establish without a precise context. We make use of use cases to represent them all. When a use case represents a semantic node (a node in a diagram that, if absent, will obscure the interpretation or understanding of the diagram), have no CRT parameters attached to it, or is replaced by a set of activities in the same diagram, we can simplify its graphical notation ("dashed contour" or "transparent background color") to alleviate its visual load on a diagram.

Concepts and graphical notations	Comments and examples

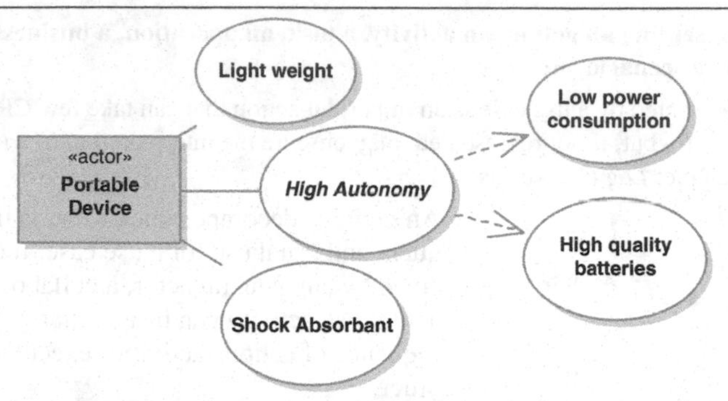

High Autonomy is converted into two goals connected with <<include>> relationships. The <<include>> stereotype is suppressed as the line style (dashed) and the orientation identify unambiguously this relationship.

***Extension: Communicate* is used as an undefined relationship**

The *communicate* relationship can be used as an undefined relationship that can mean *involve, concern, dependent, have property, obey to the rule, satisfy*, etc. This fuzziness is an advantage as it avoids spending too much time looking for a precise meaning at the requirements analysis phase.

Example: The *communicate* relationship in the previous extension is transformed into *satisfy* relationship (A portable device must satisfy Lightweight condition).

Include and *extends* relationships

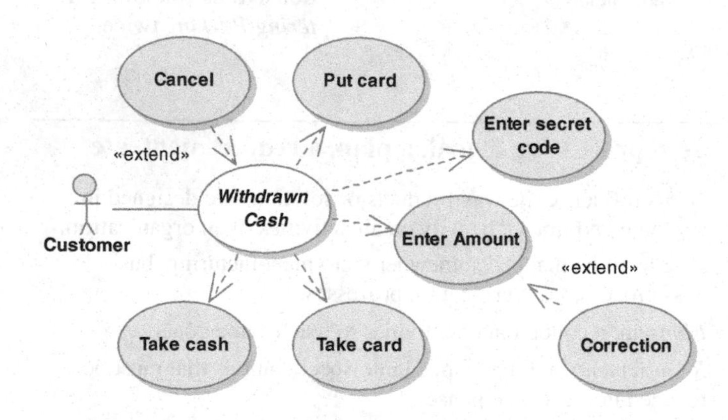

These two powerful concepts structure and connect use cases together and support decomposition mechanism.

An <<included>> use case is embedded inside another use case. It is activated or involved each time the including use case executes.

An <<extended>> use case is an activity that is executed under some conditions. Even if the probability of calling this activity is very low, from a development viewpoint, as CRT parameters must be deployed to handle this case, its importance is equal to that of other regular activities.

Example:

In this example, the *Customer* is connected to *Withdraw Cash* via a communicate relationship. *Withdraw Cash* is transformed into an "empty" use case as the main activity is transformed into five more elementary use cases and replaced by ALL decomposed activities (this transformation is not allowed for nonexhaustive decompositi on as we still need

(*cont.*)

Table 6.3.1.1 (Continued)

Concepts and graphical notations	*Comments and examples*
compute residual CRT parameters from the original use case). *Correction* and *Cancel* are software pieces that must be designed even if they are "exceptionally" activated. *Cancel* can be operated in any of the five decomposed activities as it is defined at their root. *Important Note*: Sometimes, the *extend* relationship is interpreted as an equivalent of a *switch* or *case* instruction in programming language. This fact is only coincidence. The use case diagram is not a flow chart or an activity diagram (it is not a dynamic diagram), so, this interpretation does not hold in all cases.	

some way inside the whole development process. Goals may appear as required attributes of the end product but the resulting state of a system is the fruit of hard labor and intensive design activities. Instead of involving ourselves in interminable debates, a pragmatic approach is using use cases to represent these concepts but, if the Developer thinks that a given use case does not correspond to a real activity or replaced by some activities elsewhere, its graphical representation can be alleviated and it will be considered as a semantic node.

(a) When a use case represents a semantic node, have no CRT parameters attached to it, or is replaced by a set of activities in the same diagram or conditions the way the whole process is engineered, we can simplify its graphical notation ("dashed contour," "transparent background color," or "abstract" representation) to lighten its visual load on a diagram.

2. The *communicate* relationship can be used as an undefined relationship that can mean *involve, concern, dependent, have property, obey to the rule, satisfy*, etc. The fuzziness of the model relative to the meaning of this relationship is not really important at this phase and does not impact greatly on the estimated volume of activities necessary to reach CRT parameters.

   (a) To alleviate the visual load, all the communicate relationships in a use case will be shown as full line without any stereotype and navigability. The relationships *include* and *extend* are shown with dashed lines, the *include* relationship without its stereotype, and the *extend* relationship may be shown with or without its stereotype. The *extend* arrow is oriented from the extended use case towards the extending use case, so this fact distinguishes it from the *include* relationship noted with a similar arrow with a reverse direction.

6.3.2 Conducting the Requirements Analysis Phase

The extended conventions allow us to represent all types of requirements: functional, nonfunctional, goal type, etc. inside the same formalism (UC diagrams) and thus extending the expressiveness of this diagram and reducing the textual part to its minimum. To illustrate the requirement analysis process, let us return to early requirements of a RMSHC (Fig. 6.3.2.1) and analyze this textual document, paragraph by paragraph, presuming that this text comes from the Client and is dissected by the Developer.

The textual analysis of Table 6.3.2.1 allows us to identify all actors/objects, use cases, and their relationships with this early requirements specification. Once all the inquiries are elucidated with the Client, a complete set of diagrams

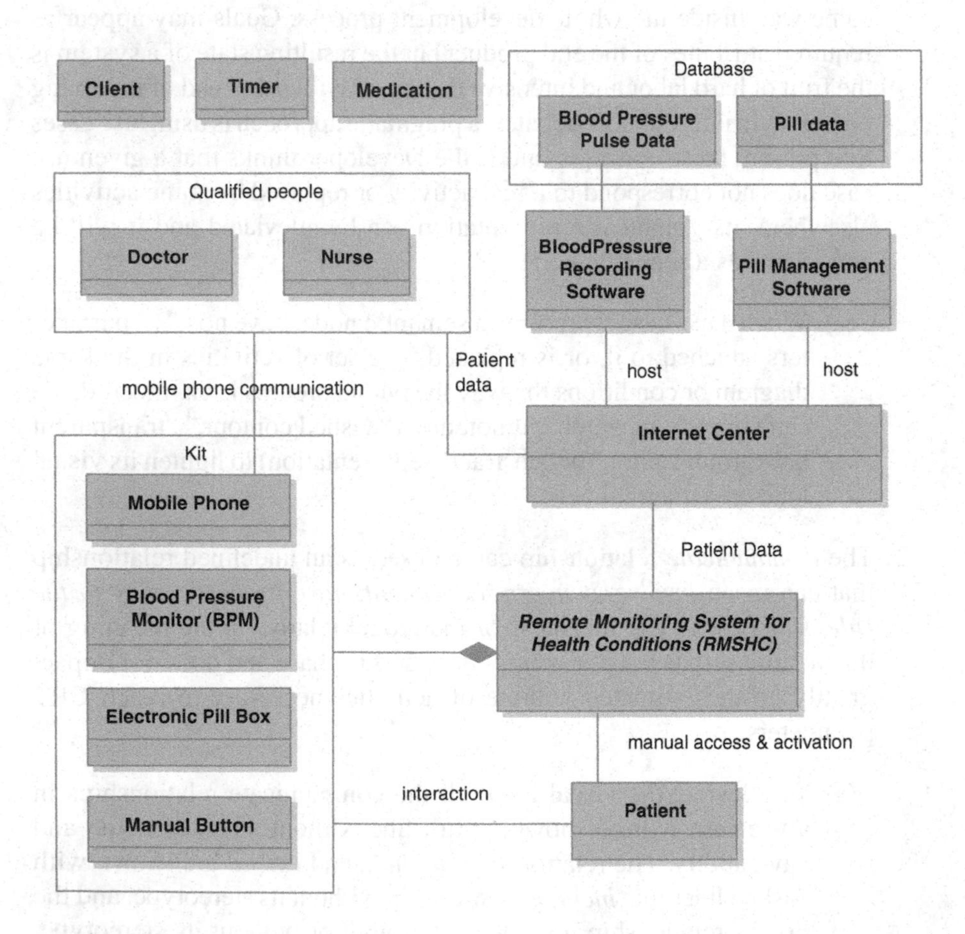

Fig. 6.3.2.1 Class diagram of the RMSHC system. This class diagram allows us to create quickly separate instances to be used as nonhuman actors in subsequent use case diagrams to support several views of a system

Table 6.3.2.1 Example of Text Content Analysis in order to establish use case diagrams structuring requirements. Fuzzy clauses must be elucidated with the Client. A more complete table template is shown next. Diagram reference contains diagram numbers (the same concepts may be displayed in multiple diagrams)

Step	*Initial text*	*Concepts*	*Reformulation*
1	A Client desires to manufacture a	Actor/Object	Client
1a	system that limits a patient's	UC/Activity	Manufacture
1b	health risks, reduces the time and frequency of check-ups in a hospital, and may occasionally	Actor/Object	System: Remote Monitoring System of Health Conditions (RMSHC)
1c	serve as pharmaceutical field	Goal	Limit patient health risks
1d	tests.	Actor/Object	Patient
1e		Goal	Reduce check-ups time in hospital
1f		Goal	Reduce frequency of check-ups in hospital
1g		Goal	Serve occasionally as pharmaceutical field tests
2	This technique can bring peace of	Actor/Object	Technique (abstract)
2a	mind to people giving them the	Goal	Bring peace of mind
2b	insurance that they are constantly monitored by qualified people.	Goal	Give insurance of good health monitoring
2c		Actor/Object	Qualified People (abstract, to be clearly identified)
2d		UC/Activity	Monitor Patient Parameters
3	The system is a kit containing a	Actor/Object	Kit
3a	wristwatch BPM (Blood	Actor/Object	BPM (Blood Pressure Monitor)
3b	Pressure Monitor), a pill box, a	Actor/Object	Pill Box
3c	mobile phone that transmits	Actor/Object	Mobile Phone
3d	*automatically* patient data to an	Relationship	Component of RMSHC
3e	internet center where doctors and	UC/Activity	(Phone) Transmit patient data
3f	nurses can monitor the patient at will.	Constraint	(*automatically* is a fuzzy term) Transmission starts each time data is available or at regular intervals? *(to be elucidated)*
3g		Actor/Object	Internet Center
3h		Actor/Object	Blood Pressure Pulse Data
3i		Actor/Object	Data Management Software (implicit specification)
3j		UC/Activity	Manage all Patient Data (implicit)
3k		Actor/Object	Doctor
3l		Actor/Object	Nurse
3m		Relationship	Doctor and Nurse are "Qualified Person" (See 2c)
3n		UC/Activity	Monitor Patient Conditions

(*cont.*)

Table 6.3.2.1 (Continued)

Step	*Initial text*	*Concepts*	*Reformulation*
4	The BPM measures at	UC/Activity	Measure systolic pressure
4a	determined frequency systolic	UC/Activity	Measure diastolic pressure
4b	and diastolic blood pressure and	UC/Activity	Measure pulses per minute
4c	pulse readings.	Constraint	*At determined frequency (unspecified for the moment. Limits have to be elucidated for the design phase)*
5	The patient can activate manually the BPM with only one	UC/Activity	(Patient can) Activate measurements manually
5a	push button command to inflate,	Actor/Object	Push Button
5b	measure and display.	UC/Activity	(BPM) Inflate
5c		UC/Activity	(BPM) Measure (See 4, 4a, 4b)
5d		UC/Activity	(BPM) Display
6	The request can be sent by the doctor or the nurse through the telephone.	UC/Activity	(Doctor, Nurse) Communicate orally (via Patient Mobile phone)
6a		UC/Activity	Request manual activation (BPM)
7	The pill box is of electronic type. It must keep track of the time a pill is removed from the box.	Attribute	(Pill Box (already defined in 3b). *of electronic type*
7a		UC/Activity	Keep track time pill removed from Pill Box
7b		Actor/Object	Pill Management Software
8	If the patient forgets to take a pill, a reminder is sent through the mobile phone.	UC/Activity	Forget to take pills (this use case is abstract as it does not correspond to any CRT parameters but explains only the process)
8a		Actor/Object	Pill Data (implicit)
8b		UC/Activity	(Pill Software) Check Pill Data against Medication (implicit)
8c		Actor/Object	Pill Software (implicit)
8d		Actor/Object	Medication (prescribed by Doctor)
8e		UC/Activity	(Doctor) Prescribe Medication (implicit)
8f		UC/Activity	Enter pill data to Internet Center (Nurse or Doctor?)
8g		UC/Activity	(Pill Software) Elaborate Electronic Reminder
8h		UC/Activity	(Pill Software) Send Reminder by Mobile Phone
8g		UC/Activity	(Nurse) Contact Patient forget taking pills

Step	*Initial text*	*UC concepts*	*Reformulation*	*Diagram reference*
				UC1, UC5, etc.

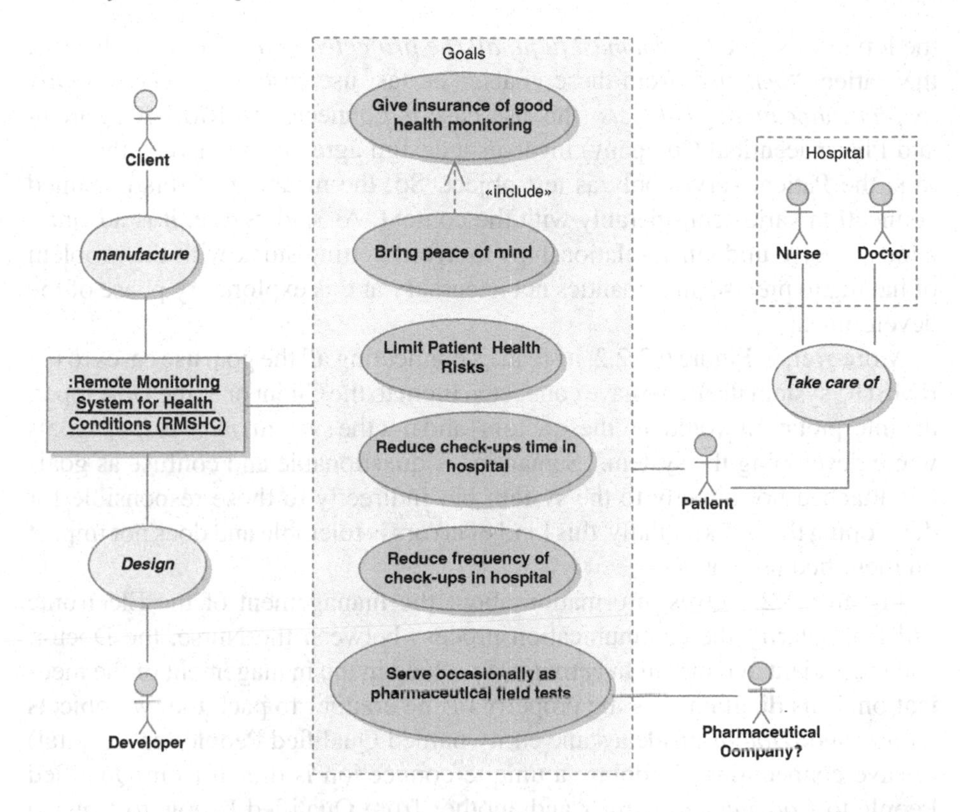

Fig. 6.3.2.2 Use case (UC) diagram depicting goals enumerated for the RMSHC. Except RMSHC system, human actors are created directly from the UC diagram (not from classes) to show another possibility of generating actors (but we suggest to create all actors by instantiating them from classes). Patient is connected to the contour of goals and therefore is connected to all goals, idem for RMSHC. Pharmaceutical Company crosses the contour to reach individual use case is not connected to other use cases

can be built to compute CRT parameters with all use cases identified through this Requirements Analysis phase. When establishing diagrams, each concept identified in Table 6.3.2.1 can figure more than once to support many views of the system.

In Figure 6.3.2.2, the RMSHC is instantiated from a class diagram (its name is preceded by a ":" and underlined). All human actors can be either created from scratch in the UC diagram or instantiated from a class. To implement the second choice, a class diagram must be created to define all classes and afterwards, objects can be instantiated from these classes at will to support all UC diagrams.

The goal *Give insurance of good health monitoring* is included to *Bring peace of mind* to show a possibility of making hierarchies of goals if needed. Goals are shared mainly by two actors, the Patient and the System. If the interpretation at the left side is "*the System must fulfill all the projected goals*," at the right side, the Patient "*benefits*" from these goals. The last use case *Serve occasionally as pharmaceutical field tests* (this use case is connected to RMSHC, Patient and Pharmaceutical Company) involves a design agreement between three actors, the Patient serves only as test object. So, the meaning of this unnamed connection varies considerably with the context. As said before, it is advantageous to have undefined relationships to avoid getting stuck with the problem of having to precise all semantics not necessary at this exploratory phase of the development.

Moreover, in Figure 6.3.2.2, instead of connecting all the goal use cases to the RMSHC system itself, we have connected them to the Client or to the Developer, the interpretation would be "these actors (and not the system) *must respect* goals while developing the system." Semantics is questionable and confuse as goals are attached not directly to the system but indirectly to those responsible for developing them. Essentially, this kind of errors is tolerable and does not impact on identified use cases.

Figure 6.3.2.4 gives information about the management of the Electronic Pill Box, clarify the communication process between the Nurse, the Doctor, and the Patient and highlight activities involved in the management of the medication. This diagram uses the property of the contour to pack the two objects *:Nurse* and *:Doctor* inside a same entity named Qualified People (or Hospital) to save connections. So doing, a unique connection is drawn from Qualified People to *Communicate orally* and another from Qualified People to *Consult Patient Data*. All connections can go through the contour of Qualified People to reach individual objects *:Nurse* or *:Doctor*. Two *extend* relationships are defined within *Communicate orally* to show that this communication is either for requesting a manual activation of the BPM (Fig. 6.3.2.3) or urging the Patient to take the forgotten pills. *Forget Taking Pills* is an abstract use case attached to the object Patient. This possibility shows that use cases with zero CRT parameters can be represented as semantic nodes in order to explain the context and give a coherent interpretation.

In Figure 6.3.2.4, several activities are made abstract as they do not impact directly on the determination of CRT parameters but are there to explain a context justifying the usefulness of some software pieces. For instance, if the Patient forgets to take pills, the Electronic Pill box cannot register any pill taken out of the box, the Pill Management Software will detect this anomaly. Two actions are therefore programmed, the first one is sending a Reminder through Mobile Phone, and the second is sending a warning to the Hospital that urges the Nurse to contact personally the Patient if the electronic Reminder has no effect. The presence of abstract use cases is thus essential to capture the meaning of the whole process.

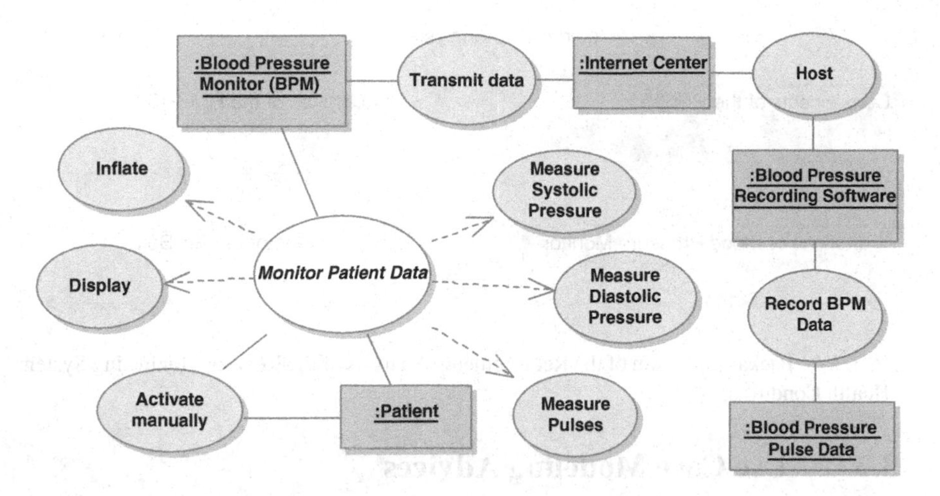

Fig. 6.3.2.3 Activities of the Blood Pressure Monitor (BPM). The central use case Monitor Patient Data is turned into abstract as it is replaced entirely by included components. The Activate manually needs the intervention of the Patient

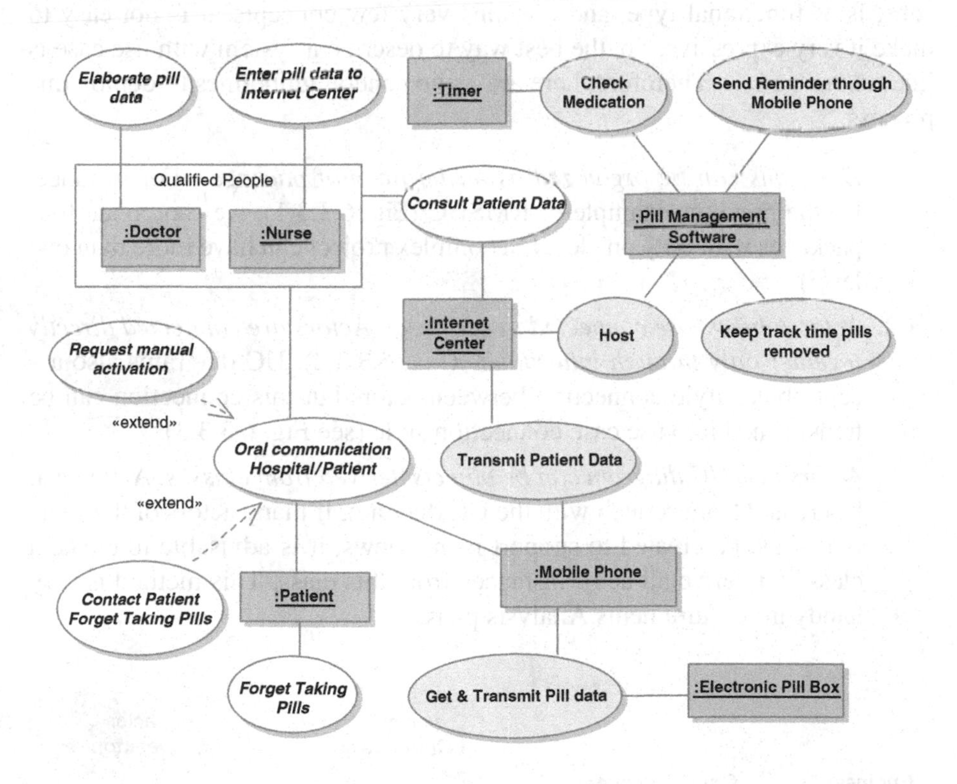

Fig. 6.3.2.4 Use case diagram showing activities related to the management of the Electronic Pill Box

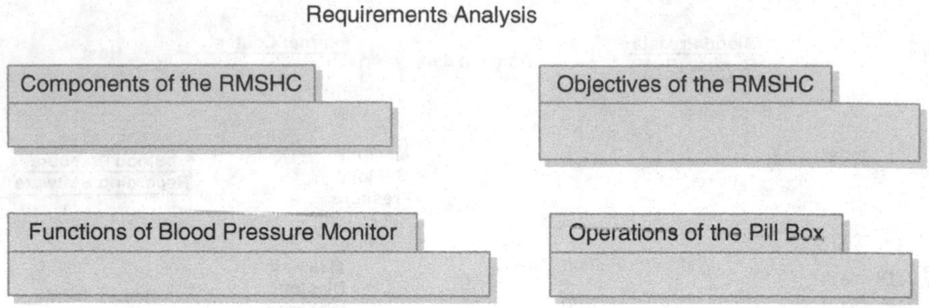

Requirements Analysis

Components of the RMSHC

Objectives of the RMSHC

Functions of Blood Pressure Monitor

Operations of the Pill Box

Fig. 6.3.3.1 Package diagram of the Requirements Analysis of the Remote Monitoring System for Health Conditions

6.3.3 Use Case Modeling Advices

UC diagram is a very simple formalism and it addresses to the first phase of the modeling process, even qualified by Developers as phase 0 outside the technical model itself. As it could eventually be read by nontechnical persons, is of functional type, and contains very few concepts, it is not easy to make it very expressive. So, the best way to describe a system with use case is like *telling a story*. Therefore, here are some general guidelines to deploy this process:

1. *Diagrams can be organized as a hierarchy of packages.* For instance, for the previous example of RMSHC (Fig. 6.3.3.1), we can create four packages with only one level (a complex project can have more than one level).
2. *Actors/objects are connected to use cases. Actors are connected directly together only through inheritance* (Fig. 6.3.3.2). UC diagrams also accept object style connection between actors but this connection can be transformed into use case connection style (see Fig. 6.3.3.3).
3. *Actors in a UC diagram can be objects derived from classes.* Actors can be created from scratch with the UC diagram. If many actors of the same type must be created to support many views, it is advisable to create a class and derive all actor instances from this class. This method is very handy in Requirements Analysis phase.

Fig. 6.3.3.2 Actors are connected directly through inheritance only

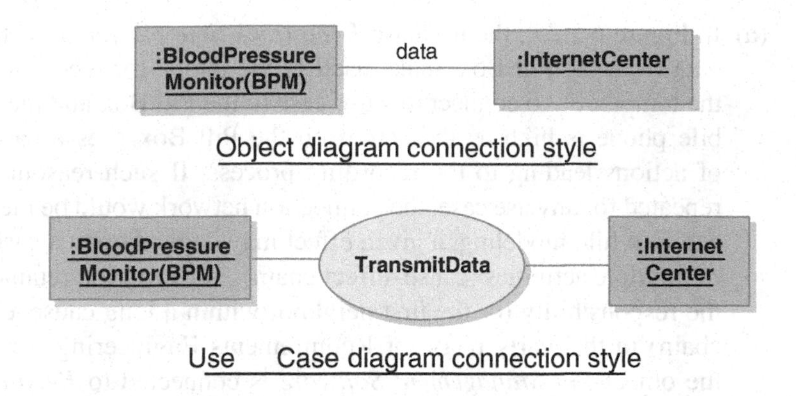

Fig. 6.3.3.3 When actors are connected together, we have an object style connection. In a use case diagram, a use case is inserted between two actors meaning that actors are involved in the activity

4. *Distinguish goals, rules, and competencies from activities.* Goals, rules have no CRT parameters. Conversely, activities have CRT parameters, so goals and rules can be made abstract to distinguish them from regular activities.
5. Use inheritance for actor roles to reduce use case repetition or connection complexity (Fig. 6.3.3.4).
6. *It is not necessary to connect a use case to all actors involved.* The use case model is essentially based on responsibilities, competencies, goals, and activities of main actors. So, only main actors responsible of the activities are relevant. If there are several, the first neighbor will be chosen. A UC diagram will become very complex if each use case is transformed into a "bubble" of a DFD diagram.

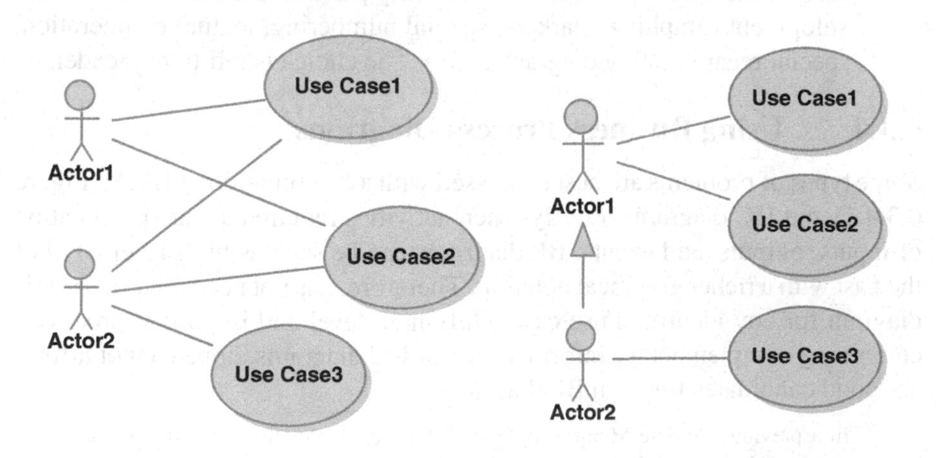

Fig. 6.3.3.4 Inheritance between roles. The left diagram is replaced by the right diagram

(a) In Figure 6.3.2.4, the use case *Keep track time pill removed* takes data from the Pill Box, make used of the phone for recording, so the temptation to connect this use case to the Pill Box and the Mobile phone is high as the action on the Pill Box fires a cascade of actions leading to the recording process. If such reasoning is repeated for any use case, the connection network would be inextricable. While modeling, a given effect may depend upon a cascade of multiple activities (cause–effect chain). The criterion retained is the responsibility (or the first neighbor within a long cause–effect chain) in this early phase of Requirements Engineering, so only the object *Pill Management Software* is connected to *Keep track time pill removed* as the latter is a component of *Pill Management Software* itself connected only to its first neighbor and responsible host *Internet Center*. As said earlier, the utmost goal is to identify activities with CRT parameters by telling an understandable story.

7. Be careful in adding new relationships to a UC diagram. Normally, we can stereotype any relationship between actors/actors, use cases/use cases, or actors/use cases, but *the tendency to transform a UC diagram into an entity relationship diagram of structural flavor must be avoided.* To explain the composition of the system, a true class diagram may be used to create classes with aggregation/composition and possibly early inheritance relationship between classes.
8. *Well define the frontier of the system to be developed.* As said earlier, diagrams are targeted to tell a story. But a system is generally part of a context and interacts with other systems. The story needs all of them. It is recommended to choose a way to identify parts devoted to the current development (graphical markers, special numbering, textual enumeration, special recapitulating diagrams, etc.). The choice is left to the reader.

6.3.4 Using Business Process Diagrams

Some types of problems are best expressed with a combination of BPDs (Figure 6.3.4.1) and UC diagrams, mostly when activities require a clear specification of inputs, outputs, and events. BP diagrams can be seen as high-level DFD of the past with a richer graphical notation. Therefore, it is not necessary to use this diagram for any identified use case. Only high level and important processes containing complementary information that UC diagrams alone cannot afford, are good candidates for such BP diagrams.

In a previous Remote Monitoring System for Health Conditions (RMSHC), a good candidate for a BP diagram is the Pill Management Software Component (actor/object). At this Requirements Analysis phase, the Developer must evaluate

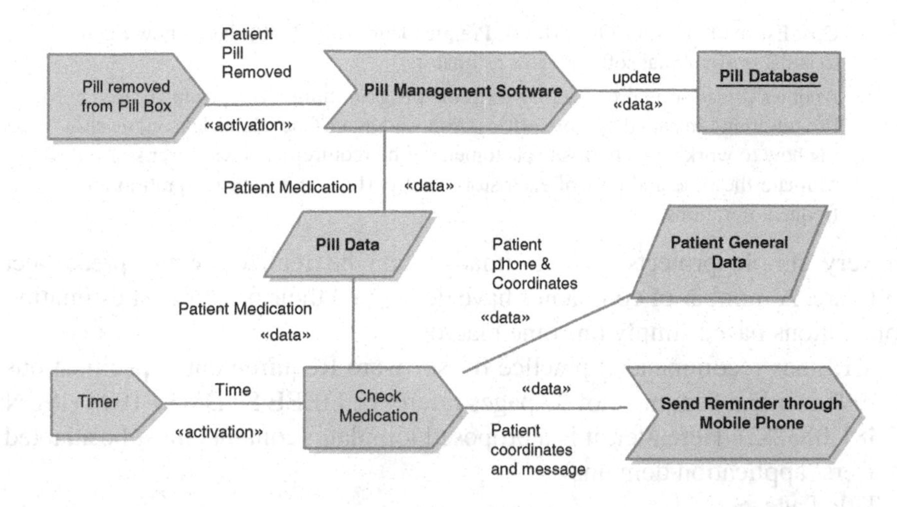

Fig. 6.3.4.1 Business Process Diagram showing the functional aspects of the Pill Management Software Component of the RSMHC project. Data and control flows are stereotyped to indicate the nature of the information and names given to flows are application dependent

information that must accompany the BP diagram to avoid entering into an early design of the system itself.

As said before, the BPD has the DFD as ancestor, so business processes can be cascaded at will. Flows are of data or control types. Stereotypes are used to typify them in the modeling domains and names given to these flows are application dependent. Business processes can be used in the design phase too. At this stage of the project development, its role is essentially to show input and output parameters of important processes in the system for communication purposes with stakeholders only.

6.4 System Requirements Specifications

We bypass the Cost Resource Time Estimation process that alone can require a whole book to pack project experiences of project leaders and discuss about tools used for this purpose. Risk factors can be taken into account at this step. A project estimate is not just a matter of picking the use cases to be included but counting in efforts for auxiliary tasks such as project management, testing, general tasks, documentation, deployment, contingency reserve, etc. The cost is not the same for neophyte and professional companies that have built, in the past, similar systems or solved analogous problems. The former can charge the Client for the experience, but the latter can charge more for training their own staff.

The original COCOMO (Constructive Cost Model) model was first published by Dr. Barry Boehm in 1981, and reflected the software development practices of the day. Emphasis on reusing existing software and building new systems using off-the-shelf software components was found in improved COCOMO II (Software

Cost Estimation with COCOMO II, Prentice Hall, July 2000) that is now ready to assist professional software cost estimators.

Another direction is proposed by Extreme programming (XP) group [Extreme Programming Installed by Ron Jeffries, Ann Anderson, Chet Hendrickson] teaching how to work with an on-site customer, define requirements with user stories, estimate the time and cost of each story, and perform constant integration and frequent iterations.

In very simple projects, cost estimation can be realized with spreadsheet software. A number of companies have developed their proper cost estimation applications based simply on some macros.

IEEE has recommended practice for software Requirements Specifications in 1998 inside a document of 37 pages referenced IEEE STD 830-1998 (ISBN –7381-0332-2). Hereafter, it is a proposed template (contents must be adapted to users' application domains):

Title Page

Project Title, Project Number, Date Signed, Date of Delivery, Authors, Version, Total Number of pages, etc.

Document history

This document is needed if there are many versions, many persons involved at various steps for a large project. The current document may be part of a more important document.

Table of contents

All chapters listed from here. Executive summary would be the first item.

Executive summary

Overview of the project, its impact on the market, its impact on the organization, its overall cost, etc. Brief summary of all important results, numbers, or statistics.

Introduction

Give the context of the project, stress its importance

Overview of the architecture of the product or of the solution proposed

As we are still in the Requirement phase, the architecture must be sufficiently vague so as not to become a constraint, unless this constraint is requested by the Client. The architecture of the RMSHC corresponds to the class diagram of Figure 6.3.2.1.

Context of the Application and User Interface

Describe in detail the context of the application and define the typical Customer of the product, the Person or the Organization targeted by the Solution. This definition allows us to define at the design phase exactly what kind of interactions the system has at its user interface.

Requirements for each component or unit enumerated

Component1/Unit 1

Use case and business process, or relevant UML diagrams can be used to describe the structure (avoid to formalize as a constraint) and the expected behavior of the unit.

...

Component N/Unit M

User interfaces specifications are considered as parts of Components or Units.

Project Cost breakdown

Large projects need details about components or subunits.

Schedule

Detail schedule and order of sub components delivery.

Product or solution validation process

Define all test scenarios, inputs and outputs, acceptance conditions, and metrics.

References and complementary documentation list

6.4.1 Requirements Engineering Model in the Model Driven Architecture

Owing to some specific constraints of the Client specification, it is not evident that the Requirements Model is always platform independent. Moreover, platform independence is not a binary concept. There could be different levels of platform independence, so what we really put in a model must be practically and explicitly defined in the model itself and annotated in the project so there must be no misunderstanding of the way developers interpret this independence concept.

Is the decision of building all databases in the RMSHC project with relational technology (written in final and contractual document as a constraint) really platform independent? It would be hard to demonstrate that the relational rules do not impact on the way schemas of entities and relationships are designed at the "logical" phase.

Owing to the "gray" interpretation of independence in PIM concept, if a backtracking is made to retrieve the utmost goals of MDA which is reuse of design patterns, separation of the logical part of a development, separation of concerns, and possibility of evaluation of the correctness of embryonic designs, there is certainly some relationship between models and domains with a background canvas of reusing development knowledge and intellectual assets of a company. So, seeing models as a succession of *cohesive models* (a model is cohesive if, taken at any level, forms an understandable, intelligible system) could be another view of the development roadmap that can be superposed on the binary concept of the PIM and PSM. Therefore, a complex PIM may be layered and compartmented. The same subdivision may exist in the PSM as well.

From the perspective of the Developer, the Requirements Engineering Model (REM) is oriented towards the Client needs. The reuse at the Developer side means a core development that can be shared by more than one application of a unique Client. Moreover, the Requirements Model is established with the goal of evaluating cost, resource, and time, is not supported by a technical and crystal clear interpretation, so *it would be difficult to make now a perfect correspondence between the Design Model and the Requirements Model* as we cannot really transform something very technical in something that is not. This is the reason why we consider the Requirement model as separate and disconnected to the series of Design models to facilitate the design process. From a practical viewpoint, we instantiate a new project name, recreate all classes. The connection between the two models is the "realize" relationship that indicates which part of the design will realize which use case or actor in the Requirement Model. This "link part" can be considered as a small independent "Link Model" (Fig. 6.4.1.1) that maps elements of the Requirements Model to components of Design Model or Implementation Model.

This absence of correspondence does not mean that the Developer cannot practically pass the validation final tests included in the contractual commitment signed at the end of the requirements engineering phase. The REM is used as a canvas for the design phase.

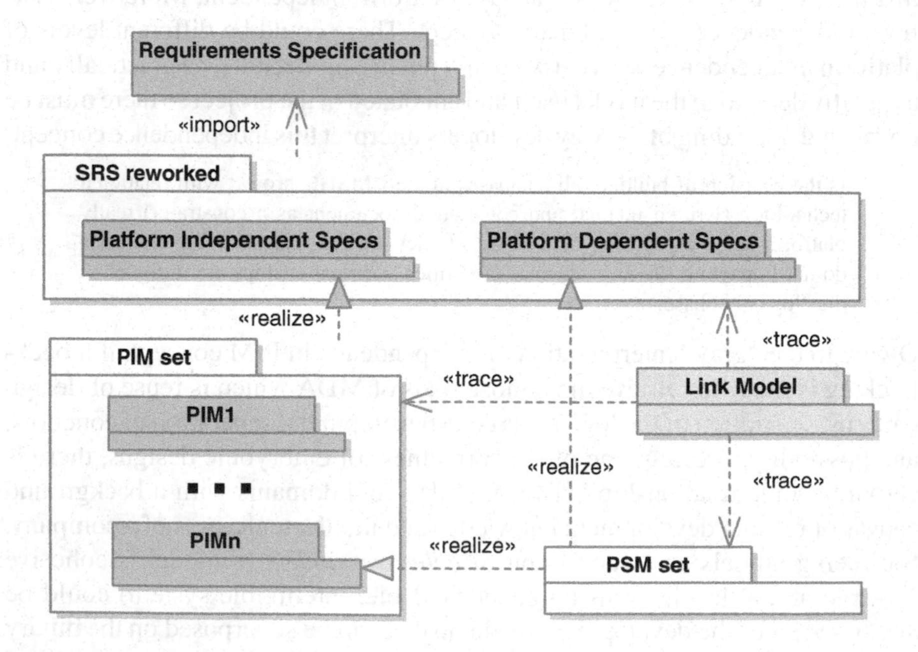

Fig. 6.4.1.1 A link Model traces all mappings between the SRS elements and components of PIM set and PSM set

The layering process using platform dependent criterion is often "horizontal" (some components are implementation dependent among other components that can be made "logical") in the Requirements Model and *vertical* (successive model refinements) in the design phase. This is the reason why, at the beginning of the design phase, the system can be partitioned into at least two parts to separate the requirements specification into two distinct sets of constraints, a *platform independent set* and a *platform specific set*. As the partitioning process into models is not limited uniquely to the criterion based on the platform, inside the PIM or the PSM, we can define submodels or layers, even layers with compartments to isolate reusable parts of the development, thus propelling forwards the MDA objectives as far as possible to optimize reuse.

6.4.2 Dealing with Platform Dependent Components in PIM

At the early phases of design, components that will be imported "as is" are identified, their interface completely specified, and their exact operations schematically represented by activity diagrams or any other dynamic diagrams (state machine, sequence, etc.). The component may itself be platform independent (reuse of abstract development patterns) or platform dependent. In the latter case, there could be many attitudes:

1. *Transform a PSM into a PIM*: Create a new interface of the component for the development making abstraction of all implementation details
2. *Find PIM part of the PSM*: Use implementation independent parts of the components
3. *Come back to old habitude of text comments*: Clearly identify part of the PIM that must deal with some sort of implementation constraints
4. *Delaying*: Drop the whole development part that is "infected" by the platform-dependent component and delay its development to the next PSM phase
5. *Long-term planning*: Find all platform-dependent components of the market, analyze their specifications, extract the core logical concept, and build a virtual platform-independent component as if the Developer must develop this component himself (the best but time consuming attitude).

6.4.3 Various Steps in the Design Process

Figure 6.4.3.1 summarizes steps over the development roadmap.

Concerning the opportunity of automating some steps in this long process, the answers are based upon our current knowledge, research results onmodeling

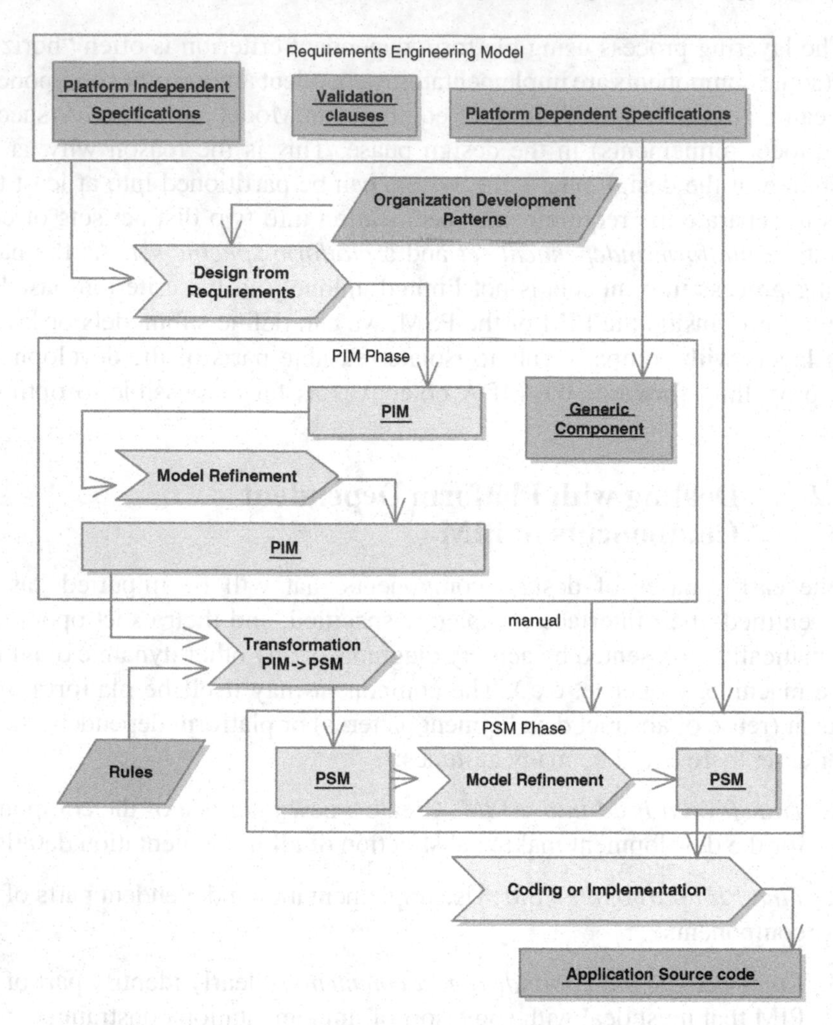

Fig. 6.4.3.1 Design process roadmap represented by a business process diagram combined with an object diagram (this diagram can be drawn with an activity diagram too). Starting from the Requirement Engineering Model, the first PIM is designed then refined. The transformation PIM → PIM is Model Refinement. The passage from PIM to PSM is called PIM → PSM transformation. Many PSM model refinements may occur. When exiting the last PSM, a coding process transforms finally the PSM into a running system

tools, development of new concepts (like MDA), and evolution of modeling standards. Obviously, what we can do today is more than in the past and the future should be better, but, we cannot automate everything, we cannot automate all along the development roadmap, and we cannot have the same automation result for all application domains. Parameters that influence this process are:

1. *Activities, goals identified in the Requirements Engineering phase are not based on highly technical diagrams.* The structure is thus sufficient to understand what is to be done in a project. So, is it really convenient to design transformations and rules for such imprecise and incomplete inputs? When designing, we transform the REM into PIM. Development patterns can be reused. It would be hard to automate completely this process as it needs human skill, trade-offs. Moreover, automating a process is worth the investment when the same or similar situations happen more than once. Domains like administrative applications, e-commerce web sites, comes back very often with similar requirements, so development patterns that can be adapted quickly to generate working designs. But, in more general cases, the best we can really do would be reusing existing components and deploying development patterns, not really automating the process with transformations and rules.

2. *PSM is commonly assimilated to code.* It is particularly true if the Developer starts coding after building his model with UML diagram (his PIM). Things are worse if the code is written directly from Requirements Model without any design (surprising fact but this practice is more common than expected). In medium size projects, there are many steps covering the platform-specific stage. The coding phase is only the last step, the early phase is the transformation from PIM to PSM. Some intermediate phases can go between, involving middleware, platform-dependent utilities, sideline developments, etc.

   (a) If we build for instance a real-time system, in the PIM, all objects are considered as independent and their reaction to events, inputs are instantaneous. But, at the first stage of the PSM, if we decided to allow only one processor for handling all tasks in the real-time system, we are facing the problem of task scheduling that must require a dynamic study. So, scheduling is typically PSM bound and this phase is platform dependent as a technological choice has been proposed for the implementation. Only the transformation of last PSM to PSM is coding.

3. If the passage from Requirements model into PIM models is not straightforward, *the passage from PSM to application source code (last PSM) should be partially possible by defining rules, code patterns*, etc. This automation is possible with standard like the UML that imposes consistent rules for creating models and diagrams. The transformation of a PSM design into code and running application is possible nowadays with very simple, demo applications. This code automation is on the way but true applications still need some research

efforts to create dynamic diagrams that can express unambiguously algorithms. We ignore all automatic phases after the source code generation (compilation, interpretation, intermediate code, linking to running code, etc.) as they are low-level processes and are currently handled by standard development platforms offered by main computer software players.

6.4.4 Hierarchical Decomposition or Bottom-up Method?

Cutting up a complex object into smaller chunks is a fundamental problem of many disciplines. At the starting point, large system must be divided into manageable subsystems. Each subsystem can be considered practically as a smaller system that can be completely specified with the same modeling concepts (goals, context, interface, inputs, outputs, etc.), exactly as the original system.

> A cordless phone consists of a base unit that connects to the landline system and also communicates with remote handsets by radio frequency or higher bands (900Mz, 2.4 GHz, and 5.8 GHz). The handset and the base can be studied as two separate systems. Their functionalities are nearly identical so a common set can be identified. Special functions that allow the connection of the handset to its base will be described as input/output function from the modeling viewpoint. A small part of the project will be devoted to the integration of the handset to its base.

When all loosely coupled parts are separated and specified, each part can then now studied with apparently two opposite methodologies: *top-down* hierarchical decomposition and *bottom-up* component building. In fact, they are perfectly complementary approaches. We cannot really start with bottom-up as long as all the required functionalities are not still identified correctly by a top-down approach. If we start by building components (bottom-up) without having previous knowledge of what problems to be solved, possibly the results could be very interesting but probably, we have to find applications or customers who want to integrate our innovations.

Bottom-up building process makes use of either elementary objects or off-the-shell components available in the market and integrates them in systems. To master the whole behavior of the final system, structural, functional, and dynamical characteristics of each component/object must be known in its utmost details. Documents, diagrams, and all artifacts are normally available either in paper or electronic form. A development database containing all their descriptions and characteristics is a must for serious designers. The structure and the behavior of a mechanical assembly, an electrical circuit, a software component, or a whole organization containing many objects/components/agents can be studied globally only if we have access to individual properties and behavior of each constituting element.

If the previous required RMSHC system needs a BPM (Blood Pressure Monitor) and if this Developer decides to acquire this component as a COTS (Component Off The Shelf) obtained from OEMs (Original Equipment Manufacturers) to include it into his design, detailed technical specifications and undisclosed information are generally available from 3rd party developers or OEM.

A processor included in a system as an intelligent device or controller is an electronic component. Even if the internal structure is not disclosed, all the available specifications (hardware, software, limit conditions, environmental parameters, instruction set, timing of all signals, etc.) give to the integrator all the necessary information to predict the behavior of a system having this processor as component.

The behavior of an electrical circuit (for instance RLC or active filters, etc) is known through differential equations deducted from either passive or active component characteristics.

Kinematics and dynamics of a robot tool tip can be computed if we know all dimensions, masses of all articulations and tools.

We can evaluate the success of a university course and find out what to enhance only if all parameters are known (provenance of students, individual academic results, quality of the course, quality of the professor, quality of supporting resources, biases caused by circumstantial group dynamics, etc.).

To summarize, the behavior of the whole system is made from the behavioral contribution of all its constituting elements or components. When designing a system, we start identifying all functional requirements by a top-down approach, find elements/components, or build assemblies to satisfy those requirements by a bottom-up approach, then specify the structure and study the dynamic behavior of the whole knowing in principle all about parts. Classical systems engineering and industrial manufacturing process breaks down large, complex systems into component parts then rebuild them so that the characteristics on the whole can be controlled efficiently. The automobile industry is probably the most visible systems engineering success following this design principle. At each stage of rebuilding the whole from parts, full testing and dynamic studies must be conducted thoroughly. If the quality of each step is not fully controlled, initial deterministic systems may exhibit chaotic behavior, but this is another problem.

6.5 Designing Real-Time Applications with MDA

The special characteristic that makes the software industry stand apart from other domains is its aptitude to adapt to fast technological changes. The MDA is not a new concept (it has been applied already to define ISO-OSI standard of network communication) but has recently been an incentive for generalizing the software development process, or more generally the design process of all systems, software or multidisciplinary projects. Roughly speaking, the technique stresses the importance of separating the PIM from other PSMs to save the maximum knowledge and experience assets of a company that must daily deal with very fast platform changes and that which wants to effectively deploy reuse. Moreover, the MDA methodology offers to create a formal architecture of high level for reasoning about system and for communication purpose.

> In this view of the development roadmap, the previous blue prints of the RMSHC studied with class, use case, and business process diagrams are parts of a new development component named REM.

A PIM provides formal specifications of the structure and behavior (functions and dynamics) of any element of a given system but abstracts away any technical or implementation details. Implementations in many platforms can therefore be realized from one PIM. The essence of PIM is reuse of known patterns of design, validation of the logical architecture early in the process, delay of problems like scheduling, system integration, or interoperability to PSM. Several PIM may exist in a complex project. The passage from one PIM to another is called *model refinement* or *model transformation*.

The PSM can be viewed (in this proposal) as a PIM plus incorporation of platform-dependent elements (agents/components/objects). They are still not a code. Those platform-dependent elements will be studied with the same diligence as for PIM, except that the whole project is now targeted to be implemented in a defined environment.

In PSM, emphasis is put on technical aspects of project design that is influenced somewhat by project management aspects. UML diagrams can be deployed along this long design phase. UC diagram can still be used for goal, responsibility, capacity, activity, and resource specifications if necessary. At the starting point of the design phase, package diagrams are used to structure tasks necessary to make a Project Assignment.

A project is like a human or social system (temporary organization) (Gareis, available at: http://www.p-m-a.at) that afford resources, manage them in time and possesses algorithms to deal with unexpected problems or to resolve conflicts that may appear during the development process. When executing the design phase, there are two projects that are running side by side, the *Client project* and the *Developer project*. The Developer solves the problem of the Client, builds the Client system while managing its proper organization to afford necessary resources at appropriate time to help Client project to get closer and closer to the Client goals and expectations. So, project management (this "art of getting things done by others" is absolutely necessary for midsize and large projects) is the *Developer system* and the system to be built is *Client system*. Even if the role of the Developer is constantly managing development, he must adapt or match the structure of his organization to the nature or the domain of the Client system as all client systems are different.

6.5.1 Project Breakdown Structure into Packages: Re-Spec Phase

There are many ways to start designing a system. The choice depends upon system complexity, problem domain, and quality of the Requirements Engineering phase, experience, and skill of the design staff relative to the product to be designed. The general guidelines are:

1. Well study the SRS (System Requirements Specifications)
2. *Operate first a horizontal repartition.* If the system contains obvious independent subsystems, break the project into several smaller independent subprojects. Structural diagrams representing a coarse structure of the system to be built, being established at the Requirements Engineering phase, may serve as a staring point. Identify all subprojects by their package names with a package diagram. A series of packages (smaller independent sub projects) can be created at the end of this process.
3. *Operate a vertical decomposition.* When entering each subpackage created at the previous phase, operate a vertical decomposition to find subpackages. It would be important to make a clear difference between a horizontal repartition and a vertical decomposition. A vertical decomposition supposes that the package at the top contain all subdecomposed packages and the suppression of a subcomponent impairs the existence of the top project. In a horizontal decomposition, if a subhorizontal project is interrupted; all remaining projects can be continued normally. Sometimes, due to the tight coupling of horizontal subprojects, it would be difficult to make such a distinction. But, developers must always make such an inventory. Think of *collaboration* for a horizontal repartition and *aggregation/composition* for vertical decomposition.
4. *Repeat horizontal and vertical decomposition again* for complex system to isolate more elementary subpackages. This two-step (horizontal then vertical) decomposition process is conducted until a high degree of coupling between objects/agents/components is detected. Leaf packages reached at this time are *monolithic projects* and contain small subsystems that can be now studied with conventional object methodology.
   (a) It is worth noticing that elements inside packages may interact together via loosely coupled interactions. For instance, in a Car, the *braking system*, the *steering system*, and the *powering system* interact together in a driving activity, but they are

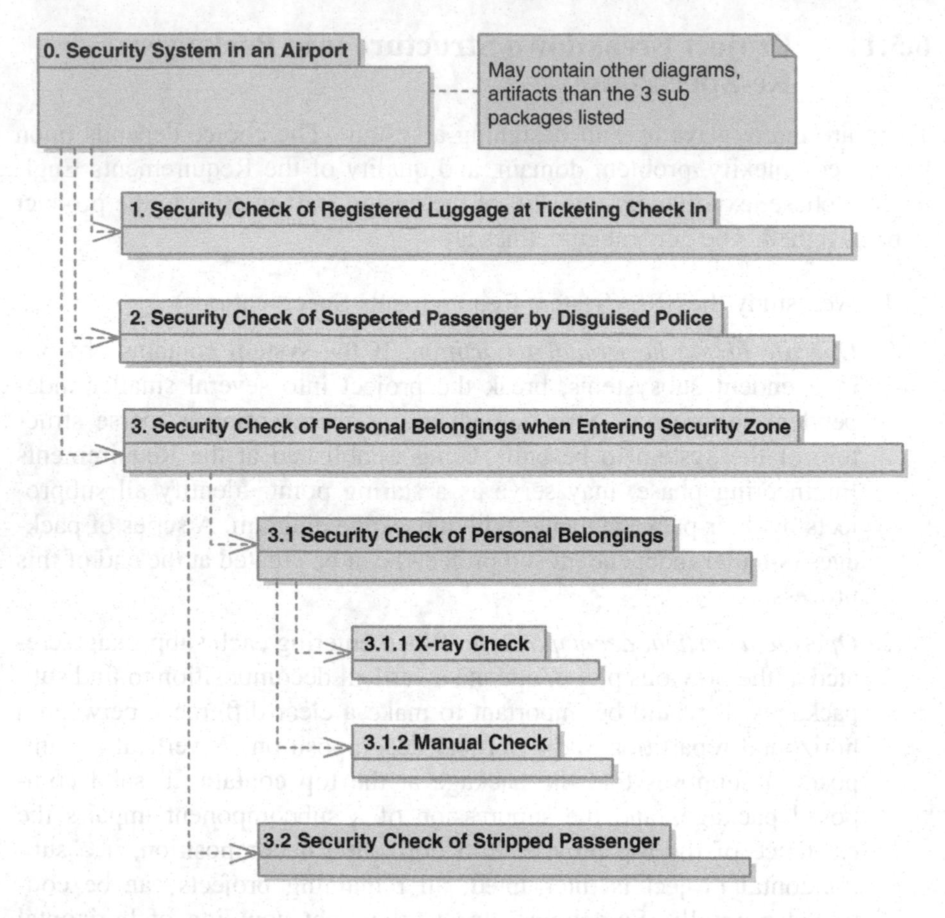

Fig. 6.5.1.1 Decomposition of a Security System in an Airport. The topmost imports three packages 1, 2, and 3. Package 3 is decomposed at second level to 3.1 and 3.2, and 3.1 is decomposed at third level to 3.1.1 and 3.1.2

nevertheless considered as subsystems (packages) that can be studied independently.

(b) The problem of detecting the crucial moment when we must terminate high-level package decomposition is a matter of developer experiences. Leaf package can contain packages for organizing the design process. Leaf package is a kind of development unit (Fig. 6.5.1.1).

5. *Distribute original specifications among packages and respecify if necessary*. Frequently, specifications must be reworked when the original system has been broken down into packages. Each package can be considered as a *work unit* for an individual or a group of developers for a while. A design respecification may be needed to have a better view of

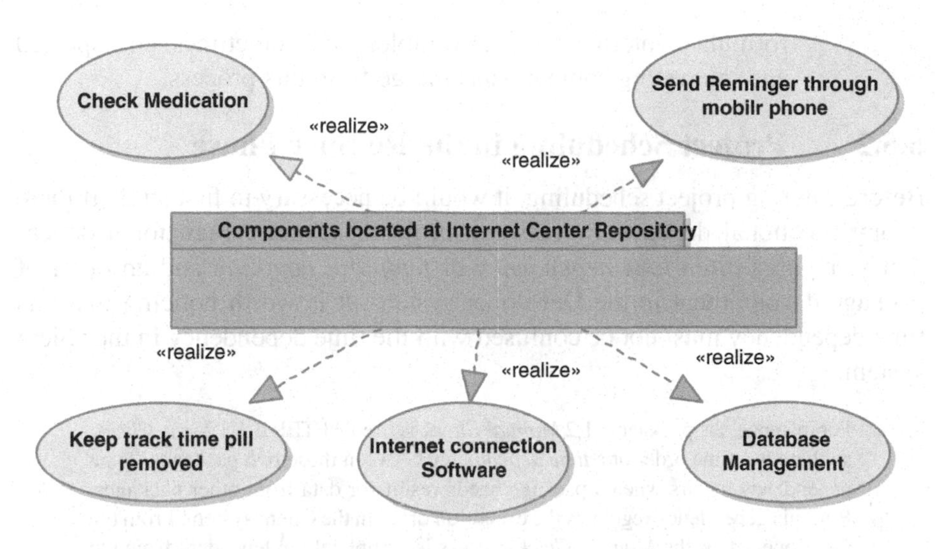

Fig. 6.5.1.2 Re-specs with use cases and packages illustrated via the RMSHC example

the initial system, the responsibilities of all subpackages, and a clarification of objectives and contents of the work unit. This respecification facilitates the design and the communication of participating development staffs or individuals. Metrics and evaluation criteria can be defined if necessary. This re-specs phase reuse SRS with additional use cases with *realize* relationship stressing the responsibilities of the current package (see Fig. 6.5.1.2) and its relationships with other packages. The repartition of packages can let people think erroneously that they are independent, but generally, we can have a tight coupling between packages.

(a) In Figure 6.5.1.1, package 3.1 *Security Check of Personal Belongings* must contain some diagrams that link packages 3.1.1 *X-ray Check* and 3.1.2 *Manual Check* together. For instance, a diagram may be used and stored in 3.1 node to show that *Manual Check* is used only when the agent making *X-ray Check* is not confident with what he has observed on the screen. At the beginning of the project break down, the exact connection cannot be known until the two component packages are thoroughly studied. Dynamic diagrams that link packages together are generally built when leaf packages are completely designed and junction points determined.

(b) This Project Breakdown and Re-Specification phase (Design Specifications by opposition to Requirements Specification) is important as it divides the project (if the nature of the system allows such a division) into several manageable subphases. Re-specs may formulate intermediate deliverables and sometimes unsuspected and interesting components emerge from this process.

6.5.2 Project Scheduling in the Re-Spec Phase

Before entering project scheduling, it would be necessary to first study if there is any functional dependency between project packages. Functional dependency imposes often *time dependency* or *temporal coupling* and an order of package development in the Developer system. It is worth noticing that this time dependency must not be confused with the time dependency in the Client system.

For instance, the package 3.1.2 *Manual Check* is fired AFTER 3.1.1 *X-ray Check*, so there is a time order or a *time dependency* between those two packages. Time dependency occurs when a package needs results or data from other packages. But, this dependency regulates the execution order in the Client system. From the Developer view, the *Manual Check* process is completely independent from the X-*ray Check*, so those packages can be studied in any order.

Project dependency occurs, for instance, in the construction domain. We cannot build anything when the foundations are not in place. Painters would be the last equipment to be hired.

Project scheduling is simple if the Developer side has only a single project. A high number of companies have an organizational structure that can accept multiple projects to run simultaneously. So, scheduling is a complex optimization exercise of matching this organizational structure (with existing projects running) with the schedule of a new Client project. Multiproject management (Speranza and Vercelis, 1993; Silver et al., 1998; Jin and Levitt, 1996) must deal with complexity and uncertainty. The breakdown of the Client project into more manageable parts plays in this context an essential role for workflow optimization. For a survey of this project scheduling please consult Herroelen et al. (1998).

6.5.3 Starting a Real-Time System with the UML Sequence Diagram

There are many possibilities to start designing a system. The choice of the starting diagram depends upon the domain, the complexity, the experience of the developer, and his biased inclination towards using a particular UML diagram. When developing database schemas, developers start almost with class diagrams. For a real-time system, if the developer knows the application domain, he can start with a class diagram directly, but, if he is a neophyte in a new domain or has never developed any similar project, it would be more appropriate to start with a sequence diagram, an interaction overview diagram (an activity diagram that packs fragments of sequences

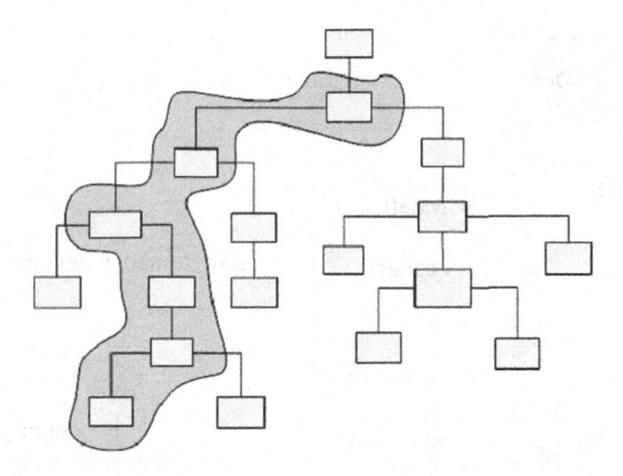

Fig. 6.5.3.1 Example of call graph showing all execution paths in a program. A sequence of execution or *scenario* is represented as a path through this graph. The whole graph packs an algorithm and contains all possible scenarios. Each node represents an operation X of an object Y or a collaborative task of multiple objects

inside nodes), or an activity diagram. Hereafter are some rationales about this suggestion.

The graph of Figure 6.5.3.1 represents an example of *call stack* of a single program without any parallelism consideration. When we start building a system containing many subsystems, all algorithms of all subsystems are to be discovered. An algorithm packs all possible execution paths of a subsystem into a call graph. If the whole system supports some parallelism, we have more than one algorithm (more than one call graph) running at a time. If they are all independent, there is no communication between graphs and therefore, there is no data dependency or precedence rules. If algorithms are not independent, a synchronization algorithm must be designed to coordinate dependent graphs.

The image of the algorithm can be displayed in an *execution graph*. The *call graph* is an implementation image of the execution graph. The execution state of a system can be viewed as a marker that jumps from node to node in this structure. Each node of this graph is an *operation X defined in an object Y* (instantiated from its corresponding class) or a *collaborative task involving many identified objects*. When starting the execution of an algorithm, we start at the top of the graph. According to run-time conditions, one path is chosen at any branch derivation. When reaching a leaf node, the lowest and most elementary operation is executed. When this operation is finished, the graph backtracks to its last higher level, and so on, but not to the highest root where program termination condition occurs. The marker will stay somewhere in the graph waiting for other run-time conditions that allow other paths of the graph to be explored.

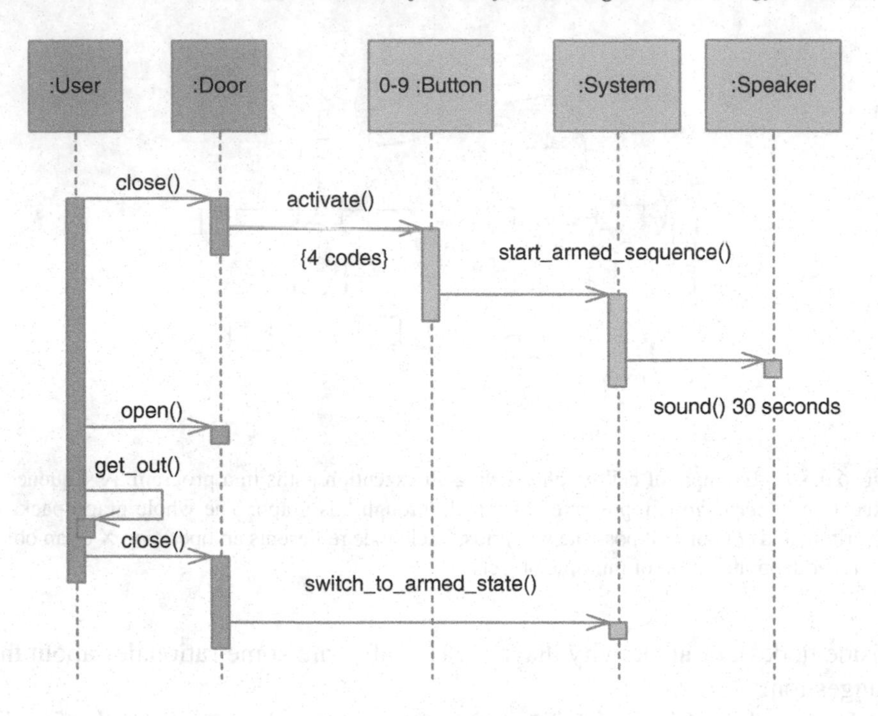

Fig. 6.5.3.2 User's Instruction "How to arm a home alarm system" expressed by a sequence diagram

A *scenario* or a *sequence* is a particular path in a graph (or a subgraph) extracted at any two levels, not necessarily at the root of the graph. Each application defines some typical scenarios and explains them as *Users' instructions.*

> When using a mobile phone, the instructions we must learn are "how to power on, how to power off, how to change batteries, how to charge, how to make a call, how to answer a call."

Therefore, when starting a real-time system study, the same Users' instructions sequences can be used to identify objects (with their corresponding classes) and messages (object operations). User's instructions stop all interactions at the system interface and consider the system as a monolithic object. Design sequences penetrate inside the boundary of the system and try to find out all internal objects. Figure 6.5.3.2 is a sequence diagram that explain to a User (User's instruction sequence) how to arm the Alarm. Figure 6.5.3.4 is the same sequence at the interface, but it goes inside the system and identifies internal objects (Developer sequence). The two sequences are identical if we stop at the interface of the system, but the Developer sequence explores the system internals to find out all constituting objects. By adding sequence over sequence, all objects are identified and a class diagram can be built at

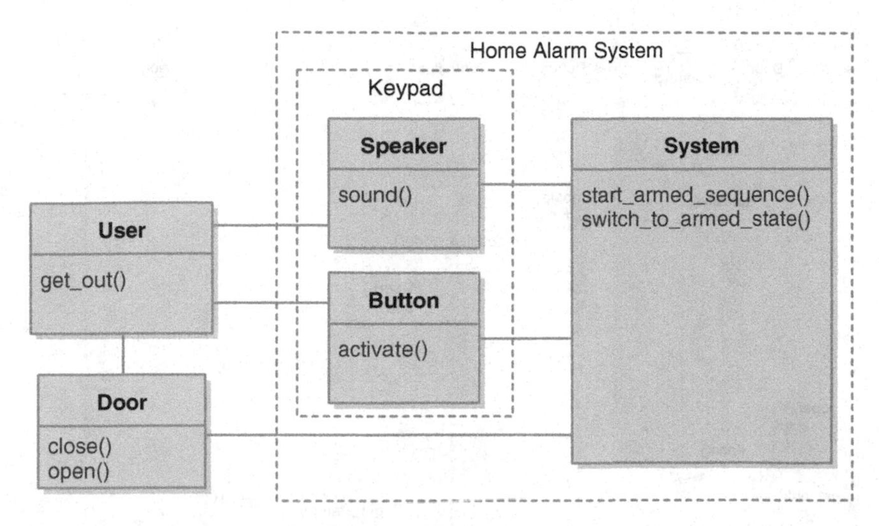

Fig. 6.5.3.3 Corresponding classes identified while drawing the sequence diagram

the same time to capture all operations (messages in the sequence diagram) (Fig. 6.5.3.3).

The number of sequences in a complex system is very large and sequence diagrams are time consuming to draw. They are good for capturing the attention of an audience, but not very handy for exploring the algorithm of a complex system. So, this diagram is handy for presentation, for starting a system investigation of its first elements and interactions, but once main components of the system are identified, other diagrams (state machine or activity diagrams) are more appropriate for finishing the algorithm. In other words, sequence diagrams are powerful for expressing the chronology of interactions between several objects in an expressive form but are not really targeted to find out systematically all interactions of a system.

The sequence diagram of Figure 6.5.3.2 is focused on the User's interaction at the interface level. The Alarm is seen as a monolithic system with high-level operations. Only the keypad is visible to the User and the sound is perceived through the Speaker integrated in the keypad assembly. *0–9:Button* is a shorthand notation that packs 10 buttons 0–9 of the class Button. The User acts on the Door and the Door, in turn, acts on the System (The design phase will reveal the existence of an intermediate object that is the microswitch in the door frame). Other sequences will be necessary to explain other functions (disarming, detecting intrusions, changing code, etc.).

The sequence diagram of Figure 6.5.3.4 starts as the sequence of Figure 6.5.3.2 but crosses over the frontier of the interface to explore internal components. This fact shows the difference between a sequence established at the Requirement specification and a sequence of the Design phase.

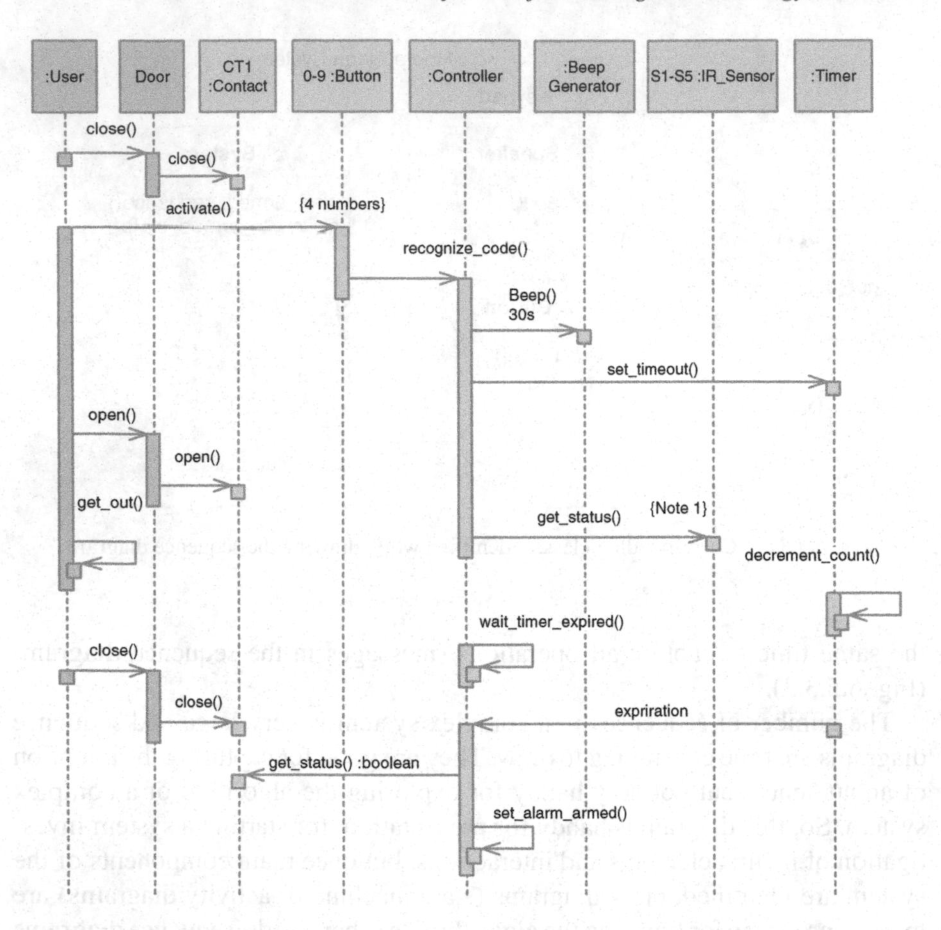

Fig. 6.5.3.4 Sequence diagram for developing the Home Alarm System. The frontier of the System is crossed over to explore internal components. Successive decompositions are necessary to identify more elementary objects. (Note1: If Infrared Sensor is activated, the alarm cannot be armed since there is somebody inside. But this sequence is not represented)

6.5.4 What Does a Logical Model in Real-Time Applications Mean?

Sequences used to depict objects in design phase are nearly identical to user functions but they cross the system frontier to find out all objects necessary to design the system. Real-time models deal directly with buttons, switches, sensors, controllers, etc. so apparently, it is somewhat questionable what a real-time logical model really means? To apply the MDA concept, the logical design must be considered as a *high level and logical model*. To reach these objectives, it would be necessary to understand specific characteristics of a logical model and to observe the following recommendations when elaborating a logical real-time model:

1. *All objects are made independent at the logical phase* to isolate the roles performed by each object in the system. Timing constraints specified at the logical phase are those needed for the operation of the object itself, leaving aside all other constraints proper to a specific implementation.

   (a) The Timer operates independently from the Controller in this logical model, even if, at the implementation phase, the Controller and the Timer are mapped into a unique microcontroller. The Beep Generator can be implemented by a subroutine of the microcontroller at the implementation phase as well, but at this logical phase, its functionalities are separated to evidence its role.

   (b) Microcontrollers (microprocessors equipped with parallel, serial interfaces, timers and a small amount of RAM, programmable ROM memory) are used at the implementation phase as intelligent devices. To minimize system cost, generally, they are used as all purpose devices. In this respect, a scheduling problem is superposed to the proper timing of the system defined initially. For instance, if we make use of the integrated timer of the microcontroller both for the Timer function, the Beep Generator, and real-time interaction with users at the keypad interface. As all these functions are activated simultaneously, there is an interference of the Beep Generator timing with the timing of the Timer, and with the keypad busy wait loop. At this logical phase of development, this fact must be ignored and all functional devices are considered as independent. The scheduling problem is a design problem at the implementation phase but not at the logical phase.

2. *Even if a real-time system makes use of real world components, their functionality is made generic at this logical phase of design*. Primary functions of devices like controller, sensor, switch, button, display, motor, actuator, sensors, etc. are well known, so a library of classes containing such logical components are very useful and is recommended to build a library of generic components. If more specific devices are needed, they can be derived from generic devices.

   (a) An actuator is a special mechanism designed to act upon an environment. A motor associated with a harmonic reducer (high ratio hollow shaft gear reducer) is an actuator used to move an arm of a robot relative to another mechanical part. Actuators can be of hydraulic or pneumatic types. So their classes must be defined individually to take into account their specific features. Actuators and sensors are complementary devices in real-time applications. Computers, microprocessors, or microcontrollers are intelligent controllers in this context.

3. *At logical level, the model often follows the natural order of events and its chronology. The way real systems implement the capture of events is not of concern at the logical phase, so we do not have to reproduce exactly the implementation arrangement.* In a real-time system, we have to manage passive components like a switch, a button, and a passive sensor with an intelligent element as a Controller. There are two methods for repeatedly checking whether an event occurs: a busy wait loop or an interrupt. In a busy wait, a loop is created to scan all passive devices. From a modeling viewpoint, the busy wait loop reads the status of the passive component (switch or button contact), hence the arrow is oriented from the active component to the passive component requesting its status. It is an acceptable solution if the Controller is idle most of the time. As an alternative, an interrupt signal can be sent to a Controller requiring it to process an *interrupt handler*. A real interrupt layout needs hardware to generate electronic signals. With the interrupt sending a signal to the Controller, the arrow is reversed and oriented from the passive towards the active components. Figure 6.5.4.1 shows how to represent these two different cases.

4. *When modeling components that are highly implementation dependent, we can consider them as black box components. All internal operations*

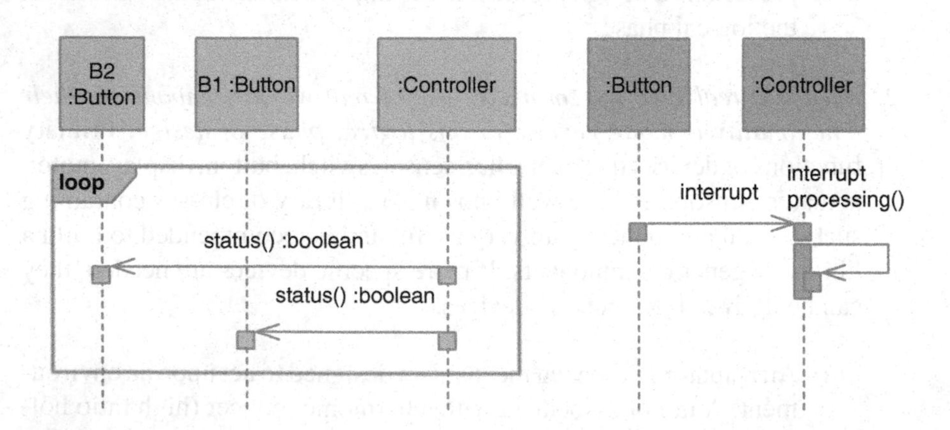

Fig. 6.5.4.1 The status of a passive component in a real-time system can be detected by reading the status of the component with a busy wait loop or mounting appropriate hardware to send an interrupting signal to the Controller (arrow direction must be reversed). When making a logical model, this decision can be postponed to the implementation phase, so the two representations are considered as equivalent at the logical phase

are brought to the interface and self messages abstracting actual internal structures can be used to postpone their development to the implementation phase. Self messages are those that are redirected to the same object (User getting out of the house, Timer decrementing its count, controller setting its alarm armed in Figure 6.5.3.3). In the execution of a self message, it appears as if the operation is self sufficient and does not need any collaboration of other objects to perform the self message. But, more frequently, the presence of several self messages is symptomatic of a system not insufficiently broken down or not decomposed at all. At the limit, an active system without any interaction with its environment has only self messages. Self messages can be very useful in the context of logical model development as they can be used to postpone implementation-dependent components to next steps.

5. *A logical model must be compatible with implementation that can be derived later from this first design model*. The logical model is the first piece of reuse. The realization of previous points will contribute and reinforce this goal, guarantee that no knowledge is lost, and proven practices can be reused in a regular and systematic manner. Design quality and productivity are enhanced accordingly. It is highly desirable to have a smooth transition between PIM and PSM. The term "smooth" means that classes are imported, derived, and reworked. The transition between SRS calls upon a "mapping" process. A mapping is generally understood as a nonsmooth process.

6.5.5 Starting Real-time Systems with the UML Interaction Overview, Activity, or State Machine Diagrams

It is not mandatory to start a system with a sequence diagram as shown in the preceding section; the UML has in its arsenal several diagrams that support other starting methods. Despite its high expressiveness, a sequence diagram relates a story in a precise context and, with the way it was designed, it would be difficult to display more than one scenario. Even a long sequence requires some work to organize it into smaller chunks of sequences.

The design of a whole system needs full algorithms with many scenarios fitted together in a complex network ruled by run-time conditions, the last UML 2 standard proposed the interaction overview diagram as a complementary formalism to be used jointly with the sequence diagrams for building a more complex dynamic structure involving many scenarios. Theoretically, by linking all interaction overview diagrams together, we must be able to build the whole algorithm. But, it is widely recognized that a sequence diagram is interesting

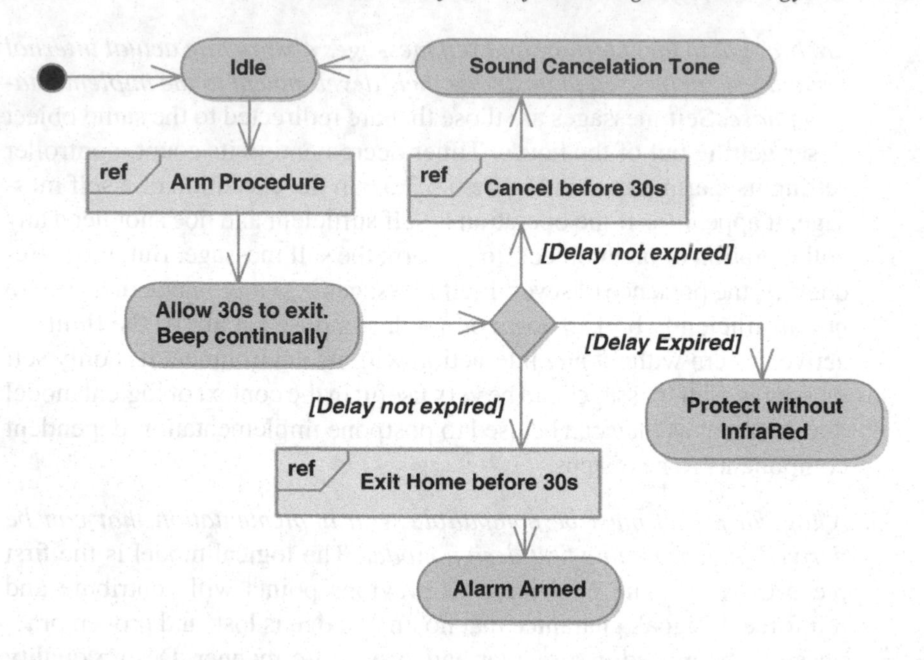

Fig. 6.5.5.1 Interaction Overview Diagram packing many references to sequence diagrams inside an activity diagram. Arrows are control flows and activities are mixed with "ref" fragments that must be detailed elsewhere

to start investigating a system, but we rarely see anyone making use of this formalism to develop a full dynamic study of a medium-size system. Probably, the sequence diagram is very time consuming to establish and needs space to align all objects horizontally. However, it is intuitive and can easily catch the attention of nontechnical persons.

The interaction overview diagram (Fig. 6.5.5.1) is the third diagram of the Interaction Diagram Suite and is a variant of the Activity Diagram (Fig. 6.5.5.2). The focus is put on the flow of control where nodes are Interaction (Unit of observable behavior) or InteractionUse (a reference to Interaction). A sequence is a piece of behavior and simultaneously an interaction so nodes can be mapped into sequences through *ref* fragments.

Figure 6.5.5.3 shows the use of the State machine diagram as a starting method for studying a system. Class and object diagrams can be built simultaneously to support the structural view. To summarize, there are many ways to start designing a system when studying real-time systems:

1. *Use sequence diagrams only* (appropriate only for very small system).
2. *Use sequence diagrams associated with interaction overview diagrams.* Interaction Overview diagrams layout high-level algorithms and sequence diagrams fill nodes referenced with "ref" fragments. Control

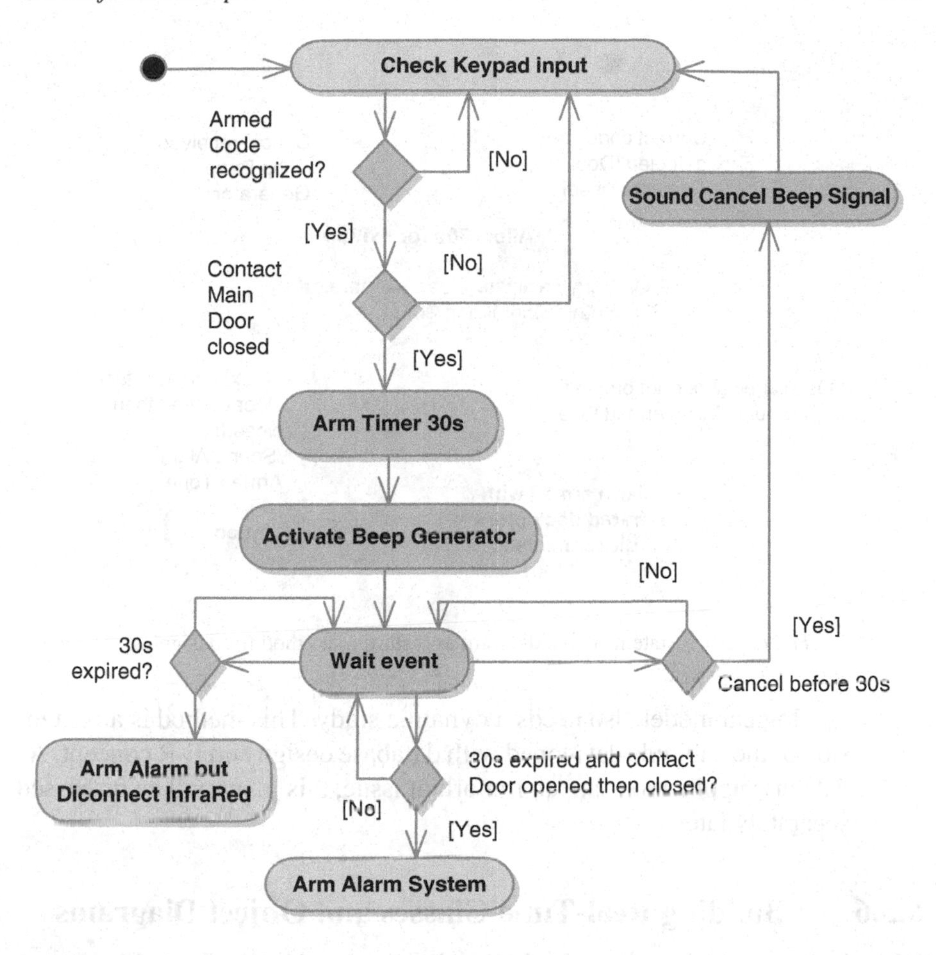

Fig. 6.5.5.2 Activity diagram as a starting method for the design. A class diagram can be built simultaneously to store operations (activities). Each test is also an activity. This diagram is only a chunk of an algorithm and is targeted to show a possible way of starting a design

structures like "alt," "loop" fragments, etc. can be used as well to enhance expressiveness.

3. *Use activity diagrams associated to structural diagrams* (class diagrams to register operations/attributes and object diagrams to list all objects that collaborate inside execution scenarios).
4. *Use state machine diagrams* replacing activity diagrams in the previous method. Class and object diagrams continue to support structural views. Electrical engineers are very fond of state machine diagrams as they are used in the past for studying sequential logic circuits.
5. *Use structural diagrams only*, relaying dynamic studies to the implementation model. This method is not really appropriate for real-time systems

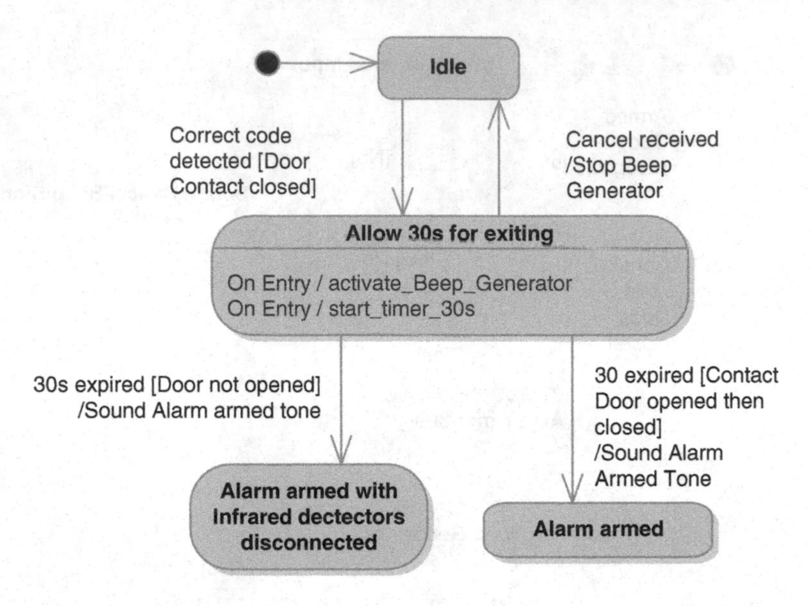

Fig. 6.5.5.3 State machine diagram as a starting method for designing

as a logical model also needs a dynamic study. This method is an extension of the habitude developed with database design and E-R concept. As database applications are an important issue, this point will be discussed separately later.

6.5.6 Building Real-Time Classes and Object Diagrams

Historically, real-time systems are built using low level languages (often assemblers) and are targeted for dedicated hardware and specialized and light weight (operating system) kernels. Efficiency, performance, and other criteria proper to embedded or mobile systems are of primary concern. Software modularity and reuse are secondary factors. Recently, with the complexity and the integration of real-time systems into a more connected world, things have moved towards the design of modular and reusable real-time components. Finding the right abstraction to build extendible and reusable components requires experienced object-oriented designers. From a software viewpoint, reuse was a reality.

> Standard object libraries (C++, C#, or Java) are components, interfaces, algorithms, containers, and data structures that can be used and reused in building object software. To get the current date of the day, we have just to instantiate an object from the class Date belonging to the standard library. Storage structure like B-tree does not need any programming effort.

A common piece of software is merely an art of associating interacting objects mostly derived from standard libraries. If necessary, new objects can be instantiated from existing classes of standard libraries with additional algorithms and operations. If they are catalogued and systematically reused, they constitute effectively company software assets. Inheritance explores similarities and enforces the whole process.

Embedded and multidisciplinary systems throw a more stringent challenge to the industry of embedded components. Hardware and software cannot be dissociated and classes are not limited only to software classes. The description of multidisciplinary component classes is therefore necessary.

> A Vending_Machine class can be designed and specialized later to sell drinking bottles, chocolates, candies, or coffee. Instances of these classes all need coins, product selection function, commands sent to actuators to perform appropriate actions according to targeted sub domains. The challenge would be how to develop the most universal component that can be reused in many contexts, in many domains, roughly over space and time continuums. Only actuators must be dissociated as they must be adapted to the final product sold. Commonalities grouping or gathering are the master concern to construct new classes that describe multidisciplinary components.
>
> Robot articulations and visions systems are nowadays COTS (Components Off The Shelf). They can be purchased separately in the OEM (Original Equipment Manufacturer) market. The application is not defined in the components themselves. It is up to the final integrator to decide upon the application purposes.
>
> Generic components are not limited only to things but may contain humans, agents and organization. For instance, in military applications, tactical intervention units can be seen as well defined multi-agent components with a given composition, defined operations and responsibilities. With internal knowledge and the nature of the missions, they can adapt operational processes to achieve some goals defined only at runtime. The same concept may be seen in Public Security Domain (Fire Intervention Unit, Crisis Intervention Unit), Medical Domain (Reanimation Unit, First Care Unit, etc.) or in Financial Domains (Tax Recovery Patrol).

Generic task models (Brazier et al., 1998) will help considerably deploying system reuse. Diagnosis, analysis, design, and process control, scheduling, dynamic scheduling, etc. are, for instance, current tasks in everyday life. Doubtlessly, experienced designers would be more predisposed to extract common behaviors in a specific domain than newcomers. New domains cannot be mastered immediately. So, the deployment of reuse concepts is the concern of experienced developers, but at the starting point, there must be awareness, conscience, and organization determination and resource devoted for exploring this avenue. We mean resource since, at some period of the reuse process, the organization has the impression that it does not solve customer problems directly but something with very long-term benefits. But, this way of doing things is *highly strategic* and will make the difference between two performers in a domain. The MDA supports this generic step by inserting a first generic design as a starting model or by importing/merging elements of a Common Generic Model (CGM) available for all applications. Therefore, genericity will condition the way we design our classes in the logical model (Fig. 6.5.6.1).

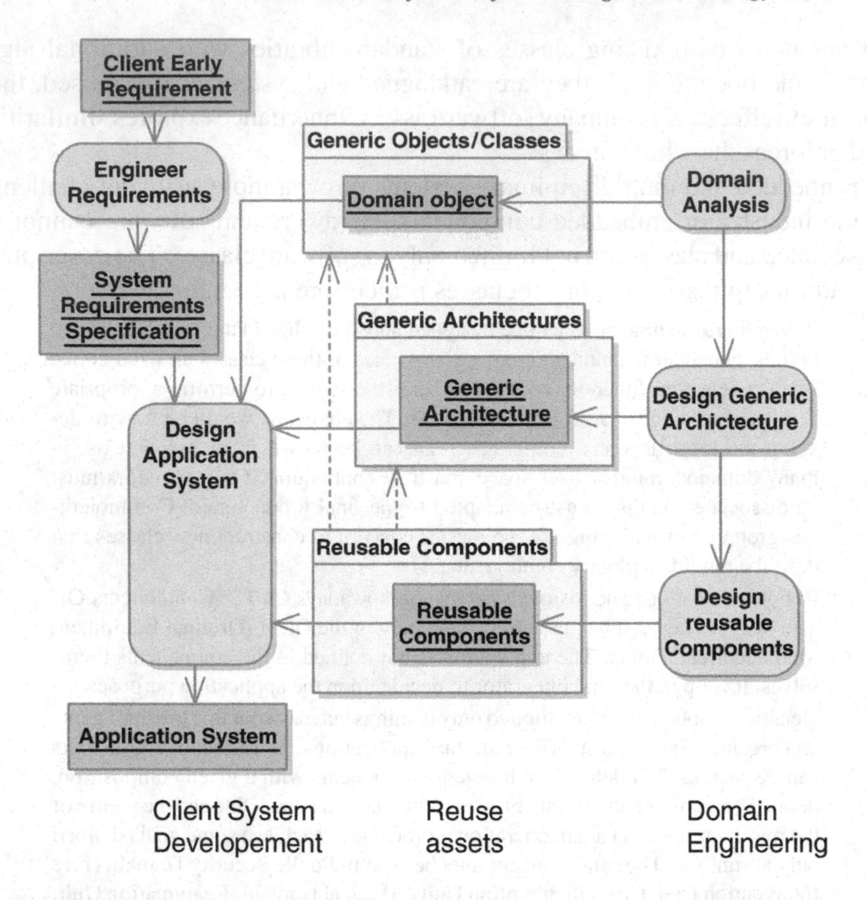

Fig. 6.5.6.1 Domain engineering with two activity flows (Domain Engineering and Client System Development) that may be conducted in parallel or not. Relationships between packages are import/merge. Objects and Components are underlined. Flows connected an activity to an object are object flows. Flows between two activities are control flows. Designing an application consists of choosing domain objects, domain generic architecture, domain components, or COTS, taking into account SRS, then generating final application system

> It is worth noticing that Standard Template Library (STL) represents already a breakthrough in genericity in C++ programming language. Comprising a set of data structures and algorithms, STL provides components available to many apparently different applications. Adopted by the ANSI/ISO C++, STL is an important tool for every C++ programmer.

Problems to be solved in a domain evolve daily with circumstances, with new actors so the dynamics of systems appear each day under a new appearance but the number of objects in a domain is nearly static. The key factor of reuse is thus how to build all *objects/classes of the domain while solving the problem*. Subsequent problems will add new actors, and new operations to the existing system.

This methodology of development will naturally manage reuse by creating two applications:

1. *Domain Engineering of reusable components and architectures*. Objects in this domain have general features (attributes and operations). They are integrated inside general architecture
2. *Client Application*. The client application imports, merges objects of the domain and mixes them with specific objects of the application.

For the first systems, Developer can start a Client application then select what he can keep in the Domain for later use. It is not necessary to have a domain fully stuffed before starting a Client project. A rich domain attests, however, the experience of the Developer in the domain.

In the case of an Alarm System (Fig. 6.5.6.2), a full library contains all current sensors and devices with defined features like Alarm Controller, Backup Battery, Keypad, Siren, Motion Detector (for human), Special Motion Detector (for human having pets), Door/Window Contact, Central Monitoring Station Model (with list of operations), Control Panel Container, Smoke Detector, Glass Break Detector, Panic Button, Pressure Mat (for under rugs), Closed Circuit TV, Alarm Screen, LCD Display, etc.

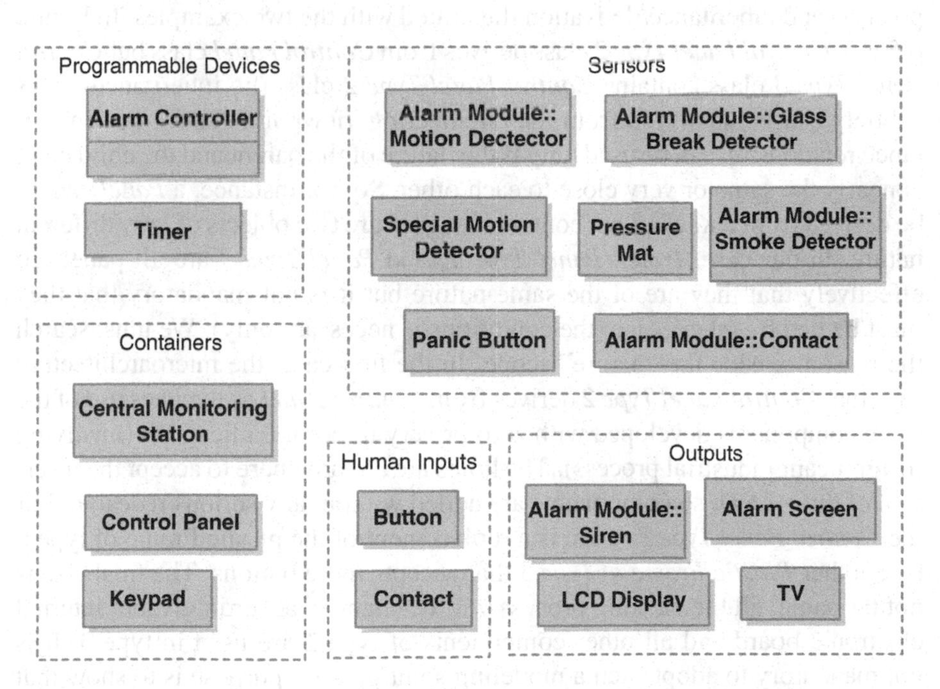

Fig. 6.5.6.2 Classes of Alarm System Domain (Generic Classes package). Some classes may appear more than one time according to the classification used. Classes appearing in this package are nearly not connected together, except some evident inheritance. Panic Button and Button can derive from an abstract Generic Button

Real-time systems are naturally multidisciplinary systems (an *electronic* Controller containing a *software* program, an *electrical* Siren, a *mechanical* Panic Button, a *mechanical container* as a Control Panel, etc. are all objects instantiated from their corresponding classes). They deal directly with real components so they can appear to some, as an implementation model and therefore lose them for some genericity. In this domain, practitioners are not disposed to go back more in genericity for not losing contact with reality. The application is till considered as logical if there is no specific implementation constraint or if the type of constraint imposed at Requirements does not contradict or destroy genericity. At this logical phase, each element of the system is considered as working independently and there is no scheduling to deal with timing constraint.

Figure 6.5.6.3 illustrates what is to be stored in the Generic Architecture package of an Alarm System. It mimics the concept of Library found in the software programming language. The number of entries is unlimited and depends upon the experience of the Developer. It is not necessary to create all generic architectures in the domain to be able to start applications. This library will grow naturally with time and the diversity of applications developed for clients.

The two important relationships used in this package are aggregation/composition and inheritance/derivation illustrated with the two examples. In Figure 6.5.6.3, *Control Panel Type 2* class derives from *Control Panel* class but *Control Panel Type 3* class contains *Control Panel Type 2* class (no inheritance). This subtlety can be answered from the engineering viewpoint. The single inheritance relationship can be used only if the nature of the parent and the child class is nearly the same or very close to each other. So, for instance, a *Panel* cannot be derived from a *Keypad* but contain it as they are two objects of very different nature. In our case, *Panel*, *Panel Type 2*, and *Panel Type 3* are all panels so effectively that they are of the same nature but it is not mandatory that they must be tied by inheritance (the conditions is necessary only). We must search the reason outside the "nature" scope. In the first case, the microarchitecture says that *Control Panel Type 2* derives from *Control Panel* as it makes use of the same components developed with zero or very minor modifications (involving insignificant industrial process). The holes are already there to accept the insertion of three LED; so components are added without any serious redesign. For the *Control Panel Type 3*, there is a replacement of the plastic Frame of type 0 by another *Plastic Frame of Type 3* that accepts more buttons. The final size is not the same, all the packing process will be changed accordingly, but internal electronic board and all other components of type 2 are used in type 3. It is not mandatory to adopt such a modeling solution. Our purpose is to show that Developer can adopt proprietary rules to model the architecture of his system as long as he has logical reasons to justify the solution retained and diagram elements can be easily understood by the whole group.

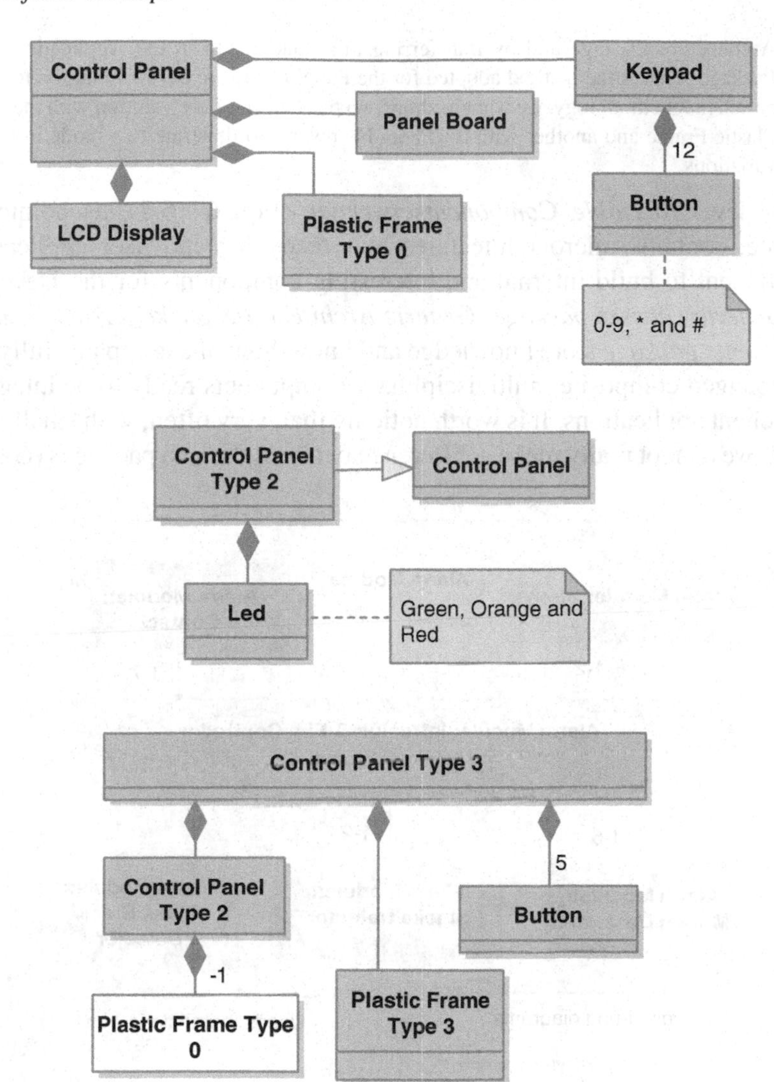

Fig. 6.5.6.3 Contents of Generic Architectures package (Alarm System). A Keypad may be defined originally as containing 10 Button objects. Structural relationships structure classes, but generic architectures remain generic in this package and are not targeted to any particular applications (or at least, they are targeted to support a whole family of applications). Two relationships (aggregation/composition and inheritance) are mainly used to build microarchitectures (see text for comments). In the third case, Plastic Frame Type 0 with multiplicity of –1 must be removed from the Control Panel Type 2 (proprietary notation)

> Please note that our use of a multiplicity -1 to exclude some elements of an assembly is not recognized (for the moment) in the current UML standard. Properties modified or generated with the presence of the subtracted element must be suppressed. We can avoid the presence of the aggregation/composition with multiplicity -1 by suppressing Plastic Frame Type 0 in the Control Panel in Generic

Architecture Package and by transferring this relationship in the Application Package. That is the method adopted for the Panel Firmware but, voluntarily, to enrich the methodology, we want to show two possibilities (one solution with the Plastic Frame and another with the Panel Firmware) to illustrate two modeling variations.

A third level *Reusable Components package* (Fig. 6.5.6.4), as pointed by its name, contains microarchitectures, not ready for end user applications, but sufficient to build internal and interesting components for the Developer side. *Generic Classes package*, *Generic Architectures package*, and *Reusable Components package* store knowledge and know-how of a company, fully contain debugged composite multidisciplinary components ready to be integrated inside client applications. It is worth noticing that, very often, with small applications, we cannot really make a difference between the two packages (*Generic*

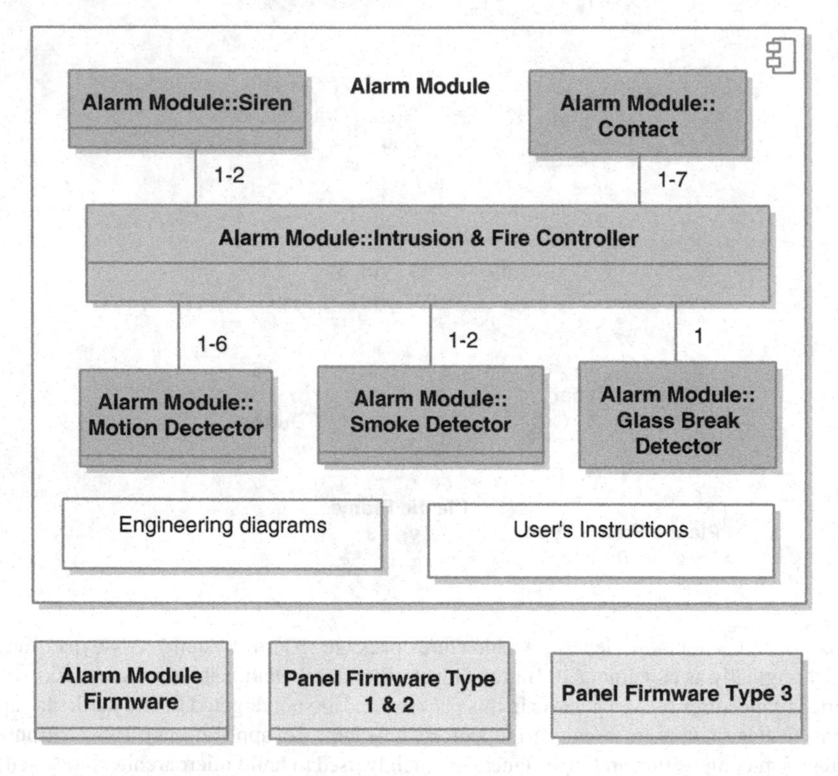

Fig. 6.5.6.4 Contents of Reusable Components package (Alarm System). This package contains monolithic components with well-defined interfaces, and well-specified characteristics, artifacts, and documentation. Component classes can be derived as objects/components in any user's application. Classes are connected together by associations whose real meaning can be understood only when examining Engineering Diagrams (artifacts). Each component is an elementary system defined at a smaller scale than the Client application, but its internal structure and behavior are well known. The last three classes "Firmware" give all firmware compatible with the component

Architectures and *Reusable components*). From the development viewpoint, *Generic Architectures package* contains simply elementary *constructs* but not *components*. The term components implies monolithic constructs with well-defined interfaces, well-specified characteristics, and well-documented OEM pieces that can be easily integrated inside real application. Moreover, components can be sold often "as is" in the OEM market. So, in the presence of some confusion, try to answer "whether elements of Reusable Components package can be or not easily converted into COTS." If the company decides to change its policy and sell COTS instead of addressing to the end user market, this package will be of topmost importance.

This hypothetic example illustrates the long process of designing applications based on reuse at several levels (Classes, Generic Micro-Architectures, and Components). The contribution of original development for any new application is reduced to its minimum. The maximum reuse portion is null for the first time, and will be reached after several developments. A 100% reuse occurs when a Customer buys a standard product developed for another customer in the past. It is not always evident that client applications can really make use of reusable components defined in the corresponding packages. For an unusual client application, it is possible that the Developer must redefine entirely new classes, generic architectures, and builds new components (Fig. 6.5.6.5).

6.5.7 Completing Dynamic Studies with UML State Machine and Activity Diagrams

One of the objectives of the MDA is the possibility of verifying early in the logical phase nonplatform-dependent algorithms that govern the way objects behave individually and in collaboration with other objects in the execution of a given task. Each task has its proper set of participating actors and each actor participates with a subset of operations.

> In a system like the Home Alarm, tasks are for instance Idle (alarm disconnected, wait for rearming), Monitoring with infrared sensors, Monitoring without infrared sensors (alarm armed, watch various sensors).
>
> With the first task *Idle*, the Keypad is monitored to wait for the rearming code, and the main door switch acts as a condition for rearming. Other alarm sensors do not participate in this task.
>
> With the second task *Monitoring with infrared sensors*, the Keypad wait for command whereas the Alarm Module monitor all sensors when the occupants are out of the protected volume. All objects participate in this collaboration. The third task *Monitoring without infrared sensors* excludes all infrared sensor objects.

Before digging into the way we study the behavior of systems, let us synthesize some results scattered in the preceding chapters.

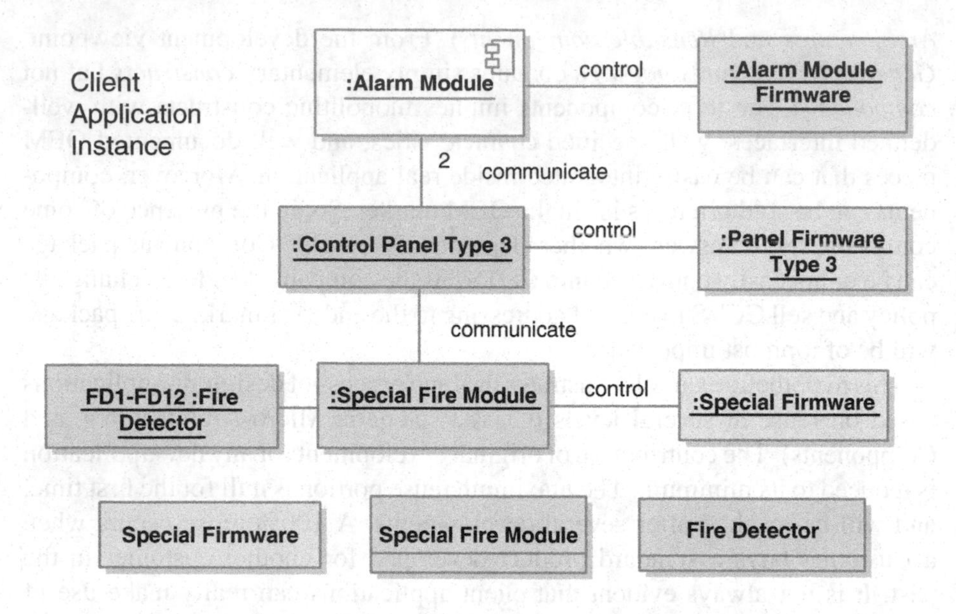

Fig. 6.5.6.5 Instance of Customer application (Alarm System). A Customer application contains objects instead of classes. They combined objects instantiated from various packages (Reusable Components, Generic Architectures, Generic Classes) taking into account Customer Requirements as inputs. This diagram illustrates the reuse of a Control Panel Type 3 to control two Alarm Modules and the engineering of a Special Fire Module specifically for a Customer. This special module is developed with special firmware and will be redirected to Generic Classes package for reuse

1. *State and activity are close concepts.* As stated in fundamental concepts, activities and states are very close concepts, state machine diagrams, or activity diagrams that can be used to study the behavior of objects. It is not uncommon to mix them.

2. *Dynamic behavior of a simple and passive object is often trivial.* Some objects are trivial so they do not need any dynamic investigation. For instance, the dynamic behavior of a switch or a button contains only *on()* and *off()* operation and *status* as attributes is not worth a study.

   (a) A passive object was defined in Fundamental Concepts as an object that does not require service from other objects and does not take any initiative to modify its environment. Passive objects can be alive objects; in this case, they may have a very rich behavior or state evolution. So the rule is not always true.

3. *Complex objects cannot be studied dynamically*. If some objects have a trivial behavior, complex objects cannot be studied dynamically if they are not decomposed correctly. So, *dynamic studies can be conducted on sufficiently elementary objects that are not trivial*. Algorithms can be built on those objects then assembled or combined to get the algorithms of complex objects. Dynamic studies are both top-down (to identify tasks and to find out all constituting elements of a system) but the process of building algorithms, therefore object dynamic behaviors is bottom up.
   (a) In the Alarm System, if we say that the Controller of the Control Panel will undergo three states Arming process, Wait for arming delay expired, and Disarming Process, this assertion cannot help us build the firmware as the system is still considered at its high level. A more thorough decomposition, highlighting the main door contact, the programming buttons 0–9, the sound beeper, and the LED lights will dictate in what order commands must be entered into the system. The resulting algorithm can be used to make the firmware of the Control Panel.
4. *The state of a whole system is often considered as the resulting states of all of its collaborative objects and semantics can be fuzzy*. As stated in the Fundamental Concepts chapter, the state of a whole system cannot always describe the states of all of its components. Sometimes, a synthetic term is hard to find for characterizing the state of a whole from those of its parts.
   (a) A teacher is exposing his subject in a classroom. Saying that "the attention is moderate" cannot make any people figure out that only half of the class is listening, the rest are either somnolent, working, or gaming on their laptops.

The process of studying the dynamic behavior of a system can be split into several steps:

1. *First, identify individual tasks, simple tasks, or activities*.
   (a) Individual tasks are those executed by a single object by opposition to a collaborative task.
   (b) A simple task is a task executed with the collaboration of passive objects only.
   (c) A task or an activity is not really an object operation although at low level a task is implemented by an object operation or method. An elementary operation is tied to the nature of an object, what it can do, but an operation is not goal oriented. For instance, a controller can

occasionally make a conversion because it is an intelligent object and has internal resource to realize this conversion but this operation is not considered as a task or an activity. If a controller scans buttons to detect a command, it is executing a simple task.

2. *Do not consider parallelism at the beginning of the dynamic study.* If several tasks must executed in parallel in a real context, it is imperative to first study characteristics of each task separately. If a given object system requires simple forms of parallelism, this constraint must be inherent to the nature of the logical task but not a precocious temptation towards implementation. For instance, when activating Ctrl-Alt-Del with the keyboard, we send a command with three simultaneous actions.
3. *With an object diagram, show all objects that participate to the task execution.* Classes are used essentially in real-time systems to store attributes (structural, functional, and dynamic). The use of associations in class diagrams are not recommended since the object diagram (or a communication) is more expressive to represent collaboration scenarios for executing identified tasks. Objects will be connected with links (equivalent to associations in a class diagram).
   (a) If a microwave oven has several buttons Start, Stop, Clock, Cook, 0–9 etc. a class diagram will represent only a class Controller connected to a single class Button, so the semantics is unclear. In an object diagram, all objects are represented with their links to the Controller.
4. *At each phase of a task, an object contributes to the task with its proper operation and its state will evolve through the execution of this operation.* If the algorithm is complex, it happens that an object may contribute to the global task with more than one operation. In this case, it is worth considering the several phases in the participation of this object, each phase will be studied separately with its own dynamic diagram.
   (a) For instance, in the process of arming and disarming an alarm, the Controller of the Control Panel will be solicited with three operations *Arming process, Wait for arming delay expired*, and *Disarming Process*. It is worth considering the three separate dynamic studies as they call for three separate operations.

Figure 6.5.7.1 shows the state machine diagram of an arming process that brings an alarm system from the *unarmed state* to the *armed state* (the reverse process is not represented). The state displayed must be the state of a unique object,

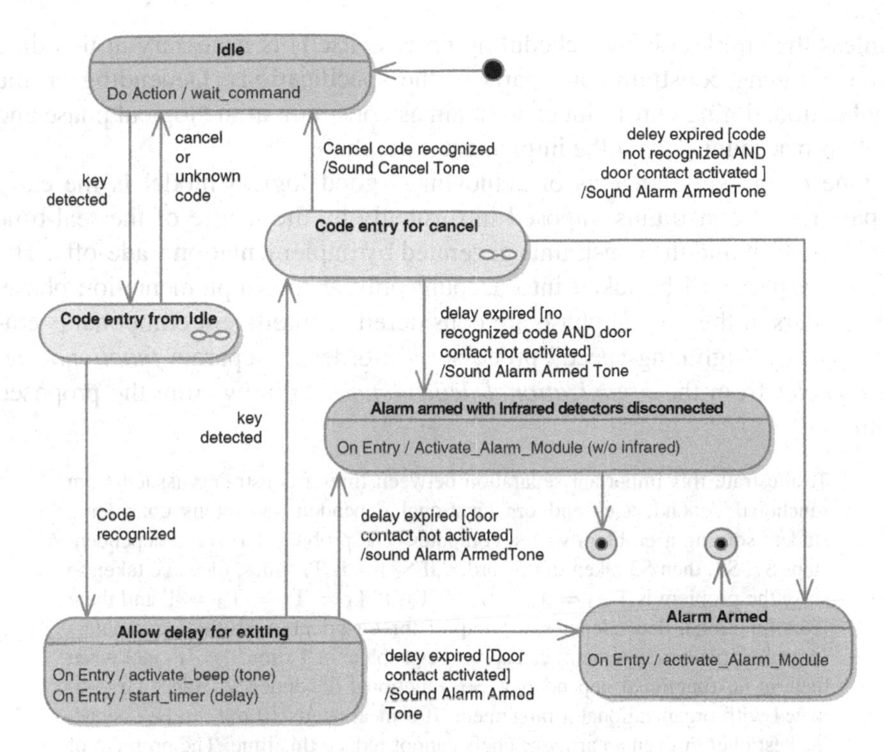

Fig. 6.5.7.1 State machine diagram of the Panel Controller showing activities accompanying thc arming process. This state machine diagram supposes a better decomposition of the alarm system and therefore, it is more detailed than the state machine diagram of Figure 6.5.5.3 established at the start of the design process. Code Entry states are composite states and must be decomposed in other diagrams. From Alarm Armed states, we can come back to Idle state with Alarm disarmed, but corresponding transitions and states are not representcd

in our case, the Panel Controller that is the intelligent element of the system. As said before, the way passive components (button, contact, etc.) interact with active components (controller, etc.) is not specified (polling or interruption) at this logical phase. Events sent from the passive to the active components are a valid approach with the principle of induced energy exposed in Section 6.1.3. When the owner acts on a button, he transmits induced energy to the button that closes the contact and sends an event to the controller.

States displayed in Figure 6.5.7.1 are states of the *Control Panel Controller*. Another controller in the *Alarm Module* is responsible for the monitoring of all sensors. The *Alarm Module Controller* must be studied separately. The two systems are considered as independent even if at the implementation step, we merge the two controllers into one unique active device. This example illustrates the first logical model built directly from specification. The design breaks the system into its utmost independent functions and assigns each important and intelligent task to a separate and independent "virtual" controller. No scheduling (unless the problem is the scheduling process itself) is necessary at this time even if timing constraints are parts of the specifications. Depending on the application, timing constraint can remain as constraint at the logical phase and took into account only at the implementation phase.

One of the key concepts of achieving a good logical model is the early separation of constraints imposed intrinsically by the nature of the real-time problem itself and the constraints generated by implementation trade-offs. The second aspect will be taken into account only at the implementation phase. Controllers in the logical phase are considered as intelligent components empowered by "lightning speed" processors in order to separate *functional dependencies* from the *organizational dependencies* coming from the proposed solution.

> To illustrate this important separation between timing constraints issued from functional dependencies and organizational dependencies, let us consider a student solving a problem with a computer. The problem has three dependent steps S1, S2, then S3 taken in this order. If S_i needs T_i times, the time taken to solve the problem is $T_{123} = T_1 + T_2 + T_3$. If $T_1 = T_2 = T_3 = T$ and there is no functional dependencies, a group of three students with three computers (organizational arrangement) can solve the problem in T time $T = T_{123}/3$ when there is no functional dependency. So, functional dependency cannot be optimized with organizational arrangement. If a meal needs 10 min. to be cooked, the best chef or even an army of chefs cannot reduce this time. The problem of the logical phase is to determine that the meal needs really 10 min. to be cooked and to demonstrate that it is the fastest delay.

6.5.8 Assigning Task Responsibility

A task requires some work, some time or can bring a small change/move of the microworld built around a given system. A task can trigger another task. The notion of unit task is dependent on the way a system is decomposed and approached. A unit task or "elementary task" will be affected to the most elementary object/agent/component identified in this system so the concept of "element" is very elastic; it varies with the context and concern.

Facing task responsibility, we have some specific recurrent problem with object modeling. First, when a task is elementary and reach sufficient low level (implementation level), it is executed by a single object and become an operation or method of the object, sometimes; it is difficult to identify the goal of the task, particularly when this operation is defined only to support a collaborative task (for instance, if an LED is flashing, without a high-level identification, it is difficult to know why it is flashing). In the opposite direction, if we can identify easily the object/agent/component responsible for elementary or individual tasks, the task analysis of high level and collaborative tasks is more complicated and may be affected by many factors:

1. *The responsibility is collective.* As the task is of collaborative nature, the part of responsibility of each participating object/agent/ component is modulated by some criteria retained for the task (execution time, frequency of implication, who is the task inventor, who finances the task, who is nominated or elected to be the task organizer, etc.) Reasons are not only technical but can bear some social, political, or human facets. If the gaming rule is well defined, they act as constraints and technically there is no difficulty of assigning tasks to objects/agents/components. The problem is more stringent when there is no clear rule or when some reasons enumerated in the next points impair the analysis process.
2. *An unclear task analysis pleads for a fuzzy responsibility assignment.* When defining a task, the decomposition is not conducted to reach individual and elementary object/agent/component, so the task is ill defined at the lowest level so it happens as if, from the development viewpoint, a flaw in the way the system is designed and decomposed. People have a very limited view on the way the task must be executed.
3. *Inefficient or inexistent dynamic study.* Tasks may be defined but the algorithms are not established. Objects/agents/components may execute tasks in a contradictory, unplanned order at run-time or worse, some people are doing things that they are not allowed to do or things that are out of their roles in the organization, etc. In the latter case, we have both a dynamic and structural flaw while designing the system. Design errors impact on responsibility attributed to objects that participate into the collaboration.

Our purpose is to point out that technically, we must build entities like a group, a bureau, a committee, etc. that constitute role aggregates to support a task or an operation. Those aggregates will have complex tasks as operations. They contain roles with a subdivision for responsibilities. In some circumstances, it may happen that individual object/agent/component cannot entirely take the responsibility of a given task even if it is the instigator, the trigger, even sometimes the author if the task calls for a collaboration of numerous stakeholders at run-time and worse, there is no clear definition of the notion of responsibility in the system. Political and social systems are generally subject to poorly designed criteria, technical products are easier to handle from this viewpoint.

> Tasks are tied to structures in object technology as we must find a class to host any task. If a task is itself so important, for instance *fight_terrorism()*, a full featured class can be built to host just this operation. A thorough decomposition will identify all subtasks needed to execute this high-level operation and all organizations inside a given country that take part in the responsibility of conducting this process. For continuation, read points 1, 2, and 3.

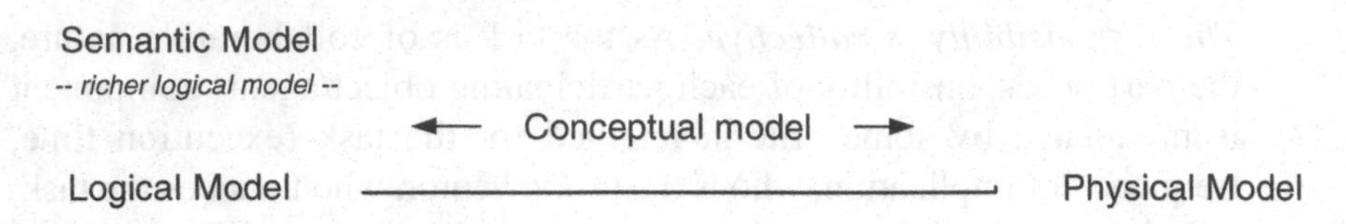

Fig. 6.6.0.1 Physical, logical, conceptual, and semantic models on a user perception axis of low or high level of data representation

6.6 Designing Reusable Database Models

Our goal is not explaining how to transform, with object technology, conceptual models currently used in the domain of relational database. The literature in the computer science domain is rather tremendous about modeling databases, relational or object, with entities-relationships or UML diagrams (Fig. 6.6.0.1). Commonly, a *logical model* of data exposes the logical view of data organization without any reference to performance or storage issues. The *physical model* is platform dependent, it shows access paths, files, indexes, and storage organization. Between these two poles, we find conceptual models that can fluctuate between these two poles according to its "conceptualized" degree. Too much optimization storage constraints will draw the conceptual model towards the physical side. Conceptual models provide the mapping from the logical to the physical models. A *semantic model* provides more semantic content (by authorizing relationships like inheritance and aggregation) than a common logical model. So, semantic models appear as richer logical models. It can cover the conceptual zone partly.

Object or relational explains the way data are globally organized. Roughly, relational means a spreadsheet organization with multiple tables and object means a graph-based representation of data. Some modern databases mix both object and relational organizations. Both the models are useful. If we make a description of a car, it is more natural to use a graph-based description with the whole car at the root of a tree structure. Leaves are made of most elementary mechanical, electrical pieces entering into the constitution of the car. But, if we manage a store of car spare parts, it is more natural to have a table view of all pieces, their part number, their price, etc. The application itself must dictate the kind of physical model that must be built. If we have to manage the stock of parts, a physical model based on tables is the ideal answer. But if we must make a description of the car and explain, from an engineering viewpoint, how this car must be modified to work both with electricity and gas, an object description will be the best approach. Saying that the best database system is object (or relational) from an absolute viewpoint is simply nonsense. The model used cannot be disconnected from its application. Before going further with the discussion, we suppose that "fundamental concepts" of Chapter 5 are fully assimilated.

6.6.1 Categorization and Classification: Reuse of Static Structures

Objects are real world. There is no class in real world. Class is an intellectual invention. A classification system is a hierarchical structure of well-defined classes nested in a series of inheritance relationships. A classification provides a powerful cognitive tool that minimizes the cognitive load on individuals by embedding information about reality through the class structure. Class membership recognition can be seen as a relatively simple pattern-matching. Relationships link classes together through *hierarchical* (inheritance) or *lateral* relationships (normal association). The overall structure of a classification system serves as a repository for the storage of information about classes, and information about their interactions. By capitalizing on the hierarchical and lateral relationships between classes, this structuring process reduces the load on our memory.

Categorization is the previous step before *organization*. Categorization groups entities on the basis of similarities based on some selected criteria but a system fully categorized may lack meaningful, informative relationships so it does not necessarily constitute an organization. A system is organized according to a set of criteria. If we change the criteria, we obtain another organization.

All humans are theoretically equal regardless of race, color etc. but the poor are adversely affected due to their inequality before the lawyers. The affluent have therefore more opportunities of evading the rigor of the law. So, if the initial group is split into two different groups, the organization structure is therefore different as we introduce another classification criterion. Good classification criteria give good answer to systems.

This observation of the destruction of a structure when the organization criteria are changed is a destabilizing factor. From an operational viewpoint, it is like relationships between classes are dynamic, so they can change at runtime. *Databases are built fundamentally on static relationships*. The reuse of databases schemas is possible only if the set of relationships remains static and invariable with time.

6.6.2 Reusability Issue of Some Conceptual Database Models

We start with an example of a database used for instance for identifying an object Person in a classical approach that shows the weaknesses of such approach regarding the reuse mechanism (those databases can be very helpful instead in regular situation).

A *Person* class (Fig. 6.6.2.1) is declared currently in the past with all attributes that allow a security agent or a policeman to control fully the identity of this person. With this design, the following situations need a database update: change of the

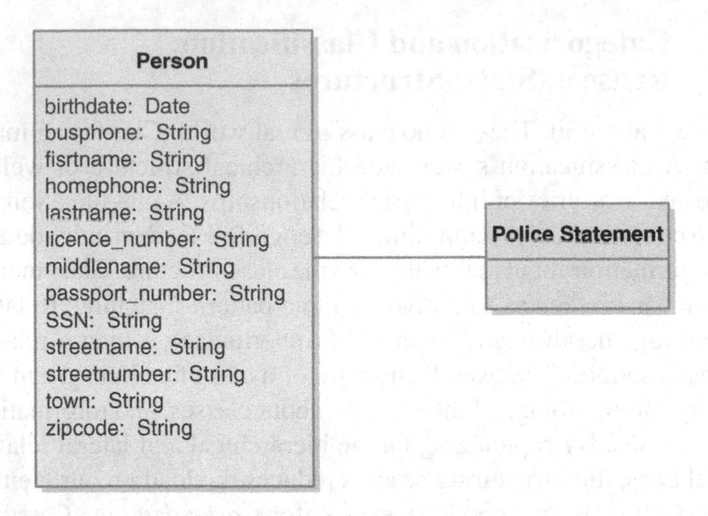

Fig. 6.6.2.1 Classic conceptual Person class often seen in database model

passport number when renewing, moving to another address, the town decides to change the street name, etc. Moreover, if the names of the person in the Passport or the Driver License do not correspond exactly (for instance abbreviations of the middle name) to the names registered in this database, the person cannot be identified correctly and he can occasionally have a lot of trouble having to justify this discrepancy.

This classic design is indeed oriented towards the final application and mimics all information that a person must fill in by hand, for instance, in a police statement. In the worst case, if the physical model has only one statement, we can put all fields inside one unique class. In Fig. 6.6.2.2, there is a slight effort of conceptualization as two classes are identified (class *Person* with all his personal information and another class *Address*). Even qualified as conceptual, this kind of model is very close to the *physical model* on the Perception Axis of Fig. 6.6.0.1.

The poor reusability comes from many facts:

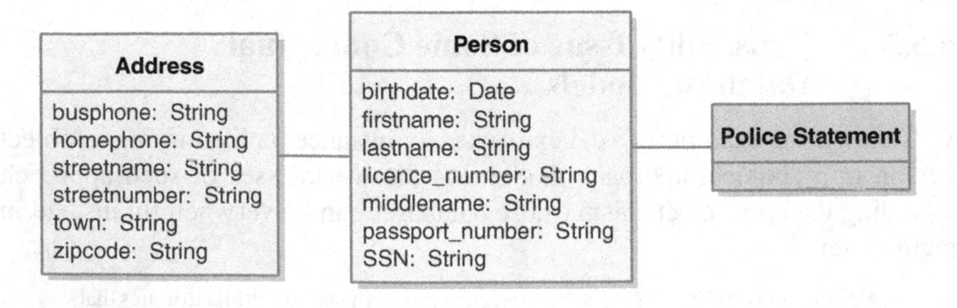

Fig. 6.6.2.2 Slight correction of the model of Figure 6.6.2.1

1. *Design is oriented towards physical support.* If the form reproduces the physical support, all small change to this support will call for a conceptual model alteration.

2. *Insufficient decomposition so information is packed into a same class.* Generally, a correction of the model of Figure 6.6.2.1 proposed is splitting the Person class into two classes *Person* and *Address*, so we now have three classes (Fig. 6.6.2.2). The social security number and the passport number are still parts of the class Person attributes. Phone numbers are parts of Address class.

3. Relationships specific to an application are not separated from more stable ones. Unfortunately, we cannot prove this point with this unsophisticated example.

6.6.3 Conceptualizing Domain and Modeling Reality: Database Models

The key concept of reuse is still, as in real-time systems, domain conceptualization of data instead of jumping directly to, in an application, to its final view of data. Moreover, if we mimic what really happens in reality, we can expect that the next application will be based on the same reality but with a different context. If we push the concept a little further to its final retrench, the number of objects in each domain is roughly static but the way they interact may vary greatly with time, so it would be interesting to identify classes of the domain then enrich them with attributes and operations each time we have a new problem to be solved. Possibly, new problems may call for creation of new classes, but this process is much slower than the process of enriching existing classes with new attributes and new operations.

Moreover, a uniform view of the whole application can be got by intimately connecting classes created in a database with real-time classes. Modern applications are complex and involve real-time applications coupled with some forms of data handling, so these two aspects are inseparable. For instance, in an airport, when designing registering systems for passengers and their baggage, and security systems for controlling passenger access operations, databases are intimately connected to real-time systems so we can model the whole chain of applications with the same object formalism.

To temporarily summarize previous ideas and make a step for designing reusable data models, here is the proposal:

1. *Identify classes of the domain* that match some information required by the physical model while avoiding to make classes uniquely for the current application. The validity of the class is weighted through its survival

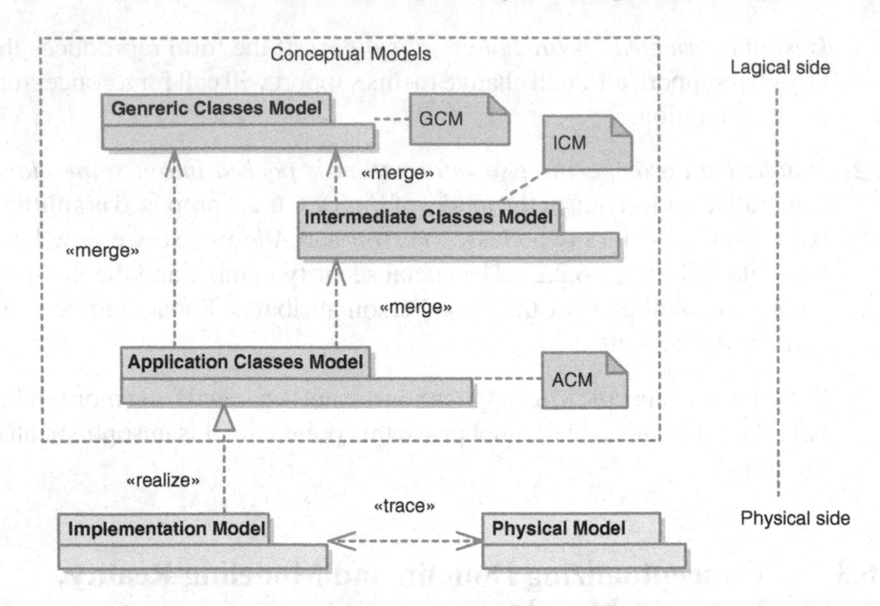

Fig. 6.6.3.1 Organization of data models, From (GCM) Generic Classes Model to the Physical Models. GCM, ICM, and ACM are all conceptual models. *Implementation Model* contains data tables, views, transactions, etc. in relational database. Physical Model contains files, index files, data repositories, computer nodes, geographic repartition of data, etc.

resistance to *space* (can the database be exported to another province or another country without modification?) and *time* (is the database structure susceptible to be modified in a near future?).

2. *Have at least two packages: one for generic classes and one oriented towards the application model.* Generic classes are part of the *Generic Classes Model* (GCM) and the conceptual model designed for the application is the *Application Classes Model* (ACM). It is not excluded that *Intermediate Classes Models* (ICM) can be built to host more structured components. The ICM corresponds to Generic Architecture package in real-time systems. Figure 6.6.3.1 illustrates the package reflecting the model organization.

3. *Use real world classes only*. Real world classes have the best chance to be there the next time another application arises. Very sophisticated or complicated classes are generally not real and thus have to be avoided.

4. *Attributes created for GCM classes are universal and tightly tied to the intrinsic nature of objects*. This is the key factor of success in the reuse process. If an attribute is not tightly tied to the nature of the class, it is probably an attribute of another real class that must be identified as well.

5. *Use only evident inheritance or aggregation/composition relationships* for classes of the domain. These relationships must be evaluated against its resistance to space and time.

6. *Specific associations are drawn only in the ACM (or ICM) to avoid spoiling the GCM.* As said before, the ACM is application oriented and the GCM is domain oriented; hence, user defined relationships must be put in the right packages.

6.6.4 Generic Classes Model

To start building a reusable library supporting the uniform methodology, we illustrate the contents of some useful and universal domains that can be reused later in our case studies. Some classes are still not broken down into their utmost elementary components but the goal is to explain the methodology.

First, the GCM top package contains four individual packages *Humans, Locations, Identifications,* and *Immobilizations* (Fig. 6.6.4.1). The number of packages is limited for explaining only the methodology. When entering the Humans package, we have the most important class that describes all particularities of a human that have a chance to last in space and time.

If we put identification card numbers, addresses, or phone numbers inside this class, we bring it to the physical side and the class will have less chance to be reused in other context. A person changes his address, and his phone number very often. The driver license number and the SSN numbers are more stable in time, the passport number does expire in some countries. The current practice of inserting those numbers in a Person class as identification strings does not hold as all identification cards of a person are separate objects, managed by independent or governmental organizations. A Person has an *association* with its Passport, his SSN card, or his Driver License instead of holding them as attributes. If those data are tied to the class Person, we violate the reuse concept both in space and time. Identification data are subjected to modification in time and the choice of card identification is a cultural matter. In most undeveloped countries, the notion of SSN identification does not even exist. Moreover, identification methods vary with time and circumstances. Recently, with terrorist threat, sophisticated biological signatures are proposed and used to identify humans. By constituting a separated package, we can add all identification means and connect them to a class Person in an application context. When dealing with multimedia attributes, we create at a logical level, attributes like *picture or image, soundtrack, videosequence*, etc. leaving all choice of implementation (LOB: Large Objects, BLOB: Binary LOB or CLOB: Character LOB, etc.) to be specified in implementation models.

Moreover, an identification card or a passport is effectively a separate object, very different from a person. A passport may have a first name and a last name

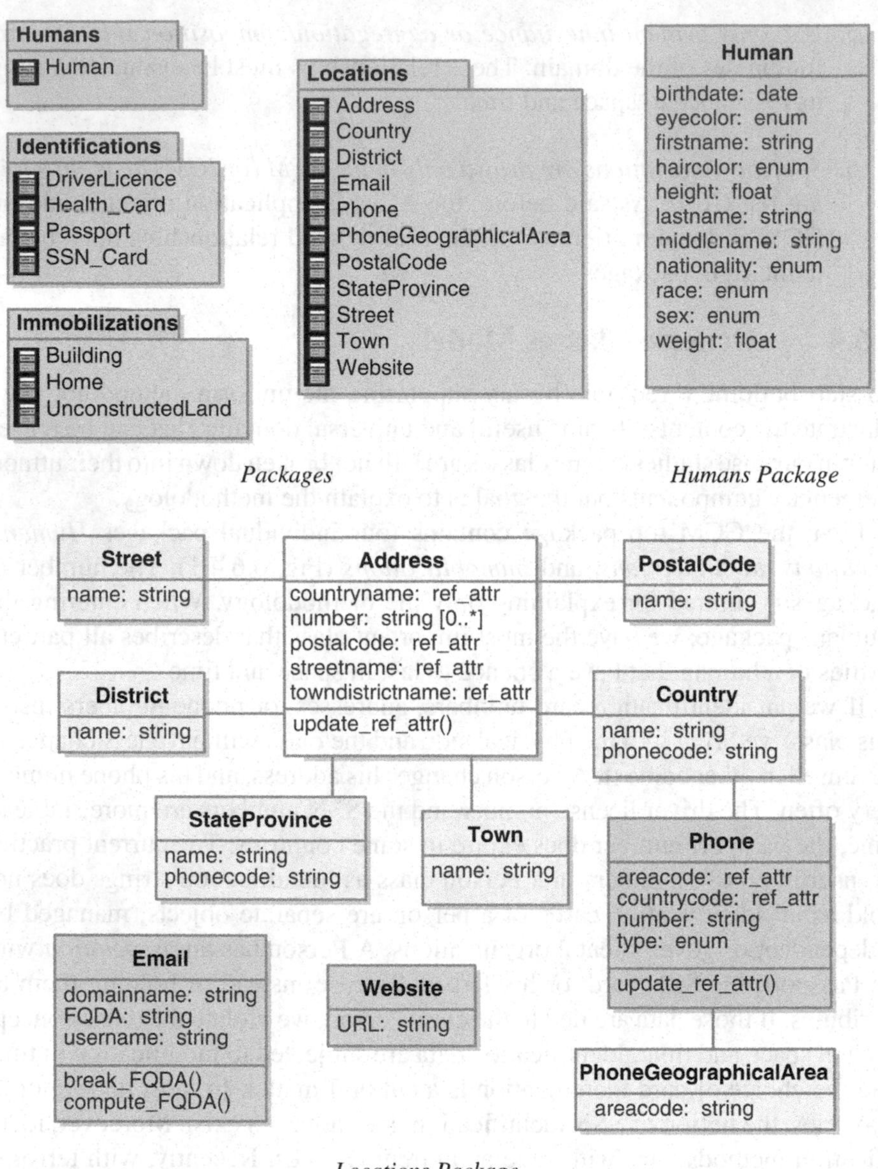

Fig. 6.6.4.1 Elementary GCM (Generic Classes Model) elements of Humans, Identifications, Locations, Immobilizations packages illustrating the contents of Generic Classes Model. The notation Locations::Address indicates that this class is defined in Locations package and there is an individual association between Home, Building, UnconstructedLand classes (contour property) and Address class

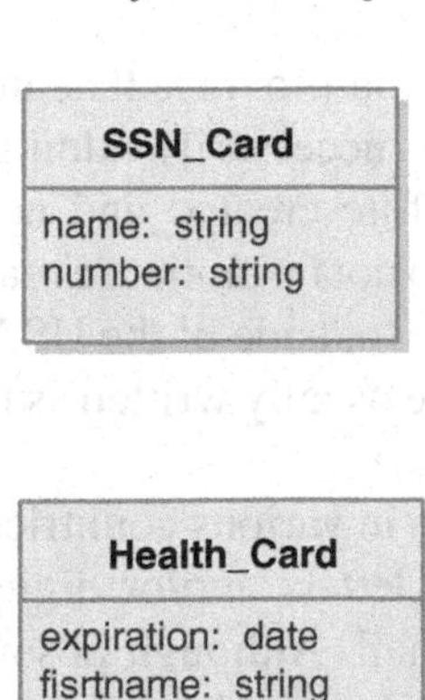

Identifications Package

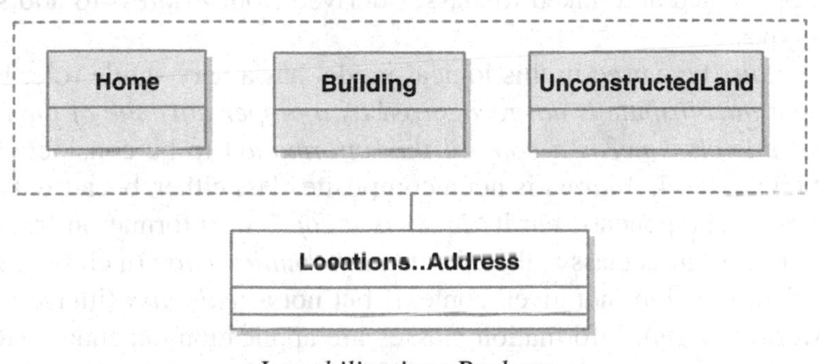

Immobilizations Package

Fig. 6.6.4.1 (Continued)

on it. Their orthography may differ from first name and last name on other documents. So representing them as various documents linked to a person allow us to detect unsuspected problems as inconsistent identifications on various documents, forgery, identification stealing, etc.

The Address class is an example of complex data specification as an address is composed of many individual fields that are dependent of other objects. In many actual designs, the resident phone number is considered as an attribute of the address and this fact is inconveniencing as, in many countries, a person may ask the telephone company to keep the same phone number when moving, if the new location is inside the same area. In this case, the phone number is not a component of an address, or a component of a person. So, the Phone is considered over here as a separated class. Area code is not specifically an attribute of a phone number but of the region. Much complication may arise as area codes usually indicate geographical areas within one country/region/province that are covered by hundreds of telephone exchange devices, not necessarily administrative frontiers. The *PhoneGeographicalArea* is therefore created and not connected to the rest because in this generic package, we do not instantiate a real situation. Nongeographical numbers, as mobile telephones (outside of the USA and Canada), do not have an area code even though they are usually written as if they do.

The same difficulty is observed for postal codes, known in various countries as a post code, postcode, or ZIP code. It is a series of letters and/or digits appended to a postal address for the purpose of sorting mail. Although postal codes are usually assigned to geographical areas, special codes may be assigned to individual addresses or to institutions with large volumes of post, such as government agencies and large commercial companies (French Cedex system).

If a real situation calls for a design for a multilingual application, cultural aspects interfere with the way the class *Address* will be made. Other classes need to be created and linked to classes derived from Address to add several address types.

The *ref_attr* type used in this logical model has a very subtle role. First, it attests that *the attribute is not recognized as a proper attribute of the current class but this class needs a copy of this information* to be considered complete at higher level. *Address* is not a composite class either, because it cannot contain other components, but it "*imports a copy*" of information from other components. *Address* class is therefore an *information class* (a class that gives pieces of information in a given context) but not a *real class* (image class of a real world objects). Information classes are application dependent (ICM or ACM) and are not part of GCM, except for some classes whose use has been so generalized that they become generic (case of *Address* class). Connections between *Address* class and Country class, for example, allow *ref_attr* to reach an instance of Country for retrieving a value of a specific attribute.

The *ref_attr* can be mapped differently from GCM to ACM. There could be a discrepancy in the way the logical model is designed and the implementation data model as well.

> At the implementation phase, if we want a simpler database system (economic consideration), a reference attribute can be mapped naturally into the type of the attribute it is referencing (for instance, *countryname* of class *Address* will take the *string* type that is the type of *name* in the class *Country*). In this case, the *ref_attr* loses one indirection and becomes a plain attribute. We have just decided to include the country as a plain attribute instead of creating a navigational structure. In this context, if a postal code must be changed for all residents of a street, each instance of the address contains a copy of the old postal code value so the update process must be fired on all Address instances. If objects are separate and linked, when making a reference to an instance of the class *PostalCode*, we must change only the name of one *PostalCode* instance.

The *Phone* class is completely disconnected from the *Address* class in this design proposal. In the past, a home phone number is often associated to an address and a mobile number is associated to a person. Very often, when editing a form, most databases give a choice of one business phone number, one home number, one fax number, and one mobile number. If, occasionally, a person wants to write a second alternate home phone, it is generally impossible to accommodate this case. Users have finally no choice, they can decide to enter extra phone numbers inside *comment* fields or equivalents (if they exist), generally neither indexable nor searchable.

The old rigid model can be made more flexible by remarking that a phone number is not attached to an address nor a person but has an independent existence (this independence corresponds to reality: when moving, sometimes we cannot keep the old phone number so this number will be affected to another person) and the relationship between a phone number and a person is simply a temporary association.

Phone numbers are not considered as attributes of class *Person*. Each newly created phone number is now tied to the class *Person* through an association, we can now instantiate an object *:Person* then a multitude of phone number objects (*firstnum:Phone . . . lastnum:Phone*, etc.) for this person, and the number of phone numbers can now vary from one person to another according to their specific needs. So, a person can possess an infinite number of addresses and each property, and an infinite number of phones.

Attributes of a class characterize this class. The UML metamodel is designed with an ownership relationship between a class and its attributes. The ownership relationship is often misleading. For instance, when putting an address as attributes of a person, we declare that the property lying at this address is permanently or statically tied to this person for an undetermined time frame. Modern life tends towards very dynamic situations. A person owns a property. A home owns an address (true, in old design since *address* is attribute of *Home* and a home belongs to a person, therefore, address is an attribute of *Person*; questionable in object design since *Address* and *Home* are two objects linked by an association only and *Home* is not permanently linked to *Person*). Parents own their children or husband their spouse. A car owns all its parts (true), a company owns all its departments (true), a company owns all its employees (false, they work for the company), a person owns all his money (arguable, he has all his money and can decide how to use them; very different from the "car owns all of its parts" example because money cannot define functionalities and abilities of a person). In an unknown case, it would be more cautious to use an association instead of an ownership or aggregation/composition relationship.

To summarize, a generic class targeted to be reused is likely to exist across space, resist to time erosion, if it is designed naturally as a real class of the world. Attributes must be tightly attached to this class in all (or at least "most") circumstances. A class with reference attributes (e.g. *Address* class) is less generic but it is highly useful as it collects useful information into a unique *Information Class* targeted to support a given application. Reference attributes are logical and attest that the information is loosely tied to the current class. At implementation level, reference attributes can be temporarily transformed into real attributes if this complexity is not needed. In this case, we will get back into more classical database schemas. But, at the logical level, generic classes can support reuse in other contexts. While working with database system, to really catch the subtleties of all data classes and attributes, it would be interesting to break a system into its utmost elementary classes then rebuild information classes.

6.6.5 Intermediate and Application Classes Models

The ICM and ACM are application models. The ICM plays the role of microarchitectures package that can be imported or merged directly into an application defined by the ACM. To give an image of the role of the ICM and ACM, to make a building plan, an architect makes individual plans showing all type of elementary construction elements chosen for the entering into the composition of rooms, kitchens, parlors, bathrooms, etc. (doors, faucets, fireplace, bathtub, shower, dim switches, etc.). This inventory is equivalent to our GCM level. Later, he can assemble building components to make complete space arrangements (various bathroom types, various room arrangements, various kitchen sizes, etc.). This level corresponds to the ICM level. In the final ACM level, the whole building arrangement can be now described with components of the ICM level (and occasionally with GCM level if necessary).

In the previous example, from *Address* class that has no specific regional specificities, we can derive *AddressUS* class for USA, *AddressCA* for Canada, *AddressFR* for France, and for other countries in the world that can have different fields, and put them in an ICM package. Operations defined for each address class *AddressXX* are intended to verify data input for this country and their algorithms will be different from one country to another. Roughly, GCM can be seen as common to all applications, ICM common to a group of applications and ACM is targeted for a given application.

6.7 Unifying Real-Time and Database Applications

Real-time system developers and database developers adopt traditionally two different development strategies. When developing databases, we create schemas. Schemas are generally drawn with E-R tools or with class diagram of UML tools. Real-time applications are developed with functional models or object models with the UML, and implemented with development platforms based on common languages as C, C++, Basic, Assembly, C#, Java, etc. Unifying the two development platforms has been attempted in the past, mostly in the direction of object persistence, and is considered by many as a commercial failure (Keller, OOP-2004, http://www.objectarchitects.de]. We do not really address this issue. At the research viewpoint, the important question is "is it useful to unify real time and database applications?" At first sight, these two application kinds seem to be disconnected.

> A real-time application considers a system from the object viewpoint, builds real-time classes to model real-time objects. It defines attributes and operations of classes, studies dynamic behavior of objects through the execution of individual or collaborative tasks. Attributes are of three types: structural, functional, and dynamic. Real-time objects are individual objects, have identities, work and act on the surrounding world. The static structure of a real-time object is a *forest of tree structures* based on the aggregation/composition relationship as structuring criterion. To describe the operation of a *Car* and its states, subsystems extracted from the *Car* tree structure are considered as interacting objects that requires a *Driver* object as a source of events, a *Route* object as the environment that the car acts on, etc.
>
> Database management system considers other kinds of problems. A car retailer may hold a store of spare parts of the same car, but the problem that worries him is to maintain always a correct stock to support car repair operations and occasionally direct sale to other service stations. A complete database application includes usual sale operations (requisition, purchasing, enquiries, report, inventory functions, data import/export, etc.). It contains the same pieces of the preceding tree structure mentioned in the real time application. Attributes stored in the database are of structural and technical types found in the previous real-time application. Another set of attributes in Business Domain (part price, date of reception, date of sale, etc.) are added to handle financial operations. When manipulating business data, the best view is undoubtedly a *spreadsheet view* or a *table view*. It would be really laborious for instance to calculate a global stock price by exploring a tree structure. Conversely, it is also hard to join relational tables to get a description of a car from the engineering viewpoint.
>
> The most important aspect is: parts in the described database have no identification. If we have 100 carburetors, they are all identical. The part number used for the inventory is not an individual identification; it allows the system to categorize the carburetor and to make sure that it will go inside a type of car for some production period. Occasionally, we can have a distinct inventory number that seems to make the part unique, but in fact, the sale person can take any carburetor and give it to anyone.

Relational databases are said to be value based and is adapted to applications that required a spreadsheet or table view of data.

> In the Car example, a line (record) in a table inventories all car parts having the same price and the same part number. So, objects are not identified individually in this context. As many as 100 carburetors need only one record with a field that counts the number remaining in stock.

When a supermarket stores 100 cans of maple syrup, it does not make sense to identify them individually. A unique record is therefore needed to count and characterize all objects of the same kind. So, effectively, we can find several applications where real-time application and database applications concern the same objects but, as the purposes of the two applications are far from each other, they can be developed as two independent applications and there is no reason to unify real-time and database projects in this context, arguing on the single fact that they refer to the same physical object.

However, there are several applications in which we need to control the evolution of an identified object, and simultaneously, store structural attributes of this object in a database in order to retrieve its description when needed. Moreover, each object needs a *separate record* in a table. The application therefore requires a real-time object (evolutional and behavioral aspects) and its description (structural aspect).

> In the previous example of car parts database, if we suppose that people must make a data entry for each carburetor (100 entries for 100 parts), there is no clear evidence of the advantage of viewing all car part data with a table or spreadsheet view. We can instantiate each carburetor one by one, identify them individually. Objects are ready to work in a real-time environment and their attributes are simultaneously indexed for data management purposes. Here are some examples:
>
> The Canadian gun registry requires every firearm in Canada to be registered or rendered in an unusable state. This was an effort to reduce crime by making every gun traceable. In this case, each gun must have a separate record.
>
> A baggage handling service in an airport is another example of database that requires both a real-time tracing and transfer of the baggage and a database management system. The system receives information in real time from various applications that support the distribution of transfer baggage, the control of the baggage offloading points, the amount of traffic on the conveyors, etc.

The unifying process is therefore desirable when objects must be *individually registered*, and *their states monitored at real time*. Objects in this context are full-fledged and the model necessitates many kinds of attributes: *structural* for identifying the objects, *behavioral* for tracing its evolution in a real-time context. Real-time classes contain large number of real-time operations to describe the transfer, handling, distribution of the baggage, and state attributes to store states that result from those operations.

When we need to work with such an object, we instantiate a copy from its class and follow its evolution through space and time. The way we figure out the list of all objects is still a table but, at this time, the table is exceptionally crowded as each object needs one separate entry.

> If the Car store of spare parts contains 2 million parts, the number of entries is limited to the number of part numbers. This is not the case with a system that requires individual tracking. The number of baggage entries is exactly that which is registered. If a security system must track all passenger states, each passenger must have its own record.
>
> Handling baggage is a very complex scheduling problem that requires simultaneously a real-time model and a database model. Simulations are conducted to detect deadlocks, for instance, to detect the resources to be made available to

handle an airliner arrival. An interesting article describes the "Denver Airport Baggage Fiasco" in 1997 (Swartz, 1997).

To model an application that needs unification (e.g. baggage), several solutions were proposed:

1. *Handle the whole system in a real-time context with data persistence management.* Classes contain both structural (for database and real-time applications) and behavioral attributes (for real time applications). Operations embrace conventional database operations with real-time operations

   (a) Structural attributes are for instance baggage weights, dimensions. Real time operations are those that accompany the transfer of the baggage through multiple steps, its security inspection (passed, not passed, etc.), that describe its individual states (normal, damaged, lost, etc.).

2. *Handle the whole system entirely with a relational DBMS* (or eventually an ODBMS or ORDBMS) with limited real time extensions

3. *Keep the two independent tools and make use modern techniques of accessing to databases.* Nowadays, it is very easy to connect any programming platform to a database and perform conventional database operations. Techniques like JDBC, ODBC. ADO.NET etc. are common and allow a standard programming language to pass a SQL clause to ask the xDBMS to perform a database operation. So the connection between the two apparently very different platforms is not really a problem

   (b) The baggage system continues to handle two independent tools: standard real-time environment without persistence management and a relational database to store structural and state attributes of baggage. Synchronization processes must be defined along the transfer line of baggage from the check-in step towards its final delivery to the traveler. An *information class* can be added in the database to store, at every step of the transfer, state attributes of the baggage.

The third system is probably the most appealing method to connect those two applications. In the global system, now coexist two objects that must be synchronized, a real object, e.g. *bagRT:Baggage* in the memory of any real-time system along the transfer line and a record *bagDB:BagageDB* in the memory of the database system. To explain the synchronization, let us suppose, for instance, that a security system detects a firearm packed inside the real baggage, it changes the state of *security* attribute from *unchecked* to *suspected* in a local *bagRT*, fires an alarm, echoes this value immediately to *bagDB* to warn other

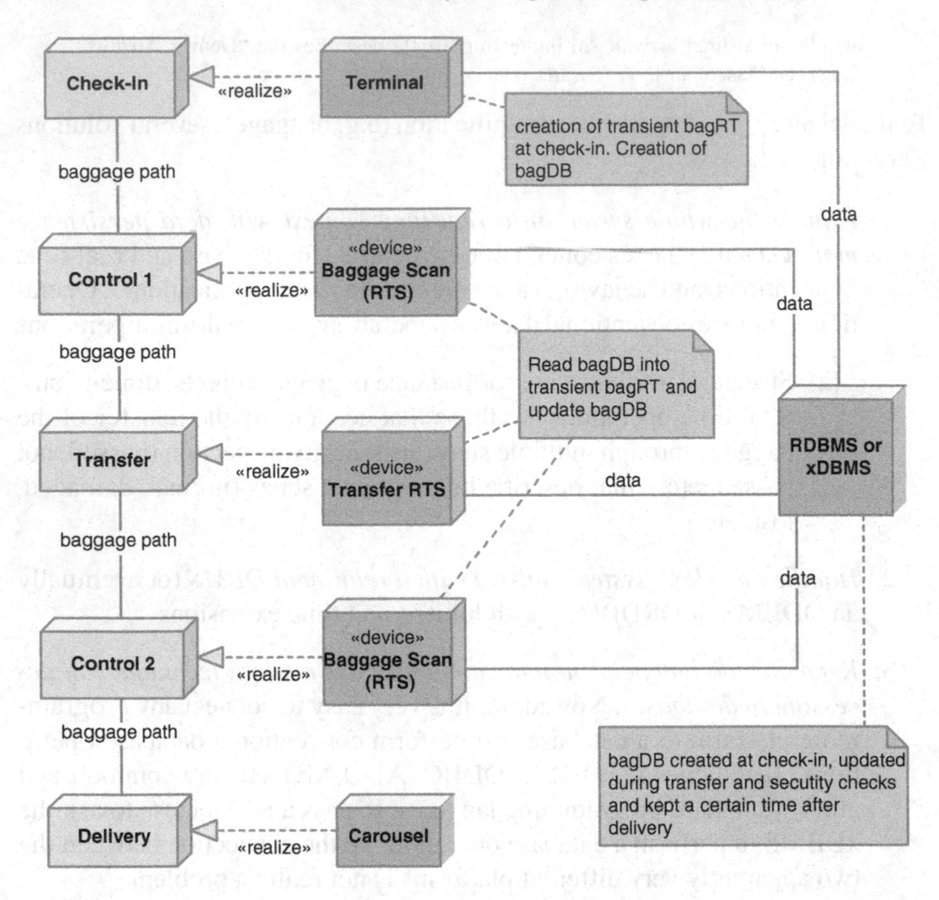

Fig. 6.7.0.1 Deployment diagram explaining that many bagRT are created temporarily (they can be made persistent locally). The object bagDB is persistent in the database and can be reached all over the network. Each real-time object is defined differently according to the nature of the device. Some of their attributes are read dynamically from bagDB if needed. bagDB is updated after visiting each node so bagDB keeps track of the current location and state of the baggage

services of the airport so that security countermeasures can be immediately deployed to intercept the real baggage corresponding to the current clone *bagRT*. Figure 6.7.0.1 represents is a deployment diagram explaining the relationship between multiple real time objects bagRT (each bagRT is created locally inside each real time application) and a database object or record bagDB (coming from an *information class BaggageDB* created in the RDBMS).

Data models used in business domains and objects in engineering worlds are generally not structured in the same way. So, the nature of the application will dictate finally what the model that must be used. This fact is certainly one of the reasons that explain why relational model with table and spreadsheet view of data still occupy a dominant position in the market. It fits perfectly to the business domain. In engineering domains or in knowledge representation, where data are structured as networks or forest of tree structures, modeling data with object style will be the best answer. However, logical models must be dissociated from implementation models. If knowledge and engineering databases must be modeled with object style, their implementation could be easily mapped into a hybrid system of type 3 mixing an object environment to a relational database with software cement (ADO, JDBC, etc.). The baggage transfer example highlights such a solution. Several models must be developed to support such an arrangement. Real-time models are based on *real-time classes.* Database model are based on *information classes.* At run-time, corresponding objects are instantiated and they require continuous synchronization to maintain data coherence.

Chapter 7

Real-Time Behavioral Study Beyond UML

The dream of a software maker is how to get running software directly from a CASE tool, to make a jump from abstraction to functionality. The passage is targeted either from the PSM to code or from the PIM to code. The later joins the former with a preliminary transformation from PIM to PSM.

> Presently, visual programming environment based on C# or Java languages allows programmers to automate many programming tasks in a graphical window-based system. List boxes, combo boxes, text boxes, buttons, etc. and dynamical components can be created graphically on a design area, then events added to realize a windows-based interface frame that needs in the past hours and hours of programming tasks linking with complex libraries. When an event are created, for instance a click on a button, only its headers *button_click()* is added and the event inserted in the event list. The programmer has to add manually codes to say to the program how to handle this event.

Presently, many tools implement automatic code generation for different programming language out of static UML class diagram. When making classes, attributes and operations are already named, their types and visibilities specified, so theoretically, CASE tools can generate programming class headers and variables in any language.

For complex or user-defined data types or data structures, there must be some means that allows the developer to specify the complex structure in terms of existing types, by associations or by any other methods. Therefore, we suppose that this feature is not a theoretical issue when the complex structure can be mapped into elementary types available in the target language.

As for the code inside functions, programmers still need to fill in the blank if the corresponding diagram cannot establish the algorithm unambiguously, and if activities are not sufficiently decomposed to reach elementary actions and class/object operations. Operations in the PIM or PSM are considered as elementary if they correspond to action/activity not decomposable at their respective level. At low level, an elementary action is translated into one line of code in high level language (Basic, C, C++, C#, Java, SQL, etc.) or eventually into a code snippet. With a low-level language like assembler, an elementary action will be translated into more than one instruction, e.g. an assembler routine

D. M. Bui, Real Time Object Uniform Design Methodology with UML, 381–416.

or assembly code snippet. In this case, a library of code snippets identical to those used in compiler techniques can be used for this purpose.

For instance, in a real-time system, instructions like

```
Timer.Start()
Button.On()
Timer.Delay = 10000
```

are considered as elementary operations.

When all the previous conditions are validated, there is effectively a possibility to get a run-time environment out of the UML diagrams. If the operations are not elementary, in the PSM, we can store *high-level language source code snippets* (sometimes called *code patterns*) that can be exported directly from UML diagram specifications towards target language code as contents of class operations. This method is not recommended as it is like making code at diagrammatic level or bringing the low level up to the UML level.

> In some specific situation, code snippets are necessary for instance for importing libraries, defining macros, etc.

Programming languages are a mature domain. To really be able to support the automatic code generation between higher models and code, it is necessary to have a means of specifying unambiguously the behavior of system. The Behavior has been addressed with three important concepts in UML: interactions, activities, and states. They are related to other concepts: use cases at the Requirement Analysis, business processes at Design, and operations inside classes at Design and Implementation. "Behavioral" is a global concept merging the two subaspects, functional and dynamic, together. Functional aspects are preponderant when dealing with operations defined in classes, analyzing elementary actions, activities, processes, use cases, etc. Dynamic aspects are more important when discussing control structures, control flows, or workflow patterns. When reasoning with states, the two aspects are mixed in an inextricable way. To contribute to the effort of designing an unambiguous diagram for automatic code generation, in this chapter, we will expose a Petri-like network SEN (State-Event Network) (Bui, 2006). The SEN diagram proposes a unifying formalism melting together interactions, states, and activities that are finally three different views of the same reality. If interactions, states, and activities are used in the PIM and PSM, we need an intermediate model between PIM/PSM and the programming model that is code. Figure 7.0.0.1 locates the proposed model in the whole architecture.

The second aspect that will be discussed pertains to the safety of systems. As computers are ubiquitous in any applications, their failure (software and control aspect) is very costly, in terms of monetary loss and/or human suffering (aircrafts, railway traffic, safety in security systems, industrial plants, military

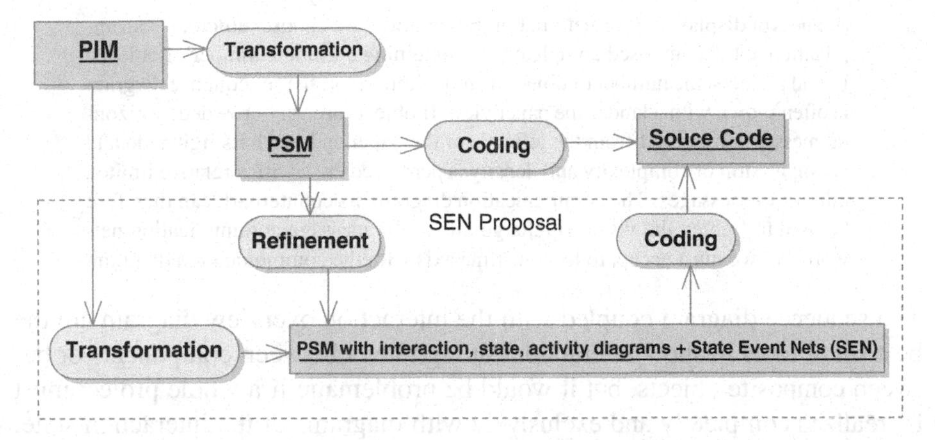

Fig. 7.0.0.1 The Dynamic Model developed with SEN (State Event Net) acts as intermediate model between PSM and Code. Three development paths are displayed: a standard PIM- > PSM- > Code, three other possibilities add the SEN to other dynamic diagrams as intermediate model before coding

weapons, banking systems, commercial worldwide transactions, etc.). Very often, the subtle design fault is at the origin of a catastrophic failure and the problem would be how to isolate such a design flaw that is often masked by the fact that the whole system seems to perform well and pass all functional approval tests. Is it one of the reasons why we still need a "black box" inside all aircraft when it passes all regular and maintenance tests? Some failure may arrive as a combination of circumstances. Its occurrence is so complex and unpredictable so that normal humans cannot anticipate. Complexity and unpredictability could hide design problems or design management issues. The problem is very complex because we can build a system that fulfills all SRS clauses, and however, this fact does not mean that the system is safe. Safety–critical situations may be disregarded in the specs even at the starting point of the development process. But, as demonstrated later, safety problem can be evidenced by developers. So, we will discuss all those aspects that are very important in real time embedded and embarked systems and propose a combinatorial methods called "Image attribute analysis" that, coupled with the SEN diagrams, can be an interesting research avenue for investigating safety-critical systems in a deterministic way. The first thing that must be done is to track all design faults and to realize the safest design.

7.1 Sen or State-Event Net Diagram

The expressiveness of UML dynamic diagrams is unquestionable, but there are some problems probably due to their high expressiveness.

Let us take the sequence diagram that represents lifelines in vertical direction and messages between objects in horizontal direction. If this diagram shows objects, messages, and sequences at the same time, real system may suffer from the richness of displayed information. For presentation, we choose judiciously an application not decomposed completely or containing a limited number of objects. In real projects, the number of objects is more important, so the sequence diagram is often shown with a landscape paper view. If objects are very crowded, horizontal messages may start from the left side of the paper and reach its right side and an impression of complexity and density is perceived even with a relative limited number of messages. The communication diagram, its counterpart, can therefore be used to lighten the visual surcharge due to the message communication network, but we must accept to lose the time axis with the communication diagram.

The sequence diagram coupled with the interaction overview diagram are the best set of tools to show interactions at high level between components or between composite objects, but it would be problematic if a whole project must be realized completely and exclusively with diagrams of the interaction suite. Moreover, drawing an interaction diagram is a time-consuming task. If sequence diagrams can be used to display some important sequences of the system, building complete algorithms is time consuming. It is interesting to notice that class diagrams already contain operations that are images of the messages in the sequence diagram, so information is somewhat redundant. The power of the sequence diagram is showing messages exchanged between objects, so a sequence diagram is naturally tied to collaboration. From this remark, sequence diagrams are more predestined to collaborative tasks than for internal algorithms of objects (messages will be mostly "self messages" in this case). Interesting works have been done in the fields of formalizing scenarios (Alur et al., 1996; Uchitel et al., 2003).

The state machine diagram is recommended for representing the state evolution of an object in reaction to external events or internal stimuli. To represent the collaboration and synchronization among several objects, statecharts provide constructs like AND/OR states. State charts may have substates, of a higher level state, all active at the same time. Substates can communicate with each other. Orthogonal regions of the chart accept events sent to the object and regions can generate events to other regions. Guards are used as preconditions for state evolutions. Contour properties added by Harel increase the expressiveness of statecharts, so modern statecharts offer powerful facilities to succinctly specifying state evolution. Statecharts are therefore an image of the algorithms and is therefore an interesting candidate for formal methodologies (Golin Reiss, 1989; Harel and Gery, 1996; Gnesi et al., 1999; Harel et al., 2005).

In Wohed et al. (2005), the activity diagram has been analyzed with YAWL (Yet Another Workflow Language) (Aalst and Hofstede, 2005). Workflow control patterns (Aalst and Hofstede, 2002) are known issues. The syntax of the activity diagram has changed considerably passing from revisions 1.4–2.0 and for the first time, the name of Petri has been mentioned in the standard.

Research efforts are made to separately adopt interactions (in scenarios), activities, or states (in statecharts) to build tools for checking systems, for generating code, etc. In this chapter, we try to unify these concepts and propose a network based on the Petri Net. Voluntarily, all redundant information (e.g. operations already defined in class diagrams) is pruned off. All visual surcharges are leveraged to maximum in order to achieve a very compact dynamic network that can be exploited as a visual algorithm before coding. As the interpretation is different from conventional ones, we coined in 1999 (Bui, 1999) the term SEN to name this Petri-like network.

7.1.1 Mathematical Foundations of Petri Network

Petri net theory has developed considerably from its beginnings with Dr. Petri's (1962) dissertation. Later, Petri networks have been widely studied by researchers, specifically in modeling of systems (Peterson, 1981; Shlaer and Mellor, 1992; Sowa, 1984). French GRAFCET modeling, widely used in Europe for teaching, derives from Petri net as well (David and Alla, 1992; David, 1995). A Petri net is composed of four parts, a set of places P, a set of transitions T, an input function I that is a mapping from a transition T_j to a collection of places I (T_j) called "input places of transition T_j," and an output function O that is a mapping of transition T_j to a collection of places O (T_j) called "output places of transition T_j. The Petri net has its structure mathematically well defined so that it can be retrieved from its four-tuple (P, T, I, O). As multiple occurrences of places are allowed, we have "bags" of places that are generalizations of "sets."

On each place, we can store tokens. If we consider monochrome tokens, their number is written inside the place and the multiplicity of I and O functions are integers put on arrows connecting places to transitions. Remember that arrow can connect only places to transitions, never graphical objects of the same kind. By default, in the absence of numbers, places contain 0 token but arrow has multiplicity of 1. A Petri net with tokens are said to be marked with an n-vector, n equal to the total number of places. Coordinates of n-vector are positive integers.

A Petri net executes by "firing" or "triggering" transitions. A transition fires by removing tokens from its inputs places and creating tokens in its output places. The numbers of tokens removed and created are not necessarily equal but depend upon I and O functions.

The mathematical formulation of Petri net is known for a long time. But, it is not the only network that has a mathematical foundation. We suspected that all diagrams or networks must have under the cover some mathematical formulation so as to handle just their connectivity (CASE tools must handle more information than this minimum set, e.g. the location and the size of places and transitions in a paper area). However, the movement of tokens or markers and the way multiplicities of In() and Out() functions are used to relate the movement of tokens has transformed to Petri net into an executable diagram

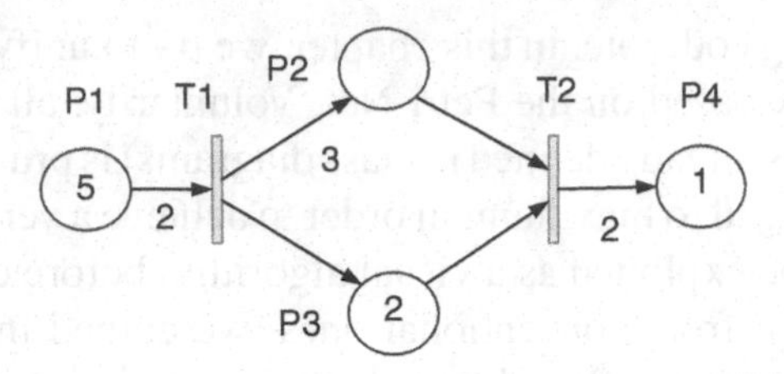

P = {P1, P2, P3, P4}
T = {T1, T2}
I(T1) = {P1, P1}
O(T1) = {P2, P2, P2, P3}
I(T2) = {P4, P4}
O(T2) = {P2, P3}
Marking vector : M0 = (5, 0, 2, 1)
After firing T1 -> M1 = (3, 3, 3, 1)
T2 not authorized to fire for lack of 1 token

Fig. 7.1.1.1 Petri net with four places P1–P4 and two transitions T1–T2. Each transition has one In() and one Out() functions. Marking vectors memorize the number of monochrome tokens inside the ordered set of places. Firing is subject to the number of tokens in the original place, and the In() function. The number of tokens generated at arrival places depends upon the Out() function

and all dynamic diagrams based on Petri formalism is doubly dynamic (current dynamic diagram like interaction, activity, state diagrams show only all dynamic paths but cannot be executed, the reader must imagine a virtual marker when examining these diagrams).

7.1.2 Using Petri Nets in Modeling

In the real world, a task can be executed if the resource is available. Some task needs more than one resource and this fact can be specified as the multiplicity of the In() function. The Figure 7.1.2.1 illustrates the way we can use a Petri net to model task execution.

Bars are used to represent primitive events with 0 execution time (*Start processing* with $\Delta T0 = 0$ and *End processing* with $\Delta T1 = 0$). *Primitive events* theoretically cannot have overlapping zone as their $\Delta T = 0$. *Nonprimitive events* with some duration can have overlapped time zone. This property will be used later to make subnets managing net hierarchies to handle complex applications.

Originally, there is no inherent measure of time in a Petri net, so we have a partial ordering of time and occurrence of events. As an event can fire when it has enough tokens and appropriate multiplicities, more than one transition can be enabled at a time and nondeterministic results could result with a complex Petri nets with many firing possibilities. Many researches on timed Petri nets,

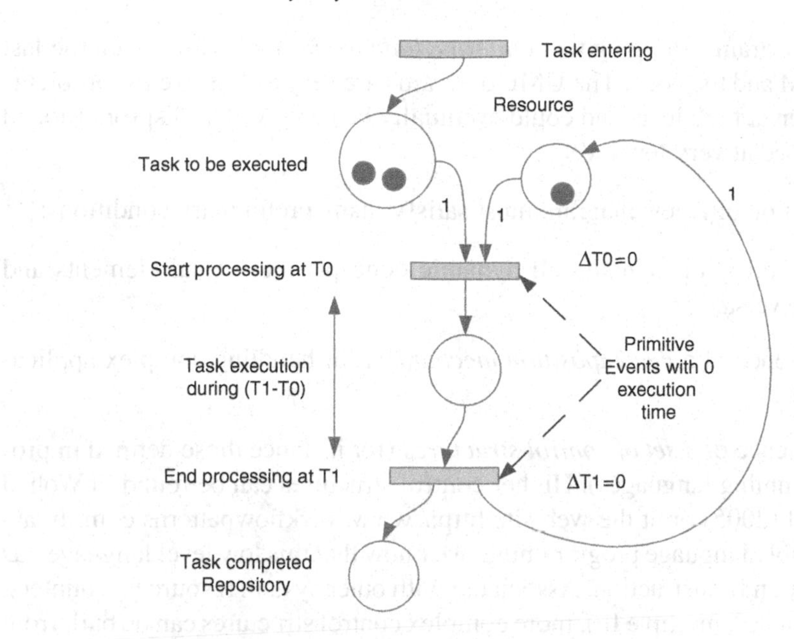

Fig. 7.1.2.1 Modeling task execution with a Petri net. Tasks need available resources to be executed. After execution of the 1st task, the resource token is returned to its original place and the system will execute the 2nd task. If a task needs two resource tokens, just mark right arrows with two instead of one. In this case, we need two resource tokens to execute one task

deterministic or stochastic nets have been made in the past. The reader can refer to a survey made by (Zurawski and Zhou, 1994).

7.1.3 State-Event Net Abstractions

As announced, the SEN diagram is a Petri-like net, so it inherits its essential characteristics (mathematical support, movement of tokens). As stated before, we need:

1. A diagram compatible with object paradigm
2. A *compact* diagram in order to express visually highly complex algorithms before attacking coding phase, a diagram for designers and programmers
3. A *nonredundant* diagram that does not express knowledge already inherent to other UML diagrams so as to leverage visual surcharge caused by redundant information
4. A *universal dynamic diagram* that can be used to express a large range of algorithms (from a simple operation defined within one object towards a complex inter operation involving a collaboration of multiple objects/components/agents)

5. A diagram belonging to an *intermediate model* located between the last PSM and the code. The UML diagrams are very expressive for problems taken at high level and could eventually be perceived as disproportioned if used at very low level

The definition of a new diagram must satisfy many preliminary conditions:

1. Possibility of mapping all dynamic concepts into visual elements and vice versa.

2. Presence of a *decomposition mechanism* for handling complex applications.

3. Presence of a *set of control structures* (for instance those defined in programming languages). Higher control structures can be found in Wohed et al (2005) or at the web site http://www.workflowpatterns.com. In assembly language programming, we know that this low level language has only an IF instruction. Associated with other system resources (counters, variables, multiple IF), more complex control structures can be built from the IF. So, this condition can be easily satisfied if we accept at the beginning that the SEN is closer to a low level programming language. More complex control structures can be added later if necessary.

4. Possibility of handling asynchronous events like interruptions.
5. Possibility of handling parallelism (e.g. UML statecharts, fork join control structures in the activity diagram).

For handling real-time system, it would be highly desirable to have a way of specifying time. Moreover, a deterministic interpretation of token movements is an advantage.

In the "Fundamental Concepts" chapter, we have already mentioned a close relationship between action, activity, task, event, condition, state, and message. These concepts lie on a same semantic continuum and the way we interpret a concept depends upon the observer viewpoint, its position on a cause–effect chain, the formalism used to interpret concepts. For instance, when an object (transmitter) executes an action, the receiver object catches it as an event. Where the state of the receiver commutes to could be viewed as a condition.

> For instance, saying that "car A hits car B" relates the story as an action, saying that "B is hit by A" describes the story as an event suffered by B and saying that "B is completely unusable after the shock" refers to the final condition of B.

Action and activity are very close semantically as activity can contain many actions. State is bound to activity or action. "A plane is flying" is both an activity and a state. In a SEN, all dynamic abstractions are represented by places. In contrast with the Petri net, transitions are not named and they serve uniquely

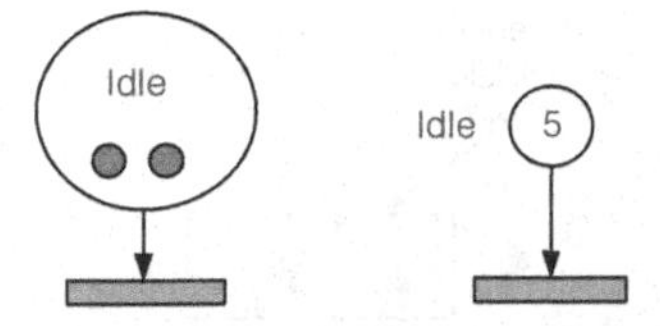

Fig. 7.1.3.1 Various graphic representations of SEN places. A SEN place represents all dynamic concepts (event, condition, state, action, activity). It can be reduced to a point with a name to save space on a diagram. The number placed near the dot give the number of tokens available in the place

to make control structures and to regulate the circulation of tokens in the SEN (except for referencing when a SEN must be partitioned into chunks in multiple sheets). Events are now considered as tokens that must appear in places. So, when an object executes an action in its SEN, it can generate a token in the SEN of another object (or in its proper SEN) and the way the two SEN are connected together mirror object interactions.

The SEN can specify any identified entity of the structural view. Thus, a system may be taken at any level of its decomposition; it can be the whole system, an external human actor, a component, an agent, a composite object, or a simple object. For graphical representation, the most "economic" visual representation would be only a dot for a place with the name written near the dot.

7.1.4 SEN Subnets

A diagram without natural support for complexity reduction through hierarchical decomposition could not be a main support for design. A SEN network supports two kinds of subnets. When entering/exiting a subnet through bars, it is defined as a *nonprimitive event* represented by a rectangular box. A primitive event is an event without zero duration and theoretically, two primitive events could never concur, only nonprimitive events can. SEN subnet with a "place" form is drawn with any means that can specify this fact (normally, we draw it with a double line and an oversized place to work in a monochrome context). Graphical representations for subnets do not add any new concepts besides place/bar/arrow; they pack only subdiagrams. There is no conservation of the number of tokens entering and exiting the subnet, we cannot know how long a subnet will keep the tokens as long as it is not decomposed and studied thoroughly.

7.1.5 SEN Control Structures

Elementary control structures in UML 2 are:

1. *Sequence of control flows*

2. *Decision node* defined in IntermediateActivities package. Decision node is a kind of IF/CASE with one incoming edge and several outgoing edges.

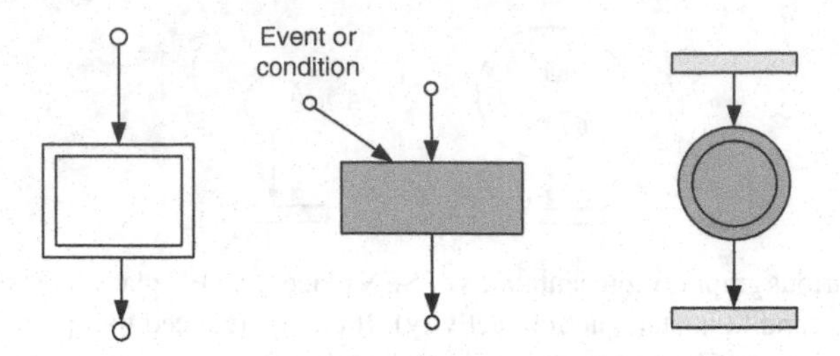

Fig. 7.1.4.1 SEN subnets. A nonprimitive event with a box materializing some duration and delimited with two primitive events can be used as SEN subnet. It must be connected to places. It can accept input conditions or events causing subnet entering. An oversized place with a double line is used to represent a subnet of "place type." It must be connected to bars

Edges can be either all object or all control flows. Each token arriving at a decision node can traverse only one outgoing edge. Tokens are not duplicated. Guards of the outgoing edge are evaluated to determine which edge should be traversed. The order in which guards are evaluated is not defined.

3. *Fork node*. A fork node has one incoming edge and multiple outgoing edges. Edges can be all object or all control flows. Tokens are duplicated across the outgoing edges. If an outgoing edge fails to accept a token, the latter remains in an FIFO queue, and the rest of the outgoing do not receive a token. The Fork Node may have guards on outgoing edges; in this case, "the modelers should ensure that no downstream joins depend on the arrival of tokens passing through the guarded edge. If that cannot be avoided, then a decision node should be introduced to have the guard, and shunt the token to the downstream join if the guards fail."

4. *Join node*. A join node synchronize multiple flows, has multiple incoming edges and one outgoing edge. "If a join node has an incoming object flow, it must have an outgoing object flow, otherwise, it must have an outgoing control flow." "If all the tokens offered on the incoming edges are control tokens, then one control token is offered on the outgoing edge." If *isCombinedDuplicate* attribute is true, then objects with the same identity are combined into one unique token.

5. *Merge node*. This node brings together multiple alternate flows. As a difference with the join node, it is not used to synchronize concurrent flows but to accept one among several alternate flows. Edges can be all object or all control flows.

Hereafter, control structures are implemented in SEN and explained.

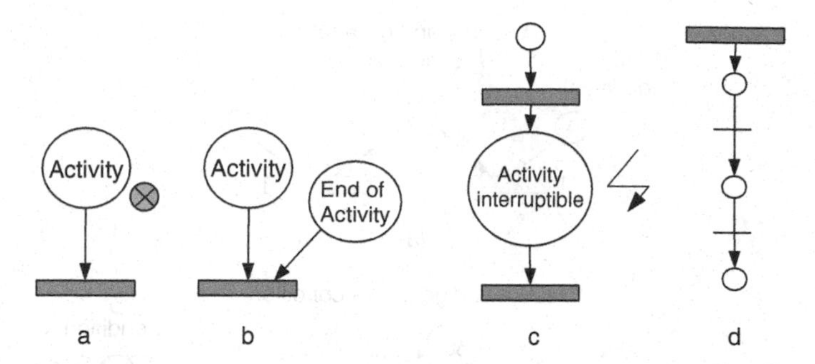

Fig. 7.1.5.1.1 Various End of Activity representation and Interruptible Activity. A global constraint can be used to characterize the whole project and force each activity to reach its end before releasing the token. In (a), a "circle with the X sign" means "End of Activity" and makes the option explicit. (b) gives an equivalent of (a) with a heavier notation. In (b), the end of the activity acts as a condition; when the activity reaches its end, the activity generates itself a token in the place "End of Activity" and the bar can fire. In (c), a lightning sign is used to express an interruptible activity. In (d), sequence of control flows without any condition can be represented with a simplified notation in which bars are reduced to a single dash put on the flow joining two places (normal bars can coexist with this simplified notation)

7.1.5.1 Sequence of Control Flows. As the SEN derives from the Petri net, places cannot be connected together directly with control flows as in the activity diagrams. We must therefore alternate places and bars. To reduce to the maximum the burden of drawing a bar after each place in the case sequential actions must succeed in time, a lightweight line can be put on the control flow to represent the bar. Moreover, each action or activities are supposed to execute to their end before releasing the place, by default, unless an interrupt forces them to leave the place prematurely. The end of process can be considered therefore as a condition not always represented for the alleviating the SEN diagram visual surcharge. In fact, in the presence of the possibility of interruption, a process can be interrupted before reaching its normal end, so the developer must weight by himself the default option (global interpretation constraints for the whole project) that can lower the burden of adding special signs or conditions.

7.1.5.2 Decision Node. The IF/CASE or exclusive choice (XOR-Split) is represented in the SEN as two or more bars with the condition (also called guard or predicate) in a place. When a predicate is evaluated to true, it is equivalent to generate a token in the corresponding place. The two bars' configuration corresponds to an IF and multiple bars corresponding to a CASE with one choice.

Conditions specified for outgoing arcs may overlap; in this case, the arc with the highest preference is selected. At design time, this preference can be specified by adding extra places filled with sequential numbers. When a condition is

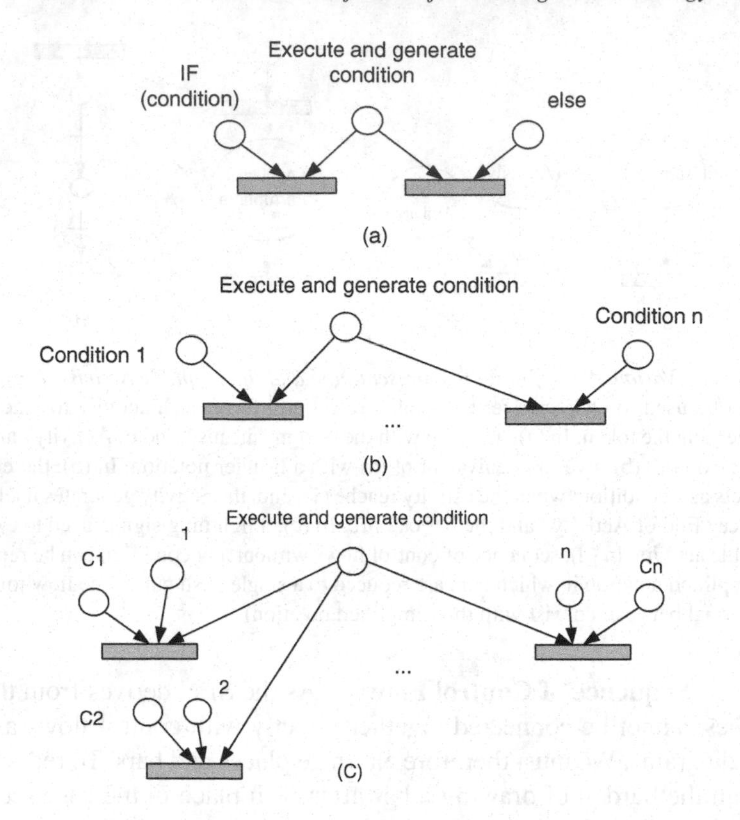

Fig. 7.1.5.2.1 Decision nodes (a) simple IF, (b) CASE with one outgoing edge, (c) handling complex case with overlapping conditions and possibility of parallel split. In (c), conditions specified for outgoing arcs may overlap; in this case, the arc with the highest preference is selected. At design time, this preference can be specified by adding extra places filled with sequential numbers (external to the place for distinguishing them from the token multiplicity). When a condition is evaluated to true, the corresponding sequential number is synthesized dynamically at this node. Conventionally, an invisible process (not represented) selects always the lowest value and affects a token to the place holding this value. If two conditions is verified at the same time (e.g. C1 true and C2 true) and the two numbers are identical (either 1 or 2), then two tokens are generated and the system behave as a parallel split on arcs 1 and 2. When C1 = C2 = ... = Cn = true and all arcs have the same number, we have a parallel split on all arcs

evaluated to true, the sequential number is synthesized at this time and we consider conventionally that an invisible process (not represented) always selects the lowest value and affects a token to the place holding this value. With this convention, two outgoing arcs evaluated to true with the same number generate an equivalent of a parallel split with two arcs. So, there is a possibility of simultaneous generation of tokens on a subset of outgoing arcs having the same number with overlapping conditions. Therefore, by extending this idea, the parallel split is a special case of the decision node with all predicates evaluated to true and all numbered places having the same number.

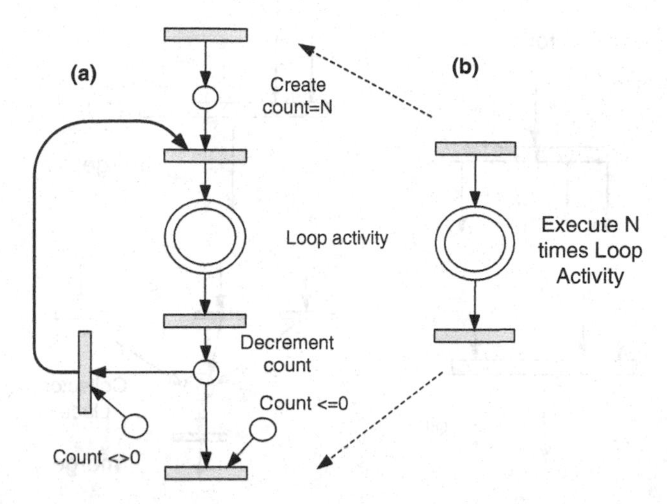

Fig. 7.1.5.3.1 Example of DO loop represented by a SEN at two levels, first at programming language level and at design level (as a sub SEN). (b) integrates all loop controls into the DO loop sub SEN. Conditions count $<>0$ or count $<=0$ come from Decrement count operation

7.1.5.3 Loop Nodes. Loop nodes can be easily made with the IF node by creating counters, extra conditions. We can define shorthand graphical notations if needed but as they are not really of theoretical concern but only conventions based on existing ones, we focus our attention only on elementary constructs. Figure 7.1.5.3.1 gives an example of the DO loop as taken at two levels: in programming language and at design time. In the later case, the DO loop is considered as a sub-SEN that does not need to be decomposed as all programmers are able to write a DO loop.

7.1.5.4 Fork, Join, and Merge Nodes. A fork node has one input on the bar and two or several outputs leaving the bar. It is also known as a parallel or AND split. A number of tokens duplicated will be equal to the outgoing edges. If conditions are used on edges outgoing from a fork, developers must ensure that no downstream join starves for token retained by conditions; there must be some mechanism for ensuring that the downstream join do work properly.

If a fork node is of configuration *1 to N*, N tokens will be generated after traversing the fork bar (if we suppose that all outgoing edges have individually 1 as multiplicity). If a join node is of configuration *M to 1*, the bar must collect M tokens (if we suppose that all incoming edges have individually 1 as multiplicity) before constituting a unique token on its outgoing edge. It is not the case for the Merge node. In the SEN, to insure a symmetrical representation of fork and join nodes, the join node is drawn as a mirror of a fork, without any added condition. The merge node has the same representation as the join node with a small M sign added to the bar to release the synchronization condition. Each

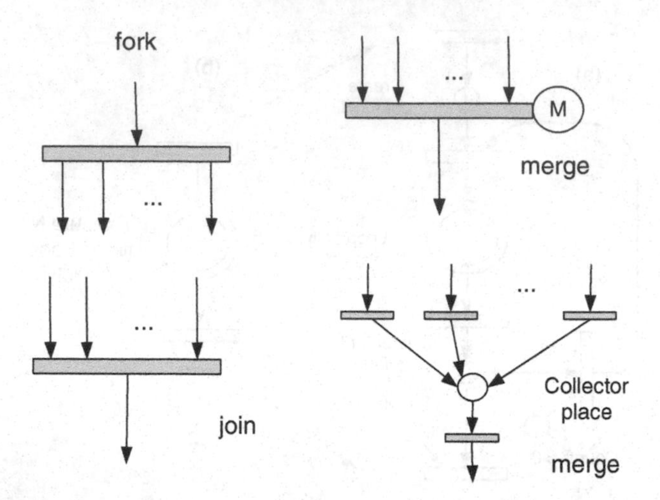

Fig. 7.1.5.4.1 Fork, Join, and Merge nodes. Fork and join are supported naturally by Petri Nets. The merge bar is a shorthand notation for a more complete arrangement shown with a token collector place

token entering one of the incoming edges of the merge node is propelled to the outgoing edge (if multiplicities are 1 and 1 at both sides). In other words, there is no conservation of tokens through fork and join nodes but there is conservation of tokens through the merge node.

7.1.5.5 Pseudoplaces. Pseudoplaces can be defined by developers, if needed, and provided that their meaning is unambiguous. As place is a unifying concept, it can thus be used to define Initial, Final, and Flow Final in Activity Diagram or Initial, Final in State Machine Diagram.

7.2 UML Diagrams Mapped into SEN

As the SEN is targeted to be an intermediate model before coding, it is therefore essential to demonstrate that it has the capacity of mapping all high-level constructs used in UML diagrams. We do not really make a full demonstration (that could be very boring for the reader) but will take most representative cases.

7.2.1 State Machine Diagram Mapped into SEN

Figure 7.2.1.1 illustrates the way the state machine diagram packs two general states and a transition between these two states. Each state is defined with its *On Entry* action, *Do* action, *On Exit* action. Transition is defined with its complete characteristics (trigger, guard, action). Actually, this definition overlooks all the timing of various actions involved. When the object receives the event that triggers the state change, a series of actions is scheduled within the state

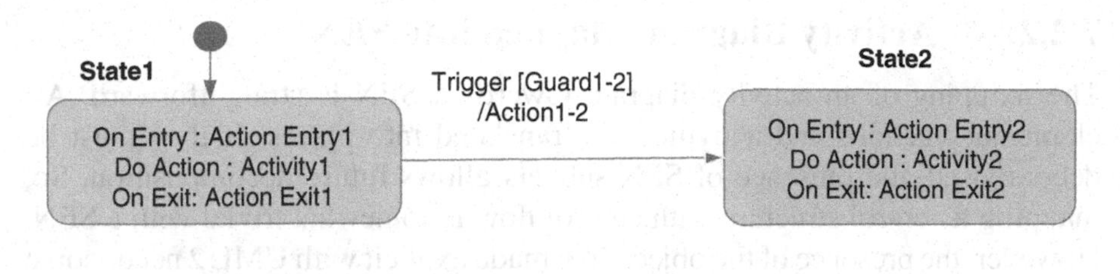

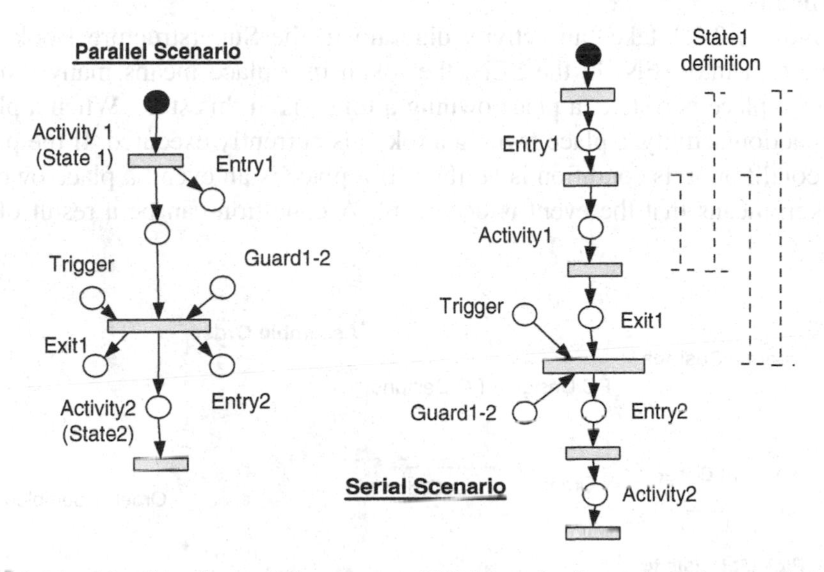

Fig. 7.2.1.1 Mapping of a state diagram into SEN. Many details overlooked on the timing of actions inside the state machine diagram appear now with the SEN. The actions Entry2, Action1–2, and Activity2 may start in parallel in the first scenario or they can be executed serially in the second scenario. Many other scenarios mixing parallel and serial actions are possible. The definition of a state is "elastic" in serial cases as we can include Entry or Exit actions

change: the *action1-2* for passing from the *state1* to *state2*, the *Exit1* action for leaving the *state1* and the *Entry2* action executed for entering the *state2*. These three actions may occur either in parallel, either sequentially (the order must be specified in this case) or possibly in a mixed mode with some time overlapping. Moreover, the notion of "state" has now been replaced by "state component", in our case, by *Activity1* and *Activity2* that defined *state1* and *state2*. But, when actions are executed serially or with some time overlap, it is not clear that states must be defined when *Action1-2* starts, when *Entry2* starts or when *Activity2* starts. Modelers that must translate the state diagram into code must therefore decide when a state begins and when it ends. But, the SEN can already highlight early all these timing subtleties. The SEN appears as a magnifying glass over a state diagram.

7.2.2 Activity Diagram Mapped into SEN

The mapping of an activity diagram towards a SEN is straightforward. All elementary actions and activities are translated into places. If they must be decomposed, the presence of SEN subnets allows future decomposition. So, mapping a control structure with control flow is somewhat trivial with a SEN. However, the presence of the object flow made explicit with UML 2 needs some comments.

Figure 7.2.2.1 takes an activity diagram in the Superstructure Book and converts it into SEN. In the SEN, the token in a place means many things. When a place is a state, a place owning a token is in this state. When a place is an action/activity, a place owning a token is currently executed. If the place is a condition, this condition is verified. If a place is an event, a place owning a token means that the event is occurring. A condition can be a result of an

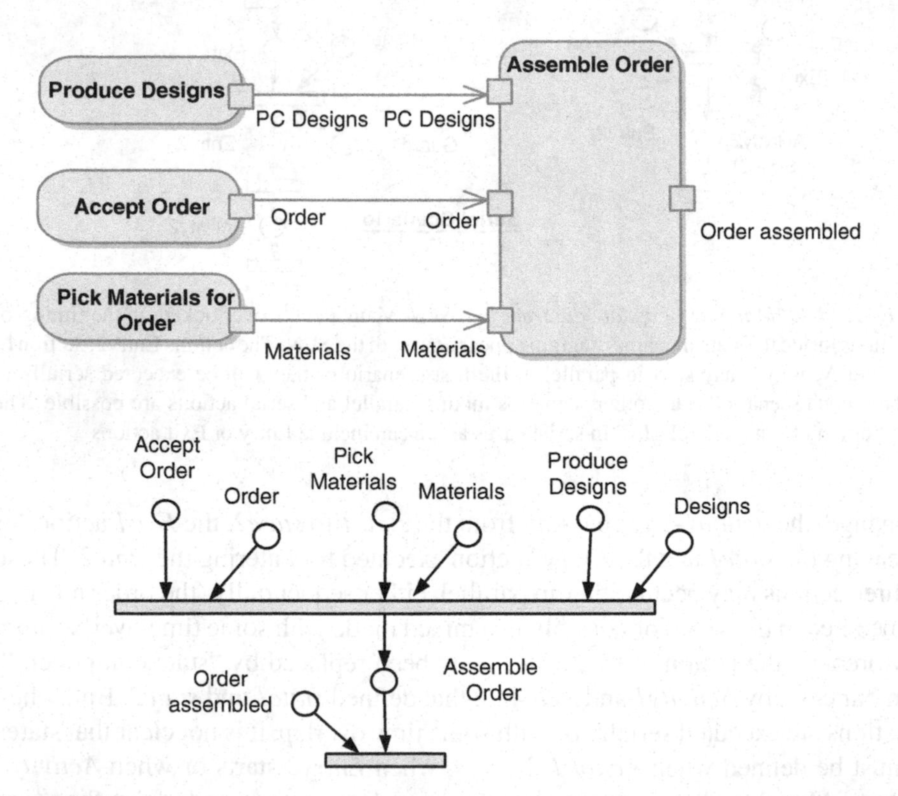

Fig. 7.2.2.1 Conversion of activity diagram with object flow into SEN. Example 12.126 of Superstructure Book. When each activity ends, it produces the necessary object so a token will be made available. We suppose that all activities must end before their tokens are made available. If needed, an end of activity sign (Fig. 7.1.5.1) can be added to make the interpretation more explicit

action/activity but this dependency between places are either not represented or are rather implicit (the same consideration is observed for an activity diagram, in which a guard that characterizes an outgoing edge may result from the execution of an upstream action located in a decision node or in an action node). So, a monochrome token cannot be really used to represent objects circulating in a SEN network. Following this logic, an object node is also a place, with a name that represents the object type. If a token is found in an object place, the interpretation is: object is available and made ready by the upstream activity and can now be used to feed as input to the next activity.

7.2.3 Sequence Diagram Mapped into SEN

If the state diagram and the activity diagram generally describes internal algorithms of systems or algorithms of many systems taken as a whole (previous activity diagram), intrinsically, sequence diagrams highlight the communication between objects, their interaction. As said earlier, arrows in the SEN syntax participate to the interpretation of the control flow, are used to support control structures. Dashed arrows exchanged between objects/systems are interobject communication. All arrows are of the same nature in a SEN, the distinction allows us to distinguish *intraobject interactions* or *interobject interactions*. In the activity diagram, some places are interdependent in the SEN and, normally arrows can be represented to highlight this dependence. As the interpretation is straightforward, they are generally removed. The Figure 7.2.3.1 shows this dependence. The same consideration happens for places in two different SEN, but now, the communication path must be made explicit to show the synchronization between many SEN places.

This synchronization concept with dashed arrows can be applied to other diagrams as statecharts (with regions) or parallel execution in the case of activity diagram (for instance, communication between places inside a fork/join region).

The interaction suite contains a second diagram, very closed to the sequence diagram, named *communication diagram*. This diagram shows all objects that collaborate together to perform a given task. Sequence of messages can be read with a numbering scheme affected to messages. This kind of diagram is halfway between structural and behavioral concepts. The structural part can be assumed by an *object diagram*. The behavioral part can be assumed by a sequence diagram in which the sequence is more explicit and easier to grasp. This diagram can be considered as a kind of "dynamic" object diagram.

7.2.4 Exception Handling

A token does not stay in SEN place unless there is a condition to force it to remain there (for instance a global *end of process* condition that forces all actions

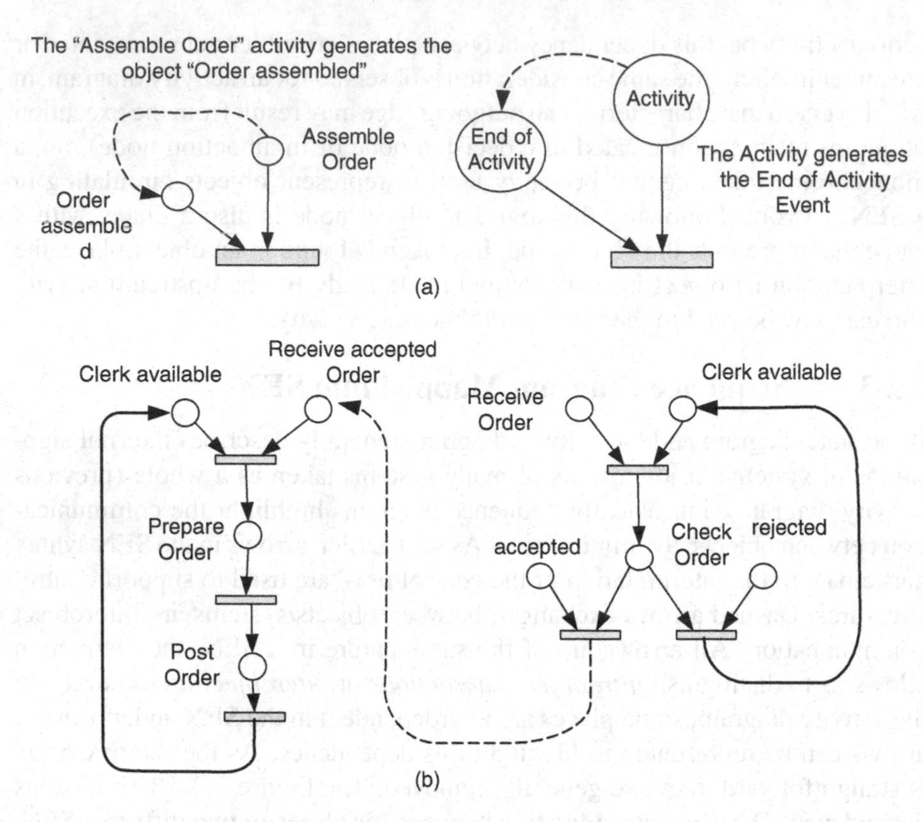

Fig. 7.2.3.1 Intraobject interactions and interobject interactions. (a) Shows the dependence of SEN places when interactions occur in the intra object mode. In (b), many SEN are synchronized through the same kind of interactions made explicit with dashed arrows. Intraobject interactions are generally not represented in SEN except if necessary

or activities to reach their end before exiting the place). Other conditions that retain token are, for instance, "next event not occurred", "condition not met", etc. on the bar connected to an outgoing edge.

When an interrupt occurs, it acts as a normal event on the bar connected to an outgoing edge. However, we must decide if this interrupt must follow the global policy (waiting for the action/activity to terminate) or force a precocious interruption of the currently executing task and entering an exception subnet. As the term "interrupt" means stopping a current process, it seems more normal to opt for the second possibility but it is not impossible that the first possibility still prevails in some situations. So, it is the burden of the modeler to decide on the appropriate scenario. If a global *end of process* condition is specified and we decide that the interrupt must stop the current activity, a lightning sign can be used to cancel this global condition locally.

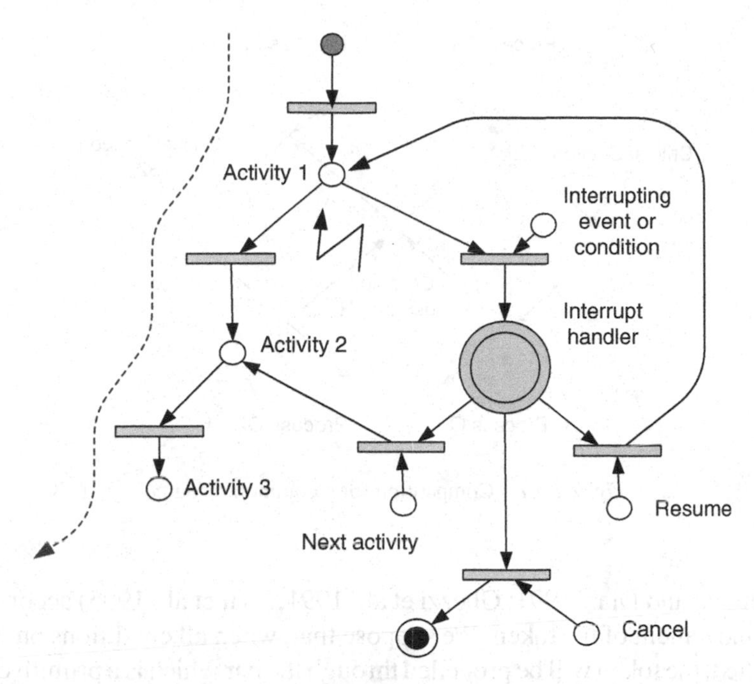

Fig. 7.2.4.1 Exception handling in a SEN. The exception is considered as a condition or an event that can force the place to stop its execution and go into an exception handler that is considered as a subnet

Figure 7.2.4.1 shows a standard way of representing an interrupt in a SEN. We suppose that globally, an *End of Process* condition is imposed on places. A lightning sign put on *Activity1* allows this activity to be interrupted by the *Interrupting Event or Condition.* Special processing after the interrupt will be defined in Interrupt handler that is supposed to return a token after ending its job. The token will have three choices: go to a *Cancel* state stopping everything, *Continue* with the next process, or *Return* to the interrupted process with other parameters.

7.2.5 Competition for Common Resource

If two persons need to print their job and the printer can print only one job at a time, the first job will monopolize the printer and the second job that arrive epsilon time after the first one must wait. A SEN can easily represent this case and many other cases of resource contention (Fig. 7.2.5.1).

7.3 Timing Constraints with SEN

For research on the use of Petri nets for controlled discrete Event Systems, please refer to survey article by Zhou and Dicesare (1993) and Holloway et al. (1997). Our goal in this direction is limited to the specification of time constraints

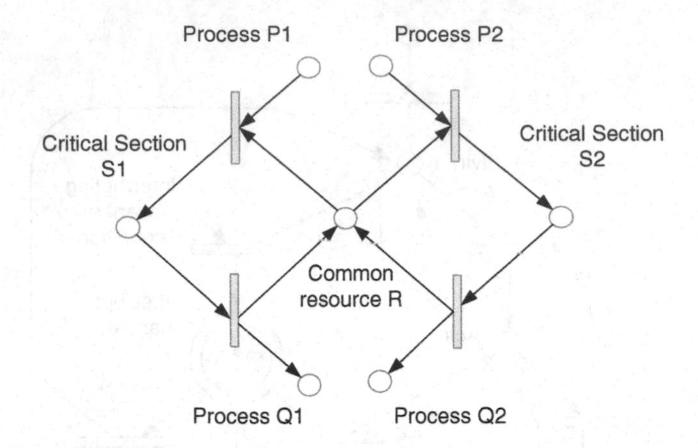

Fig. 7.2.5.1 Competition for a common resource

(Berthomieu and Diaz, 1991; Ghezzi et al., 1991; Tsai et al., 1995) accompanying the movement of the token. We suppose that, when all conditions on the bar are verified, the token will be propelled through the bar which is a primitive event with zero time duration. This attitude does not mean that the timing of the system is oversimplified. Actually, elementary actions are already detailed at each step to reach the most elementary action chunks (see the decomposition of UML state transition in Figure 7.2.1.1 which isolate various actions *Do, On Entry, On Exit* in the states and action accompanying the transition). Secondly, in a SEN, timing constraints are reported in places instead of arcs. When a timing constraint is satisfied, a token is generated in the corresponding place and the SEN evolves normally to the next state. If the timing constraint is not validated, the net must provide an exception to explain explicitly the way the system must handle such a case.

The dynamic analysis of a system can be done in two phases, the *functional analysis* and the *timing analysis*, in this order. We must assert the correctness of the functionality earlier without any timing constraint consideration. The timing analysis is then performed to assert the correctness of the timing behavior. This correctness is obtained if the system complies with the timing constraint specifications. We address to the first step of timing analysis: the *timing specification.*

> The timing analysis needs a reachability analysis. A place is reachable if it can get a token, even under restricted conditions. A SEN can be reachable in absence of timing constraint and not reachable after timing constraints are imposed. In complex cases, computer simulation can be used to test this reachability.

Abadi and Lamport (1994) mentioned that “the old fashioned methods handle real time by introducing a variable, which we call *now,* to represent time. This

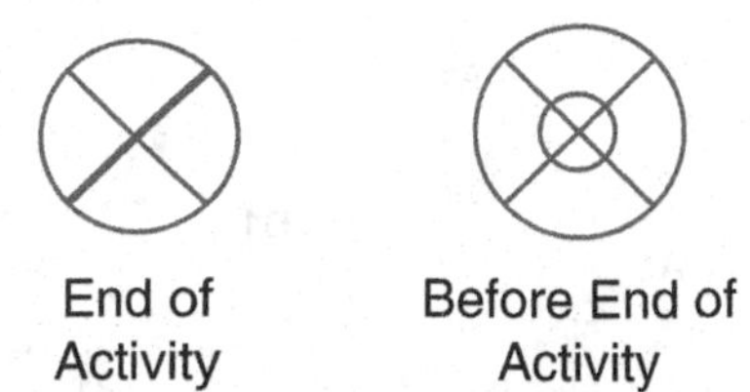

Fig. 7.3.0.1 Representation on the End of Activity (EOA) and Before End of Activity (BEOA) conditions. The EOA has been discussed in Figure 7.1.3.1. BEOA is the opposite logical condition, so a token is present in this place when an activity starts and disappears when it ends

idea is so simple and obvious that it seems hardly worth writing about, except that few people appear to be aware that it works in practice." Berthomieu and Diaz (1991) have analyzed concurrent systems whose behavior is dependent on explicit values of time. The number of scenarios for time specification is very large but fundamentally we can identify some current time constraints.

A process *duration* (or *end* variable if we suppose that this variable is reset each time a given process starts) can be bounded by a low limit and a high limit, for instance, $low \leq duration \leq high$ generates three cases to be considered. When process *duration* must exceed a low limit ($low \leq duration$) or when process *duration* must not exceed a high limit ($duration \leq high$), two cases have to be considered.

Another kind of specification relates a given point of time to another point of time (Fig. 7.3.0.2). For instance, if a process A must terminate before another process B, but their respective durations are not specified; Figure 7.3.0.1 elaborates a *Before End of Activity* (BEOA) condition that is the negation of *End of Activity* (EOA) condition already defined in Figure 7.1.5.1. Figure 7.3.0.3 shows an example of specification.

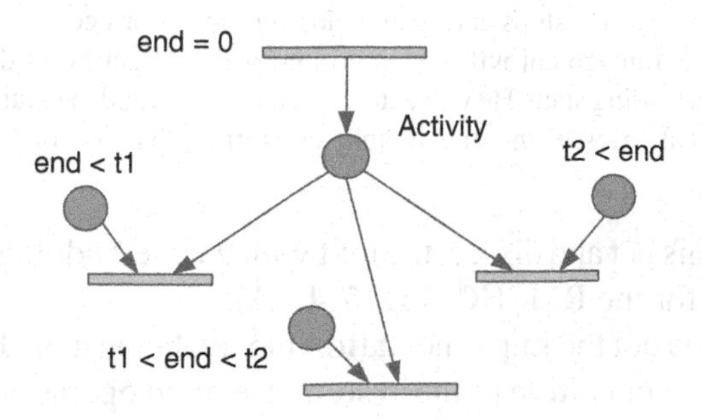

Fig. 7.3.0.2 Timing specification with variable "end". Timing limits can be specified with an end variable reset to zero when the activity starts. With two limits t1 and t2 ($t1 < t2$), the system may show three different behaviors considered in the SEN as three conditions

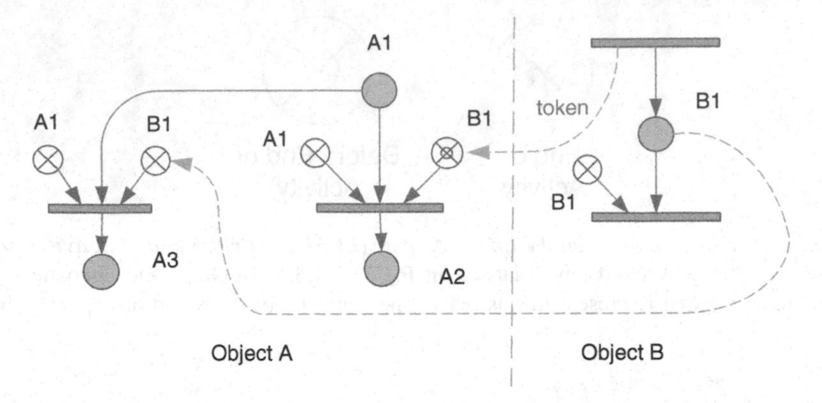

Fig. 7.3.0.3 Timing dependence between two objects A and B with End of Activity (EOA) and Before End of Activity (BEOA). Dashed arrows show the timing dependence of the two SEN but they are they can be omitted in real situations. The same name used for all the places is sufficient to connect these diagrams. An EOA place receives a token when an activity reaches its end. A BEOA must receive a token when the activity starts (when B1 receives a token, a second token is given to the BEOA named B1). When an activity ends, its corresponding BEOA disappears. If A1 finishes before B1, then A1 goes to A2. If A1 finishes after B1, then A1 goes to A3

7.4 Case Study with SEN

A Tamagotchi (or Tamaguchi) is a digital pet (virtual digital companion) created in 1996 by Ali Maita and sold by a Japanese Toy Manufacturer Bandai Co., Ltd. Hereafter, we describe an imaginary and simple toy that simulates the working principle of a Tamagotchi to study the dynamic behavior of this system with a SEN. Hereafter is the description of an oversimplified Tamagotchi (Table 7.4.0.1).

> When starting a Tamagotchi cycle, he enters first in his Happy state. But this state lasts only during TA seconds called Autonomy Time. At the expiration of TA, the Tamagotchi gets hungry and cries. A person must put the Tamagotchi at the table so he can eat. He stops crying and eats during TF or Feeding Time. At the end of TF, the Tamagotchi will cry again and we must get him off the table to avoid the overfeeding state. He will enter his Happy state and the Autonomy Time reapplies. In all cases, if the Tamagotchi cries more than TC or Critical Time, he will die.

We can analyze this pet and dissect the text with the methodology of text analysis already exposed for the RMSHC (Fig. 7.4.0.1).

Figure 7.4.0.2 is not the implementation model, but just the logical model explaining the overall operation of the system. Detailed operations setting various states of the Tamagotchi are not mentioned. Dashed arrows represent communication messages between SEN belonging to different objects. In this diagram,

Table 7.4.0.1 Text Content Analysis of the Tamagotchi specification

Step	*Initial text*	*UC concepts*	*Reformulation*
1	When starting a Tamagotchi	Actor/Object	Tamagotchi
1a	cycle, he enters first in his Happy state	State/Activity	Stay happy
2	But this state lasts only during	Actor/Object	Timer
2a	TA seconds called Autonomy Time	UC/Action	Arm Timer with TA (Autonomy Time)
2b		UC	Decrement TA
3	At the expiration of TA, the	Event	TA expired
	Tamagotchi gets hungry and cries	UC/State/Activity	Get hungry and cry
4	We must put the Tamagotchi at table so he can eat	Actor/Object	Person (who feeds the Tamagotchi)
4a		UC/Action	Put the Tamagotchi at the table
4b		UC/Activity	(Tamagotchi) eat
5	He stops crying and eats during	UC/Action	Stop crying
5a	TF or Feeding Time	UC/Action	Arm Timer with TF
5b		UC/Activity	Decrement TF
6	At the end of TF, the Tamagotchi	Event	TF expired
6a	will cry again and we must get	UC/State/Activity	Get overfed and cry
6b	him off the table to avoid the overfeeding state	UC/Action	Get Tamagotchi off the table
7	He will enter his Happy state as and the Autonomy Time will apply		(No new abstractions. Come back to 1a)
8	In all cases, if the Tamagotchi	Constraint	In all cases (when crying)
8a	cries more than TC or Critical Time, he will die	UC/Action	Arm Timer with TC (Critical Time)
8b		UC/Activity	Decrement TC
8c		Event	TC expired
8d		UC/Action	Tamagotchi dies

we have voluntarily represented sub-SEN places of Person and Timer on the same diagram to show the real interactions but it is not necessary to do so in a real project.

As said earlier, Figures 7.4.0.2 and 7.4.0.3 build the first logical model of the Tamagotchi. This model explains and identifies all high-level operations and evident attributes (e.g. *Tamagotchi state* that is an *Enum* parameter taking the following values: *Happy, Hungry, Overfed, Died*). Now, if we try to simulate the Tamagotchi with a computer or if we want to make an electronic pet, a button will be added to allow the Person to interact with the Tamagotchi (Fig 7.4.0.4). The separation of class and operation diagrams is an advantage as we can store all

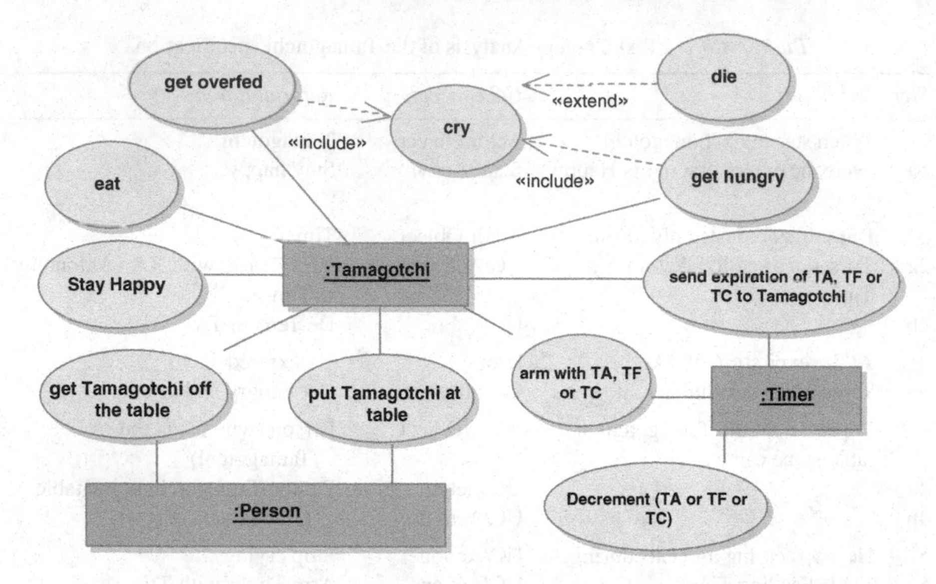

Fig. 7.4.0.1 Use Case (UC) diagram representing activities of the Tamagotchi, Person, and Timer. The object (or actor) Timer is part of the Tamagotchi but this fact is not explicit in the UC diagram. When getting hungry or overfed, the Tamagotchi cries so this action is included inside "*get hungry*" and "*get overfed*" use cases. The Tamagotchi can cry to death if he is not saved in time

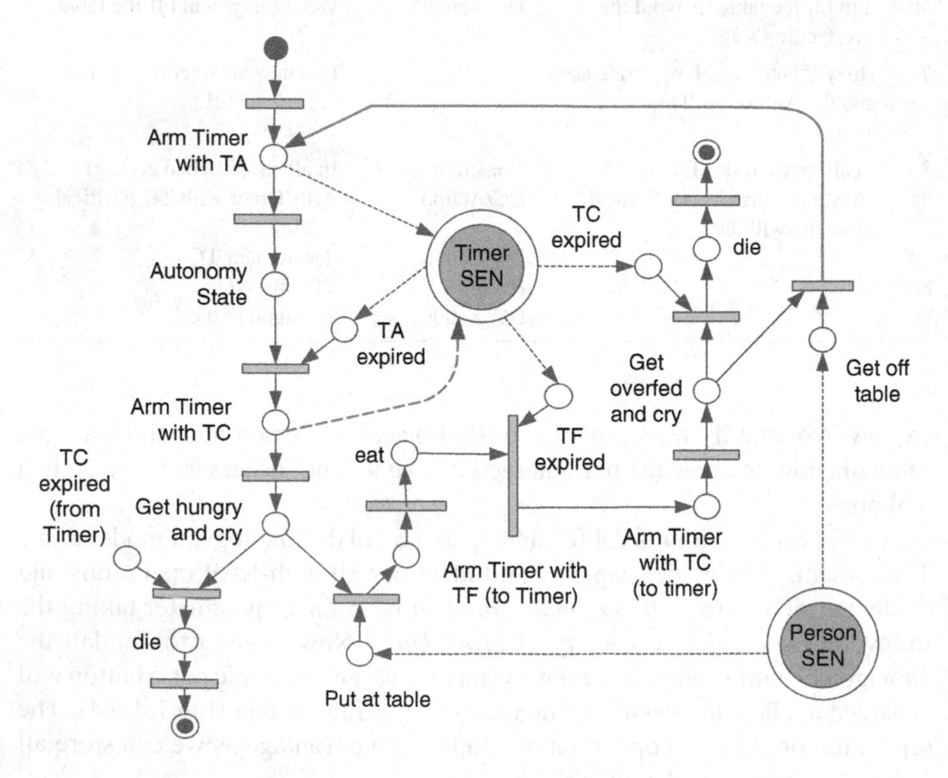

Fig. 7.4.0.2 SEN representing the state evolution of the Tamagotchi

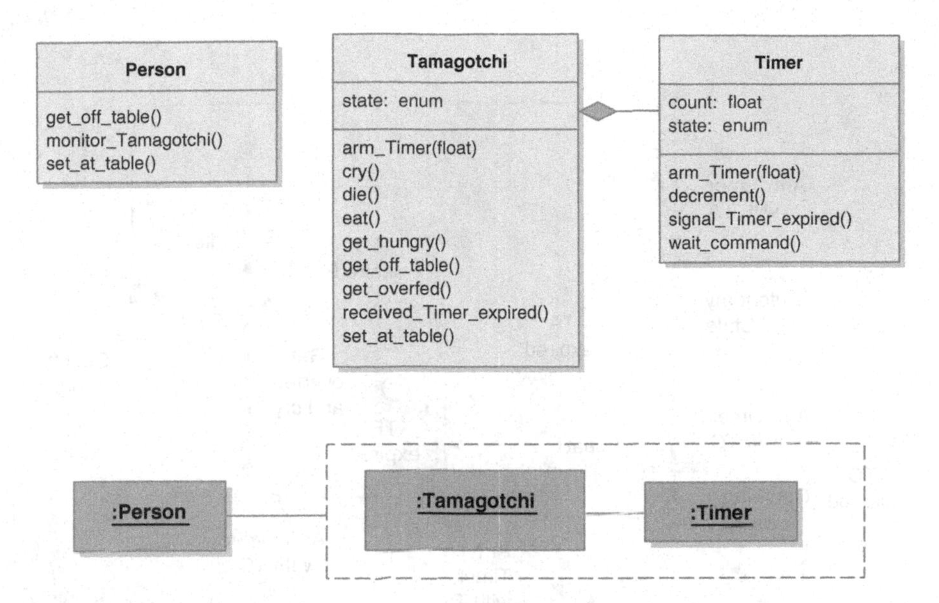

Fig. 7.4.0.3 Class and object diagrams of the Tamagotchi. In class diagram, classes are shown with their attributes, operations, and relationships. Object diagram shows communication links between the object *:Person* and the whole Tamagotchi system made of the *:Tamagotchi* himself and its internal biological *:Timer*. The model is still a logical model and does not contain any implementation details

classes of all applications inside the class package but instantiate several object system for various implementations of the Tamagotchi. For instance, when making an electronic pet, we can add a sound system to simulate various states of the Tamagotchi. New objects and new interactions are therefore concerns of subsequent models.

7.5 Safety-Critical Systems

Wikipedia.org defines *safety-critical system* as *life-critical system* and as "a system whose failure or malfunction may result in death or serious injury to people, loss or severe damage to equipment, or environmental harm." By mentioning "damage to equipment," human life may be implied indirectly and the effect cannot be immediate (environment harm). So, this concept involves a wider class of applications than expected. Safety is always considered with respect to the whole system, including software, hardware (electronic, mechanical, biological, chemical, etc.), users, and operators.

Traditionally, safety critical software has been associated with embedded, embarked, and real-time systems (military, transport, communication, medical, security, nuclear power, etc.). Its scope has expanded recently into many other contexts (internet, household apparatus, etc.). An aircraft with hundreds of passengers is totally dependent upon fully operational and safety-critical systems.

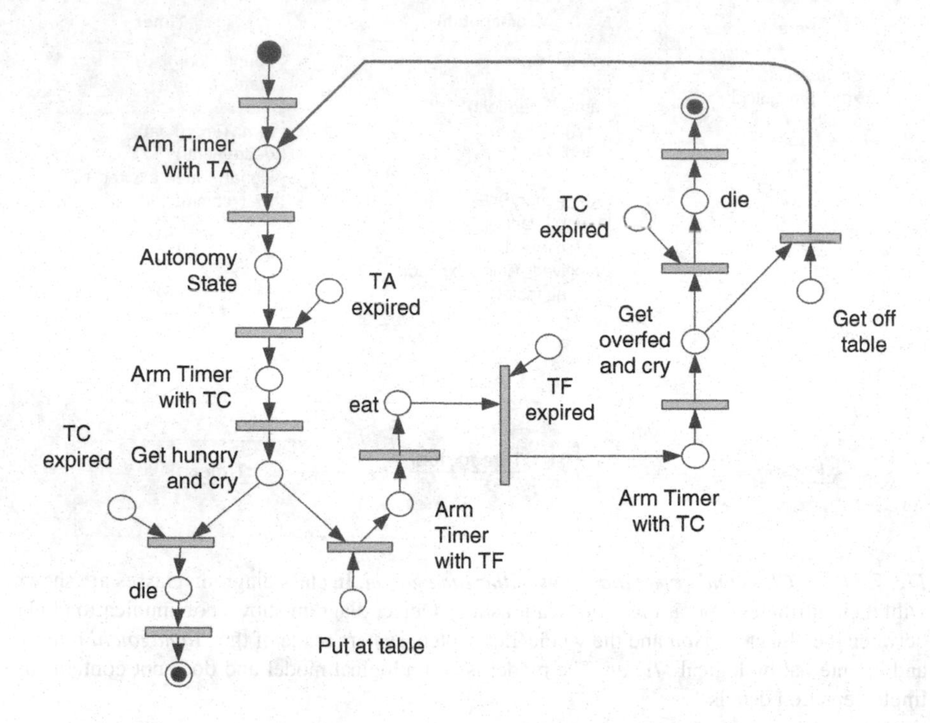

Note 1: All timer expiration signals come from the Timer and all Arming Actions are directed to the Timer
Note 2: "Put at table" and "Get off table" are operations executed by Person object.

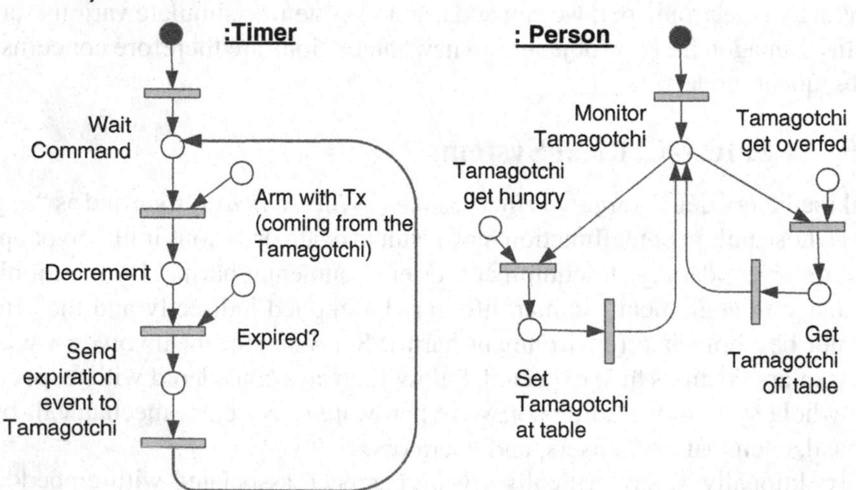

Fig. 7.4.0.4 SEN versions of the Tamagotchi application with separate SEN. Cross-interactions between objects are not shown but they are parts of class and object diagrams

Railway signaling systems preventing trains from colliding, or more generally traffic regulation, are real time and safety critical. Medical systems are safety critical as they are directly responsible for human life. Recent accidents of collapsing buildings in 2001 have thrown civil engineering structures into safety-critical class systems. Errors in software calculation or software simulation could be fatal. Many systems in a car are now software based (ABS brakes) and could potentially fail and create accidents.

The questions when developing safety-critical systems are "Can we present hazards and reduce their occurrence?" and "Can we verify that a given developed system is safe?" The vast majority of software safety is entirely in the hands and conscience of the software developers and suppliers. The first step in developing a system is performing a preliminary hazard analysis, to determine whether the system could present a hazard to safety. If the answer is yes, we must conduct a more detailed hazard analysis and safety engineering (Leveson and Stolzy, 1986; Villemeur, 1991; Isaksen et al., 1996; Lutz, 2000).

From these articles, a *hazard* is a potentially dangerous situation that can cause an accident. A hazard may come from an internal error or fault. A *failure* is seen as an event that leads to the accident. The *risk* is associated to the probability of the hazard.

Security is a concept related to the confidentiality of a system, malicious interference, unauthorized access, and privacy. In internet communication systems, there is a close relationship between security and safety; a hole in the security is a hazard for safety.

When discussing safety of software, software can become important when it is used to control potentially dangerous hardware. Moreover, safety can be dependent on human factors (a bad man–machine interface can cause misinterpretation and hazard; a patient could be given the wrong medication even with the best hospital database, so control mechanisms must be implemented to avoid even human error). Even if the software is bullet-proof and it can be established that there is no software error when an accident occurs, a badly designed system is implicitly a faulty design as it can create potentially hazardous situations for normal humans with normal reactions. The large number of states found in software usually makes exhaustive testing impossible. There is no proportionality between the hazard and the disaster it can cause. A very small omission or error may have a strong impact on safety. So, software safety is an important issue and no details can be forgotten in a development if safety is the main concern. In a multidisciplinary system, it would be more difficult to assert the safety factor when dealing with various components coming from many disciplines. This fact pleads for a uniform way of analyzing and designing a whole system (that is the subject of this book) so all factors can be judged and analyzed on the same conceptual and logical basis regarding safety independently from the discipline a piece of hardware or software. The safety consideration starts at the beginning of the development process (specification) then takes into account all the design, implementation, and testing. Special tools must be developed to assert that safety has been observed correctly. A well-designed system is the first step towards safety.

7.5.1 State Space Search with Combinatorial Method: Image Attribute Methodology

Let us consider the example of the Tamagotchi. States of the Tamagotchi have been proposed previously without any systematic methodology. The developer thinks that the electronic pet can be in the five states:

1. Normal and Happy (Normal = Happy)
2. Get hungry and Cry
3. Eating when been sit at table
4. Get overfed and Cry at table
5. Died

He then proposes the algorithm of Figure 7.4.0.3. The questions are "Are all states enumerated?" "Are there forgotten states?" We cannot demonstrate that the states enumerated are all possible states that can be observed with the Tamagotchi.

If we consider the following internal and binary variables of the Tamagotchi: *happy, hungry, cry, eat, be at table, die*, we get six binary variables with a possibility of $2^6 = 64$ combinations. So, as a first observation, the behavior of a system depends upon the *change of its internal variables*.

But, the Tamagotchi interacts with an internal Timer and an external Person. So, his state or his state change is influenced by events sent by other objects (e.g. Timer expiration event) or the actions performed by other objects on him (when the Person puts the Tamagotchi at the Table, he stops crying and eats). As a second observation, the behavior of a system depends upon *inputs* to this system (Fig. 7.5.1.1).

More surprising, when analyzing many systems, we observe that the state of an object can be conditioned by the actions this object performed on its environment or the events it creates to influence objects in its neighborhood. If the Person performs an action on the Tamagotchi, he/she supposes it will change its behavior later and stop crying, so the person has internally an "image" of the expected behavior of the Tamagotchi. If this image does not correspond to

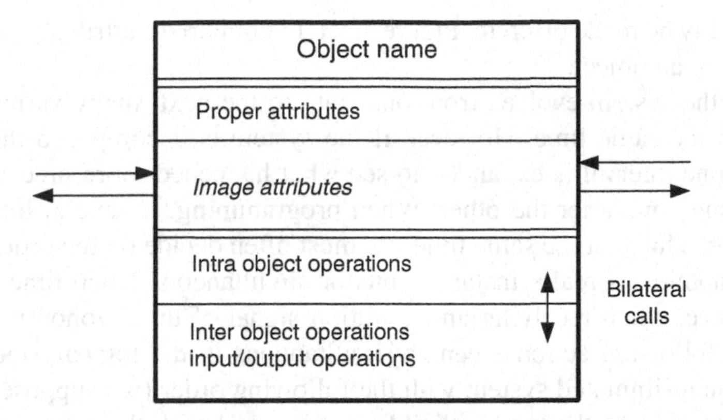

Fig. 7.5.1.1 Attributes and image attributes of an object. Proper attributes are those that describe the behavior of the object, e.g. state changes, within intraobject operations. Image attributes are those that object models its environment that impact on its proper behavior. If the object needs to communicate or exchange messages with its environment, image attributes can be built from input/output types and values. An object has internal resources (input operations) to capture all incoming events, can change its state if it is subjected to external actions. It also has internal resources (output operations) to act on the environment or send messages/events to surrounding objects. It has image attributes to compare the actual behavior of other objects to its proper references

the reality (the pet does not stop crying), in this case, more appropriate actions could be programmed. In the context of this problem, the person does not have any means to afford more appropriate actions as the person is designed with only two actions (put the pet at the table or get the pet off the table). Many real-time systems have a richer set of actions and algorithms to deal with real situations. A closed loop can therefore be established between the controlling and the controlled objects so the actual behavior can be adjusted in real time to the image of the "correct behavior." For an open loop system, the behavior of a system is effectively dependent on the actions performed blindly, for instance, if we make a redirection on a phone, we suppose it will redirect all calls and we adjust our behavior accordingly (even if this redirection is not effective). So, as a third observation, the behavior of a system depends upon *outputs* from this system.

To study a system and its behavior, it is necessary to enumerate all attributes, proper and image attributes, determine their type, and their variation interval. For analog variables, the simplest case occurs when an analog variable can be made discrete. For instance, the comfort temperature of our home is between 20°C and 24°C, under 20°C, the temperature is qualified as cold and above 24°C, the temperature is qualified as hot. Most analog systems present some threshold values settling the accepted tolerance even if the analog variables cannot really be made discrete. Figure 7.5.1.1 summarizes attributes and image attributes of an object.

When the system evolves from one state to the next, many variables may change at the same time. However, if the system is decomposed thoroughly and the time interval is expanded to see what happened more precisely, variables change one after the other. When programming, if several instructions must be scheduled at the same time, we must often decide on the order to issue them as nothing is really instantaneous or simultaneous when time expands. For instance, when receiving an expiration signal of the autonomy time TA, the three following actions seen as simultaneous in the logical model is sequenced in a simulated system with the following order (we suppose the time required to execute them is negligible if compared with the time scale of the application).

Set hungry attribute to ON
Set cry attribute to ON
Perform Cry() operation

The next step is making an inventory of all proper and image attributes of the Tamagotchi. As explained in the theoretical part of the UML, an association between classes is translated into links between objects. Very often, in an object diagram, a unique link is drawn between two objects to mean that they can communicate. When analyzing with the *Attribute Image method*, each component (input or output) of the amalgamated link must be separated and attached to a separate image attribute.

Table 7.5.1.1 lists nine Boolean or enum variables with a maximum combination value of 3072. This astronomical value has in fact few effects on the way we set up the method as a lot of combinations are impossible. For instance, if the Tamagotchi dies, all values or the remaining variables are insignificant, or the Tamagotchi cannot be simultaneously happy and crying.

Instead of enumerating all the variables in a disordered way, we can explore them systematically by building successive real and functional sequences with a SEN diagram as shown in the Table of sequences of Figure 7.5.1.2. These sequences, packed together, will constitute the global algorithm of the Tamagotchi. Unsuspected or transient states can therefore be evidenced.

As the number of variables is important, we make three vectors, the first one with an ordered set of regular attributes and others with image attributes of their respective object to save space in Table 7.5.1.2. Each line of this table corresponds to a fundamental abstraction (action, activity, event, state, etc.) of the dynamic view.

When the table of sequences contains the same set of values for all attributes, but the interpretation is first an action then later a state (e.g. 21 and

Table 7.5.1.1 Table of attributes and image attributes of the Tamagotchi. Six are proper to the Tamagotchi, two enum variables connect the Tamagotchi to its internal Timer object, and one enum variable connects the Tamagotchi to the Peron object

N	*Variable*	*Type*	*Values*	*Comments*
Tamagotchi attributes				
1	Happy	Boolean	1–0	Normal state when entering
2	Hungry	Boolean	1–0	
3	Cry	Boolean	1–0	
4	Be at table	Boolean	1–0	
5	Eat	Boolean	1–0	
6	Die	Boolean	1–0	
Tamagotchi image attributes (facing Timer object)				
10	Arm Timer with T_x (output)	Enum	Idle, TA, TF, TC	Idle is a state meaning that the Tamagotchi issues no command
11	Expired signal (input)	Enum	Idle, TAX, TFX, TCX	Idle, TA expired, TF expired, TC expired
Tamagotchi image attributes (facing Person object)				
20	Act on the Tamagotchi (input)	Enum	Idle, PAT, GOT	PAT: put at table, GOT: get off table
Total number of potential combinations: $2^6 \times 4 \times 4 \times 3 = 3072$				

21a; 31 and 31a), same numbers with different lettering scheme were used to account for this particularity. The explanation is "when entering a state, some actions are executed to set values inside the state and these values stay unchanged."

As loops are unavoidable in the table of sequence, the same point must be identified with the same number (without parentheses when the line is defined for the first time, with parentheses later, e.g. 4, 6a, and 10a). They correspond to the same place on a SEN diagram. When many elementary actions occur (e.g. 8, 21, 31, etc.), they are considered in this logical model as "without duration" and executed in parallel unless there is some logical dependency between these elementary actions. Figure 7.5.1.2 is a new SEN obtained after this combinatorial analysis with the image attribute method. Now, we can claim that there are no forgotten states or actions and the design is safe at this point.

7.5.2 Developing Safety-Critical Systems

Despite the large number of potential combinations, the *method of image attributes* is deterministic, combinatorial, and can be applied surprisingly in

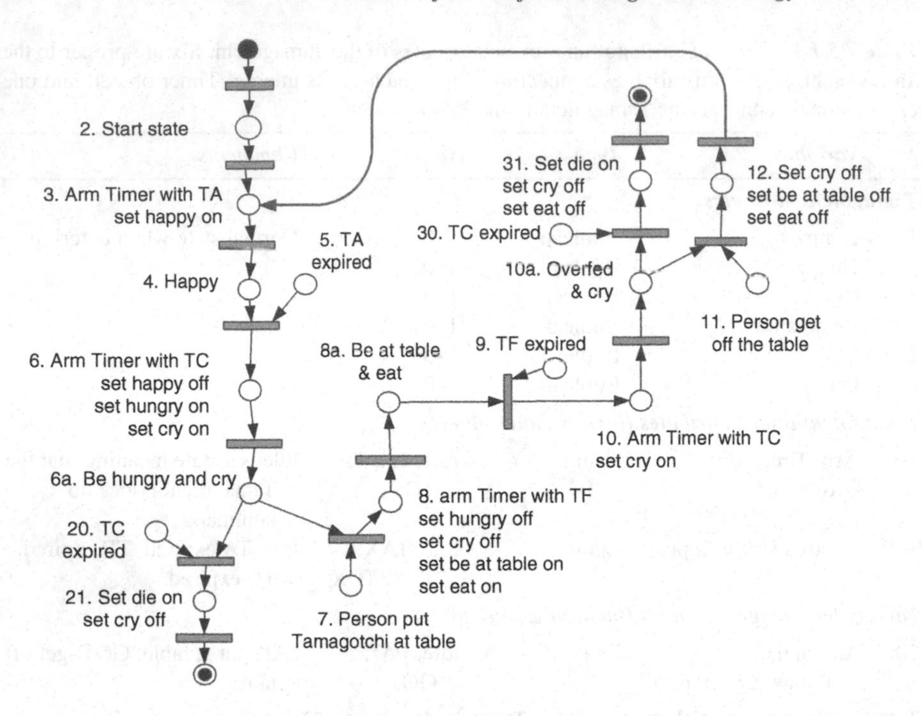

Fig. 7.5.1.2 SEN of the Tamagotchi established after an extensive study with the method of image attributes. This diagram is more precise then Figure 7.4.0.3 executed without any systematic investigation. Notice also the forgotten action labeled #12 undetected in the previous SEN. From the safety-critical viewpoint, we need just a minuscule error to get a big accident. (21a) and (31a) make a distinction how the Tamagotchi died (hungry or overfed)

several apparently complex situations. Actually, the method explores only sequences that need to be implemented. Along these sequences, all attributes and image attributes are highlighted, their values identified, and their effects in the system evaluated. By examining all sequences (to be implemented) obtained from evolutions of their attribute values, minor details and most elementary actions can be identified. Moreover, interactions of this object with surrounding objects can be evidenced by image attributes. Designing errors are naturally avoided as all elementary evolutional steps can be seriously tracked and ordered. If conducted correctly, it eliminates virtually all hazards coming from forgotten states in a design if this design can be studied in a deterministic way. Moreover, this method already offers testing steps to attest the functionalities of a system.

When developing a complex system, the problem is not to inventory all the possible scenarios but, owing to the development cost, only very well-identified functionalities and their corresponding scenarios (or sequences) need to be designed. Subsequent functionalities can be added at a later stage if needed. The system can be made safe if it cannot reach any state not listed in the table

Table 7.5.1.2 Table of Sequences built from variations of state values of Tamagotchi regular attributes and image attributes

N	*Concept identification*	*Value sets* *[happy, hungry, cry, be at table, eat, die]* *[Arm Timer T_x, expired signal]* *[action of Person on Tamagotchi]*	*Comments or Names*
Main sequence			
1	Initial state	[all undefined]	Starting point
2	State	[0, 0, 0, 0, 0, 0] [idle, idle] [idle]	Initialization
3	Action	[1, 0, 0, 0, 0, 0] [TA, idle] [idle]	Arm Timer with TA (transient action so TA is removed immediately) Set happy on
4	State	[1, 0, 0, 0, 0, 0] [idle, idle] [idle]	Tamagotchi happy
5	Event	[1, 0, 0, 0, 0, 0] [idle, TAX] [idle]	Event TA expired received
6	Action	[0, 1, 1, 0, 0, 0] [TC, TAX] [idle]	Arm critical time TC Set happy off Set hungry on Set cry on
6a	State	[0, 1, 1, 0, 0, 0] [idle, idle] [idle]	Be hungry and cry
7	Event	[0, 1, 1, 0, 0, 0] [idle, idle] [PAT]	Person puts Tamagotchi at table
8	Action	[0, 0, 0, 1, 1, 0] [TF, idle] [idle]	Arm Timer with TF Set hungry off Set cry off Set be at table on Set eat on
8a	State	[0, 0, 0, 1, 1, 0] [idle, idle] [idle]	Be at table and eat
9	Event	[0, 0, 0, 1, 1, 0] [idle, TFX] [idle]	Event TF expired received
10	Action	[0, 0, 1, 1, 1, 0] [TC, idle] [idle]	Arm Timer with TC set cry on (still eating to overfed)
10a	State	[0, 0, 1, 1, 1, 0] [idle, idle] [idle]	Overfed and cry
11	Event	[0, 0, 1, 1, 1, 0] [idle, idle] [GOT]	Person gets Tamagotchi off the table
12	Action	[1, 0, 0, 0, 0, 0] [TA, idle] [idle]	Arm Timer with TA Set cry off Set be at table off Set eat off Set happy on
(4)	State	[1, 0, 0, 0, 0, 0] [idle, idle] [idle]	(Come back to state 4)
Cry to death, been hungry			
(6a)	State	[0, 1, 1, 0, 0, 0][idle, idle] [idle]	Be hungry and cry
20	Event	[0, 1, 1, 0, 0, 0] [idle, TCX] [idle]	Event TC expired received

(*Cont.*)

Table 7.5.1.2 (Continued)

N	*Concept identification*	*Value sets [happy, hungry, cry, be at table, eat, die] [Arm Timer T_x, expired signal] [action of Person on Tamagotchi]*	*Comments or Names*
21	Action	[0, 1, 0, 0, 0, 1] [idle, idle] [idle]	Set die on Set cry off
21a	State	[0, 1, 0, 0, 0, 1] [idle, idle] [idle]	Died been hungry
Cry to death, been overfed			
(10a)	State	[0, 0, 1, 1, 1, 0] [idle, idle] [idle]	Overfed and cry
30	Event	[0, 0, 1, 1, 1, 0] [idle, TCX] [idle]	Event TC expired received
31	Action	[0, 0, 0, 1, 0, 1] [idle, idle] [idle]	Set die on Set cry off Set eat off
31a	State	[0, 0, 0, 1, 0, 1] [idle, idle] [idle]	Died been overfed

of sequences. So, it must be designed to fulfill a given set of functionalities (that could be extended later) and, at the same time, it cannot deviate from a set of paths already drawn in the execution tree. In the vocabulary of Petri nets, only a given set of places are reachable, all other places are forbidden because they are not studied and verified. Hazards or potentially dangerous situations are eliminated with this development attitude.

For instance, in electronic engineering field, if a processor must not accept interruptions and there is no exception handler written for managing interruptions, hardware interrupts must not be left uncontrolled. Their voltage must be fixed to some secure value and the Enable Interrupt bit must be masked to avoid any hazard that could come from unexpected interrupts (coming from electric perturbations or strong electric or magnetic fields). When analyzing the working model of a processor, all these factors are listed as attributes and they will draw the attention of the developer.

7.5.3 Method of Image Attributes Applied to Human and Social Organizations

There are many subtleties on the way we search the image attribute. If we take this term at its original meaning, "image" does not mean reality but an "opinion" of this reality. In real-time systems, image is reality since we are an external observer, we master all the objects in the system and their behaviors are generally deterministic. The question is "could this method be applied to human and social organizations?"

The behavior of an object depends upon the image of the environment (surrounding objects) that this object is building in its microworld. Very often, the image we forge about other people or things does not agree with reality. This image may be imperfect, oversimplified, or biased. Moreover, if an object has some form of intelligence, the image it makes of itself may not correspond to the real and objective image (that is never reachable in fact). Images are only "projections." Even so, we make projections of our proper personalities.

There is a difference in the way the images are constructed in human organizations and real-time systems. In a constellation of real-time objects devoted to accomplish high-level tasks, we search only image attributes of the first neighbors of a given object, searching their interactions at first level of interactions. This attitude is sufficient enough for many problems. In human organizations, the images go beyond the frontier of this first neighborhood. In human systems, we are accustomed to make images of everything, to judge everything and the microworld that conditions our behavior, which often oversteps our bounds. For good planning, the image of a good manager is that of someone having an accurate image on everything (human and situations). "Is it really useful or not?" We do not want to explore it yet.

> Image attributes are fundamental for deploying an advertising campaign for a company. The type of image this company wishes to convey should be elaborated from an understanding of its current image and its competitors' images. Using a series of "image attributes" (low prices, courtesy, high tech, fast delivery, etc.), the company calculates the relative importance of each attribute by establishing a mathematical model between these attributes and their impacts.

Images induce perception and dictate behaviors. While reasoning on human organizations, decision-making software can still use object technologies in a creative way by introducing a modeling concept based on the *method of image attributes* explained for real-time systems. The behavior of an object or agent is not dependent only on its original characteristics. Its behavior is highly conditioned by the interaction network it maintains with surrounding objects. To fully quantify these interactions, image attributes can be created inside each object showing how it models its microworld. If we can put values inside attributes, or at least enumerate them, computation models could then be built to study and anticipate object/agent reactions. From the object viewpoint, there is a departure from the way object technology usually handles objects. It is a really interesting research issue. A class must now have multiple compartments, both for attributes and operations. Attributes proper to an object will occupy only one compartment. Others are reserved to model the image it builds on other objects. In the simplest case, we have only the first neighbors interacting directly with this object. In a complex case, we must add more than the first neighbors. Operations available in the object are now dependent on attributes and image attributes and the behavior of the object can now be studied as usual considering that image attributes acts on the object in the same way.

This suggestion unifies the way the behavior of an object is studied and modeled with the UML and object tools. From the outside, an object/agent taken in a social or human context has the same "presentation" and characteristics as all objects encountered in programming environment. They have attributes, can work with operations, and have states. The difference is not seen on the presentation but by the way this object has been designed internally. It contains information on itself and information on the whole system as its second nature.

Chapter 8

Case Studies

8.1 Design of an Inclined Elevator or Track-Lift Tram

Our purpose aims to illustrate the Uniform Methodology by choosing a system made of objects taken from various disciplines. A simple inclined elevator going between two levels has all ingredients of such a system as it contains mechanical components (cabin, door, mechanical brakes), electromechanical components (buttons, microswitches, contact sensors, motors), intelligent components (microcontroller, electronic command for variable speed motors), biological (humans making use of the system), and computer software (program that give intelligence to the controller).

This simple and classical system is easily understood by everybody so it does not add other difficulties to the understanding of the methodology itself. Moreover, this system has inputs, outputs, close-loop control, hardware, software, and safety features (we can stop the tram with the Stop button in an emergency situation, a stalling motor triggers a breaker to go off) so we can expose all exception handling of this system and give a complete example of testable PIM model. On the other hand, we add some supplementary constraints in order to illustrate how to distribute constraints among development phases.

8.1.1 Early Requirements

A company LiftTram wants to manufacture an inclined elevator. Track-lift trams need an even slope. If the slope is not even and the terrain is rugged, a pair of cables can be used instead of the pair of tracks but fundamentally the two designs are based on the same engineering principle which includes a cabin running on wheels on either fixed tracks or static cables. The cabin is powered by a motorized drum hoist that drives the cabin up and down the tracks/cables. Two to four aerial cables (providing high safety at any charge) are used to move the cabin. Redundant limit switches are used at the high and low landings to ensure smooth stops.

The power is provided by a high efficiency three phase gear-motor with an internal brake (Motor Brake). If power is accidentally off, this brake locks

D. M. Bui, *Real Time Object Uniform Design Methodology with UML*, 417–465.

the motor shaft in place. The variable frequency motor drive allows gradual acceleration and deceleration.

The cabin must have a large surface of glass panels to allow a clear view of the surroundings and the cabin must be designed in such a way that it can be customized to meet the client needs.

The cabin is equipped with mechanical brakes to lock it to the tracks in the event of slack aerial cables (Slack Cable Brake). If the tension on aerial cables is removed, tension detectors activate the Slack Cable Brake, lock the cabin to the tracks, and stop the motor.

The cabin is supported on the tracks with castors; two to four of them have each a centrifugal overspeed safety brake (Overspeed Brake) that will be activated if the speed reaches 15% over the rated speed.

A control unit is available inside of the cabin. This allows the operator to stop anywhere, besides automatic stops at the starting and the arriving points. The cabin has safety switches at the front and back to stop the cabin in the event of an obstruction on the tracks (Track Jam Brake).

If for an unknown reason, the battery of limit switches fail to stop the cabin at the high landing, a detection of a stalling motor (Current Surge Detector) stops the motor. At the low landing, the Slack Cable Brake has the same effect by activating the Slack Cable Brake.

Manual commands can be issued inside the cabin and the same commands are duplicated outside the cabin at each landing (three sets). Essentially, each security brake category has its corresponding lights red and green (the red light means "brake activated" and the green light means "brake released"). Each brake has a corresponding set and release button. If a brake is set manually, a signal is sent to the motor to stop it. We have also general "all brakes activated" and "all brakes released" buttons with their corresponding lights. The button "all brake activated can serve as emergency stop while the cabin is in motion."

A Cabin Door Lock is a magnetic switch used to prevent the cabin from moving in the event that the door is not completely closed. This lock also prevents the door from being opened while the cabin is moving. The door itself has a pair of monitoring lights.

To move the cabin, the operator glides a cursor up or down proportionally with the speed. When arriving at landings, the motor brake stops the cabin. To move the cabin in the opposite direction, the operator moves the cursor in the neutral zone before sending a release brake. "Release Brake" signal cannot be activated if the cursor is not in the neutral zone.

The following free form diagrams (Figs. 8.1.1.1 and 8.1.1.2) locate main components of the system.

Complementary information can be obtained at ASME (American Society of Mechanical Engineers). This organization (*http://www.asme.org*) has published

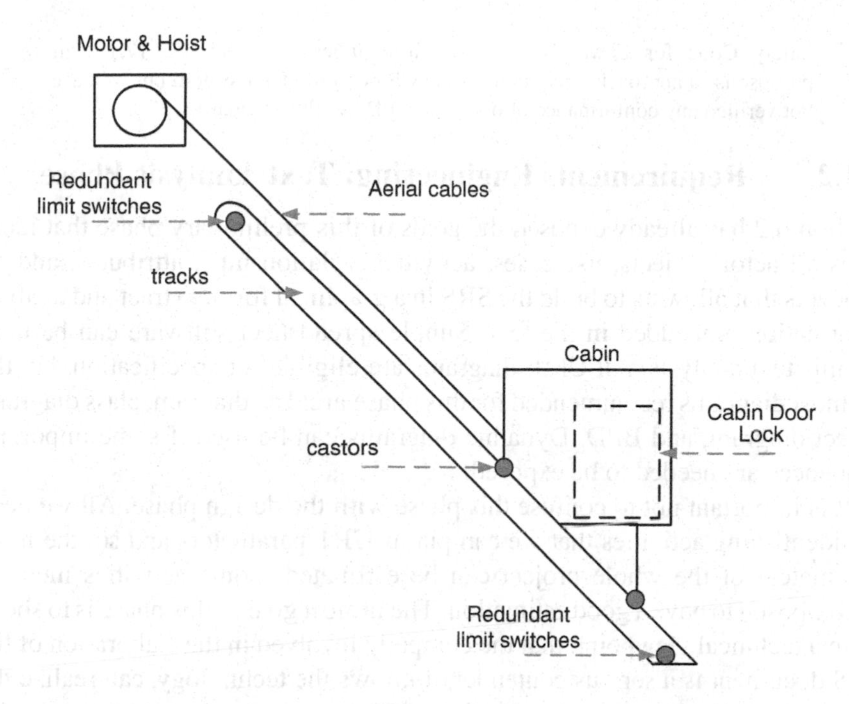

Fig. 8.1.1.1 Free form diagram showing main components of the inclined elevator

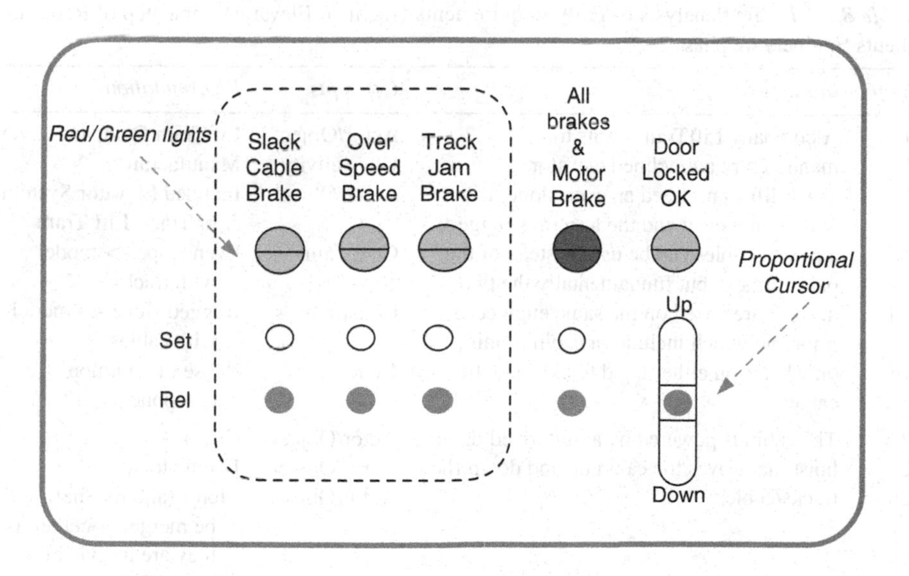

Fig. 8.1.1.2 Control Panel of the inclined elevator

Safety Code for Elevators and Escalators under the reference 17.1. Our purpose is to approach a real case to show how to model a system but we have not verified any conformance of our logical PIM to this standard.

8.1.2 Requirements Engineering: Text Analysis Phase

Section 6.2 has already exposed the goals of this preliminary phase that identifies all actors/objects, use cases, activities, relationships, attributes, and all concepts that allow us to build the SRS in a graphical form, extract and analyze information embedded in the text. Simple spreadsheet software can be used for this text analysis. All UML diagrams are eligible for specification, but the topmost diagrams recommended for this phase are: UC diagram, class diagram, object diagram, and BPD. Dynamic diagrams can be used if some important sequences are needed to be exposed.

It is important not to confuse this phase with the design phase. All we need are identifying activities that we can put as CRT parameters and so, the main parameters of the whole project can be estimated. Some activities must be decomposed to have a good estimation. The utmost goal of this phase is to show, from a technical viewpoint, that the company involved in the elaboration of the SRS document is a serious contender, it knows the technology, can realize the project, and has given a good estimation of CRT parameters, without necessarily exposing all its secret know-hows (design knowledge).

Table 8.1.2.1 Text analysis of Early Requirements (Inclined Elevator). First step of Requirements Engineering phase

Step	*Initial text*	*Concepts*	*Reformulation*
1	A company LiftTram wants to	Actor/Object	LiftTram Company
1a	manufacture an inclined elevator.	UC/Activity	Manufacture
1b	Track-lift trams need an even slope. If the slope is not even and the terrain is rugged,	Actor/Object	Inclined Elevator System or Track-Lift Tram
1c	a pair of cables can be used instead of the pair of tracks but fundamentally the two	Constraint	Even slope -> model with tracks
1d	designs are based on the same engineering principle which includes a cabin running	Constraint	Rugged slope -> model with cables
1e	on wheels on either fixed tracks or static cables	Goal	Reuse of common components
2	The cabin is powered by a motorized drum	Actor/Object	Cabin
2a	hoist that drives the cabin up and down the	Actor/Object	Drum Hoist
2b	tracks/cables	Actor/Object	Motor (and its Shaft will be merged together as they are indivisible)
2c		Actor/Object	Guiding Tracks/Cables
2d		UC/Activity	Power (Motor power Drum Hoist)

(*Cont.*)

Table 8.1.2.1 (Continued)

Step	*Initial text*	*Concepts*	*Reformulation*
3	Two to four aerial cables (providing high	Actor/Object	Aerial Cable
3a	safety at any charge) are used to move the	Multiplicity	2..4 (Aerial Cable)
3b	cabin. Redundant limit switches are used	UC/Activity	Pull (Cable pulls Cabin)
3c	at the high and low landings to ensure	Actor/Object	Limit Switches Low
3d	smooth stops	Actor/Object	Limit Switches High
3e		Actor/Object	Low Landing Position
3f		Actor/Object	High Landing Position
3g		Goal	Smooth stopping
4	The power is provided by a high efficiency three-phase gear-motor with an internal	Attribute	3-phase (attribute of Motor)
4a	brake (Motor Brake). If power is	Actor/Object	Internal Motor Brake
4b	accidentally off, this brake locks the motor	UC/Activity	Lock if power off
4c	shaft in place	Actor/Object	Motor Shaft (merged into Motor)
5	The variable frequency motor drive allows gradual acceleration and deceleration	Actor/Object	Variable Frequency Drive
5a		UC/Activity	Accelerate motor gradually
5b		UC/Activity	Decelerate motor gradually
5c		UC/Activity	Keep speed
6	The cabin must have a large surface of	Actor/Object	Glass Panel
6a	glass panels to allow a clear view of the	Multiplicity	* (many)
6b	surroundings and the cabin must be designed in such a way that it can be	Relationship	Contain (Cabin contains Glass Panels)
6c	customized to meet the client needs	Goal	Clear view of the surroundings
6d		Goal	Customization to meet Client needs
7	The cabin is equipped with mechanical	Actor/Object	Slack Cable Brake
7a	brakes to lock the cabin to the tracks in the	Multiplicity	* (many)
7b	event of slack aerial cables (Slack Cable	UC/Activity	Lock Cabin to Tracks
7c	Brake). If the tension on aerial cables is removed, tension detectors activate the	Event	Slack tension (Aerial Cable)
7d	Slack Cable Brake, lock the cabin to the	Actor/Object	Tension Detector
7e	tracks, and stop the motor	UC/Activity	Stop Motor
8	The cabin is supported on the tracks with	Actor/Object	Castor
8a	castors; two to four of them have each a	Multiplicity	4 (Castor on Tracks)
8b	centrifugal over-speed safety brake	Actor/Object	Overspeed Brake
8c	(Overspeed Brake) that will be activated if	Relationship	Integrate (to Castor)
8d	the speed reaches 15% over the rated speed	Multiplicity	2..4
8e		Event	Overspeed detected
8f		Actor/Object	Overspeed Detector

(Cont.)

Table 8.1.2.1 (Continued)

Step	*Initial text*	*Concepts*	*Reformulation*
8g		UC/Activity	Activate (Overspeed Detector Activate Brake)
8h		Condition	Speed 15% over rated spccd
9	A control unit is available inside of the	Actor/Object	Control Unit
9a	cabin. This allows the operator to stop	UC/Activity	Stop (Cabin) anywhere
9b	anywhere, besides automatic stops at the	UC/Activity	Stop at Low Landing
9c	starting and the arriving points. The cabin	UC/Activity	Stop at High Landing
9d	has safety switches at the front and back to	Actor/Object	Front Safety Switch
9e	stop the cabin in the event of an obstruc-	Actor/Object	Back Safety Switch
9f	tion on the tracks (Track Jam Brake)	UC/Activity	Stop if obstruction
9g		Actor/Object	Obstruction Object
9h		Actor/Object	Track Jam Brake
9i		UC/Activity	Activate (Safety Switch activates Track Jam Brake)
9j		Relationship	Contain (Cabin has Safety Switches)
10	If for an unknown reason, the battery of	Condition	Limit Switches fail
10a	limit switches fail to stop the cabin at the	Actor/Object	Current Surge Detector
10b	high landing, a detection of a stalling	UC/Activity	Detect surge
10c	motor (Current Surge Detector) stops the motor. At the low landing, the Slack Cable	UC/Activity	Stop Motor at high landing
	Brake has the same effect by activating the Slack Cable Brake	UC/Activity	(Slack Cable Brake) stop Motor at low landing
11	Manual commands can be issued inside the cabin and the same commands are	Multiplicity	3 (Control Unit mentioned at step 9)
11a	duplicated outside the cabin at each landing (three sets). Essentially, each security brake category has its	Actor/Object	Red/Green Lights (for each category of brakes and all brakes)
	corresponding lights red and green (the red	UC/Activity	Set lights on/off
11b	light means "brake activated" and the green light means "brake released"). Each brake has a corresponding set and release	Actor/Object	Set/Release Buttons (one set for each category of brake)
11c	button. If a brake is set manually, a signal is sent to the motor to stop it. We also in general have "all brakes activated" and	UC/Activity	Set brake on/off (one set for each category of brake)
11d	"all brakes released" buttons with their corresponding lights. The button "all brakc activated" can serve as an emergency stop	Actor/Object	"Set all brakes" Button (Emergency Stop interpretation)
11e	while the cabin is in motion.	Actor/Object	"Release all brakes" Button
11f		UC/Activity	Set all brakes on/off

(*Cont.*)

Table 8.1.2.1 (*Continued*)

Step	*Initial text*	*Concepts*	*Reformulation*
11g		UC/Activity	Set motor off (when one kind of brake is on)
12	A Cabin Door Lock is a magnetic switch	Actor/Object	Cabin Door Lock
12a	used to prevent the cabin from moving in	Actor/Object	Door
12b	the event that the door is not completely closed. This lock also prevents the door	Relationship	Contain (Cabin Door Lock attached to Door)
12c	from being opened while the cabin is moving. The door also has a pair of	Condition	Movement enable if Door Completely Closed
12d	monitoring lights	UC/Activity	Open Door during movement
12e		Actor/Object	Red/Green Lights (for Cabin Door Lock)
12f		UC/Activity	Set lights on/off (Cabin Door Lock)
13	To move the cabin, the operator glides a	Actor/Object	Cursor for speed control
13a	cursor up or down proportionally with the speed. When arriving at landings, the	UC/Activity	Move up/down proportionally
13b	motor brake stops the cabin. To move the	UC/Activity	Move to neutral zone
13c	cabin in the opposite direction, the operator moves the cursor in the neutral zone before sending a release brake.	Condition	Cursor must be in Neutral zone for releasing any brake
13d	"Release Brake" signal cannot be activated if the cursor is not in the neutral zone	Actor/Object	Operator of the Cabin

8.1.3 Building Classes for Instantiating Actors in Use Case Diagrams

As said in Section 6.3, our choice was to dissociate the requirements engineering objects completely from the design phase. The correspondence between objects, components, or functions will be established by a *mapping process*. Objects identified in a very short time frame (requirements analysis) are used to estimate the project importance and cannot be kept "as is" when addressing the complex design phase. Reuse considerations (classes/components will be merged from the organization reuse database), industrial secret cannot be unveiled at requirement analysis.

Moreover, a class identified at this phase generally has just identification. They serve to instantiate numerous objects needed to explain use cases and BPDs (or occasionally other dynamic diagrams). Attributes are absent except when their presence is unavoidable. Important operations/activities are translated into uses cases and more elementary ones are not relevant in this context. Figure 8.1.3.1 gives a list of classes identified by the text analysis process and is grouped according to their technological domains.

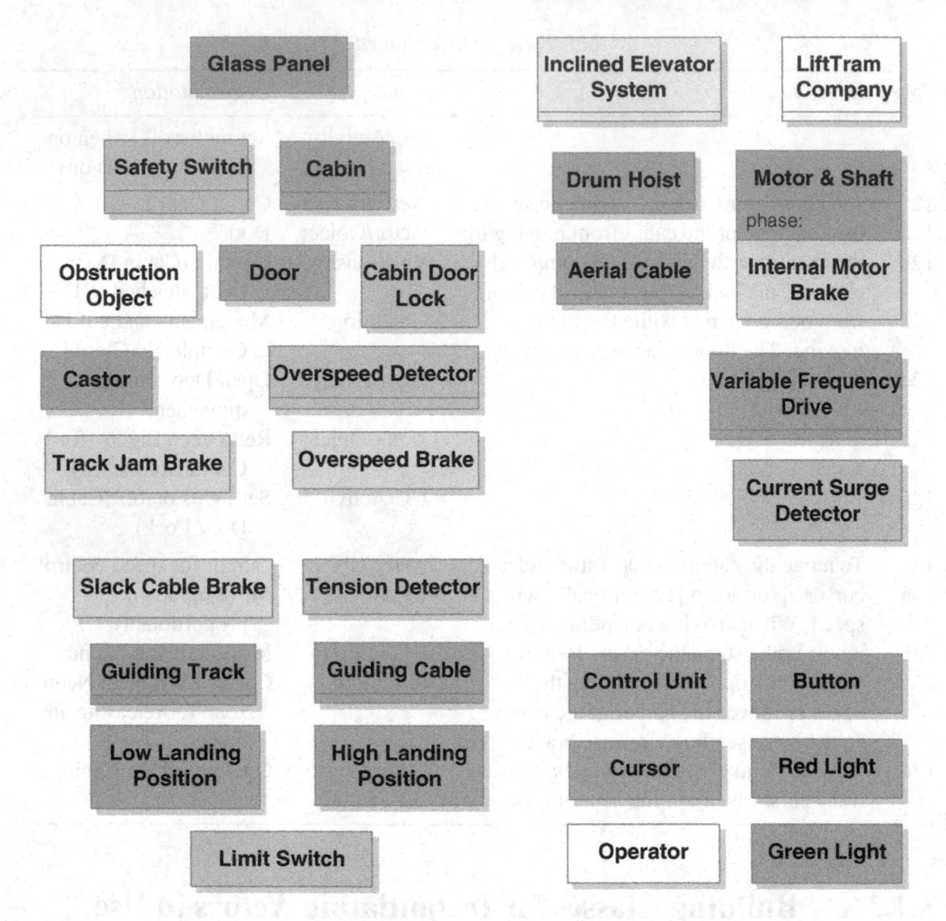

Fig. 8.1.3.1 Classes of the Inclined Elevator identified by Text Analysis of Early Requirements Specification

Classes of Figure 8.1.3.1 are generic and afford a first "evident" clustering of objects; for instance, *Safety Switch* will later give rise to *Front: Safety Switch* and *Front: Safety Switch* objects, *Limit Switch* will later give rise to *Low: Limit Switch* and *High: Limit Switch* objects. All buttons will be instantiated from a unique class *Button*. For *Red Light* and *Green Light*, we have the choice between making a unique class LED Light and putting an attribute *color* or making two different classes. We have chosen to show design variations. We make two classes as they correspond to two different electronic components (some LED integrate two colors into one light). For *Track Jam Brake*, *Overspeed Brake*, and *Slack Cable Brake*, we know from engineering experience that they are "brake type" and can be implemented later as a same brake system activated from several sensors. But, at the requirements, it is not mandatory to reveal early any technological solution. So, from a perspective of specification, it would be more interesting to specify functions (so why use cases are used) instead of specifying objects. Many functions can be implemented later by the same object and so project costs can be lowered accordingly. As a consequence of this discussion, some solutions at the Requirements Engineering phase may appear as not based on solid design criteria from a "designer perspective." But, as this preliminary phase is not a rigorous design phase and must be done in a very limited time frame, the impact of this "inconsistency" would be minor.

8.1.3.1 Requirements Diagrams. The previous class diagram was established without any relationships. The class diagram can be, at the limit, not joined to the SRS document as objects/actors are only relevant in UC diagrams that follow. The following diagrams reproduce rigorously clauses of Table 8.1.2.1 but mix clauses together and reorganize them in a more understandable way. If diagrams are named UC1, UC2, UC3, etc. (Fig. 8.1.3.1.1). these names can be registered in a supplementary column of this table allowing cross referencing. Each UC diagram exposes a simple or complex clause, and a new idea of the specification and we must avoid the temptation to mix all clauses together. Comments are put immediately after each use case to explain the techniques.

A use case often involves many actors or many objects. The "simple line" connection means, as said in the theoretical part, "concern" or "imply." A use case involves several actors/objects and allows us to identify the underlying activity. It is not mandatory to represent all involved actors or objects, just the main ones. The "extend" relationship shows that the reuse is optional (in our case, it will be deployed whenever possible). Each diagram tells a "scenario in a story."

The motor contains its internal brake. It powers the whole system inside the contour. The contour is a "white box" showing a drum hoist associating with 2–4 aerial cables pulling the cabin. So, a UC diagram in our case mixes intimately a classical UC diagram to an object diagram. The link between Motor/Brake is of aggregation type, the link between Hoist and Cable objects is not necessary to be defined.

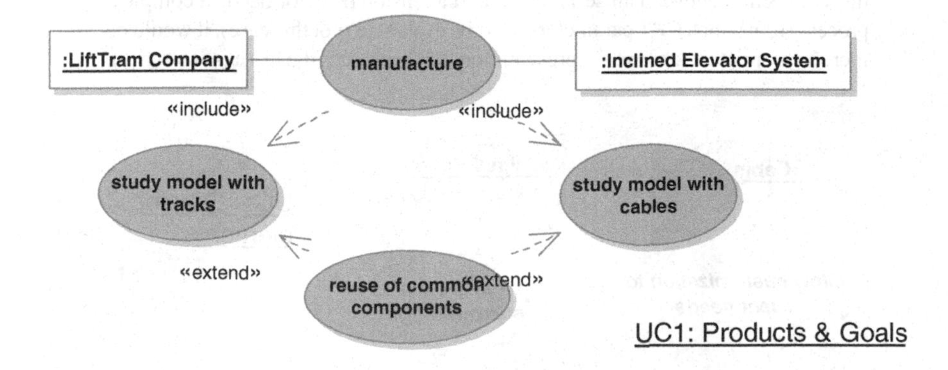

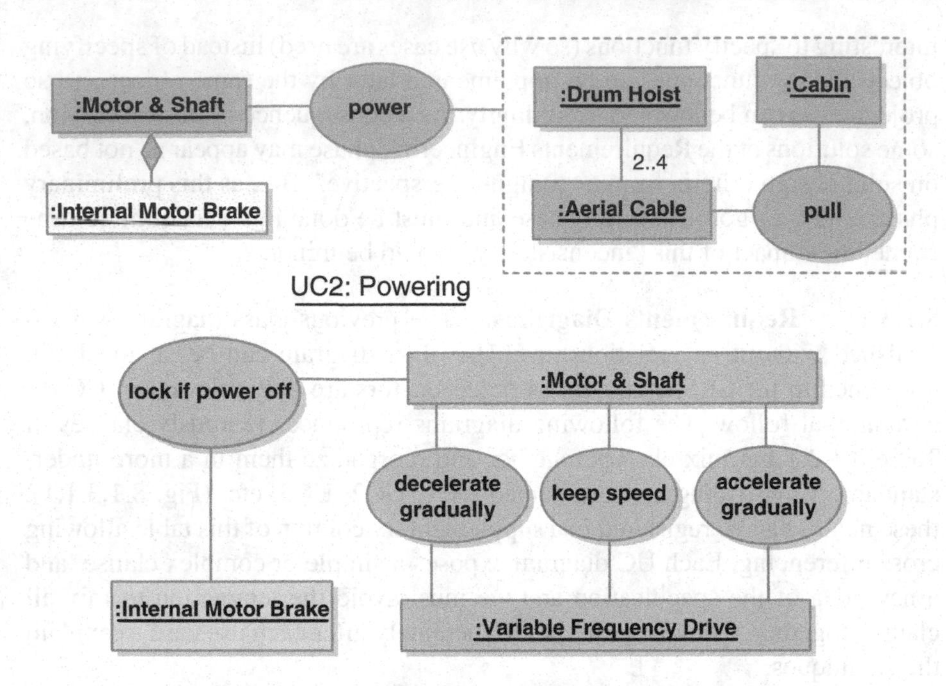

The internal motor brake locks the shaft (that is inseparable from the motor). A variable frequency drive powers the motor, accelerates, decelerates gradually, or keeps a constant speed.

The comfort elements normally will not enter into our logical design PIM but will be taken into account at the implementation phase. As said in the theoretical part, all constraints are disseminated among development models, some in PIM, others in PSM. In this diagram, we have an example of use case deployed to represent goals. In this case, we recommend making them abstract and transparent to distinguish them from regular UC that we must compute CRT parameters.

The stereotyped relationship <<equipped>> is drawn between two objects and, by this way, saves a use case. When saying for instance that "an aerial cable has a tension detector," we think of the "light" aggregation form. If mechanically, this connection between an aerial cable and a tension detector needs a complete process from which CRT parameters must be evaluated (not this case), it would be more appropriate to choose a representation that evidences the use case (activity).

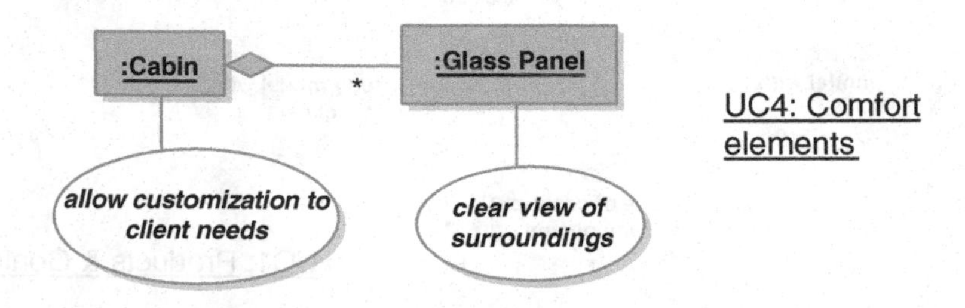

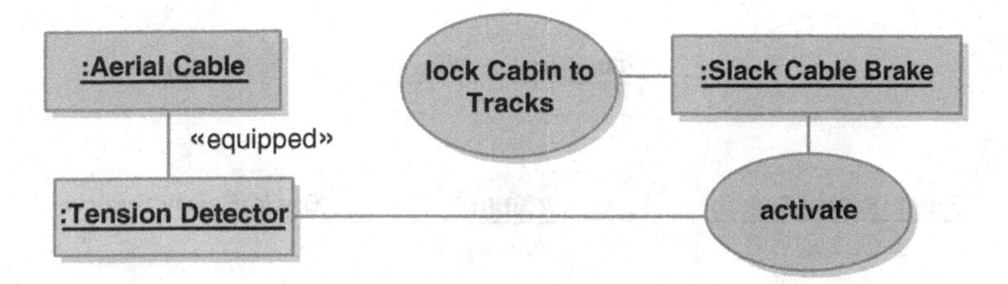

UC5: Slack Cable Protection and Tension Detector

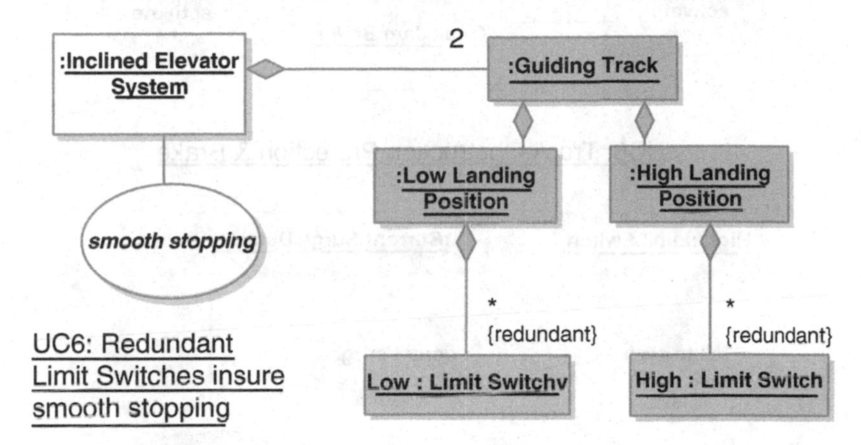

UC6: Redundant Limit Switches insure smooth stopping

Most aggregation/composition relationships can be interpreted as "is equipped with". They serve mainly to pack several components inside one logical description (See full discussion of this relationship in Section 5.5.5).

In a UC diagram, it would be difficult to represent a condition and make it visible as there is only use case and actor. We can choose to write down the condition "15% over rated speed" as comment, or in the constraint field of the use case (not visible in some tools). The simplest way is (if possible) entering it in the use case itself.

As said before, we create an object "Track Jam Brake" because at this phase, it would be irrelevant to discuss about solutions. So, objects of this kind may be introduced at requirements engineering phase to materialize a function that must be realized. "Obstruction Object" is not part of the system (it explains only the function that must be implemented) so its representation is alleviated.

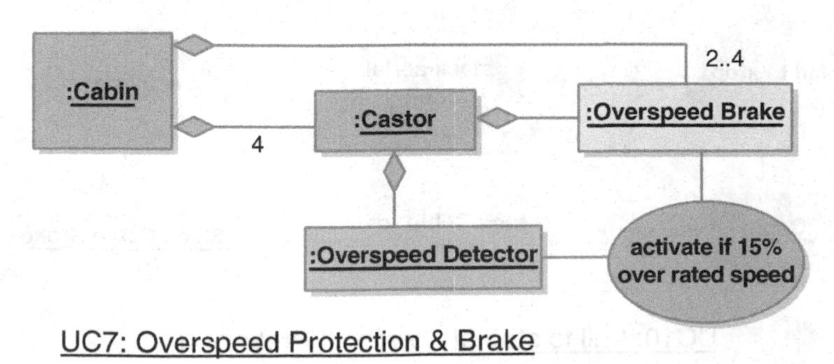

UC7: Overspeed Protection & Brake

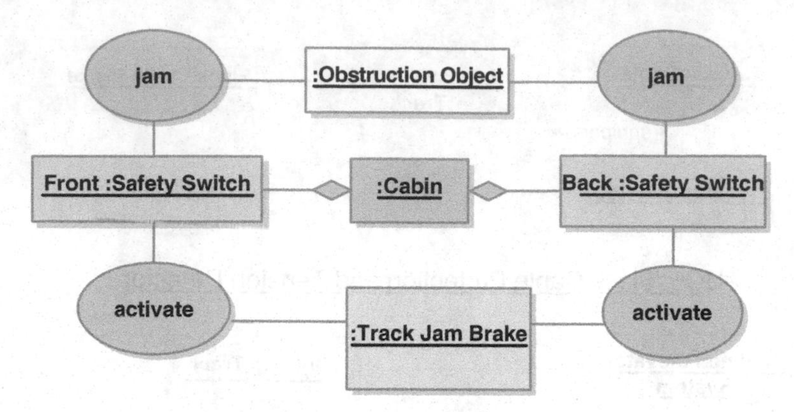

UC8: Track Obstruction Protection & Brake

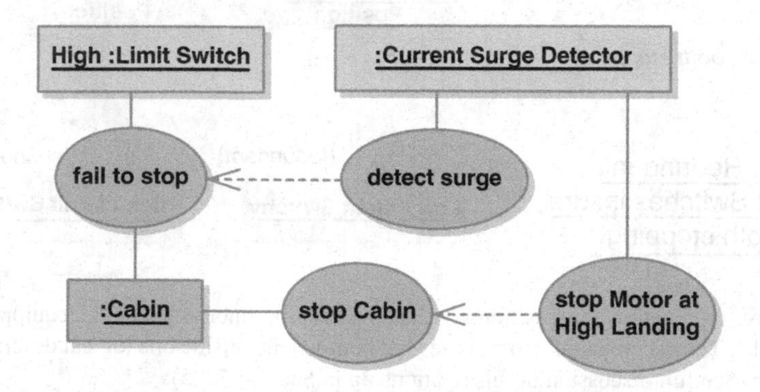

UC9: Fail to stop at High Limit & Protection

We have seen that the connection between two use cases can be <<include>> or <<extend>>, a series of actions in a cause-effect chain can be represented with the general <<dependency>> relationship generally not stereotyped. If the cabin fails to stop, a surge will occur. If the surge detector stops the motor, the

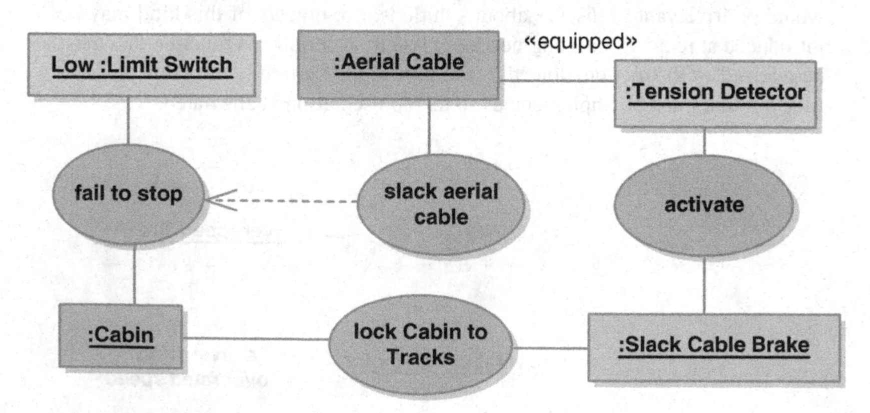

UC10: Fail to stop at Low Limit Protection

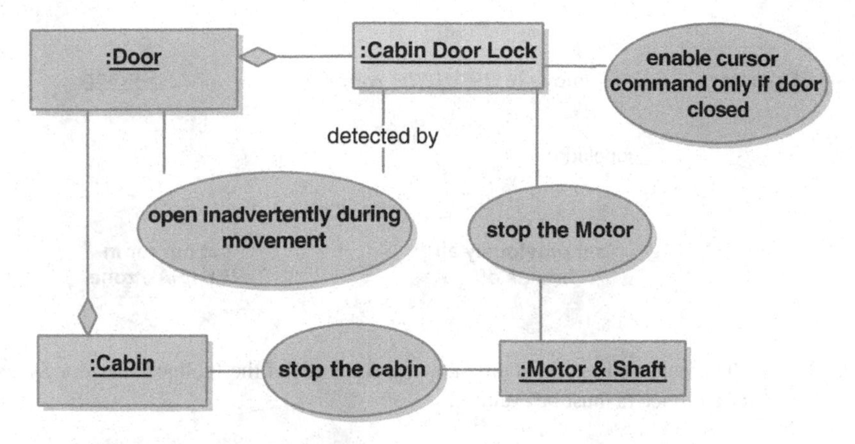

UC11: Door Lock Protection during movement. Cursor enabled only if Door locked

cabin will be stopped accordingly. The "stop Cabin" use case is tied to "stop Motor at High Landing" in a dependency relationship.

The UC12 diagram makes use of contour properties to save representation. The Control Unit has four sets of Red/Green Lights, Release/Set Buttons. Each set

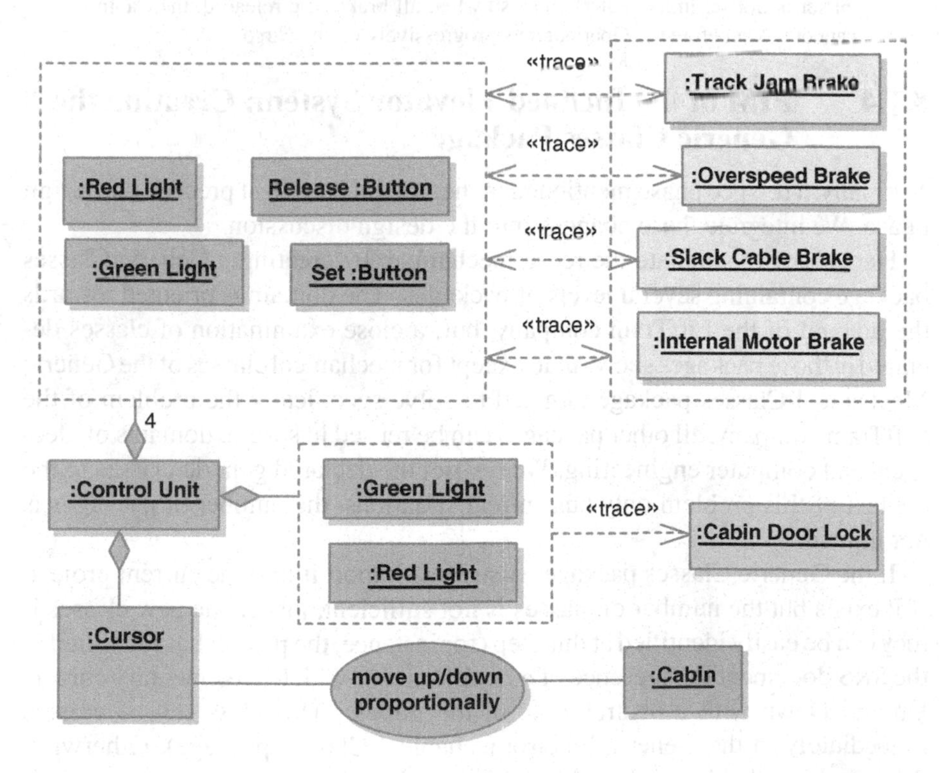

UC12: Control Unit

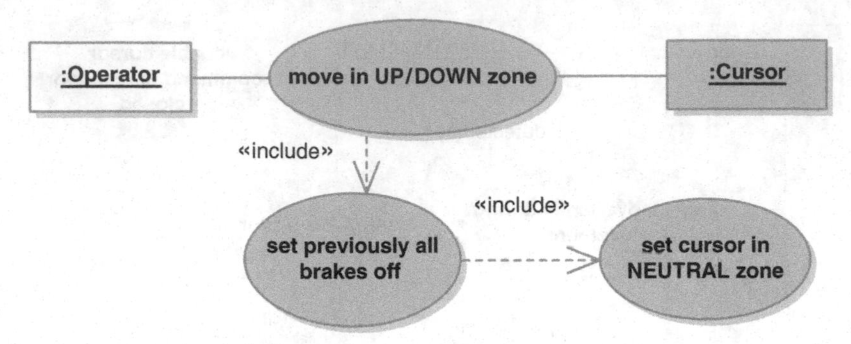

Fig. 8.1.3.1.1 Thirteen use case diagrams showing activities of the Inclined Elevator System from which CRT parameters must be evaluated

is attached to its corresponding protection. A <<trace>> relationship of bidirectional type is bound to each set as buttons are inputs and lights are outputs. Generally, this relationships is used when two actors/objects must evolve together with some dependence, but, for simplifying the representation, we cannot explicit the whole cause-effect chain.

The "include" relationship means "constraint" imposed to the use case. To move the cursor up and down, all brakes must be off. Brakes cannot be set to off if the cursor is not set in its neutral zone, so when all brakes are released, the Cabin cannot move before the Operator acts progressively on the cursor.

8.1.4 PIM of the Inclined Elevator System: Creating the Generic Classes Package

Normally, a re-spec phase mentioned in theoretical part must precede the design phase. We integrate these changes into the design discussion.

Hereafter, we simulate the reuse mechanism by creating a Generic Classes package containing several levels of packages. The domain is oriented towards the domain of the LiftTram company, but, a close examination of classes declared in those packages shows that, except for mechanical classes of the Generic Mechanical Classes package targeted to solve specifically the problem of the LiftTram company, all other packages can be reused in several domains of electrical and computer engineering. We restrict the declared generic classes to the context of this problem only but, in real situations, the number of packages is not limited.

If the Generic Classes package exists, just import it into the current project. If it exists but the number of classes is not sufficient, just create new classes if they can be easily identified at this step (for instance, the project has identified in the SRS document the presence of a double cursor with two elementary cursors Up and Down with a neutral zone in the middle. This class can be created immediately in the Generic Electromechanical Classes package). Otherwise, they can be added later when they appear in the project.

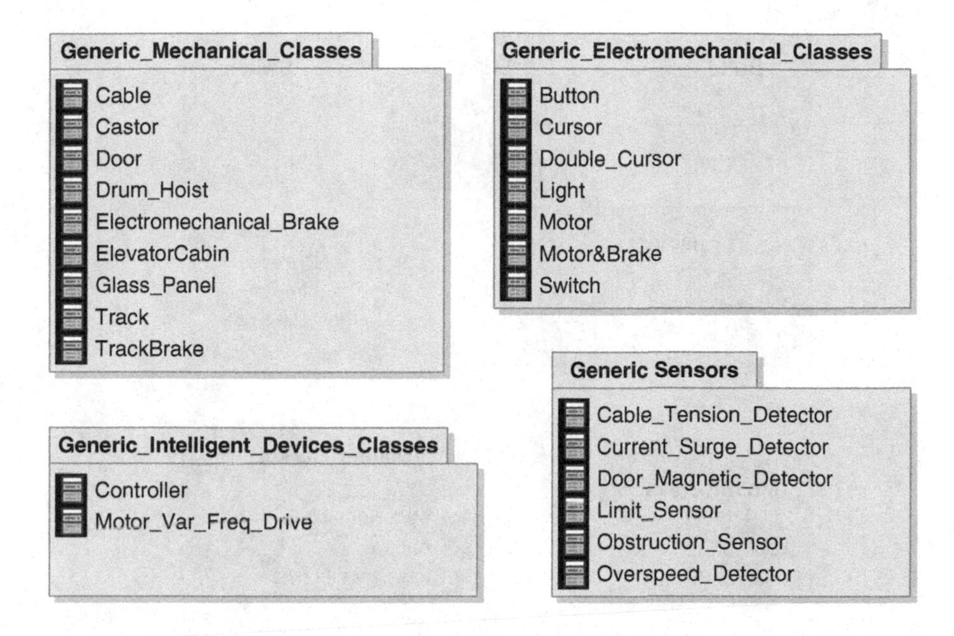

Fig. 8.1.4.1 Generic Classes package in use in LiftTram Company

The Generic Mechanical Classes package shows classes in use in the Lift-Tram Company (Fig. 8.1.4.1). All other packages are general and can be reused outside the Elevator domain.

Generic Classes are stuffed with their logical operations to support the PIM study (Fig. 8.1.4.2). They are not yet mapped into technological components

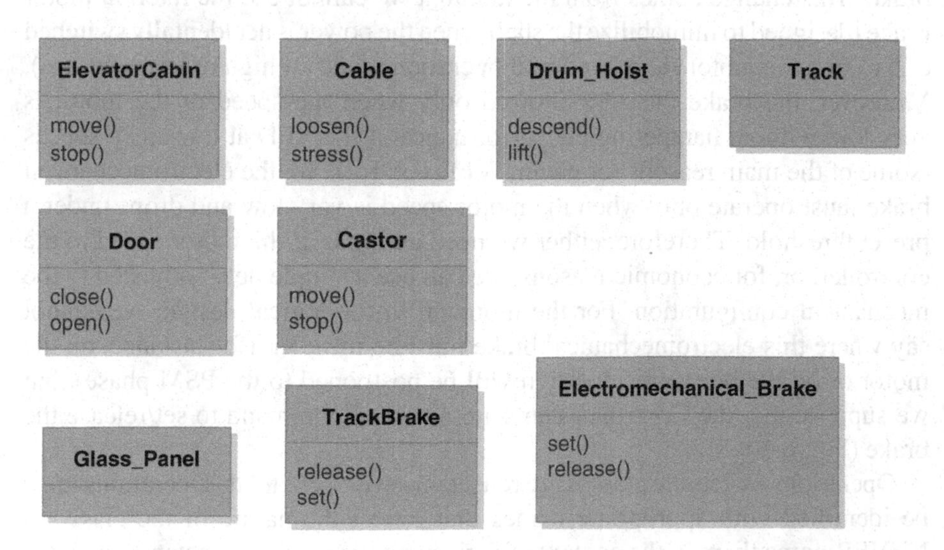

Fig. 8.1.4.2 Details of the Generic Mechanical Classes package. These classes are specific to the Elevator Domain

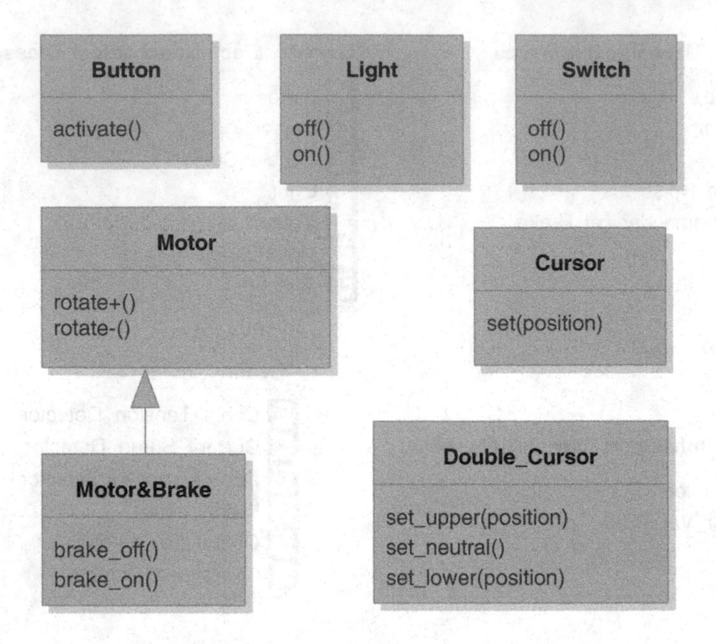

Fig. 8.1.4.3 Details of the Generic Electromechanical Classes package

so all common attributes (for instance, strength of the cable, dimensions of the Elevator Cabin, etc.) are irrelevant at this design step (of making reusable PIM).

One of the most important corrections being added to the original project and the SRS document at the re-spec phase was the creation of an electromechanical brake. This change comes from the fact that we cannot use the internal motor brake (designed to immobilize the shaft when the power is accidentally switched off) to stop the motor shaft in normal operation mode (at high or low landings). Moreover, the brake must be allowed only when the speed of the motor is very low, without hampering the performance of the VFD at low creep speeds (some of the main reasons for having VFD control). So, the electromechanical brake must operate only when the motor speed is very low and drops under a preset threshold. Therefore, either we need an input of the motor speed to the controller, or, for economic reasons, we can use a simple delay adjusted to the mechanical configuration. For the moment, in the logical design, we cannot say where this electromechanical brake must be installed (for instance, on the motor or on the hoist; this decision will be postponed to the PSM phase), but we suppose that the Controller can send a logical command to set/release the brake (Fig. 8.1.4.3).

Operations in generic classes are reused in several contexts. Operations must be identified with appropriate names that reflect the nature of the class but NEVER name them in the context of a given application. For instance, a motor has a shaft that can rotate clockwise (+) or counterclockwise (−). If, in the context of the elevator, we name operations of the motor *up()* and *down()*, we just destroy the reuse mechanism, since we interpret these operations in the context of the elevator domain. If the same kind of motor is used later to open or close gliding doors, the naming scheme, *up()* and *down()*, becomes obsolete quickly.

The Motor&Brake class is derived from the Motor class. This derivation adds two operations *brake_off()* and *brake_on()* to existing operations *rotate+()* and *rotate–()* of the Motor. We do not derive the Double_Cursor class from the Cursor class as they are very different electronic components. The double cursor rather aggregates two cursors and adds a mechanical neutral zone to the overall design. This modeling difficulty (aggregation or inheritance) has already been discussed in Section 5.5.7.

In this PIM design, the Light class is not identified as a LED Light class; despite the fact that all electronic engineers think that it is the way to implement this visual component. The same consideration has already been found in the mechanical package (Figure 8.1.4.2) in which we have not distinguished a steel Cable class used to guide the tram or a cable used to power the cabin, contrary to what was been announced in the SRS document. This genericity holds for all classes of the PIM.

Classes like Cursor and Switch arc not used in this project but are drawn to show that this package always has more classes than needed.

The same genericity has been observed in the way operation are conceived as attributes and operations (Fig. 8.1.4.4). There is no specification of the implementation types (float, Boolean, integer, etc.). Operations have parameters but no implementation types are specified. As discussed in the theoretical part,

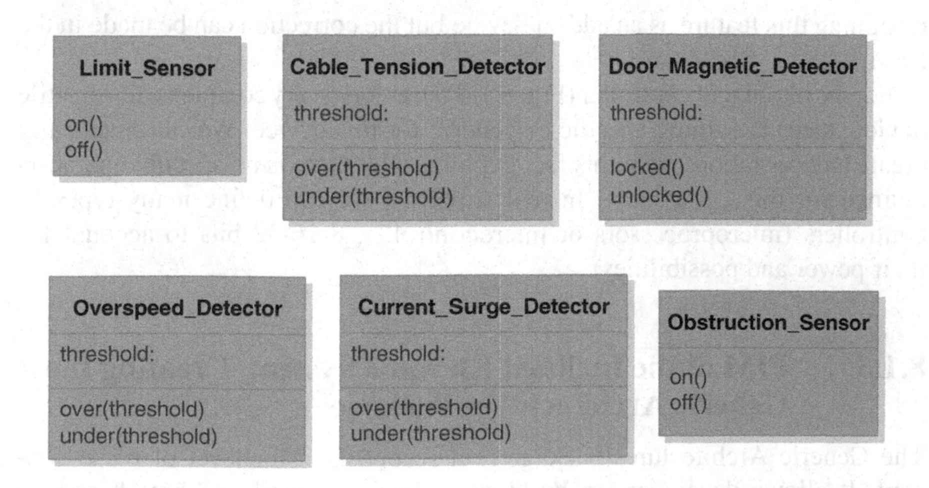

Fig. 8.1.4.4 Details of the Generic Sensor Classes package

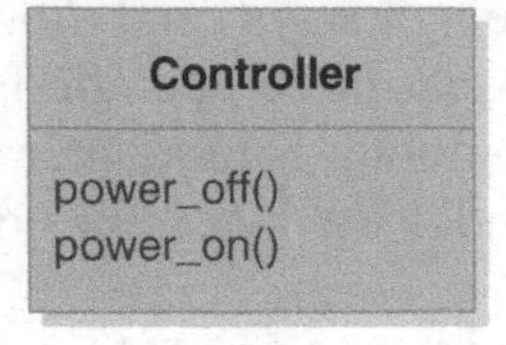

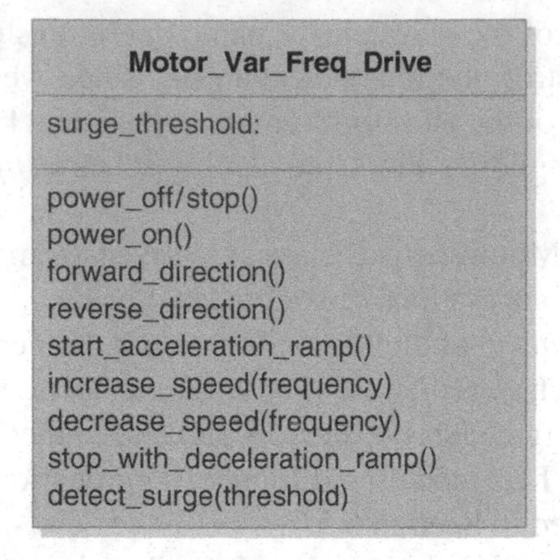

Fig. 8.1.4.5 Details of the General Intelligent Devices Classes package

to support logical models, we think that the low-level specification currently available must be enhanced in the future to meet MDA needs.

In a VFD application, the motor reacts to the varying voltage and frequency from the drive and develops a torque for the load. So doing, it draws current from the VFD. The current increases with the load. The direction of rotation can be set with *forward_direction()* or *reverse_direction()*. Stopping the motor does not mean powering it off. If the drive is currently set at some speed, it will undergo a fastest deceleration ramp to immobilize the motor shaft. We can put a power off that produces a brutal stop but this feature is not a regular command of the VFD.

The detection of the surge current has been integrated to the VFD. If such a feature is not available at implementation phase, we have the possibility of designing this feature as an add-in device but the correction can be made in the PSM.

For the intelligent controller (Fig. 8.1.4.5), as it is a very complex and versatile device, there is nothing specific or generic for this device. We can practically create any operation for it; this fact explains why there is no specific operation defined for this component. In real situation, we can define many types of controllers (microprocessors or microcontroller, 8-16-32 bits to account for their power and possibilities).

8.1.5 PIM of the Inclined Elevator System: Creating the Generic Architectures Package

The Generic Architectures package is a second layer or level of reuse concept. It allows developers to build over time structural concepts based on classes declared in previous Generic Classes package. They can be new classes

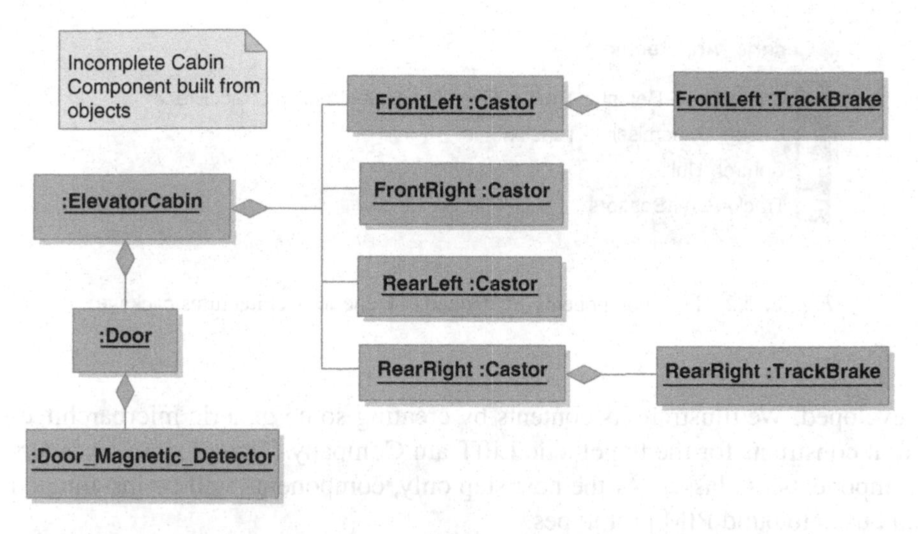

Fig. 8.1.5.1 For one shot deal project with small chance of recurrence (small scale reuse), components can be built directly from objects instantiated from the Generic Classes package. This diagram is incomplete and is targeted to show small-scale reuse process

obtained by aggregating/composing classes together, objects derived from classes and linked together into a generic component. Interfaces can be defined for components.

In UML 2, a component can be now considered as a class and thus can be instantiated. Therefore, there exist two approaches for handling real-time systems, one, with small-scale reuse and another with large-scale reuse. The first one consists of building real objects/components just after the Generic Classes package. Objects/components built under their instantiated form are "object based components" by opposition to "class based components." For instance, to build a fully equipped cabin ready to be used in the elevator, an object cabin will be instantiated, attached to an object door, an abject magnetic sensor, four objects castors, etc. If accidentally, we need duplicated components in the object form, they can be replicated. There is no inconvenience associated with this method of developing projects except that we cannot reuse microarchitectures.

Figure 8.1.5.1 shows this attitude that can be adopted for one shot deal projects with very small chance of recurrence.

For large-scale reuse, microarchitectures can now be built from classes present in Generic Classes package, and stored in Generic Architectures package (Fig. 8.1.5.2). In this package, we store composite classes and components under their "class forms," not "object forms." The number of entries is unlimited and depends upon the experience of the Developer. It is not necessary to create all generic architectures in the domain to be able to start applications. This library will grow naturally with time and the diversity of applications

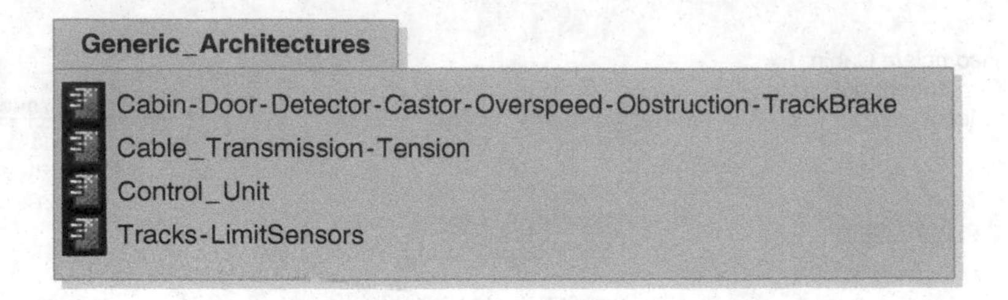

Fig. 8.1.5.2 Four components are defined in Generic Architectures package

developed. We illustrate its contents by creating some generic microarchitectural constructs for the hypothetic LiftTram Company. This process considers components as classes. At the next step only, components will be instantiated and used to build PIM prototypes.

Components defined at this stage are "class based components" (Fig. 8.1.5.3). They are built from classes. They are composite (sign "o-o" at the right low corner) and multidisciplinary as they mix inside a same entity several classes belonging to several domains. Interfaces can be defined for each component, but it is not always mandatory to do so (only when we want to afford a "black box" view of a component). In our case, components are transparent and in their "white box form." Links between components can go through the frontier of a component and reach internal classes. The constitution into components allows us to handle complexity only, not necessarily to encapsulate them all of the time.

The Drum Hoist contains 2–4 cables. Each pair of cable is controlled by a tension detector. We can build other components with other configurations then name them differently, for instance assemblies CTT1, CTT2, etc. to account for various versions of the Cable-Transmission-Tension assembly (Fig. 8.1.5.4).

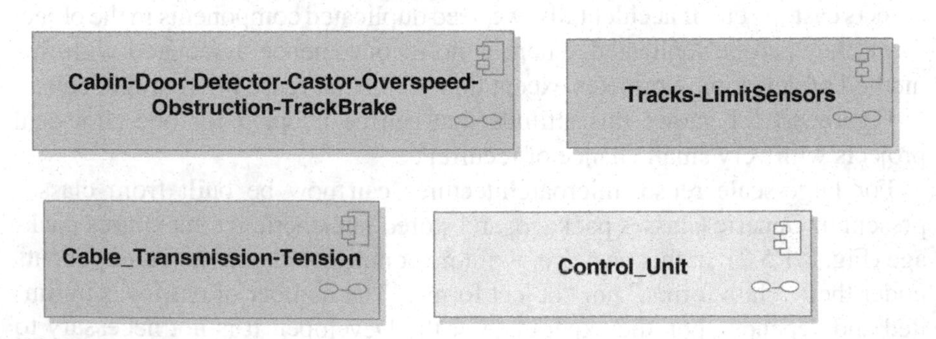

Fig. 8.1.5.3 Generic Architectures package contains the definition of four composite components

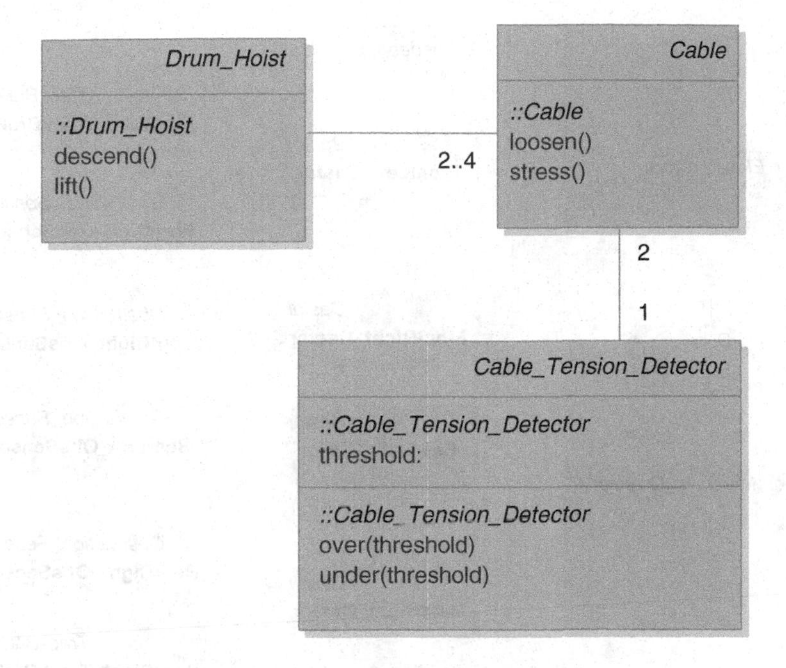

Fig. 8.1.5.4 Composition of the Cable-Transmission-Tension component

The Track-LimitSensors component (Fig. 8.1.5.5) packs inside a same set two tracks (Left, Right tracks facing the slope) and each track is equipped with two limit sensors. This aspect shows that a component may contain classes or objects/components that are not connected together.

We have the choice of merging once the Castor class and then affecting a multiplicity 4 to this class as often done in database system. In Figure 8.1.5.6, the Castor class has been merged four times instead, to generate four different classes whose names come from their positions relative to the Cabin (we suppose

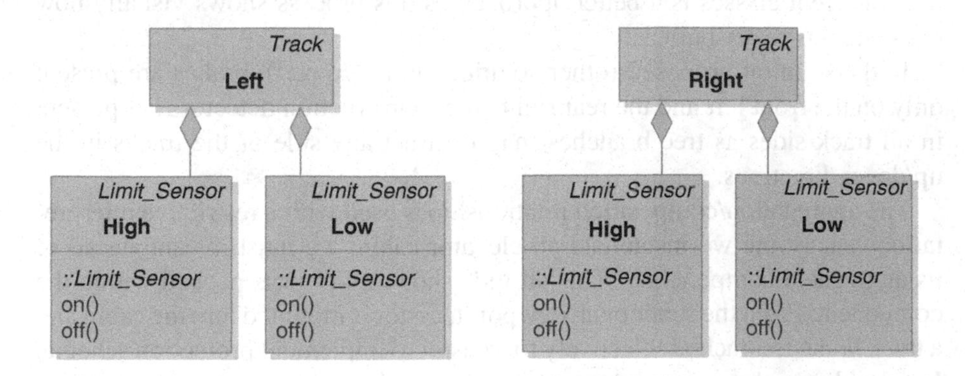

Fig. 8.1.5.5 Composition of the Track-Limit_Sensors component

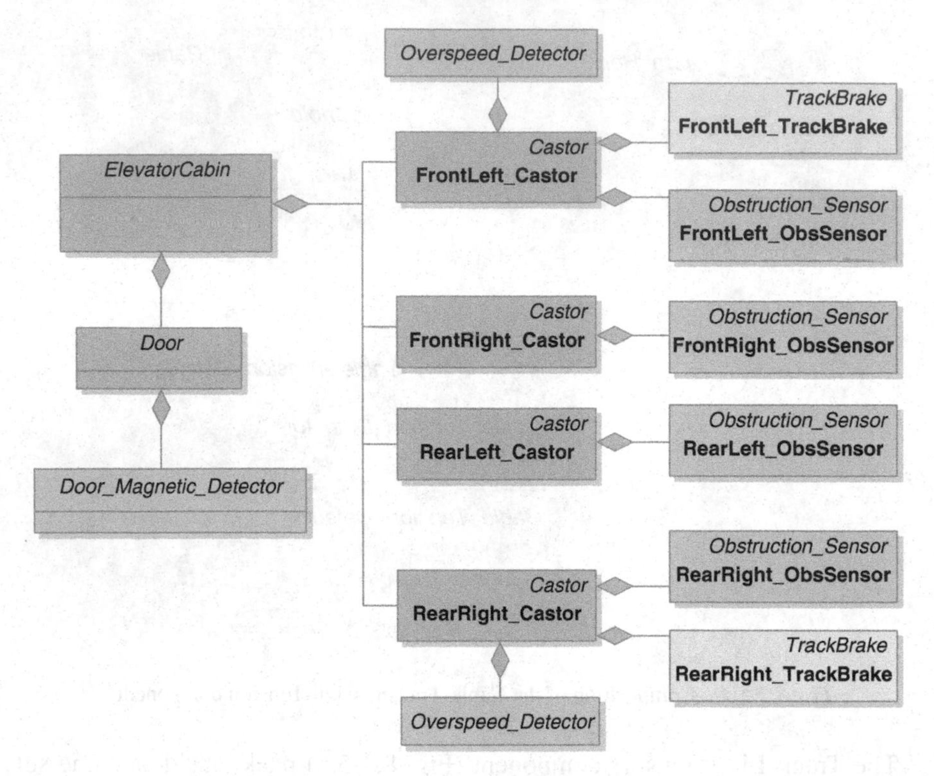

Fig. 8.1.5.6 Composition of the Cabin-Door-Detector-Castor-Overspeed-Obstruction-TrackBrake component

that we have defined a front side for the cabin, for instance the side facing the slope). In real-time system, using a multiplicity specification is not always appropriate as this process bypasses the identity of each individual object. In our case, as sensors and brakes must be mounted on specific castors, making four different classes is a better approach as this process shows visually how the system is really built.

In the solution proposed (other solutions is welcomed), brakes are present only on the front left and the rear right sides. Obstruction detectors are present in all track sides as tree branches may obstruct any side of the tracks in the up/down directions.

The aggregation/composition relationship is used with a restrictive interpretation "each time we instantiate an elevator cabin, we must instantiate all of its aggregated components." This attitude shows all objects packed within the component. From the functional viewpoint, castor is mounted into the cabin and a track brake is attached selectively to a castor to implement protection scheme. If a developer chooses to represent the aggregation with a simple association, he must maintain only his proper inventory.

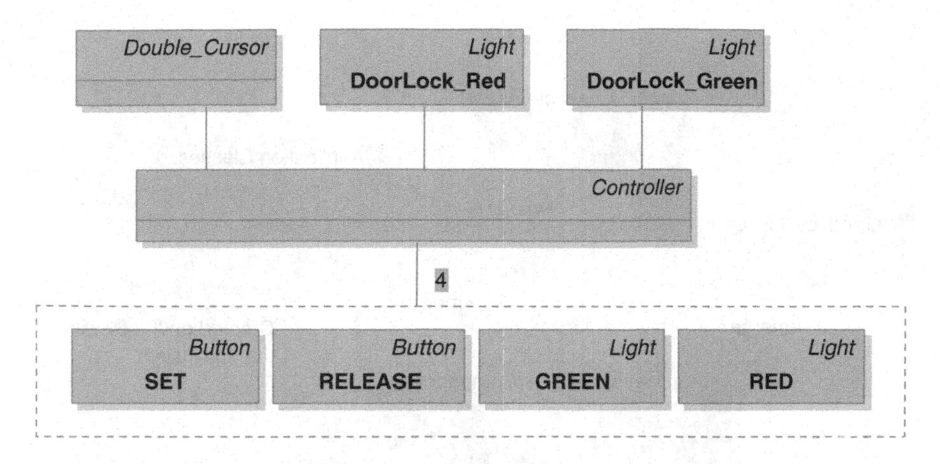

Fig. 8.1.5.7 Composition of the Control Unit component

The Control Unit component is shown with its accompanying elements. When connecting this unit in a system, connections will be drawn directly to the Controller. For instance, if the Controller must monitor the Door Lock Sensor, this signal goes through the frontier of the Control Unit component (Fig. 8.1.5.7) to join the controller directly. From the electronic interfacing viewpoint, possibly, we need more elements (parallel port, adaptation electronics, even analog–digital or digital–analog converter, etc.). These elements can be integrated in a modern microprocessor-based controller. These details are part of the PSM, not the PIM, unless algorithms to be studied involved these devices themselves. Therefore, in this logical model, we focus only on the control logic. If necessary, electronic interface classes (do not confuse the electronic interface with the "UML interface" concept) can be inserted to represent these elements. Ports, required interfaces, and provided interfaces are available embedded elements that can be added to any component to define interfacing structure (see Section 4.10).

8.1.6 PIM of the Inclined Elevator System: Instantiating a Working System

A working system can be built by instantiating objects from classes (belonging to Generic Classes package), object-based components from class-based components (belonging to Generic Architectures package) and objects directly from Application Specific Classes/Components package. The last package (Fig. 8.1.6.2) is needed because some information classes are specific to the current application, cannot be reused in any context, and must be created on the flight to support the application. Figure 8.1.6.1 explains the building process.

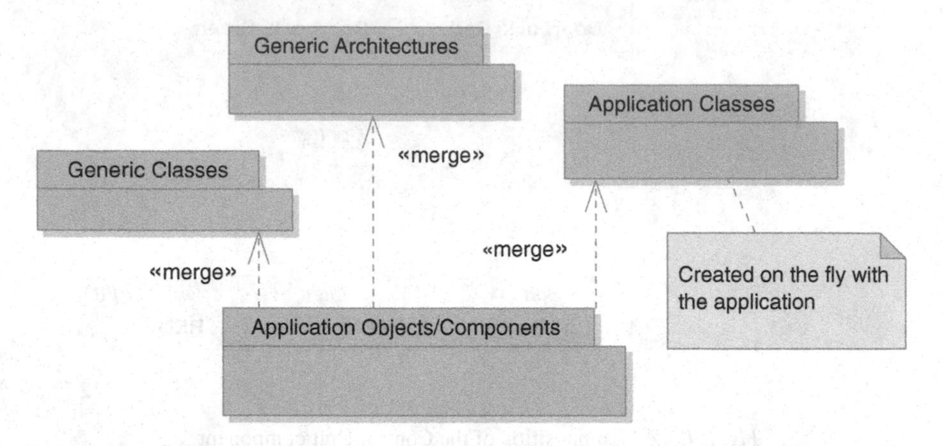

Fig. 8.1.6.1 The Package Application Objects/Components contains the PIM built from merging classes and components from other packages. Only Application Classes Package is not reusable

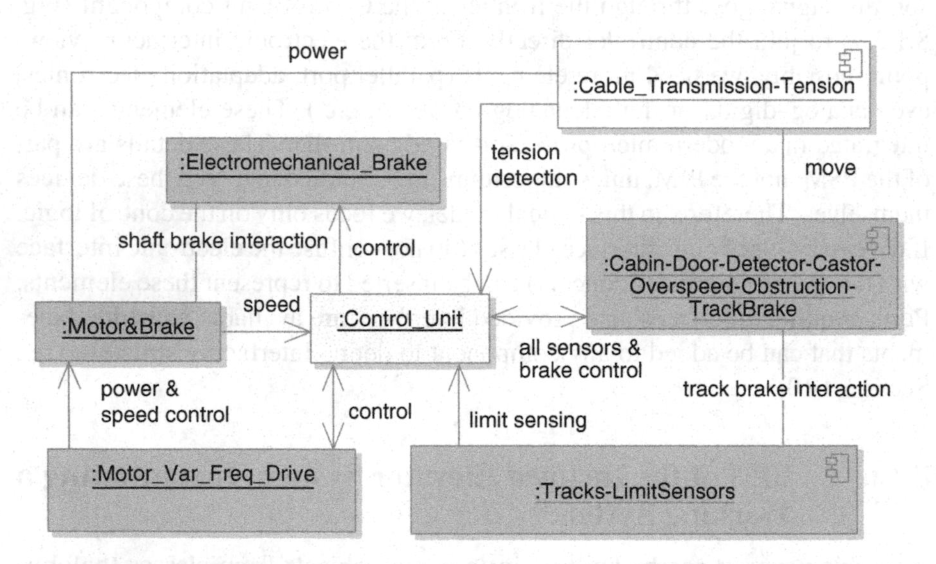

Fig. 8.1.6.2 The final system is composed of objects/components instantiated from reusable Generic Classes Package, reusable Generic Architecture Package. Links drawn between objects/components are high level identification. They show potential communication between objects/components

8.1.7 PIM of the Inclined Elevator System: Behavioral Study of the Control Unit

Each object or component inside a system has a behavior. As stated in the theoretical part, the behavior of simple and passive objects/components is trivial and so does not need any dynamical study. The behavior of the complex and/or insufficiently decomposed objects/components is difficult (even impossible) to reach. The presence of a study does not necessarily mean that it is a valid and secure design. So, the key of success for obtaining complete, reliable, and valid design passes through a thorough decomposition and identification of all elementary and nonelementary operations of objects/components (see Sections 6.5.7 and 6.5.8)

In our case, most objects of the system are passive, mechanical, or electromechanical objects, so their behaviors are trivial. For instance, a cable has two states, stressed or loosened; the door has two states, opened or closed; the Motor can rotate clockwise or counterclockwise, etc. Some components have a complex behavior, e.g. the VFD. As they are black box components, their interfaces are fully defined so there must be no problem if we observe scrupulously operating modes indicated in the instruction manual. We cannot change or modify anything when dealing with black box components.

The only device that needs a full dynamic study would be the control software implemented in the Controller. The first step is establishing a complete inventory of all inputs and outputs, possible values they take in their respective domains, as well as thresholds and actions that must be performed around those thresholds. A link in an object diagram can pack inside a unique connection a set of inputs and outputs. They must be separated individually to reach a complete investigation. A new object diagram can be drawn to identify visually all the inputs/outputs of the Controller. Table 8.1.7.1 lists all the variables of the Controller and Figure 8.1.7.1 shows the corresponding object diagram with all objects that interact with the controller. We stop the interaction at the first neighbors of the Controller.

Normally, we can connect directly the Double_Cursor to the VFD with some signal adjustment. We have chosen a full software solution that shields the operator from the real command injected to the system. This concept relies heavily upon the "intelligence" of the Controller. We do not pretend that the solution is optimal for the elevator domain. The purpose of this case study is using an imaginary system that allows us to explain a design methodology. So, specialist advices for future releases of this book are welcomed.

This shielding effect is necessary in most complex real-time systems as human operators are not always reliable, particularly in emergency situations. For instance, normally, we expect that the operator adjusts the cursor to accelerate then decelerate manually when approaching the landing zones. If for some

Table 8.1.7.1 This table lists all objects/components constituting the first neighbors of the Controller and identifies the nature of communication channels exchanged with the Controller. Signals 10, 24, 25 are analog

N	*Variable*	*Hosting object or component*	*In/out*	*Comments*
Inputs and Outputs OUTSIDE the Control Unit				
1	EM_brake	*:Electromechanical_Brake*	Boolean Output	0: No brake; 1: Brake
2	slack_cable	*:Cable-Transmission-Tension*	Boolean Input	1: Slack Cable detected
3	door_closed	*:Door_Magnetic_Detector*	Boolean Input	1: Door closed 0: Door opened
4	Overspeed	:Overspeed_Detector	Boolean Input	1: Overspeed 0: Speed under threshold value
5	Obstruction	*:Obstruction_Sensor*	Boolean Input	1: At least one of the four obstruction sensors is on 0: No obstruction
6	Trackbrake	*:TrackBrake*	Boolean Output	1: Set trackbrakes on 0: Set trackbrakes off If the power is off, normally brakes is on by design for security
7	Highlimit	*High:Limit_Sensor*	Boolean Input	1: High limit redundant sensors reached
8	Lowlimit	*Low:Limit_Sensor*	Boolean Input	1: Low limit redundant sensors reached
9	Direction	*:Motor_Var_Freq_Drive*	Boolean Output	0: Forward 1: Reverse
10	ramp_speed		Output Value	Set ramp speed (valid for two directions)
11	stop_decelerate		Boolean Output	1: Stop with decelerated ramp
12	Surge		Boolean Input	1: Current surge detected
13	power_on_off		Boolean Output	1: Power on (Motor) 0: Power off (Motor)
Inputs and Outputs INSIDE the Control Unit				
14	slack_light	*Slack:Light*	Boolean Output	1: Red 0: Green
15	slack_set	*Slack_Set:* Button	Boolean Input	1: Test slack brake security
16	slack_ release	*Slack_Release:Button*	Boolean Input	1: Release slack brake

(*Cont.*)

Table 8.1.7.1 (Continued)

N	*Variable*	*Hosting object or component*	*In/out*	*Comments*
17	OV_light	*OV:Light*	Boolean Output	1: Red 0: Green
18	OV_set	*OV_Set:* Button	Boolean Input	1: Test overspeed brake security
19	OV_Release	*OV_Release:* Button	Boolean Input	1: Release overspeed brake
20	TJ_light	*TJ:Light*	Boolean Output	1: Red 0: Green
21	TJ_set	*TJ_Set: Button*	Boolean Input	1: Test track jam brake security
22	TJ_Release	*TJ_Release: Button*	Boolean Input	1: Release track jam brake
23	DoorLock_Light	*DoorLock: Light*	Boolean Output	1: Red (door unlocked) 0: Green (door locked)
24	UP_value	*:Double_Cursor*	*Input Value*	Set UP speed
25	DN_Value		*Input Value*	Set DOWN speed
26	neutral_zone		Boolean Input	1: In Neutral Zone 0: In UP or DOWN zone

reasons, he forgets to decelerate manually when arriving, the limit sensors take over the landing operation (decelerate then apply the EM brake) even if the cursor is still at the maximum position. The only operation the operator must do is to bring back the cursor towards the neutral zone before being able to start a new cycle in other direction. Other solution like a fully automatic operation with two buttons "UP" and "Down" starting fully programmed cycles with a cursor adjusting the maximum speed is also an interesting solution as extended exercises.

Another problem detected is the potential conflict not mentioned in the SRS document. We have three sets of commands, one in the cabin and each landing zone has also an identical Control Unit. What really happened when several operators try to adjust their cursors at the same time? This hardware problem with switch logic is not solved at this PIM phase and will be postponed to the PSM phase. Voluntarily, we have introduced this point to show the difference and the role of models when evolving from the first PIM to the last PSM.

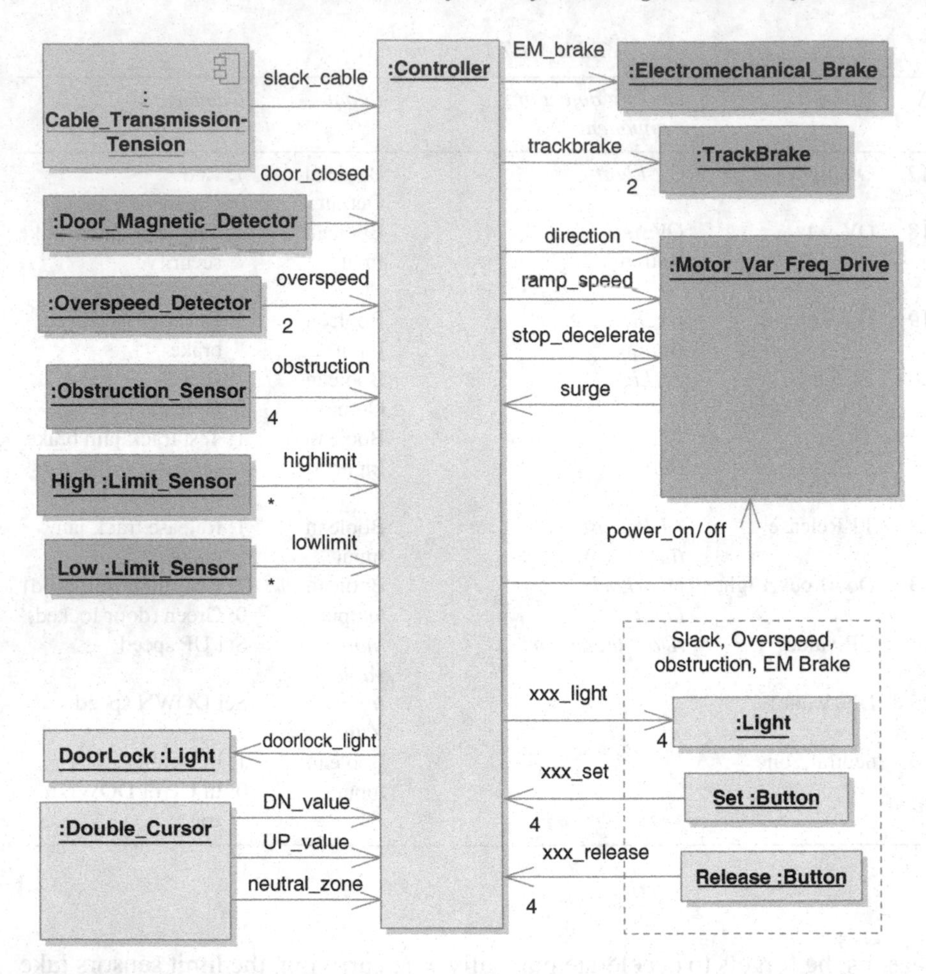

Fig. 8.1.7.1 Object diagram of the Inclined Elevator Controller with all inputs/outputs of its first neighbors

As a technological note, the Controller is able to read and issue both digital and analog signals. Microcontrollers or microprocessors boards with appropriate parallel ports, A/D and D/A converters currently support this feature.

Figure 8.1.7.1 shows a relative complex application. The best approach is top down with a full description of the natural sequence to identify in the first round high-level tasks. Progressively, tasks are decomposed to reach more elementary operations. The following figures algorithms of the controller studied with a SEN diagram. Those dynamic diagrams are first attempts and must be confirmed by more systematical tools like "images attribute methodology" exposed in Section 7.5.1.

The top-level SEN diagram of Figure 8.1.7.2 explains the overall software organization and represents a specific design of this elevator. Other designs are

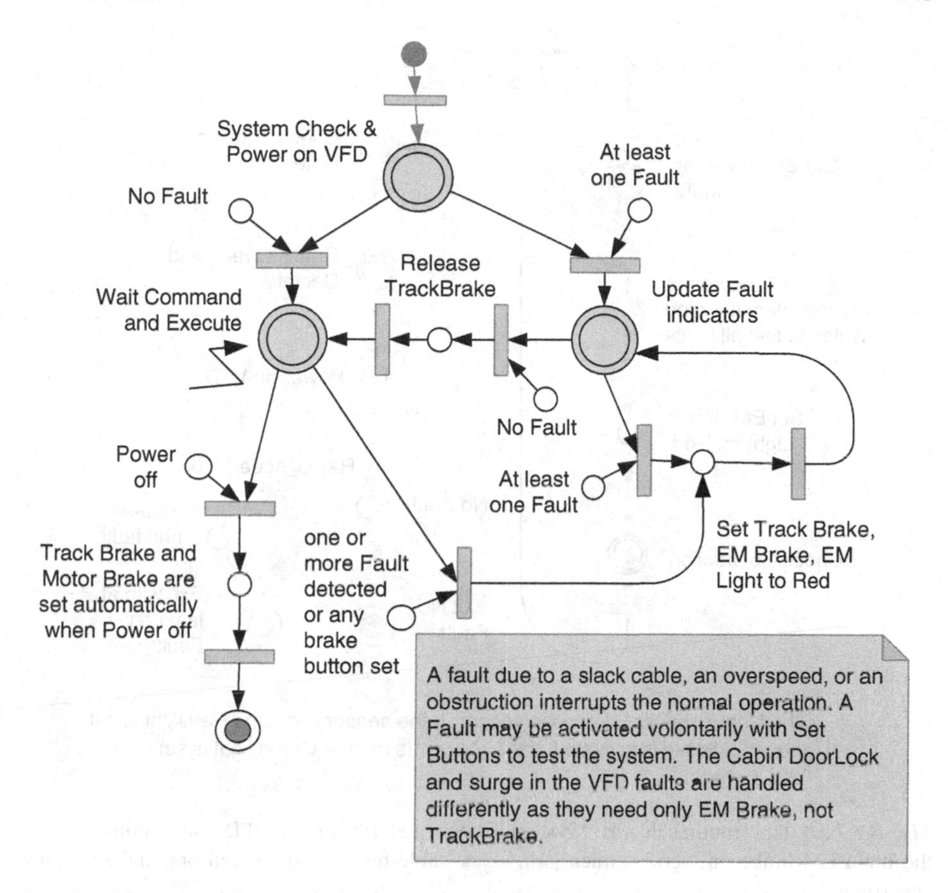
System Check &
Power on VFD
At least
one Fault
No Fault
Release
TrackBrake
Wait Command
and Execute
Update Fault
indicators
No Fault
Power
off
At least
one Fault
Set Track Brake,
EM Brake, EM
Light to Red
Track Brake and
Motor Brake are
set automatically
when Power off
one or
more Fault
detected
or any
brake
button set
A fault due to a slack cable, an overspeed, or an obstruction interrupts the normal operation. A Fault may be activated volontarily with Set Buttons to test the system. The Cabin DoorLock and surge in the VFD faults are handled differently as they need only EM Brake, not TrackBrake.

Fig. 8.1.7.2 When the power is on, the Controller undergoes a system check routine then powers the VFD. If faults are declared, the Controller requests that all faults must be cleared before enabling any movement

also possible. Our goal is to show how to model, draw hierarchical dynamic diagrams, represent interruptions, subdiagrams, not really to solve the elevator problem.

When entering a dynamic diagram, first locate the "black point" that represents the pseudoinitial state where interpretation starts. The "black point circumscribed with a circle" is the exit point representing the end of the current diagram. At the top level, this pseudofinal state marks the end of the software. At any other levels, this final state returns the control to the parent level. We can have more than one final state, in this case, they exit with different conditions. For instance, while exiting "System Check and Power on VFD" (Fig. 8.1.7.3), the subroutine exits with "No fault" or "at least one fault." Composite and decomposable nodes are double circled and normally must be detailed elsewhere. We make use of two sizes of subroutines and the decomposition of "small size" subroutines are delayed to the PSM as they can be easily understood.

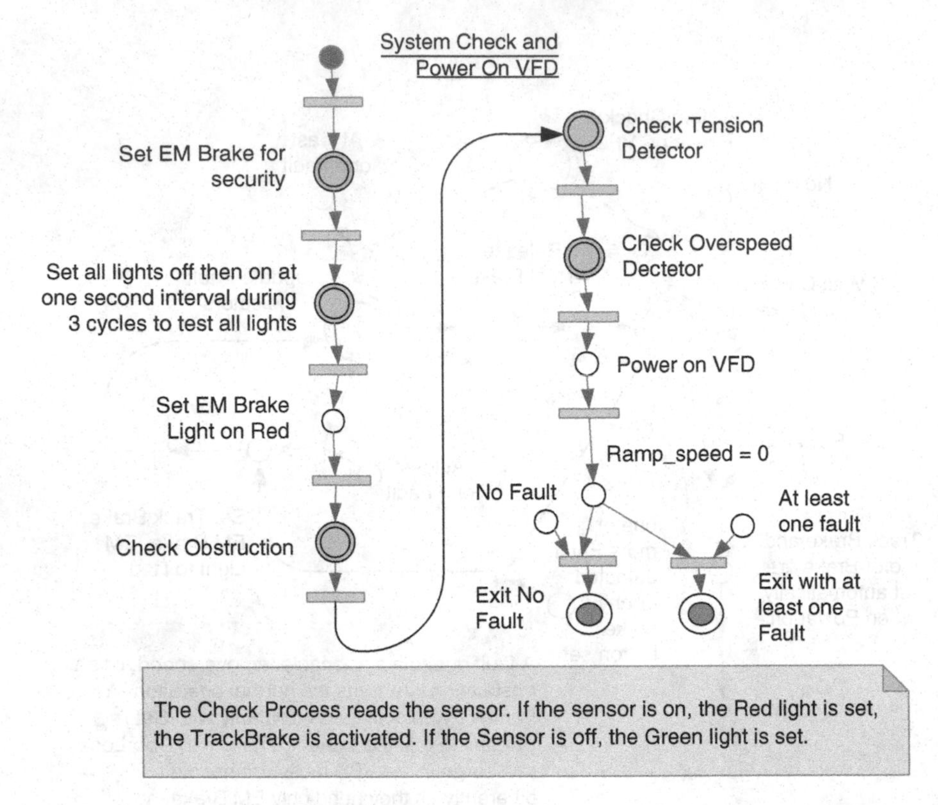

Fig. 8.1.7.3 This routine details "System Check and Power on VFD" subroutine. It sets the main EM brake on, verifies track jam, slack cable and overspeed sensors, and set lights accordingly

Normally, by default, we write as a constraint at project level that all activities, elementary or composite (subroutine), will execute to their natural end. So, they keep or "drown" the token as long as needed (see Section 7.1 for rules). Interruptible activities are marked with lightning arrows (e.g. "Wait Command and Execute"). So when the power is off (the current can be switched off by other mechanical means and not with power on/off on the Control Unit), the Trackbrake and the fail–safe internal Motor Brake (not EM Brake) are by default on and they immobilize the cabin in this circumstance.

When the Controller is powered on, it sets the EM brake immediately to stop all movements to undergo a system check. At the end of the check, the VFD is powered on. At this point, the subroutine exits with two possible conditions "with at least one fault" or "no fault". If "no fault," the system is ready to read commands from operator and execute them. In the presence of faults, all faults must be cleared (details in Fig. 8.1.7.4) before returning to normal operation. Tracks must be cleared if jammed by obstructive objects, the cable must not

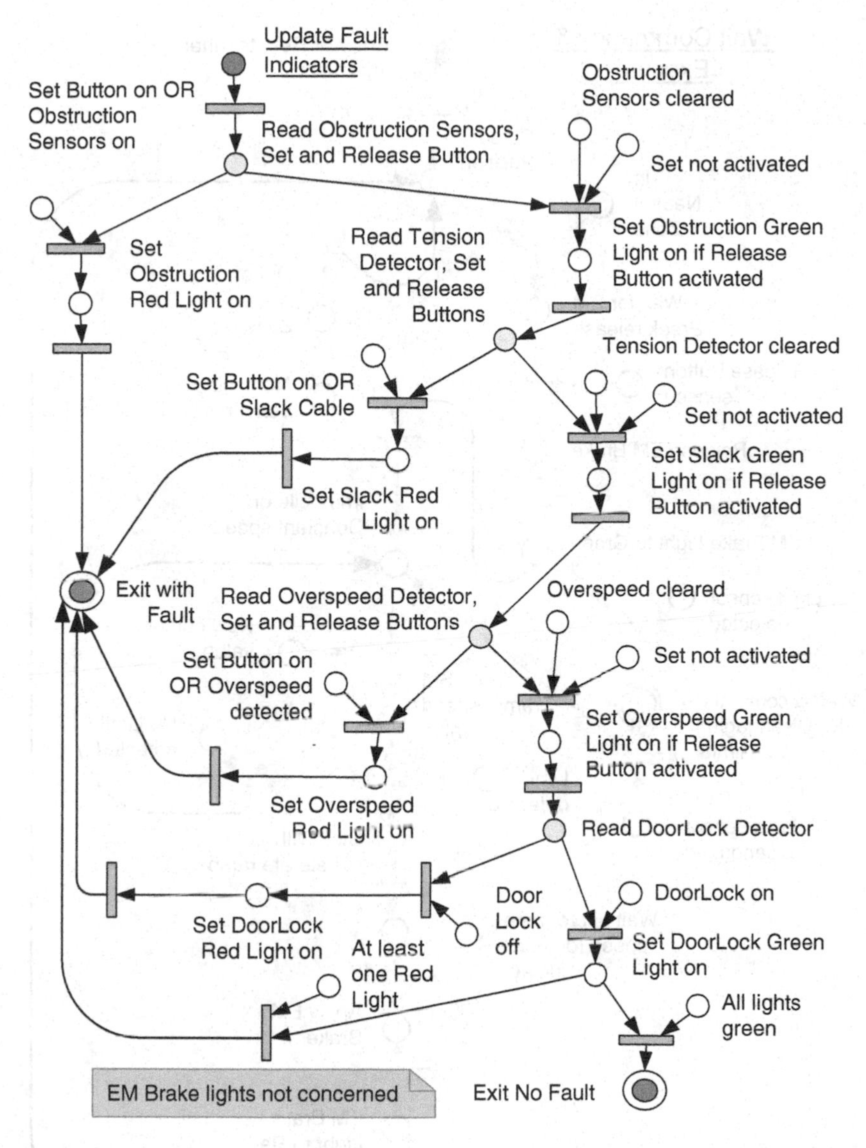

Fig. 8.1.7.4 The subroutine "Update Fault Indicators" reads all sensors and buttons then updates all signal lights (Lights are exclusive, Red or Green only)

be slack, the overspeed detector must be "under," and the cabin door must be closed before being able to issue any command. At the end of the routine, the VFD is powered on by the Controller despite the presence of faults or not.

The way the subroutine "Update Fault Indicators" is designed, the system test must be performed by maintaining the Set buttons constantly at their "on" position. If the set button is released, the set is cleared immediately as there is no

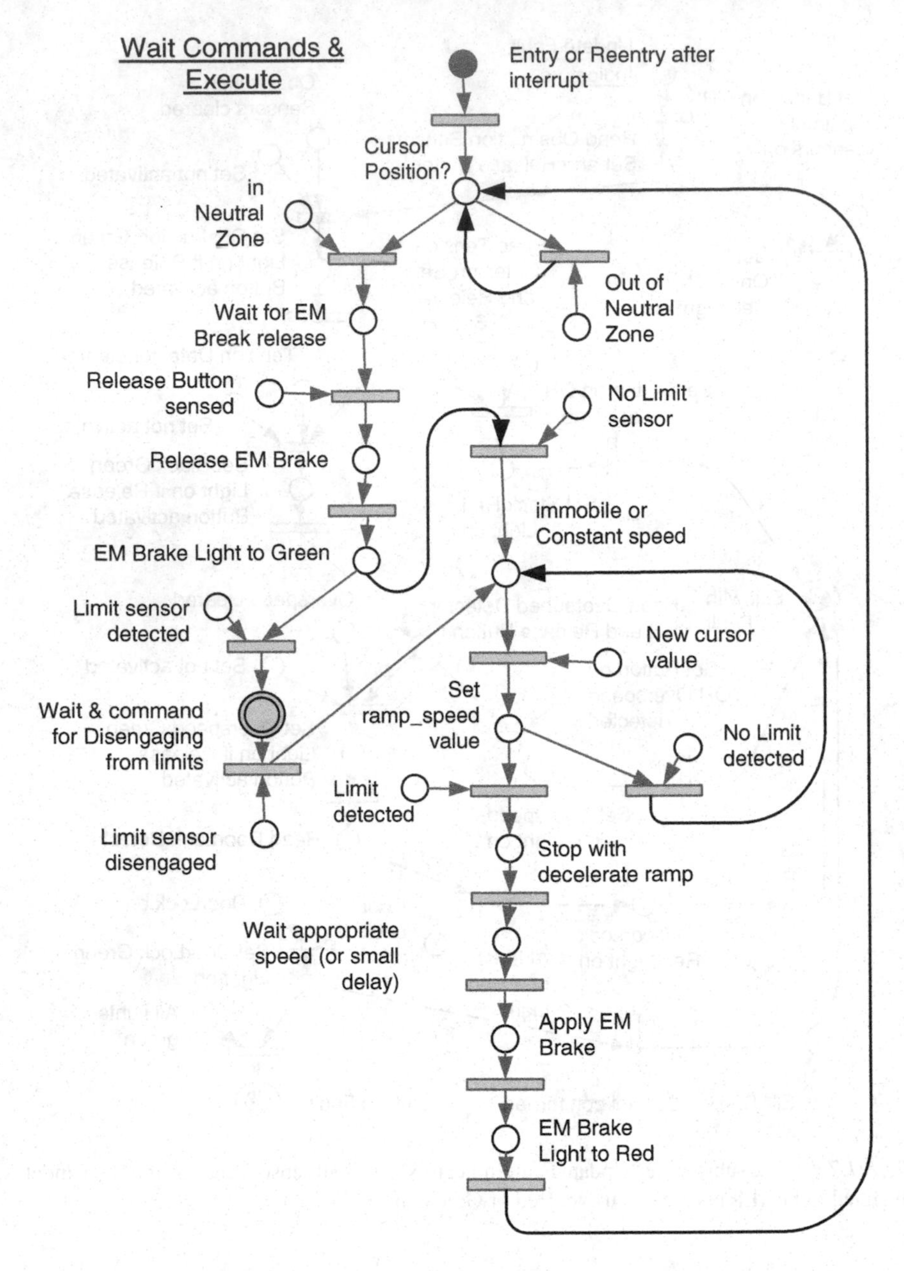

Fig. 8.1.7.5 The main subroutine supports the normal operation mode of the cabin

provision for latching the signal. Even if the sensor is in its normal state, the light does not pass to green immediately but requires a pressure on the corresponding release button, otherwise the fault condition remains. An instruction like "Set Slack Green Light on if Release Button activated" is not elementary from the programming viewpoint. But, at the design phase, we admit that it can be easily understood by any implementer.

In Figure 8.1.7.2, the "Wait command and Execute" subroutine can be interrupted anywhere, even immediately when entering the subroutine in presence of fault. Each fault arises an interrupt, triggers the trackbrake and the EM brake. So, all categories of brakes mentioned in the SRS document are designed with the combined action of two brakes: EM brake for normal operation of the cabin and EM brake combined with trackbrake for emergency situations.

The next subroutine of Figure 8.1.7.5 explains the normal operation of the cabin. When entering the "Wait command and Execute" subroutine, the EM brake avoids all cabin movement due to the incorrect position of the cursor. The trackbrake has been already released by clearing all exceptions conditions. Roughly, the operator must push on the release Motor Brake Button (EM Brake) to suppress this last brake. But this action can be realized only if the Double_Cursor is in its neutral zone.

When all brakes are cancelled, the operator can now operate the cabin by gliding the cursor in the UP and DOWN zones. At this time, it is possible that the redundant limit sensors are still present and detected by the controller. As these sensors stop the movement of the cabin when arriving at the landings, we must distinguish the disengaging procedure for quitting the landings from the brake produced when arriving at the landings; otherwise, if the limit sensors inhibit any movement, we will be not able to leave the landing zones, the batteries of limit sensors, constantly switched on.

When the cabin is moving with a constant speed, the controller continuously reads the cursor information and updates the ramp speed of the VFD at real time.

8.2 Emergency Service in a Hospital and Design of a Database Coupled with a Real-Time System

The following case is interesting as we simultaneously address two problems, real time and database, that are intimately related to each other. We are in the situation described in Section 6.7 in which objects must be *individually registered*, and *their states monitored at real time*. The model necessitates many kinds of attributes: *structural* for identifying the objects, *behavioral* for tracing its evolution in a real-time context, etc.. Real-time classes contain large number of real-time operations to describe medical operations, and state attributes to store patient states and, at the same time, the real-time system needs information in the database about patients and eventually needs to update them.

When working with such a system, the database must be built first, and then classes can be derived from the database in order to get all descriptive information. Dynamic attributes and operations are then added to make workable classes. As said in the theoretical part, when a supermarket stores 100 cans of maple syrup, it does not make sense to identify them individually. In a hospital, patients have multiple views. They are indistinguishable when we need make medical statistics but this "cans of maple syrup" view or "columns of numbers' view disappears when addressing to the patient that the practitioner must study medical information and find a good health-care plan for her or him. They must be individually monitored and treated with care and humanity. Canadian gun registry program or baggage handling service in an airport, are, for instance, problems of the same category. Owing to the limited format of this book, we cannot treat the example completely so we focus more on the database part and give very useful information to approach this double problem from the logical view. The PIM will stay valid even if at the next phase, the PSM calls for two different but communicating platforms.

8.2.1 Early Requirements

A hospital X wants to improve its emergency service and wishes to model its system to build a patient database and a decision-making system to classify patients arriving at emergency service in order to decide which patient must be examined first.

The hospital contains:

1. *Receptionists taking phone calls.*
2. *Ambulance men transporting patients to emergency or transferring them towards other destinations.*
3. *Patients who come by their own means, generally accompanied by members of their family.*
4. *Receptionists performing patient registering with an admission form. The mean time spent to register is about 2 min for a patient with an existing record and 8 min for a new patient. If a patient is unable to register, his or her family member can help for registering.*
5. *Medical assistants who verify patient records, collect, scan, and classify official certificates, letters of reference, medical prescriptions, radiographies, laboratory test results, etc. Their tasks can be performed anytime, while patients are waiting in the waiting room or later.*
6. *Chief nurses who examine patients and determine the urgency with which each patient should be examined by a practitioner. Their algorithm can be described roughly as follows. After taking the pulse and the tension, filling data in the database, they estimate the maximum waiting time allowed in minutes. The patient queue is therefore updated after each data entry (rescheduling). If a general practitioner is free before (or after)*

the expiration of the maximum delay, the patient at the top of the list will pass the examination immediately. The average duration of examination of a patient is about 12 min by the chief nurse.

7. *General practitioners are those who carry out diagnoses, prescribe drugs, radiography, blood test, or more specific tests. Medical data are collected at each step and are accessible via the medical database. Each medical examination lasts between 7 and 15 min. On the average, 11 min was retained as a mean value. Thereafter, the patient, on foot or on a stretcher, can be put under observation in a rest area between 30 min and 8 h. In this case, the general practitioner puts the patient back in the waiting queue exactly as if the later comes from the reception area. In the rest area, other practitioners can reexamine the patient for the 2nd time or the 3rd time, etc. Subsequent examinations last around 6 min. After each examination, the practitioner must write down what is the maximum waiting time allowed before the next examination (same procedure executed by chief nurses).*

8. *Nurses , for the period of observation in the rest area, take measurements on the patient like blood pressure, pulse, pain scale every 60 min, give drugs to patients, check/install medical apparatus, and update the database. This takes about 10 min for a nurse to realize this act for each visit. At night time, nurses can take over tasks of receptionists and take phone calls if needed. A chief nurse may execute all medical or administrative tasks of a nurse or of a medical assistant if necessary.*

9. *Surgeons.*

10. *Other conditions are, no patient can stay in the rest area more than 8 h. If there is an extension beyond this limit, practitioners must choose between prescribing a hospitalization in the current hospital (if beds or individual rooms are still available), transferring the patient to another hospital with a preliminary availability check, sending the patient in surgery, or sending the patient to home with a drug prescription. The practitioner can decide on a hospitalization immediately when examining the patient for the 1st time or after the 2nd, 3rd examination, etc. when he has seen laboratory or test results, or simply just by seeing that a patient is getting worse. In those cases, the patient leaves the emergency queue.*

The emergency staff counts 10 general practitioners, 3 surgeons in the daytime (7a.m. to 8 p.m.). In the evening, two general practitioners are always present and a surgeon works on call only. Initial hypotheses state that beds, nurses, chief nurses, receptionists, laboratory personnel, etc. are largely sufficient (technically speaking, they are "infinite resources"). The hospital X wants to know what is the maximum number of patients that can be admitted in a day within this medical staff, what is the maximum arrival flow manageable according to the time staying in the rest area, and finally what are important dynamic parameters of the whole process?

8.2.2 Use Case Diagrams for System Requirements Specifications Document

Hereafter, we bypass the text analysis phase to expose directly a set of UC diagrams that normally structures concepts revealed during the text analysis phase. This task affords a first attempt of structuring the description. Concepts in Requirements Analysis phase are exposed as those perceived by the hospital Client, but, at the design phase, they can be represented in a completely different manner. As said in the theoretical part, even if the use case analysis is not very accurate, it has several benefits:

1. It helps conceptualizing the problem
2. It forces to read "correctly" the project to be developed
3. It classify ideas
4. It identifies main actors and main activities
5. It organizes activities into a network of dependent activities. By laying a network of "include" and "extend" relationships, it first apprehends dynamic structures
6. It highlights goals to be satisfied
7. It puts all requirements under a graphical form
8. It tends to structure the system even if this structure is not very rigorous.

All those benefits will prepare correctly the design phase.

8.2.3 Designing the Patient Database Coupled with a Real-Time Emergency System

The challenge is to deploy reuse principle early in the PIM, despite what will be decided at the PSM level. As said before, there is two systems intermixed together in the problem of this hospital.

First, we must construct a database to store information about patients, record all medical acts performed by medical staff at every step, and store multimedia data (radiography images, letters, various documents, etc). Firstly, this database is of a special type in which all records must be identified individually. Secondly,

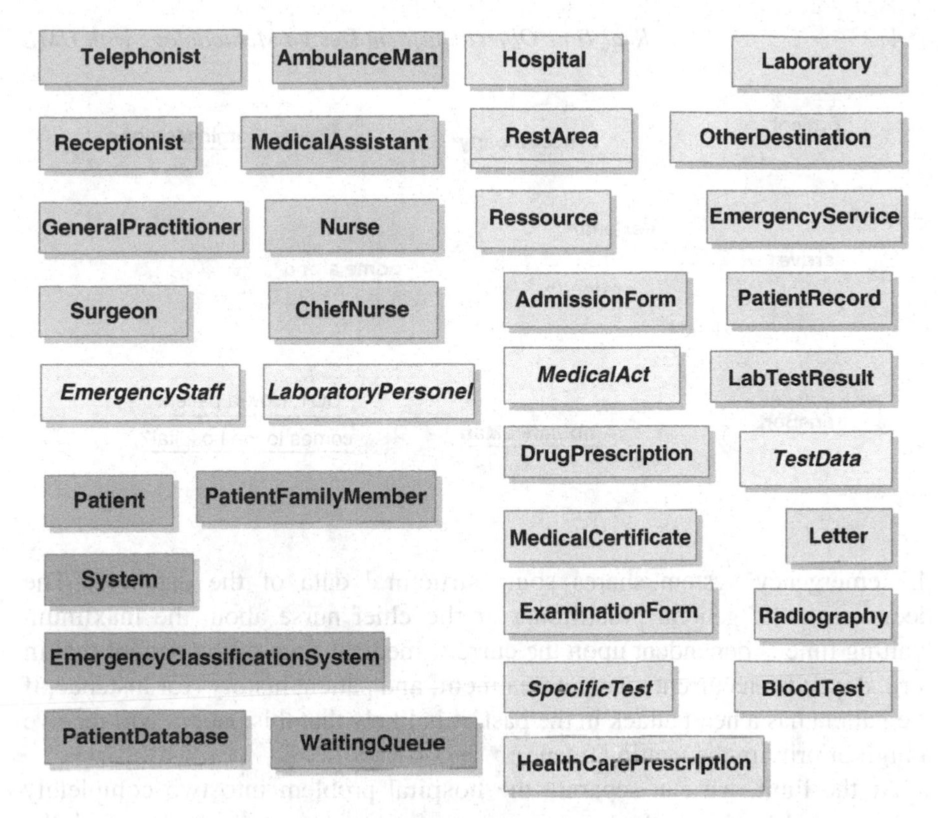

Fig. 8.2.2.1 Classes of the Hospital with the Emergency Classification Problem. They are used to instantiate objects for specification purpose only, not for design

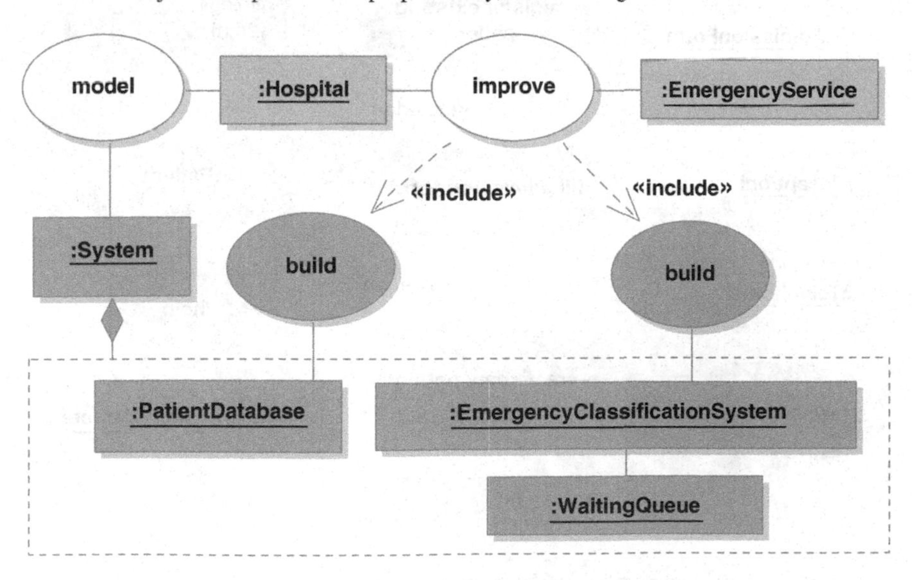

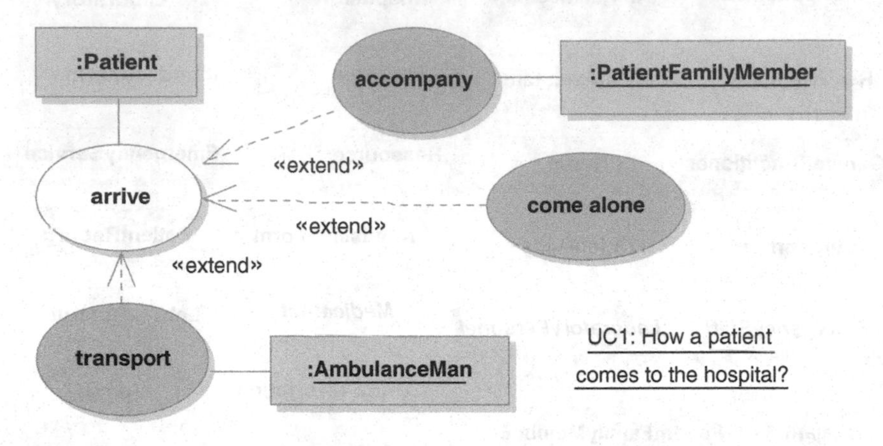

the emergency system shares some structural data of the database. The decision of the general practitioner or the chief nurse about the maximum waiting time is dependent upon the current medical state of the patient that in turn, depends on current medical treatment, and patient history (for instance, if the patient has a heart attack in the past, it is likely that this patient will receive a high priority in the waiting queue).

At the limit, we can separate the hospital problem into two completely unconnected projects: the management of the emergency system and the

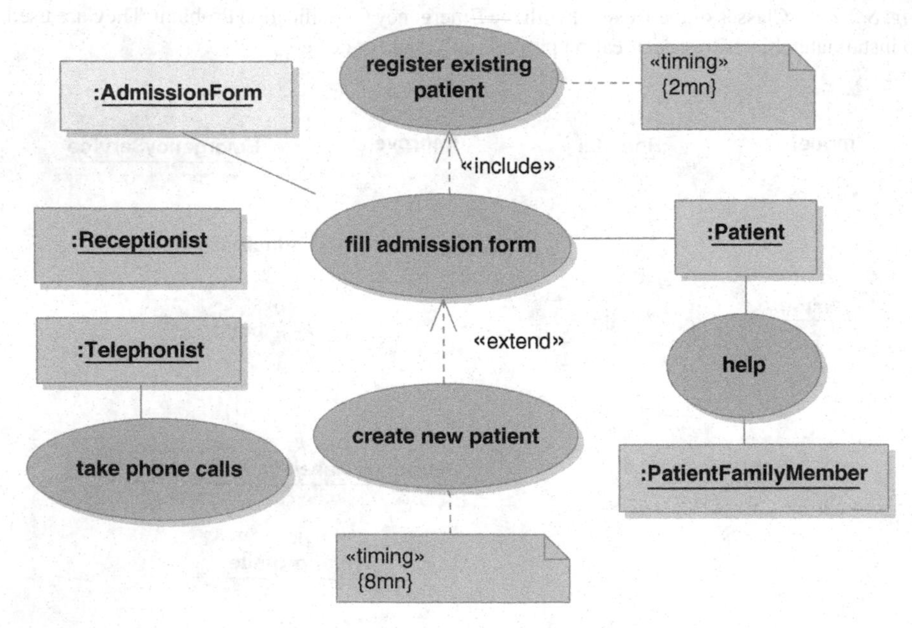

UC2: Reception & Patient Registering

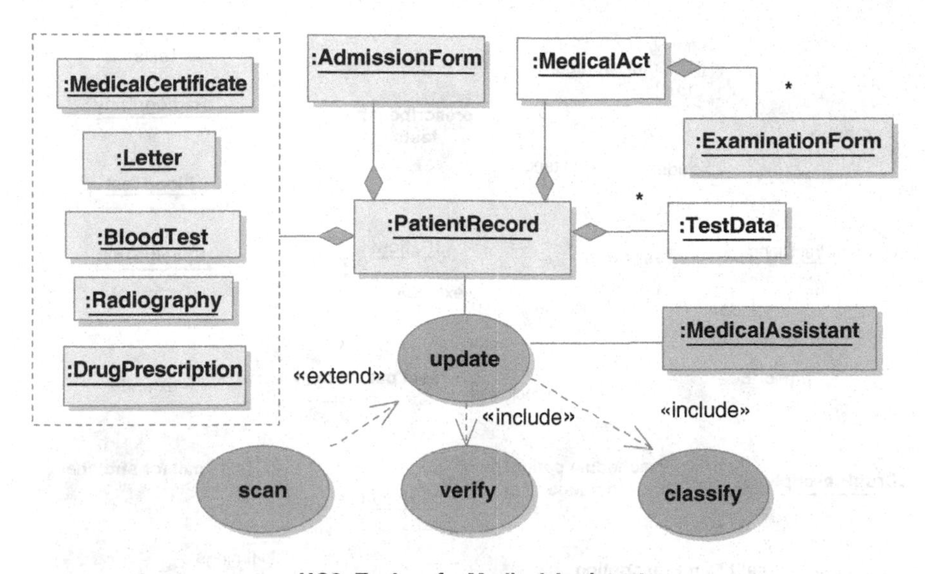

UC3: Tasks of a Medical Assistant

administrative database. In this case, we have very few information that can explain why a given patient has stayed, for instance, 5 h in the rest area, has been examined many times by the medical staff. To corroborate facts, a person must open the two applications at the same time and display information in such a way that relationships become explicit and help him in inferring new information. As this condition is very restrictive, it would be more interesting to connect systems together if, naturally, they are interrelated. Nowadays, even in

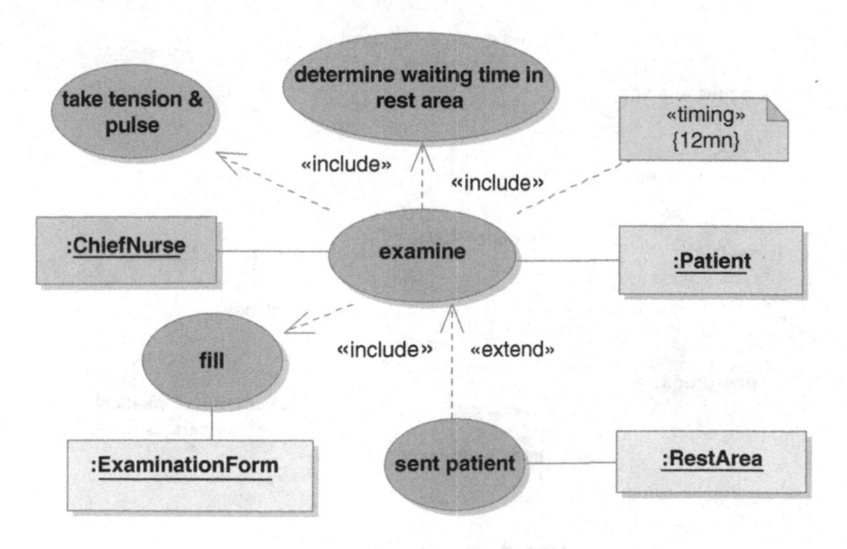

UC4: Tasks of a Chief Nurse

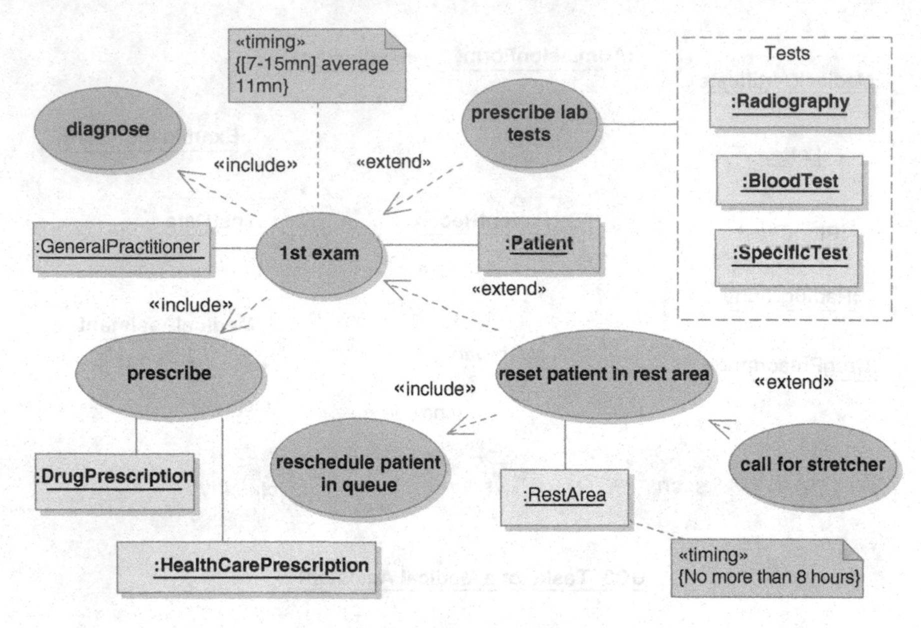

UC5: Tasks of a General Practitioner

the presence of many independent platforms, from any software environment, we can connect to any database, and gain access to its data. The problem is administrative, not technological. Some technical difficulties still remain when accessing huge amount of multimedia data through controlled channels. From a logical view, a large amount of real-time classes of this hospital are the same

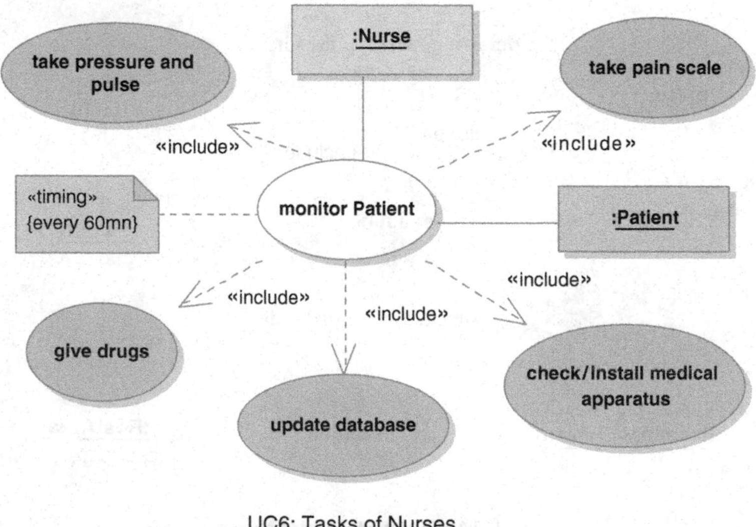

UC6: Tasks of Nurses

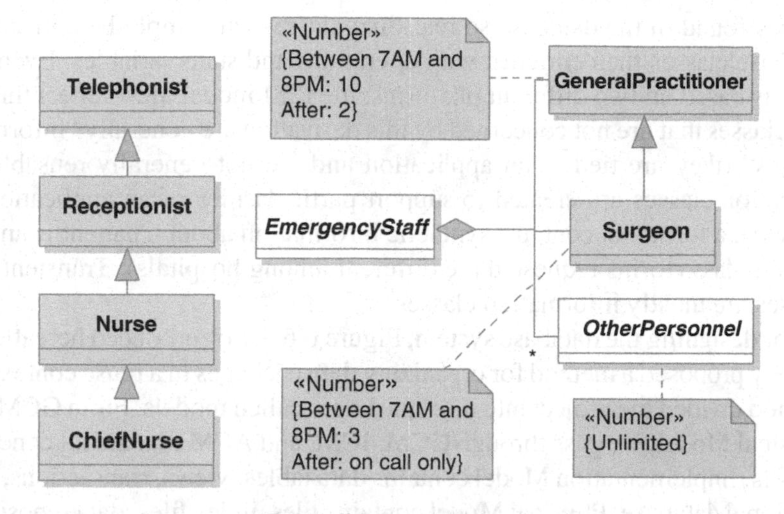

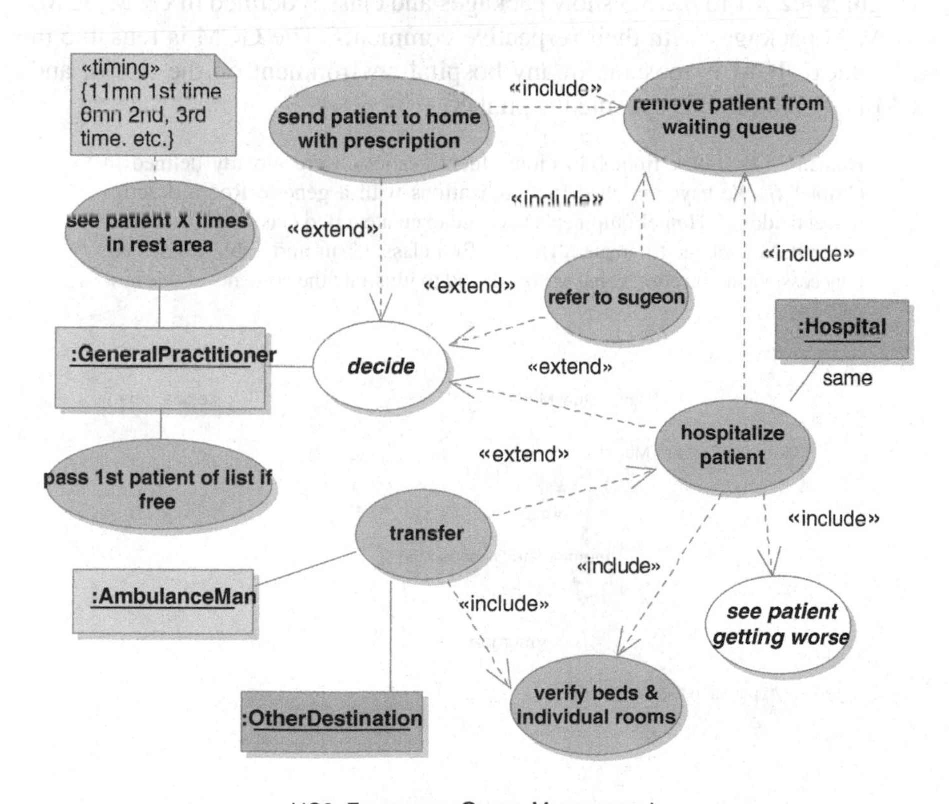

Fig. 8.2.2.2 UC0–UC8 specifying the hospital emergency problem

classes found in the database so real-time classes can simply be derived from database classes then enriched with operations and state variables. Even if the PSM is based on two different platforms, the PIM model must reflect this reality. Classes that are not concerned by this derivation are generally "information classes" (they are tied to an application and are not generally reusable). Information classes are created to support particularities of an application. For instance, a form that contains synthetic information about a patient is an information class (forms requested are different among hospitals). Transient query classes are mostly information classes.

For designing the database system, Figure 6.6.3.1 (reproduced hereafter) has already proposed a method for organizing data packages in a reuse context. This method divided the project into three packages called models. From GCM to the Physical Model, we pass through GCM, ICM, and ACM that are all conceptual models. Implementation Model contains data tables, views, transactions, etc. in relational database. Physical Model contains files, index files, data repositories, computer nodes, geographic repartition of data, etc.

Figures 8.2.3.1 to 8.2.3.3 show packages and classes defined in GCM, ICM, and ACM packages with their respective comments. The GCM is reusable in any context. ICM is reusable in any hospital environment on the world, and ACM is specific to this specific hospital X.

> Humans, Locations, Immobilizations, Identifications were already defined in Chapter 6. We have enriched Immobilizations with a generic Room description and added a HomeEquipments package to create a Bed class to be used later in the ICM package to create a HospitalBed class. Chair and Table classes are unnecessary in this context but are displayed to illustrate the contents of this new package.

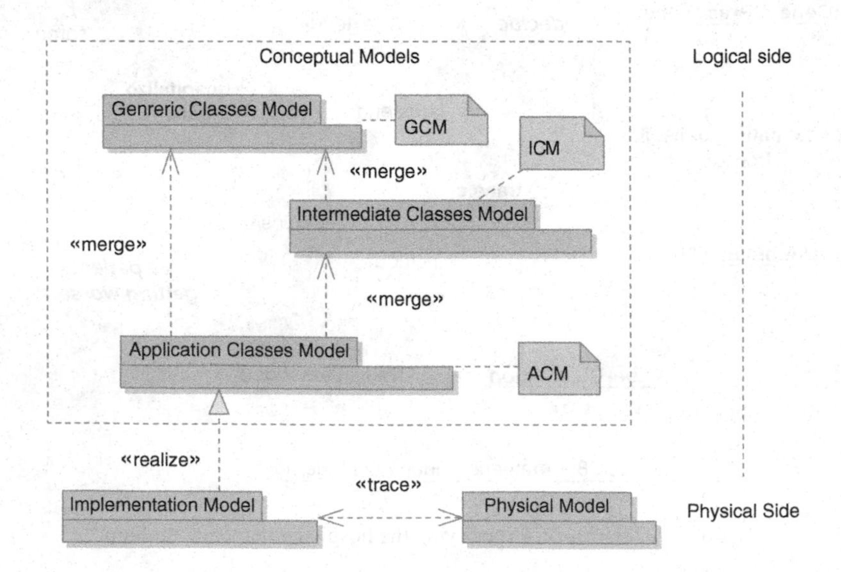

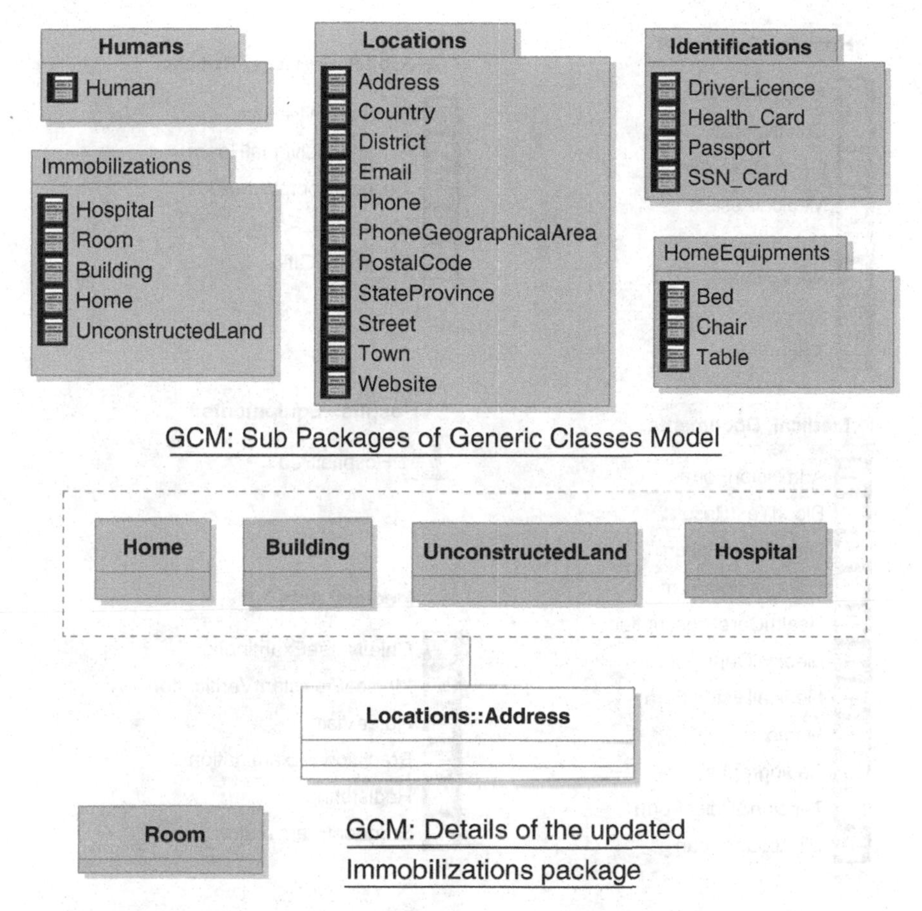

Fig. 8.2.3.1 Contents of the Generic Classes Model (GCM). Figures of this GCM suite must be completed with Figure 6.6.4.1 of Chapter 6

> The immobilizations package is the same as that shown in Figure 6.6.4.1. A Room does not bear any address so it is created outside the association. Contents of other packages are already shown in Figure 6.6.4.1. We illustrate therefore the surprising fact that a GCM created in a completely different context can be reused in the context of this Hospital. Practically, we have exported the design of Figure 6.6.4.1 to an XMI format (defined in UML) and reimported this format inside this Hospital project.

It would be interesting to note that we have not used the title "GCM package of the Hospital." Owing to its high potential of reusability, the same package GCM can be redeployed "as is" for modeling a company, a university, or a governmental organization.

The next ICM package will be a step closer towards solving the Hospital problem. There can be many ICMs before reaching the final ACM. The number of ICMs is dependent upon the complexity of the final ACM itself and how

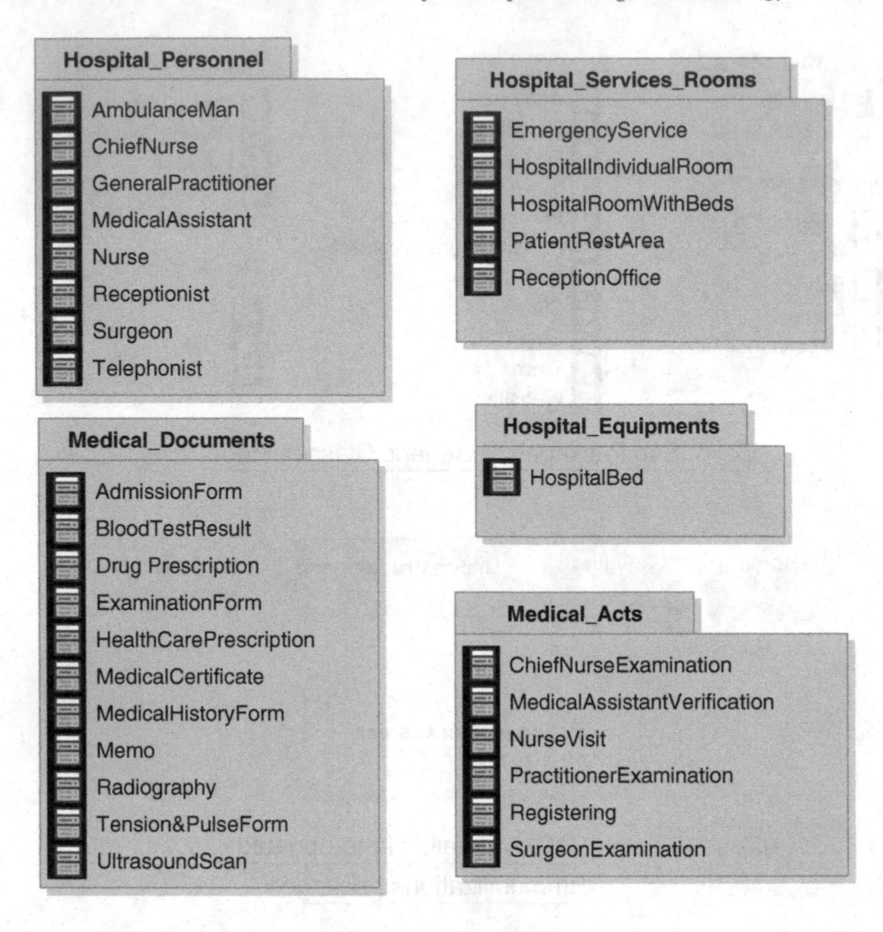

ICM: Top Level package

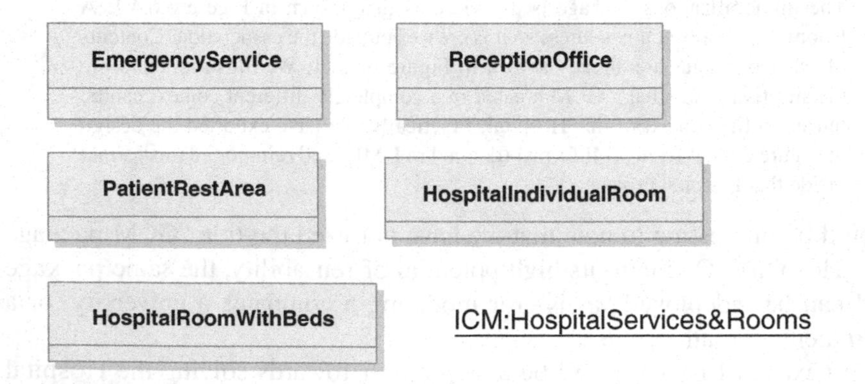

ICM:HospitalServices&Rooms

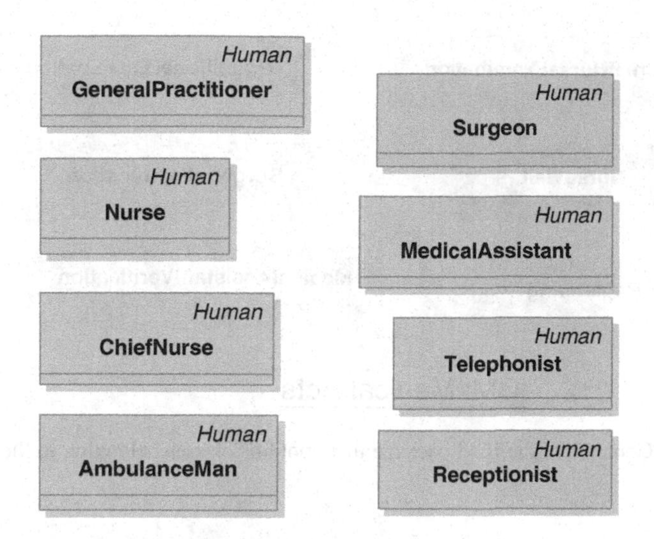

ICM: Hospital Personnel

Hospital Bed

ICM: Hospital Equipments

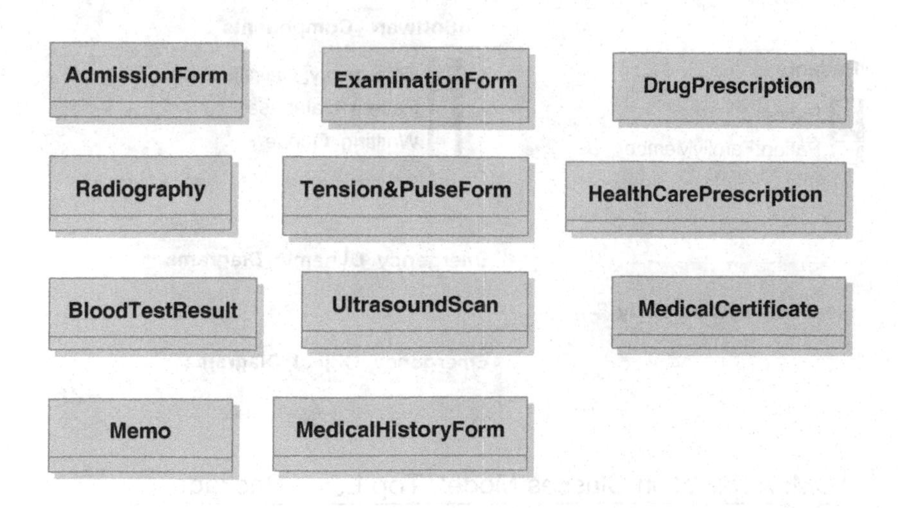

ICM: Medical Documents

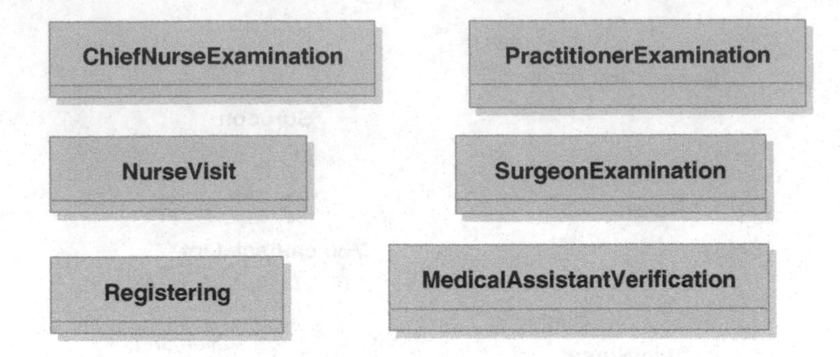

Fig. 8.2.3.2 Contents of the ICM package that contains classes belonging to the Health Care System

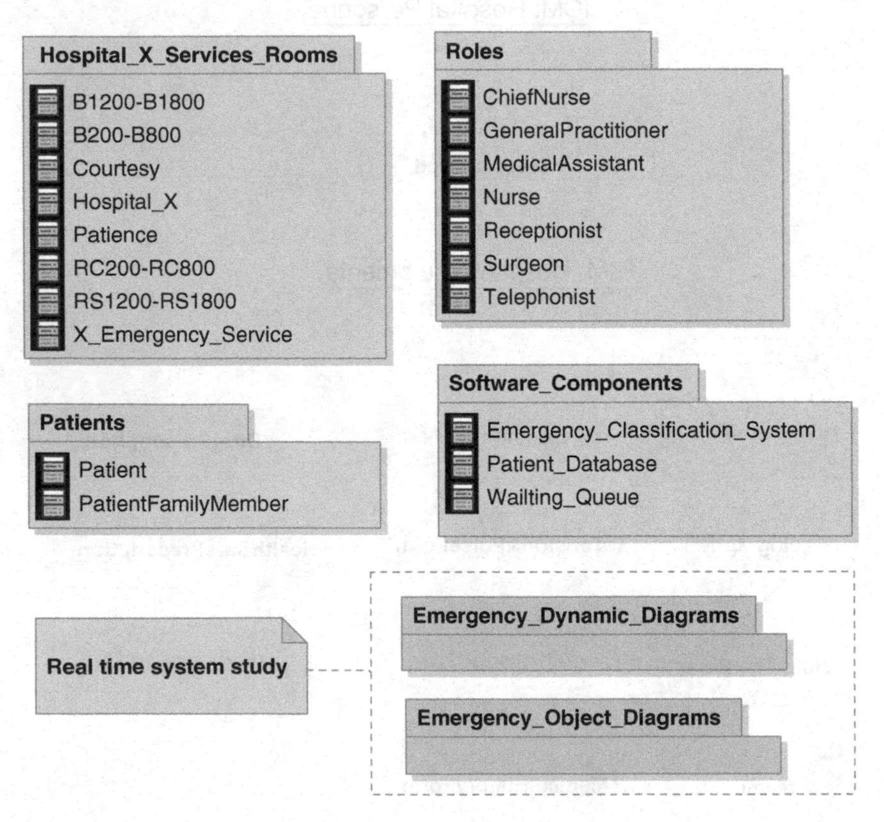

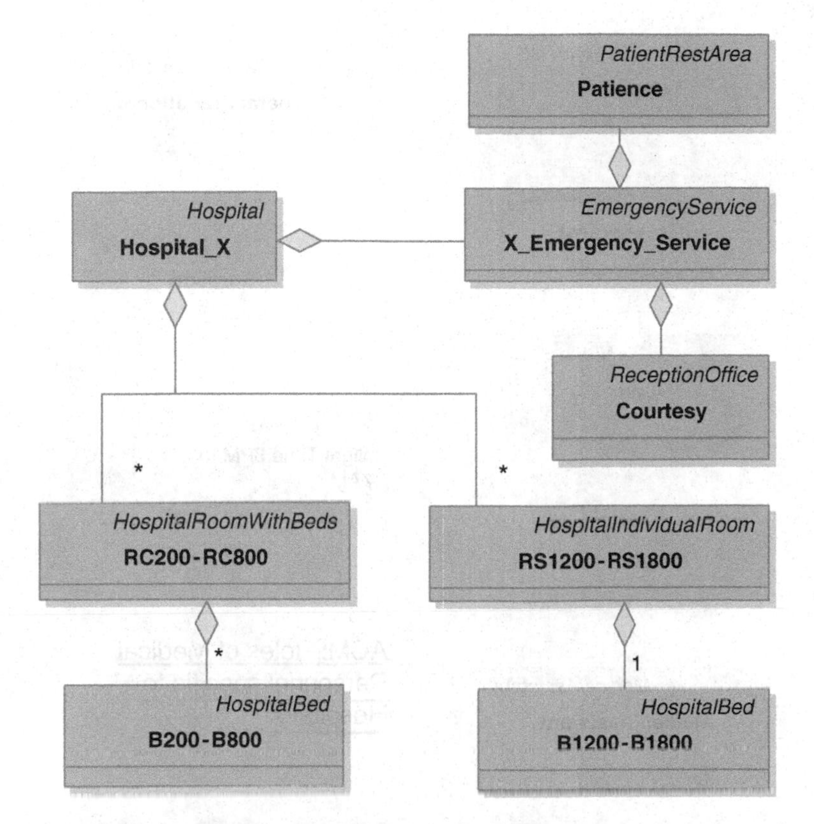

ACM:This package shows how the Emergency Service of the Hospital X is located inside this Hospital.

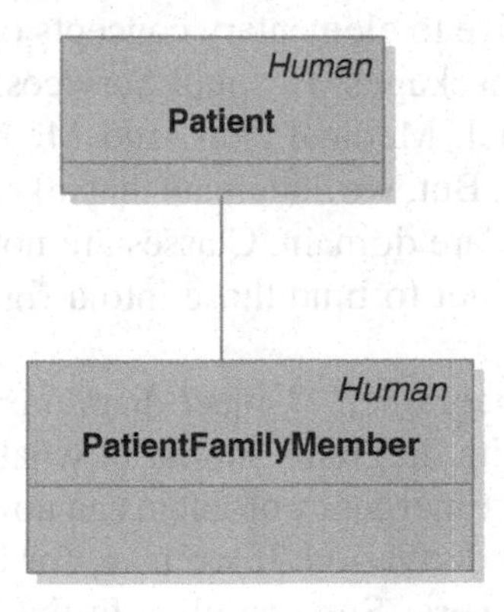

ACM: Patients and their family members

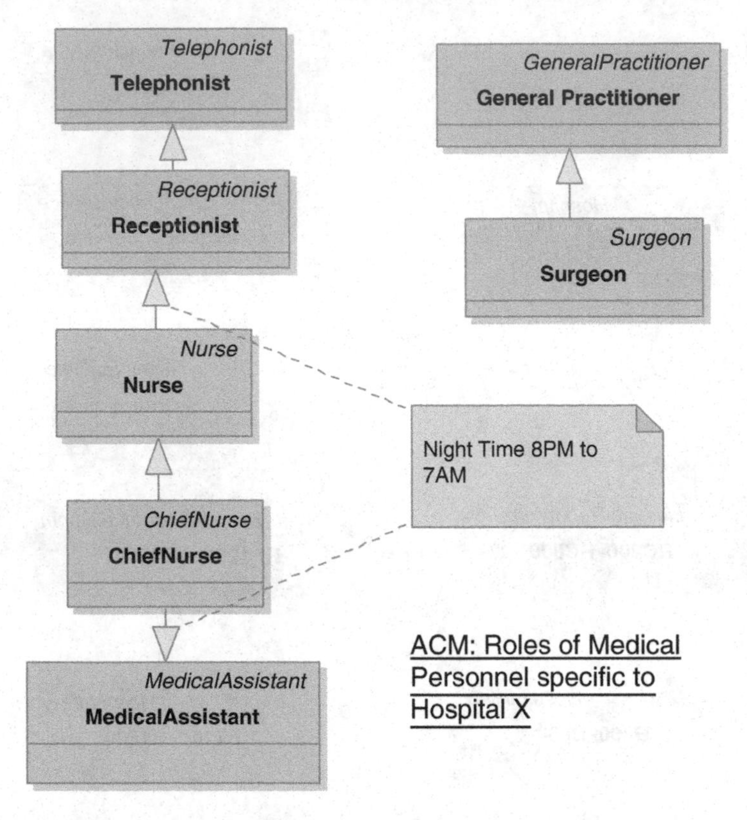

Fig. 8.2.3.3 Contents of the ACM (Application Classes Model) package. This package relates all relationships specific to the Hospital X

far concepts in the ACM are, relative to elementary concepts of the first GCM. In our case, we have created five packages: Hospital_Services_Rooms, Hospital_Equipments, Hospital_Personnel, Medical_Acts, and Medical_Documents to support the Hospital application. But, we anticipate that all classes created in the ICM can be reused in Health Care domain. Classes are not still connected together through relationships for not to bind those into a rigid structure that tends to destroy the reusability.

The subpackages "Dynamic Diagrams," "Object diagrams," and Software Components" can now be filled with diagrams similar to what we have developed for a real-time system and the Emergency problem can now be approached with all classes ready for the new derivation. If we take, for instance, a class like Surgeon, it derives from a generic Surgeon class in the ICM that packs operations very generic to a Surgeon all over the world. In the ACM packages, we put differential operations allowed in the context of the Hospital X, in country Y. The Surgeon class in ICM derives from Human class in GCM, so as a human; a surgeon will have all regular attributes common to a patient or any other humans.

With this example, we have demonstrated that databases and real-time applications can coexist in a same continuum and they are closely related in a PIM. The uniform methodology shows a high degree of reuse and creates models that defy space and time. New applications will be built from now on a common knowledge base that spares us so much energy and time. Each new problem instantiates new scenarios with new operations (occasionally new classes) but we have never to reinvent the wheel again.

References

W M P van der Aalst, A H M ter Hofstede, Workflow Patterns: On the Expressive Power of (Petri-Net based) Workflow Language, Proceedings of the 4th workshop on the practical use of colored Peri nets and CPN Tools, Vol 560, 2002, pp. 1–20

W M P van der Aalst, A H M ter Hofstede, YAWL: yet another workflow language, Information Systems, 30(4), 2005, 245–275

M Abadi, L Lamport, An old fashioned recipe for real time, ACM Transactions on Programming Languages and Systems (TOPLAS), 16(5), 1994, 1543–1571

M Alanen, I Porres, Subset and Union Properties in Modeling Languages, TUCS Technical Report 731, Turku, Finland, 2005

R Alur et al., The algorithmic analysis of hybrid systems, Theoretical Computer Science, 138, 1995, 3–34

R Alur, G J Holzmann, D Peled, An Analyzer for Message Sequence Charts, Proceedings of the Tools and Algorithms for the Construction and Analysis of Systems Conference, Passau, Germany, 1996, pp. 35–48

L F Andrade, J L Fiadeiro, The Unified Modeling Language: Beyond the Standard, Proceedings of the UML 99, Fort Collins, Colorado, 1999, pp. 566–583

F Armour, G Miller, Advanced Use Case Modeling, Object Technologies Series, Addison-Wesley, Reading, MA, 2001

C Atkinson, T Kühne, Rearchitecting the UML infrastructure, ACM Transactions on Modeling and Computer Simulation, 12(4), 2002, 290–321

C Aurrecoechea, A T Campbell, L Hauw, A survey of QoS architectures, Multimedia Systems, 6, 1998, 138–151

R Axelrod, Advancing the art of simulation in the social sciences, Complexity 3(2), 1997, 16–22

C W Bachman, M Daya, The Role Concept in Data Models, Proceedings of the 3rd International Conference on Very Large Databases, 1977, pp. 464–476

B Berelson, Content Analysis in Communications Research, Handbook of Social Psychology: Theory and Method, 2nd edn., Addison-Wesley, Cambridge, MA, pp. 488–522

B Berthomieu, M Diaz, Modeling and verification of time dependent systems using time Petri nets, Third Transactions on Software Engineering, 17, 1991, 259–273

T J Biggerstaff, An assessment and analysis of software reuse, Advances in Computer, 34, 1992, 1–57

G Booch, Object Oriented Analysis and Design with Applications, Benjamin-Cummings, Redwood City, CA, 1993

P Bradley, B Fox, L Schrage, A Guide to Simulation, 2nd edn., Springer-Verlag, New York, 1987

F M T Brazier, C M Jonker, J Treur, Principles of Compositional Multi-Agent System Development, Proceedings of the 15th IFIP World Computer Congress, 1998

P Bresciani, A Perini, P Giorgini, F Giunchiglia, J Mylopoulos, Tropos: an agent-oriented software development methodology, Journal of Autonomous Agents and Multi-agent Systems, 8(3), 2004, pp. 203–236

Bui Minh Duc, Conception et Modélisation Objet des Systèmes Temps Réel, Eyrolles, Paris, 1999, ISBN 2-212-09027-7

P A Buhr, M Fortier, Monitor classification, ACM Computing Surveys, 27(1), 1995, 63–107

L Cardelli, P Wegner, On understanding types, data abstraction, and polymorphism, ACM Computer Surveys, 17(4), 1985, 471–523

L Cardelli, A semantics of multiple inheritance, Information and Computation, 76, 1988, 138–164

P P S Chen, The Entity-Relationship model: towards a unified view of data, ACM Transactions on Database Systems, 1(1), 1976, 9–36

E F Codd, A relational model of data for large shared data banks, Communications of the ACM, 8(2), 1983

E F Codd, The Relational Model for Database Management, Version 2, Addison-Wesley, New York, 1990

A Cockburn, Writing Effective Use Cases, Addison-Wesley, Reading, MA, 2000

B Curtis, M I Kellner, J Over, Process modeling, Communications of the ACM, 35(9), 1992, 75–90

A Dardenne, A van Lamsweerde, S Fickas, Goal-Directed Requirements Acquisition, Selected Papers of the Sixth International Workshop on Software Specification and Design, 1993, pp. 3–50

R David, H Alla, Petri Nets and Grafcet: Tools for Modeling Discrete Event Systems, Prentice-Hall, New York, 1992, pp. 339, ISBN 0-133-27537-X

R David, Grafcet: A powerful tool for specification of logic controllers, Control Systems Technology, IEEE Transactions, 3,(3), 1995, 253–268, ISSN 1063-6536

P Davidsson, Multiagent Based Simulation: Beyond Social Simulation, Multi-Agent Based Simulation, Springer-Verlag, New York, 2001

T DeMarco, Structured Analysis and System Specification, Prentice-Hall, Englewood Cliffs, New Jersey, 1979

E W Dijkstra, Notes on Structured Programming, in: Structured Programming, Academic Press, New York, 1972

C A Ellis, G J Nutt, Office information systems and computer science, ACM Computing Surveys, 12(1), 1980, 27–60

J M Epstein, J D Steinbruner, M T Parker, Modeling Civil Violence: An Agent-Based Computational Approach, Working Paper 20, Center on Social and Economic Dynamics, Washington, DC, 2001

F Flores, M Graves, B Hartfield, T Winograd, Computer systems and the design of organizational interactions, ACM Transactions on Office Information System, 6(2), 1988, 153–172

W B Frakes, C J Fox, Sixteen questions about software reuse, Communications of the ACM, 38(6), 1995

E Gamma, R Helm, R Johnson, J Vlissides, Design Patterns: Elements of Reusable Object-Oriented Software, Addison-Wesley, Reading, MA, 1995

C Gane, T Sarson, Structured Systems Analysis: Tools and Techniques, Prentice-Hall, Englewood Cliffs, NJ, 1978

G Génova, J Llorens, V Palacios, Sending messages in UML, Journal of Object Technology, 2(1), 2003, 99–115

D Georgakopoulos, M Hornick, A Sheth, An overview of workflow management: From process modeling to workflow automation infrastructure, Distributed and Parallel Databases, 3(2), 1995, 119–153

N Gilbert, K G Troitzsch, Simulation for the Social Scientist, Open University Press, Milton Keynes, UK, 1999

C Ghezzi, D Mandrioli, S Morasca, M Pezze, A unified high level Petri nets formalism for time critical systems, IEEE Transactions on Software Engineering, 17(2), 1991, 160–172

S Gnesi, D Latella, M Massink, Model Checking UML Statechart Diagrams Using JACK, Proceedings of the 4th International Symposium on High Assurance Systems Engineering, Washington, DC, 1999, pp. 46–55

M Gogolla, M Richters, Transformation Rules for UML Class Diagrams, Lecture Notes in Computer Science, Springer-Verlag, New York, 2004, pp. 92–106, ISBN: 3-540-66252-9

E J Golin, S P Reiss, The Specification of Visual Language Syntax, Proceedings of the IEEE Workshop on Visual Languages, Rome, Italy, 1989, pp. 105–110

T R Gruber, A Translation Approach to Portable Ontologies, Knowledge Acquisition 5, 1993, 199–220

M Hammer, D McLeod, Database description with SDM: a semantic database model, ACM Transactions on Database Systems, 6(3), 1981, 351–386

D Harel, A Pnueli, On the Development of Reactive Systems, Logics and Models of Concurrent Systems, Springer-Verlag, New York, 1985, pp. 477–498

D Harel, Statecharts: A visual formalism for complex systems, Sciences of Computer Programming 8(3), 1987, 231–274

D Harel, On Visual Formalisms, Communications of the ACM, 31(5), 1988, 514–530

D Harel, E Gery, Executable Object Modeling with Statecharts, Proceedings of the 18th International Conference on Software Engineering, Berlin, 1996, pp. 246–257

D Harel, E Gery, Executable object modeling with statecharts, Computer, 30(7), 1997, 31–42

D Harel, H Kugler, A Pnueli, Synthesis Revisited: Generating Statechart Models from Scenario-based Requirements, Springer-Verlag, New York, 2005, pp. 309–324, ISBN 978-3-540-24936-8

G T Heineman, W T Councill, Component Based Software Engineering: Putting the Pieces Together, Addison-Wesley, Boston, MA, 2001, ISBN: 0201704854

B Henderson-Sellers, F Barbier, What is This Thing Called Aggregation, Proceedings of the Technology of Object-oriented Languages and Systems, Nancy, France, 1999, pp. 236–250

W Herroelen, B D Reyck, E Demeulemeester, Resource-constrained project scheduling: a survey of recent developments, Computers and Operations Research, 25(4), 1998, 279–302

L E Holloway, B H Krog, A Giua, A survey of Petri net method for controlled discrete event systems, Discrete Event Dynamic Systems, 7(2), 1997, 151–190

R Hull, R King, Semantic database modeling: survey, applications and research issues, ACM Computing Surveys, 19(3), 1987, 201–260

U Isaksen, J P Bowen, N Nissanke, System and Sofware Safety in Critical Systems, University of Reading, Computer Science Department, Whiteknights, UK, 1996

I Jacobson, M Chriserson, P Jonsson, G Overgaard, Object Oriented Software Engineering: A Use Case Driven Approach, Addison-Wesley, Reading MA, 1992

I Jacobson, G Booch, J Rumbaugh, The Unified Software Development Process, Addison-Wesley/Longman, Reading, MA/London, 1999

N R Jennings, On agent based software engineering, Artificial Intelligence, 117, 2000, 277–296

N Jennings, K Sycara M. Wooldridge, A roadmap of agent research and development, Autonomous Agents and Multi-Agent Systems, 1(1), 1998, 7–38

Y Jin, R E Levitt, The virtual design team: a computational model of project organizations, Journal of Computational and Mathematical Organization Theory, 2(3), 1996, 171–195

D H Jonassen, Instructional design model for well-structured and ill-structured problem-solving learning outcomes, Educational Technology: Research and Development, 45(1), 1997, 65–95

H H Kassarjian, Content analysis in consumer research, Journal of Consumer Research, 4, 1977, 8–18

G Kiczales, Aspect-oriented programming, ACM Computing Surveys, 28(4), 1996

S N Khoshafian, G P Copeland, Object Identity, Proceedings of the Conference on Object Oriented Programming Systems, Languages and Applications, ACM Press, Portland, OR, pp. 406–416

C Kobryn, UML 3.0 and the future of modeling, Software and System Modeling, 3(1), 2004, 4–8

B B Kristensen, K Osterbye, Roles: conceptual abstraction theory and practical languages issues, Theory and Practice of Object Systems, 2(3), 1996, 143–160

D Kulak, E Guiney, Use Cases: Requirements in Context, ACM Press/Addison-Wesley, New York/Boston, MA, 2000

C W Krueger, Software reuse, Computing Surveys, 24(2), 1992, 131–184

A V Lamsweerde, Goal-Oriented Requirement Engineering: A Guided Tour, Proceedings of the 5th International Symposium on Requirements Engineering, Toronto, Canada, 2001, pp. 249–262

P J Leach, B L Stumpf, J A Hamilton, P H Levine, UIDs as Internal Names in a Distributed File Systems, Proceedings of the 1st Symposium on Principles of Distributed Computing, ACM Press, Ottawa, 1982

E A Lee, What's ahead for embedded software? IEEE Computer, 33(9), 2000, 18–26

D Lefffingwell, D Widrig, Managing Software Requirements: A Unified Approach, Object Technology Series, Addison-Wesley, Reading, MA, 2000

N G Leveson, J L Stolzy, Software safety: Why, what, and how, Computing Surveys, 18(2), 1986, 125–163

A H Lewis, Human Organizations as Distributed Intelligence Systems, Proceedings of the IFAC Symposium on Distributed Intelligence Systems, Pergamon Press, Oxford, 1988

M Luck, P McBurney, C Preist, Agent Technology: Enabling Next Generation Computing – A Roadmap for Agent Based Computing, Agent link Community, Southampton, UK, 2003, ISBN-0854-327886

R R Lutz, Software Engineering for Safety: A Roadmap, Proceedings of the Conference on the future of Software Engineering, Limerick, Ireland, 2000, pp. 213–226

G A Miller, The magical number seven, plus or minus two: Some limits on our capacity for processing information, Psychological Review, 63, 1956, 81–97

M Morisio, M Ezran, C Tully, Success and failure factors in software reuse, IEEE Transactions on Software Engineering, 28(4), 2002, 340–357

O Nierstrasz, Regular types for active objects, ACM SIGPLAN Notices, 28(10), 1995, 1–15

J D Novak, Concept maps and Vee diagrams: two meta cognitive tools to facilitate meaningful learning, Instructional Science, 19(1), 1990, 29–52

J Odell, H V D Parunak, B Bauer, Extending UML for agents, Proceedings of AOIS 2000, Austin, TX, 2000, pp. 3–17

P S Pande, L Holpp, What is Six Sigma? McGraw-Hill, New York, 2002

D L Parnas, On the Criteria Used in Decomposing Systems into Modules, CACM 15, Vol 12, 1972, pp. 1053–1058

W H Pierce, Failure Tolerant Computer Design, Academic Press, New York, 1965

J L Peterson, Petri nets, ACM Computing Surveys, 9(3), 1977, 223–252

J L Peterson, Petri Net Theory and the Modeling of Systems, Prentice-Hall, New Jersey, 1981, ISBN 0-13-661983-5

C Petri, Kommunication mit Automaten, Ph.D. thesis, University of Bonn, Germany, 1962

R A Pottinger, P A Berstein, Merging Models Based on Given Correspondences, Proceedings of the 29th Very Large Data Base Conference, Berlin, Germany, 2003

C Potts, K Takahashi, A Anton, Inquiry-based requirement analysis, IEEE Software, 11(2), 1994, 21–32

M R Quillian, Semantic memory, in: M Minsky (ed.), Semantic Information Processing, MIT Press, Cambridge, MA, 1968, pp. 216–270

B Randell, System Structure for software fault tolerance, IEEE Transactions on Software Engineering, 1(2), 1975, 220–232

W R Reitman, Cognition and Thought: An Information Processing Approach, Wiley, New York, 1965

C Rolland, C Souveyet, C B Achour, Guiding goal modeling using scenarios, IEEE Transactions on Software Engineering 24 (12), 1998, 1055–1071

S Robertson J Robertson, Mastering the Requirement Process, Addison-Wesley, Reading, MA, 2000

E A Rundensteiner, L Bic, Set Operations in Object-based Data Models. IEEE Transactions on Knowledge and Data Engineering, 4, 1992, 382–398.

J Rumbaugh, M Blaha, W Premerlani, F Eddy, W Lorensen, Object Oriented Modeling and Design, Prentice-Hall, Eaglewood Cliffs, NJ, 1995

J Sametinger, Software Engineering with Reusable Components, Springer-Verlag, New York, 1997, ISBN 3-540-62695-6

R S Sandhu, E J Coyne, H L Feinstein, C E Youman, Role-based access control models, Computer, 29(2), 1996, 38–47

N Scharli, S Ducasse, O Nierstrasz, A P Black, Traits: Composable Units of Behaviour, ECOOP 2003, Springer-Verlag, New York, 2003, pp. 248–274

S Shlaer, S J Mellor, Object Lifecycles: Modeling the World in States, Yourdon Press Computing Series Archive, New York, 1992, pp. 251, ISBN 0-13-629940-7

E A Silver, D F Pyke, R Petersion, Inventory Management and Production Planning and Scheduling, Wiley, New York, 1998

H A Simon, Applying Information Technology to Organization Design, Public Administration Review, 33, 1973, 268–278

J M Smith, D C P Smith, Databases abstractions: aggregation and generalization, ACM Transactions on Database Systems, 2(2), 1977, 105–133

A Snyder, Encapsulation and inheritance in object-oriented programming languages, in: N Meyrowitz (ed.), Proceedings of the Object-oriented Programming Systems, Languages and Applications, ACM Press, New York, 1986, pp. 38–45

J Sobieszczanski-Sobieski, R T Haftka, Multidisciplinary aerospace design optimization: survey of recent developments, Structural Optimization, 14(1), 1997, 1–23

I Sommerville, P Sawyer, Viewpoints: principles, problems and a practical approach to requirements engineering, Annals of Software Engineering, 3, 1997, 101–130

R M Sonnemann, Exploratory Study of Software Reuse Success Factors, Ph.D. dissertation, George Mason University, Fairfax, VI, 1996

J F Sowa, Structures: Information Processing in Mind and Machine, Addison-Wesley, Reading MA, 1984, pp. 481, ISBN 0-201-14472-7

M G Speranza, C Vercelis, Hierarchical Models for multi-project planning and scheduling, European Journal of Operational Research, 64, 1993, 312–325

F Steimann, On the representation of roles in object-oriented and conceptual modeling, Data and Knowledge Engineering, 35, 2000, 83–106

W Stevens, G Myers, L Constantine, Structure design, IBM Systems Journal, 13(2), 1974, 115–139

J Swartz, Simulating the Denver airport automated baggage system, Dr. Dobb's Journal, 22(1), 1997, 56–62

Z Tari, J Stokes, S Spaccapietra, Object normal forms and dependency constraints for object-oriented schemata, ACM Transactions on Database Systems, 22(4), 1997, 513–569

J J P Tsai, S J Yang, Y H Chang, Timing Constraint Petri nets and their application to schedulability analysis of real time system specifications, IEEE Transactions on Software Engineering, 21(1), 1995, 32–49

S Uchitel, J Kramer, J Magee, Synthesis of behavioral models from scenarios, IEEE Transactions on Software Engineering, 29(2), 2003, 99–115

J Ullman, Database Theory: Past and Future, Proceedings of ACM SIGACT News – SIGMOD – SIGART Principles of Database Systems, San Diego, 1987

A Villemeur, Reliability, Availability, Maintainability and Safety Assessment, Wiley, New York, 1991

E Yourdon, L Constantine, Structured Design: Fundamentals of a Discipline of Computer Program and Systems Design, Prentice-Hall, Eaglewood Cliffs, NJ, 1979

M R Waldmann, Y Hagmayer, Seeing versus doing: two modes of accessing causal knowledge, Journal of Experimental Psychology: Learning Memory and Cognition, 31, 2005, 216–227

P Ward, S Mellor, Structured Development of Real-Time Systems, Vol 1–3, Yourdon Press, New York, 1986

P Wegner, Dimensions of Object-Based Language Design, Proceedings of the Conference on Object-Oriented Programming Systems, Languages and Applications, Orlando, FL, 1987

P Wegner, Concepts and paradigms of object oriented programming, ACM OOPS Messenger, 1(1), 1990, 7–87

P Wohed, W M P van der Aalst, M Dumas, Pattern-based Analysis of the Control Flow Perspective of UML Activity Diagram, Lectures Notes in Computer Sciences, Springer-Verlag, New York, 2005, pp. 63–78, ISBN-978-3-540-29389-7

W Wolf, Hardware–Software co-design of embedded systems, Proceedings of the IEEE, 82(7), 1994, 967–989

P Zave, M Jackson, Four darks corners of requirement engineering, ACM Transactions on Software Engineering and Methodology, 6(1), 1997, 1–30

A Zeller, Isolating Cause–Effect Chains from Computer Programs, Proceedings of the ACM SigSoft (FSE-10), South Carolina, 2002

M C Zhou, F Dicesare, Petri Nets Synthesis for Discrete Event Control of Manufacturing Systems, Kluwer Academic, Dordrecht, 1993

R Zurawski, M Zhou, Petri nets and industrial applications: a tutorial, IEEE Transactions on Industrial Electronics, 41(6), 1994, 567–583

Index

Abstract class, 58, 67, 98, 105, 110, 195, 219, 265, 266, 280
Abstraction, 19, 20, 43, 48, 52–58
AcceptSignalAction, 164
ACM, 195, 368, 369, 372, 374, 458, 459, 462–464
Action, 6, 24, 25, 28, 144, 145, 148, 150, 151
Active object, 29, 290, 296, 297
Activity, 305, 308, 310, 316, 317, 325, 331, 336, 340
ActivityEdge, 153, 155, 158
ActivityNode, 152, 153, 160, 162, 163, 164, 165
ActivityParameterNode, 160, 166
ActivityPartition, 163, 164, 168, 214
Actor, 32, 182, 183, 184, 206, 308, 310, 313
ActivityGroup, 155, 163, 164, 168
Adornment, 21, 53
Agent, 13, 15, 16, 26, 27, 29, 31, 224, 238, 240, 245, 261, 291, 308, 351, 362, 363, 389, 415, 416
AggegationKind
Aggregation, 22, 23, 31, 32, 39, 47, 53–55, 84, 99, 100, 106, 111, 114, 115, 118, 128, 133, 224, 255, 268, 269, 276–279, 326, 337, 354, 355, 364, 369, 373, 375, 426, 427, 438
Alarm System, 240, 246, 292, 293, 303, 342, 343, 344, 349, 353–356, 358–361
Algorithm, 144, 159, 205, 236, 289, 291, 341, 343, 347, 349, 359, 360, 381, 385, 408, 410, 450
AND-states, 148, 232, 238
Argument, 64, 179, 273
Artifact, 97, 128, 130–134, 202
Assembly, 7, 10, 11, 118, 131, 159, 205, 298, 334, 343, 355, 374, 382, 388, 436
Assessments, 212, 219
Association, 244, 249, 270–277, 287, 288, 365, 369, 370, 373, 410, 438, 459
Association class, 69, 98, 104, 105, 107, 123, 134, 272, 273
Association end, 53, 54, 66, 69, 94, 95, 98, 99, 127, 287
AssociationClass, 90, 104, 136
Asynchronous message, 178
Attribute, 16, 54, 55, 58, 60, 66, 69, 94, 95–97
Author, 15, 16, 207, 216, 363
Automatic Registering System, 182, 183

Bag, 96, 97, 111
Baggage handling, 376, 450
BasicBehaviors, 124, 136–138, 141, 151, 152, 162, 176, 183
Behavior, 5, 11, 12, 14, 15, 25, 29, 30, 33–35
Behavioral diagrams, 82, 83, 194, 195
BehavioralFeature, 63, 64, 69, 70, 90, 91, 138
BehavioredClassifier, 123, 124, 137, 138, 145, 183
BehaviorStateMachines, 136, 141, 152
BMM, 212, 217–219, 224, 225
Book, 3, 9, 15–17, 40, 47, 48, 50
Bottom-up, 22, 334, 335
BPM, 78, 213, 216, 217, 313, 322, 335
BPMI, 78, 212, 213
Break, 174, 175, 245, 335, 337, 353, 361, 374
Bubble, 24–25, 207–208, 325
Business modeling, 78, 201, 213, 315
Business process, 25, 28, 36, 78, 79, 201–205, 211–224, 231, 237, 288, 307, 309, 315, 326–327, 329, 336, 382
Busy wait, 346

C#, 5, 97, 100, 201, 247, 269, 282, 350, 374, 381
C++, 5, 97, 100, 201, 247, 269, 350, 352, 374, 381
Call graph, 341
Cap putting machine, 10
Car, 2, 6, 10, 22–23, 29, 30, 40
CASE, 4–5, 24, 42, 49, 78, 84, 85, 149, 163, 173, 312, 381, 385, 391, 392
Categorization, 194, 365
Cause–effect chain, 248, 352, 252, 303–304, 326, 388, 428, 430

CentralBufferNode, 155, 161, 209, 211, 229
Class, 5, 10, 14, 27, 30, 31
Class diagram, 80, 81, 93–109
Class factorization, 31–32
Classification, 32, 49, 61, 80, 165, 189, 193, 259, 260, 262, 264–266, 271, 282, 353, 365, 453
Classifier, 57–64
Client-server, 27, 205, 257, 293
Closed-loop, 34, 289, 409
COCOMO, 327–328
Collaboration, 4, 22–23, 29, 60, 81, 104, 118
CollaborationUse, 119, 123–125, 127, 221
Collaborative task, 2, 23, 27, 118, 220, 234, 303, 341, 359, 362, 375, 384
Color, 21–22, 178, 189, 233, 315, 317, 365, 424
CombinedFragment, 173–177, 180
Comment, 42, 47, 54, 66, 68, 78, 81–83, 86
Communicate, 20, 22, 28, 30, 31, 34, 129, 139, 191, 193, 199, 206, 229, 231, 249, 250, 269, 296, 297, 303, 313, 316, 317, 320, 322, 334, 358, 384, 409, 410
Communication diagram, 82, 169, 171–173, 177, 179, 181, 222, 233, 243, 249, 252, 253, 384, 397
CommunicationPath, 133, 135–136
Complex system, 1, 4, 5, 29, 40, 83, 116, 128, 159, 181, 200, 241, 289, 290, 335, 337, 343, 412
Complexity, 5, 7, 16, 19
Complexity reduction, 22, 28, 31, 33, 39, 40, 184, 191, 245, 255, 256, 289, 302, 303, 389
Compliance levels, 86–88
Component, 5–8, 10–14
Component diagram, 77, 81, 106, 128–134, 194, 198, 226
Composite state, 139, 146–150, 220, 231–233, 236238, 254, 361
Composite structure diagram, 81, 113, 114, 118, 120, 121, 124–126, 131, 194, 195, 221, 226, 271
Composition, 22, 23, 31, 32, 49, 40, 47
Concept, 2, 13, 14, 20, 28, 30, 33, 42, 43, 48
Concurrency, 5, 11, 24, 33, 139, 151, 159, 301
Condition, 11, 15, 37, 52, 57, 70
ConnectableElement, 119, 120, 123, 124, 176
Connector, 21, 112–114, 120, 123, 126, 131, 202, 270, 271
ConnectorEnd, 122, 123, 271
Constraint, 4, 6, 8–13, 16, 20, 34, 39, 47, 50, 52
Content analysis, 307, 310, 312, 319, 403
ControlFlow, 153, 158, 188, 201, 207, 210
ControlNode, 155, 160, 162
Cordless phone, 334
Core concepts
Core
- abstractions
  - behavioralfeatures, 63
  - changeabilities, 62
  - classifiers, 59
  - comments, 55
  - constraints, 72
  - expressions, 72
  - generalizations, 61
  - instances, 62, 88
  - literals, 58, 88
  - multiplicities, 63, 65
  - multiplicityexpressions, 63, 65
  - namespaces, 56
  - ownerships, 54, 72
  - redefinitions, 53, 61
  - relationships, 60
  - structuralfeatures, 62
  - super, 59
  - typeelements, 59
  - visibilities, 56
- basic
  - classes, 64, 66, 68, 69
  - datatypes, 65, 67, 70
  - packages, 65, 67
  - types, 64
- constructs
  - classes, 68, 69
  - classifiers, 67
  - constraints, 72, 73
  - datatypes, 70
  - expressions, 72, 73
  - namespaces, 71, 72
  - operations, 69, 70
  - packages, 71
  - root, 60
- profiles, 72–75

COTS, 7, 335, 351, 352, 357
CRT parameters, 305, 310, 315–317, 320–322, 325, 420, 426, 430

Data abstraction, 191–193
Data modeling, 36, 39, 285
Data structures, 27, 262, 270, 350, 352, 381
Data type, 67, 70, 71, 381
Database, 13, 28, 36–39, 41, 78, 80, 98
DataFlow, 24, 201, 207, 209
Datastore, 24, 25, 158–162, 206, 207–211
DatastoreNode, 206, 209, 211, 225
DataType, 67, 70, 71, 381
DBMS, 4, 36, 38, 39, 377, 378
Decentralization, 39
Decision, 1–6, 9, 15, 17, 99, 151
Decision making, 3, 5, 8, 9, 36, 415, 450
DecisionNode, 215
Deduction, 15
deepHistory, 150

Dependencies, 26, 48, 108, 109, 119, 124, 128, 183, 193, 270, 310, 362
Dependency, 52, 107–109, 114, 270, 271, 275, 340, 341, 397, 411
Deployment, 9, 80, 87, 128, 132–135, 212, 300, 351
Deployment diagram, 132–135, 378
DeploymentSpecification, 134
Design model, 78, 330, 347
Device, 4, 5, 8, 9, 28, 36, 78, 132–134
DFD, 24, 156, 204, 207, 208, 227–230, 307, 326, 327
Diagram, 16, 21, 24, 28, 50, 52
Diagram interchange, 50, 87
DirectedRelationship, 60, 68, 71, 73, 92
Disaster control, 10–13
Do Action, 144, 237, 394
Do Activity, 144
Domain, 1, 2, 6, 9–14, 26
Dynamic studies, 241, 335, 349, 357, 359, 360

Early requirements, 311, 313, 318, 417–420
Edge, 20, 24, 141, 151, 154, 158–163, 228, 389, 390
Effect, 7, 141, 248, 297, 303, 304, 312, 322, 405, 418, 441
Element, 33, 52, 53–58, 60
ElementImport, 71
Elicitation, 308, 311
Embarked, 11, 383, 405
Embedded, 8–12, 36, 128, 139, 246, 257, 306, 350, 351, 383, 405, 420, 439
Emergency service, 448–450
Encapsulation, 30, 256, 294
Entry point, 146, 148, 234
Enumeration, 55, 67, 70, 189, 236, 260, 261, 311, 326
ER
Event, 6, 103, 143–145, 155
Exception, 70, 102, 103, 152, 163, 164, 231, 281, 397–400, 417
ExceptionHandler, 163, 164, 168
ExecutableNode, 163, 164
ExecutionEnvironment, 132–134, 137
ExecutionSpecification, 145
Exit Point, 146, 148, 150, 234, 445
Expansion region, 167, 168
ExpansionNode, 166
Expression, 5, 50, 53, 56–58, 62, 63, 65, 72, 73
Extend relationship, 184, 185, 187, 245, 317, 322, 425, 452
Extension mechanism, 51, 75, 112, 186, 211
ExtensionPoint, 183–185
Extreme programming, 328

Feature, 11, 33, 53, 57, 59, 60–64, 67
Final, 7, 16, 141, 148, 154, 156, 157, 159, 161, 162, 169
FinalNode, 155, 160, 162, 214
FinalState, 141
FlowFinalNode, 155, 162, 214
ForkNode, 155, 162
Fragment, 173–180, 222, 233, 245, 340, 348, 349
Frame, 2, 5, 16, 176, 343, 354–356, 373, 381, 423, 425
Free diagram, 19, 20
Functional decomposition, 22, 23, 25, 26

Gate, 175, 176, 179, 294, 315
GCM, 368–370, 372, 374, 458, 459, 464
Generalization, 53, 57, 60, 61, 73, 88, 90–92, 94, 107
Generic architectures, 352, 354, 355–358, 434–436, 439, 440
Generic classes, 353, 356, 358, 368–370, 374, 430–432, 434, 435, 439, 440, 459
Genericity, 351, 352, 354, 433
Global system state, 234
Global warming, 12
Goal, 4, 7–9, 23, 25, 27, 28, 43, 50
Guard, 141, 144, 148, 149, 160–163, 173, 174
GUID, 266, 267

Hair dryer, 139, 140, 235, 236
Hazard, 407, 412, 414
Hospital, 268, 313, 319, 321–323, 407, 448, 450–454

ICM, 368, 369, 372, 374, 458–462, 464
ID, 40, 266, 267
Ill structured, 8, 9
Image attributes, 409–415
Implementation diagram, 77
Import, 51, 71, 85, 88, 115–117, 128, 152, 164, 203, 330, 352, 375, 430
Inclined elevator, 417, 419, 420, 424, 425, 427, 430, 434, 439, 441, 444
Include relationship, 184, 316, 317, 430
Induced energy, 290, 296–298, 361
Induction, 15, 290
Influencers, 212, 219
Information class, 372, 374, 377–379, 439, 458
Information hiding, 25, 30, 147, 191–193
Inheritance relationship, 86, 186, 258, 313, 314, 326, 354, 365
InitialState,
InputPin, 154
InstanceOf, 114, 115, 271
InstanceSpecification, 60–62, 90, 109, 110
Instantiation, 77, 105, 108, 283
Interaction, 11, 12, 15, 23, 32, 36, 79, 80, 82, 83, 87
Interaction diagram, 82, 83, 169, 182, 186, 241, 294, 296, 384

Interaction overview diagram, 79, 82, 169, 173, 180, 244, 245, 340, 347, 348, 384
Interaction suite, 79, 165, 169, 175, 176, 181, 213, 384, 397
InteractionFragment, 173, 175–177, 180
InteractionUse, 82, 169, 173, 175–177, 348
Interface, 7, 14, 15, 25, 30, 31, 44, 80, 81, 89, 90
InterfaceRealization,
Interrupt handler, 346, 399
InterruptibleActivityRegion, 163, 164, 168, 229
Interval, 65, 202, 233, 248, 251, 264, 274, 319, 409, 410, 446
isAbstract, 59, 138
isActive,
isComposite, 141, 188, 232, 234, 260, 277
isIndirectlyInstantiated, 130
isOrdered, 63, 69, 96, 410
isReentrant, 138
isUnique, 63, 69, 96

Job, 8, 39, 41, 105, 157, 204, 205, 229, 272, 286–287, 399
JoinNode, 155, 162
Junction, 141, 149, 234, 339

Kernel, 88, 89, 93–96, 104, 109, 110, 114, 116

L0, 87
L1, 87
L2, 87
L3, 87
Lifeline, 169, 170, 174, 176–180, 202, 241, 243, 250
Link, 60, 64, 98, 99, 107–109, 111, 113, 114, 123, 135, 186, 188, 249, 255, 257, 270, 271, 294, 330, 339, 365, 410, 425, 441
Literal, 58
LiteralBoolean, 58, 91
LiteralInteger, 58, 91
LiteralNull, 58, 91
LiteralSpecification, 58, 91
LiteralString, 58, 91
LiteralUnlimitedNatural, 58
Logical model, 12, 19, 20, 255, 301, 344–347, 350, 351, 361, 362, 364, 372, 379, 403, 405, 410, 411, 434, 439
LoopNode, 155

M0, 48, 200, 386
M1, 48, 49, 95, 386
M2, 48, 49, 51, 68
M3, 48–51, 68
MABS, 13
Manifest relationship, 133
Manifestation, 108, 109, 133
Manipulator arm, 35

MDA, 9, 12–14, 26, 42–45, 78, 191–193, 198, 200, 256, 282, 283, 306, 329, 331, 332, 335, 336, 344, 351, 357, 434
MDO, 6
MergeNode, 155, 162
Message, 19, 20, 30–32, 35, 81, 82, 84, 103, 130, 132, 133, 134
Message interaction model, 293, 296
MessageEnd, 175, 176, 179
MessageEvent, 146
MessageKind,
MessageOccurrence,
Method, 5, 14, 22, 26, 28, 31–33, 49, 52, 75, 94, 100, 108, 142, 143, 151, 186, 201, 202, 204, 205, 246, 256, 258, 260, 269, 334, 346–350, 356, 359, 362, 369, 377, 381–383, 410–412, 414, 415, 435, 458
Methodology, 3, 15, 22, 24, 26, 28, 32, 39, 40, 77
Middleware, 44, 45, 257, 333
Mission, 8, 201, 212, 213, 218, 219, 261, 291, 292, 301, 302, 351
Model, 3–6, 12–15, 21, 23, 26–29, 35–38, 40
Model Element, 5, 21, 50, 54, 60, 65, 67, 80, 81, 84, 85, 87, 89, 90, 92, 93, 104, 117, 132, 136, 137, 139, 142, 144, 157, 163, 171, 181, 183, 187, 188, 192–194, 199, 241, 244, 254, 270, 271, 370
Model transformation, 13, 43, 49, 198, 336
Modular programming, 25, 27, 41
Modularization, 22, 25, 26
MOF, 48, 49, 51, 65, 74
Monitor, 5, 16, 224, 252, 297, 299–303, 313, 318, 319, 323, 324, 325, 335, 357, 405, 406, 439, 456
Multidisciplinary, 1, 4–8, 36, 78, 196, 284, 286, 289, 290, 297, 300, 301, 335, 351, 354, 356, 407, 436
Multiple inheritance, 66, 73, 74, 86, 281–286
Multiplicity, 52, 54, 55, 65–68, 88, 91, 97–99, 105, 107, 110, 126, 127, 151, 202, 273, 274, 355, 385, 386, 392, 393, 421, 422, 437, 438
MultiplicityElement, 63–65, 67, 69, 96, 123, 154

NamedElement, 55, 56, 58, 61, 64, 65–67, 70, 72, 90–92, 94, 124, 141, 145, 151–154, 160, 163, 164, 170, 175–178, 183
Namespace, 55–57, 59, 62–64, 67–74, 81, 85, 89–94, 102, 114–116, 128, 130, 132, 141, 202, 261
N-ary association, 98, 107, 272–274
Navigability, 98, 99, 107, 269, 317

Node, 26, 56, 80, 82–84, 132, 133–137, 143, 144, 150–152, 154, 155, 156, 158
Normalization, 37, 275

Object database, 37, 38, 197, 226, 267, 270, 274
Object diagram, 77, 79, 81, 109, 111, 112, 113, 115, 121, 194, 249, 293, 299, 325, 332, 348–350, 360, 397, 405, 410, 420, 425, 441, 444, 464
ObjectFlow, 153, 158, 207
Objective, 8, 26, 118, 201, 205, 212, 218, 288, 306–308, 324, 331, 339, 344, 357, 415
ObjectNode, 155, 157, 160, 161, 165, 166, 168, 209
Objects with volume, 20, 21
Objects without volume, 20, 21
OCL, 50, 52, 53, 84
Odd shape, 19, 20
OEM, 7, 41, 335, 351, 357
OID, 255, 266, 268, 275
OMG, 5, 41, 42, 47, 49, 50, 53, 77, 78, 204, 212, 213, 217, 219, 267, 310
On Entry, 144, 147, 148, 232, 237, 243, 245, 250, 251, 254, 350, 361, 394, 395, 400
On Exit, 144, 147, 232, 237, 243–245, 250, 251, 254, 394, 395, 400, 447
Ontology, 14, 37, 48, 75, 193, 194, 196–198
OpaqueExpression, 57, 73, 91
Operation, 14, 23, 30, 34, 35, 37, 48, 64, 66, 67
Optimization, 3, 6, 12, 107, 301, 340, 364
OQL, 38
Ordered set, 82, 96, 386, 410
OR-states, 148, 238
ownedElement, 53, 57, 58, 61–63, 66, 71–73, 94, 110, 116, 123, 124, 129, 146, 152–154, 183

Package diagram, 80, 81, 83, 106, 114, 115, 117, 196, 198, 324, 337
Package import, 71, 85, 115, 117
Package merge, 71, 85, 115, 117
PackageableElement, 71–73, 90–92, 110, 116, 129, 130, 145, 271
PackageImport, 71–74, 90, 92, 189
PackageMerge, 71, 85, 90, 92, 115, 116
Parameter, 14, 63, 64, 67, 69, 70, 90, 91, 100–102, 124, 138, 158–160, 162, 166, 175, 202, 209, 239, 252, 260, 403
Parameter list, 100–102
ParameterSet, 155
Part, 7, 24, 29, 31, 35, 40, 48, 58, 75, 80
Passive object, 296, 358
Petri, 83, 155, 156, 159, 168, 227, 229, 230, 244, 382, 384–388, 391, 394, 399, 414
Pill Management Software, 318, 320, 322, 323, 326, 327
PIM, 42–45, 255, 256, 285, 329–331
Pin, 151, 154, 158–160, 165, 167
Port, 122, 126, 127, 130, 131, 202, 209, 439
PrimitivesTypes, 51, 52, 54, 65, 67, 70, 84, 90, 136, 262
PrimitiveType, 52, 54, 65, 67, 70, 90
Printed Circuit Board, 158, 159
problem solving, 1–5, 8, 15
Procedure, 14, 24, 27, 29, 30, 99, 125, 152, 202, 204, 205, 348, 448, 451
Process, 4, 5, 9, 10, 11, 13, 14–17, 22
Process modeling, 24, 201, 206, 207, 212, 215, 220
Profiles, 48, 51, 72–74, 77, 78, 85, 87, 88, 187–190, 211, 212
Property, 11, 14, 32, 55, 64, 66, 68–70, 74, 90
Provided interface, 106, 127
PseudoState, 139, 141, 144, 148–150, 188, 234
PseudoStateKind, 141
PSM, 43–45, 82, 226, 255–257, 285, 329, 330
Public, 15, 16, 30, 50, 55, 56, 81, 97, 103, 118
Publisher, 15–17

QoS, 12, 13, 305
Qualifier, 2, 98, 99

Reactive system, 11, 33–36
Real time, 8, 10–13, 15, 16, 34–36, 40
Realization, 107, 109, 129, 224, 347
Realize relation, ship, 330, 339
RedefinableElement, 60, 61, 67, 90, 92, 94, 153, 160, 183
ref, 173, 175, 177, 348
ref_attr, 370, 372
Region, 83, 139, 141, 143–150, 158, 163, 167, 168, 229, 232, 234, 238, 291, 371
Relational database, 37, 39, 77, 98, 189, 267–270, 274, 364, 375, 377, 379, 458
Relationship, 1, 2, 14, 16, 20, 23, 31, 33, 36, 37
REM, 330, 333, 336
Required interface, 103, 106, 127, 130, 439
Requirement analysis, 41, 78, 80, 83, 114, 199, 221, 305–307, 309, 310, 312, 318, 382
Requirement engineering, 28, 309, 332
Re-Spec phase, 337–340, 430, 432
Reusable components, 350, 352, 353, 356–358
Reuse, 26, 40–42, 50–52, 65, 85, 88, 89, 128
Risk, 15, 26, 33, 42, 159, 211, 219, 238, 313, 319, 321, 327, 407
RMSHC, 313, 318, 319, 321, 324, 326, 328, 329, 335, 339, 402
Robot, 7, 15, 29, 35, 196, 289, 290, 335, 345, 351
Role, 15, 17, 32, 36, 35, 54, 77, 95, 98

Safety critical, 252, 383, 405–416
Security, 4, 6, 12, 19, 39, 338, 351, 367, 376–378, 405, 407

SEN, 382, 383–394, 397–406, 410–412, 444
SendSignalAction, 164, 165
Sequence diagram, 16, 79, 82, 115, 169–171
Service, 7, 10, 12–14, 16, 27, 30, 37, 122, 131, 158
Set of objects, 99, 110, 258, 259, 264, 265, 288
shallowHistory, 150
Simulation, 13–16, 195, 269, 289, 297, 301, 407
Slot, 61, 64, 95, 99, 104, 109, 113, 233, 262, 294, 295
Spreadsheet, 37, 38, 310, 328, 364, 375, 376, 378, 420
SRS, 337, 339, 347, 383, 420, 425, 430, 432, 433, 443, 448
State, 2, 4, 6, 9, 26, 28, 29
State list, 147
State machine diagram, 83, 139, 140, 143–147, 158, 168, 182, 222, 231, 232, 233, 237, 242–244, 250, 253, 254, 347–350, 358, 361, 384, 394–395
State transition, 83, 139, 140, 143–145, 148, 160, 241, 254, 400
StateInvariant, 180
Strategy, 3, 9, 41, 200, 201, 204, 211, 212, 218, 219, 239, 266
StringExpression, 50, 57
StructuralFeature, 60–62, 67, 68, 91, 93–96, 170
structured design, 25
StructuredActivityNode, 152, 153, 160, 162–165
StructuredClassifier, 119–120, 122–124
Subclassing, 264, 281
Submachine state, 145, 146, 148
Substate, 139, 145
Substitution, 109
Subtyping, 264, 281
Synchronization, 34, 151, 159, 341, 377, 379, 384, 393, 397

Tamagotchi, 402–407, 408–414
Task, 2, 6, 22–28, 29–32, 35
Terminator, 206–210
Thread, 24, 32, 201, 205, 206, 241
Tightly coupled interaction, 301–303
TimeEvent, 146
Timing constraint, 10, 144, 345, 354, 362, 399, 400
Timing diagram, 169, 174, 175, 241
Token, 24, 151, 156, 157, 159, 160
Top-down, 200, 334, 335, 359
Trace, 34, 156, 430
Tracklift tram, 417, 420
Tradeoff, 309
Transformation, 13, 43, 45, 49, 85, 151, 155, 168, 198, 201, 206, 224, 273
Transition, 83, 84, 139–144, 146–149, 160, 227, 232
Trigger, 37, 139, 142–143, 145, 155, 202, 240–242, 254, 289, 291, 362, 363, 394, 417, 448
Tsunami, 2, 12
Type, 3, 8, 9, 14, 25, 26, 36, 37
TypedElement, 58–64, 66–69, 72, 73, 91, 92, 154, 160, 170

UML 2, 78, 81, 82, 118, 136, 174, 260, 389, 412, 435
UML infrastructure, 47–75
UML superstructure, 77–189
Uniform, 3
Uniform concept, 290
Uniform methodology, 28, 39, 40, 78, 83, 128, 251, 369, 417, 465
Unifying, 39, 289, 374, 376, 382, 394
Unpredictability, 383
Usage, 12, 26, 55, 81, 93, 107, 108, 114, 128, 139, 174, 197, 311
Use case diagram, 83, 181, 182, 184–186, 199, 200, 206, 221, 222, 226, 245, 314, 315, 317, 323, 325
UseCase, 137, 183, 184, 186, 221
UUID, 266, 267

Validation, 311, 312, 329, 330, 332, 336
ValueSpecification, 57, 58, 61–64, 73, 91, 95, 110, 146, 154
Variable speed electric fan, 299
Vending_Machine, 351
Views, 29, 40, 79, 104, 200, 201, 220, 257, 308, 318, 321, 324, 349, 368, 382, 450, 458
Visibility, 55, 56, 72, 96, 97, 99, 100, 105, 261
VisibilityKind, 55, 56, 72
Visual surcharge, 20, 21, 134, 384, 387, 391

Web database, 134, 135
Well structured, 9, 269
Work flow, 158, 205, 207, 212, 213, 216, 217, 237, 340, 382, 384